LINEAR ALGEBRA

CORE TOPICS FOR THE
SECOND COURSE

LINEAR ALGEBRA

CORE TOPICS FOR THE
SECOND COURSE

Dragu Atanasiu
University of Borås, Sweden

Piotr Mikusiński
University of Central Florida, USA

 **World Scientific**

NEW JERSEY · LONDON · SINGAPORE · BEIJING · SHANGHAI · HONG KONG · TAIPEI · CHENNAI · TOKYO

Published by

World Scientific Publishing Co. Pte. Ltd.
5 Toh Tuck Link, Singapore 596224
USA office: 27 Warren Street, Suite 401-402, Hackensack, NJ 07601
UK office: 57 Shelton Street, Covent Garden, London WC2H 9HE

Library of Congress Control Number: 2022057625

British Library Cataloguing-in-Publication Data
A catalogue record for this book is available from the British Library.

LINEAR ALGEBRA
Core Topics for the Second Course

ISBN 978-981-125-854-1 (hardcover)
ISBN 978-981-125-855-8 (ebook for institutions)
ISBN 978-981-125-856-5 (ebook for individuals)

For any available supplementary material, please visit
https://www.worldscientific.com/worldscibooks/10.1142/12898#t=suppl

Preface

This is a book for a second course in linear algebra.

In order to facilitate a smooth transition to the world of rigorous proofs we mix, to a greater extent than most textbooks at this level, abstract theory with matrix calculations. We present numerous examples and proofs of particular cases of important results before the general versions are formulated and proved. We noticed in many years of teaching that a proof of a particular case which captures the main idea of the general theorem has a major impact on the depth of understanding of the general theory. Reading simpler and more manageable proofs is also more likely to encourage students to work with proofs. Students can try to prove another particular case or the general case using the knowledge gained from the particular case. For some theorems we give two or even three proofs. In this way we give students an opportunity to see important results from different angles and at the same time to see connections between different results presented in the book.

Students are assumed to be familiar with calculations with real matrices. For example, students should be able to calculate products of matrices and find the reduced row echelon form of a matrix. All this background material is presented in our book *Linear Algebra, Core topics for the first course*, but the present book does not assume that students are familiar with our presentation of that material. Any standard book on Matrix Linear Algebra will provide a sufficient preparation for the present volume.

On the other hand, since most material of this book mirrors, at a higher, less computational, and more abstract level, the content of our book *Linear Algebra, Core topics for the first course*, students who find a result from this book too abstract would benefit from reading the same material from the first course. For example, in *Core topics for the first course* there are a lot of concrete examples of Jordan forms and singular value decompositions. Getting familiar with those examples would make the theory presented in the second course more accessible.

The majority of results are presented under the assumption that the vector spaces are finite dimensional. Some examples of infinite dimensional spaces are given and a very brief discussion of infinite dimensional inner product spaces is included as Appendix D.

In Chapter 1 we introduce vector spaces and discuss some basic ideas including subspaces, linear independence, bases, dimension, and direct sums. We consider both real and complex vector spaces. For students with limited experience with complex numbers we provide an appendix that presents complex numbers in an elementary and detailed manner.

In Chapter 2 we discuss linear transformations between vector spaces. The presented topics include projections, the Rank-Nullity Theorem, isomorphisms, dual spaces, matrix representation of linear transformations, and quotient spaces. In order to keep this chapter at a reasonable size we only prove the results that are used in later chapters and give more results as exercises.

In Chapter 3 we discuss inner product spaces, including orthogonal projections, self-adjoint, normal, unitary, orthogonal, and positive linear transformation. A careful presentation of spectral theorems and the singular value decomposition constitute a substantial part of this chapter. Since we determine eigenvalues without using characteristic polynomials, determinants are used in Chapter 3 only in some examples and exercises.

In Chapter 4 we show how to obtain bases such that the matrix of a linear transformation becomes diagonal or block-diagonal. In order to construct such bases, we study factorization of characteristic polynomials. In order to give interesting examples we need to calculate characteristic polynomials using determinants of endomorphisms. These determinants are introduced at the beginning of Chapter 4 via multilinear algebra.

It is possible to obtain diagonal and block-diagonal matrices without using determinants, as shown in [2], but in our opinion the discussion of determinants in connection with alternating multilinear forms is an interesting part of linear algebra. While the theory without determinants has a certain appeal, when it comes to determining diagonal and block-diagonal forms in concrete cases one is limited to very simple examples where calculating the eigenvalues is trivial, as can be seen in books where determinants are avoided. We believe that students should have all possible instruments to solve problems and determinants are essential in determining the characteristic polynomials. It has been our experience that by presenting topics in a less theoretical way and showing more concrete calculations using determinants increase the understanding of the presented topics. Since every student taking a proof based course in linear algebra has some knowledge of determinants, the presence of determinants in this book helps students make connections with more elementary courses. Moreover, in our book we do not calculate determinants as in matrix linear algebra or precalculus, but instead we emphasize the connection to multilinear algebra. If there are reasons to dislike determinants because calculating determinants is tedious and non-intuitive, it is not a reason to not appreciate the elegance of the alternating multilinear forms and eliminate them in a first course of linear algebra with proofs.

Appendices that provide short introductions to permutations, complex numbers, and polynomials are included at the end of the book. Proofs of all results presented in these appendices are included.

A complete solution manual is available upon request for all instructors who adopt this book as a course text. Please send your request to sales@wspc.com.

Contents

Chapter 1

Vector Spaces

Introduction

If you are reading this book, you most likely worked with vectors in a number of courses and you have fairly good intuitive understanding of what we mean by a vector. But can you give a formal definition of a vector? It turns out that the best we can do is to define a vector as an element of a vector space. It may seem a silly definition, but actually it represents what is quite common in more advanced mathematics. The idea is that, when we want to describe a certain class of objects, we don't want to describe the objects themselves, but rather what we can do with them and what are the properties of those operations on the objects we are defining. Describing the objects directly limits applications of the methods we develop to the instances we describe. On the other hand, if formulate something about any object that has certain properties, then it will apply to examples that we may not even be aware of.

We use this approach in our definition of vector spaces. We define a vector space as a collection of objects not of a certain kind, but rather objects on which certain operations having certain operational properties can be performed.

1.1 Definitions and examples

A vector space is a set with an algebraic structure. Elements of a vector space can be added and scaled. Addition in a vector space $\mathcal{V}$ is a function that assigns to any $\mathbf{v}, \mathbf{w} \in \mathcal{V}$ a unique element $\mathbf{v} + \mathbf{w} \in \mathcal{V}$. An element $\mathbf{v} \in \mathcal{V}$ is scaled if it is multiplied by a number, called *scalar* in this context. The result of multiplying $\mathbf{v} \in \mathcal{V}$ by a scalar c is denoted by $c\mathbf{v}$. In this book scalars are either real numbers or complex numbers. If it is necessary to specify which case it is, we write a "real vector space" or a "complex vector space". If a statement applies to both real vector spaces and complex vector spaces, we use the letter $\mathbb{K}$ instead of $\mathbb{R}$ or $\mathbb{C}$. The formal definition of vector spaces given below is an example of such a situation. The set $\mathbb{K}$ is called a "scalar field". The phrase "vector space over $\mathbb{K}$"

is often used to indicate that $\mathbb{K}$ is the set of scalars for that vector space.

As explained in the introduction, addition and scaling in the definition of a vector space below are not specific operations, but any operations that satisfy the listed conditions. In most applications we already know what we mean by addition and scaling. In order to use the tools of vector spaces we have to make sure that all conditions in the definition are satisfied. After the definition we discuss some examples where all conditions are satisfied as well as some examples where some conditions are not satisfied.

Definition 1.1.1. By a *vector space* we mean a set $\mathcal{V}$ with *addition* that assigns a unique element $\mathbf{v} + \mathbf{w} \in \mathcal{V}$ to any $\mathbf{v}, \mathbf{w} \in \mathcal{V}$ and *scalar multiplication* that assigns a unique element $c\mathbf{v} \in \mathcal{V}$ to any $c \in \mathbb{K}$ and any $\mathbf{v} \in \mathcal{V}$ in such a way that all of the following conditions are satisfied:

(a) For every $\mathbf{v}, \mathbf{w} \in \mathcal{V}$ we have $\mathbf{v} + \mathbf{w} = \mathbf{w} + \mathbf{v}$;

(b) For every $\mathbf{u}, \mathbf{v}, \mathbf{w} \in \mathcal{V}$ we have $\mathbf{u} + (\mathbf{v} + \mathbf{w}) = (\mathbf{u} + \mathbf{w}) + \mathbf{v}$;

(c) There is an element $\mathbf{0} \in \mathcal{V}$ such that for every $\mathbf{v} \in \mathcal{V}$ we have $\mathbf{0} + \mathbf{v} = \mathbf{v}$;

(d) For every $\mathbf{v} \in \mathcal{V}$ there is $\mathbf{u} \in \mathcal{V}$ such that $\mathbf{v} + \mathbf{u} = \mathbf{0}$;

(e) For every $\mathbf{v} \in \mathcal{V}$ we have $1\mathbf{v} = \mathbf{v}$;

(f) For every $\mathbf{v} \in \mathcal{V}$ and every $c_1, c_2 \in \mathbb{K}$ we have $(c_1 c_2)\mathbf{v} = c_1(c_2\mathbf{v})$;

(g) For every $\mathbf{v} \in \mathcal{V}$ and every $c_1, c_2 \in \mathbb{K}$ we have $(c_1 + c_2)\mathbf{v} = c_1\mathbf{v} + c_2\mathbf{v}$;

(h) For every $\mathbf{v}, \mathbf{w} \in \mathcal{V}$ and every $c \in \mathbb{K}$ we have $c(\mathbf{v} + \mathbf{w}) = c\mathbf{v} + c\mathbf{w}$.

Now we present some examples of vector spaces. We do not verify that all conditions in the definition of a vector space are satisfied. While we do not expect that you will verify every condition in every example, you should convince yourself that they are satisfied. It is a good exercise to give formal proofs for some conditions in some examples, especially if they don't seem obvious.

Example 1.1.2. The set of all real numbers $\mathbb{R}$ with the standard addition and multiplication is a real vector space. Note that in this example the vector space and the scalar field are the same set.

The set of all complex numbers $\mathbb{C}$ is an example of a complex vector space as well as a real vector space.

Example 1.1.3. For every integer $n \geq 1$ the set of all $n \times 1$ matrices $\begin{bmatrix} a_1 \\ \vdots \\ a_n \end{bmatrix}$

with $a_1, \ldots, a_n \in \mathbb{K}$, denoted by $\mathbb{K}^n$, is a vector space with the operations of addition of vectors and multiplication of vectors by scalars defined by

$$\begin{bmatrix} a_1 \\ \vdots \\ a_n \end{bmatrix} + \begin{bmatrix} b_1 \\ \vdots \\ b_n \end{bmatrix} = \begin{bmatrix} a_1 + b_1 \\ \vdots \\ a_n + b_n \end{bmatrix} \quad \text{and} \quad c \begin{bmatrix} a_1 \\ \vdots \\ a_n \end{bmatrix} = \begin{bmatrix} ca_1 \\ \vdots \\ ca_n \end{bmatrix}$$

for $c \in \mathbb{K}$.

Clearly, $\mathbb{C}^n$ is a vector space over $\mathbb{C}$ and $\mathbb{R}^n$ is a vector space over $\mathbb{R}$. It is also possible to consider $\mathbb{C}^n$ as a vector space over $\mathbb{R}$.

Example 1.1.4. Let $\mathcal{V} = \{a\}$, a set with a single element $\mathbf{a}$. With the operations defined as

$$\mathbf{a} + \mathbf{a} = \mathbf{a} \quad \text{and} \quad c\mathbf{a} = \mathbf{a},$$

for any $c \in \mathbb{K}$, it is a vector space.

Note that we must have $\mathbf{a} = \mathbf{0}$, where $\mathbf{0}$ is the element whose existence is guaranteed by condition (c) in Definition 1.1.1. This is the smallest possible vector space. It is often called the *trivial vector space*.

Example 1.1.5. Let S be an arbitrary nonempty set. The set of all functions $f : S \to \mathbb{K}$ with addition defined by

$$(f + g)(s) = f(s) + g(s) \text{ for every } s \in S$$

and multiplication by scalars defined by

$$(cf)(s) = cf(s) \text{ for every } s \in S \text{ and } c \in \mathbb{K}$$

is a vector space over $\mathbb{K}$. We will denote this space by $\mathcal{F}_{\mathbb{K}}(S)$. Note that the constant function $f(s) = 0$ is the $\mathbf{0}$ of this vector space and the function $g(s) = -f(s)$ is the element of $\mathcal{F}_{\mathbb{K}}(S)$ such that $f + g = \mathbf{0}$.

This is a very important family of vector spaces. Several examples below are special cases of $\mathcal{F}_{\mathbb{K}}(S)$. If you verify that all conditions in the definition of a vector space are satisfied for $\mathcal{F}_{\mathbb{K}}(S)$, then there is no need to check them for those examples.

Example 1.1.6. We denote by $\mathcal{M}_{m \times n}(\mathbb{K})$ the set of all $m \times n$ matrices with entries from $\mathbb{K}$. If A and B are matrices from $\mathcal{M}_{m \times n}(\mathbb{K})$ such that the (j,k) entry of the matrix A is a_{jk} and the (j,k) entry of the matrix B is b_{jk} and c is a number form $\mathbb{K}$, then $A + B$ is the matrix with the (j,k) entry equal to $a_{jk} + b_{jk}$ and cA is the matrix with the (j,k) entry equal to ca_{jk}. It is easy to verify that with these operations $\mathcal{M}_{m \times n}(\mathbb{K})$ is a vector space over $\mathbb{K}$.

Note that the set $\mathcal{M}_{m \times n}(\mathbb{K})$ can be interpreted as the space $\mathcal{F}_{\mathbb{K}}(S)$ where

$$S = \{1, 2, \ldots, m\} \times \{1, 2, \ldots, n\} = \{(j,k) : j = 1, 2, \ldots, m, k = 1, 2, \ldots, n\}.$$

Example 1.1.7. The set of all infinite sequences $(x_n) = (x_1, x_2, \ldots)$ of real numbers can be identified with $\mathcal{F}_{\mathbb{R}}(\{1, 2, 3, \ldots\})$ and thus it is a real vector space. Similarly, the set of all infinite sequences of complex numbers is a complex vector space. The operations defined in Example 1.1.5 can be described in a more intuitive way:

$$(x_1, x_2, \ldots) + (y_1, y_2, \ldots) = (x_1 + y_1, x_2 + y_2, \ldots)$$

and

$$c(x_1, x_2, \ldots) = (cx_1, cx_2, \ldots).$$

In some applications it is natural to consider the vector space $\mathcal{F}_{\mathbb{K}}(\mathbb{Z})$, where $\mathbb{Z}$ is the set of all integers, of all "two-sided" sequences

$$(\ldots, x_{-2}, x_{-1}, x_0, x_1, x_2, \ldots)$$

of real or complex numbers.

Example 1.1.8. We will define a vector space over $\mathbb{R}$ whose elements are lines of $\mathbb{R}^2$ parallel to a given line. We recall that a line is a set of the form $\mathbf{x} + \mathbb{R}\mathbf{a} = \{\mathbf{x} + c\mathbf{a} : c \in \mathbb{R}\}$ where $\mathbf{x}, \mathbf{a} \in \mathbb{R}^2$ and $\mathbf{a} \neq \mathbf{0}$.

Let $\mathbf{a}$ be a fixed nonzero vector in $\mathbb{R}^2$. We define

$$\widehat{\mathbf{x}} = \mathbf{x} + \mathbb{R}\mathbf{a}.$$

In other words, $\widehat{\mathbf{x}}$ is the line through $\mathbf{x}$ parallel to $\mathbf{a}$. First note that $\widehat{\mathbf{0}} = \mathbb{R}\mathbf{a}$. Moreover,

$$\widehat{\mathbf{x}} = \widehat{\mathbf{y}} \text{ if and only if there is a real number } \alpha \text{ such that } \mathbf{y} = \mathbf{x} + \alpha\mathbf{a}.$$

In other words

$$\widehat{\mathbf{x}} = \widehat{\mathbf{y}} \text{ if and only if } \mathbf{y} \in \widehat{\mathbf{x}}.$$

Indeed, if $\hat{\mathbf{x}} = \hat{\mathbf{y}}$, then

$$\mathbf{x} + \mathbb{R}\mathbf{a} = \hat{\mathbf{x}} = \hat{\mathbf{y}} = \mathbf{y} + \mathbb{R}\mathbf{a},$$

and thus $\mathbf{y} = \mathbf{y} + 0\mathbf{a} = \mathbf{x} + \alpha\mathbf{a}$ for some $\alpha \in \mathbb{R}$.

On the other hand, if $\mathbf{y} = \mathbf{x} + \alpha\mathbf{a}$ for some $\alpha \in \mathbb{R}$, then

$$\hat{\mathbf{y}} = \mathbf{y} + \mathbb{R}\mathbf{a} = \mathbf{x} + \alpha\mathbf{a} + \mathbb{R}\mathbf{a} = \mathbf{x} + \mathbb{R}\mathbf{a} = \hat{\mathbf{x}},$$

because $\alpha\mathbf{a} + \mathbb{R}\mathbf{a} = \mathbb{R}\mathbf{a}$.

Now we define

$$\hat{\mathbf{x}} + \hat{\mathbf{y}} = \widehat{\mathbf{x} + \mathbf{y}} \quad \text{and} \quad c\hat{\mathbf{x}} = \widehat{c\mathbf{x}}.$$

These operations are well defined because

$$(\mathbf{x} + \alpha\mathbf{a}) + (\mathbf{y} + \beta\mathbf{a}) = (\mathbf{x} + \mathbf{y}) + (\alpha + \beta)\mathbf{a} \in \widehat{\mathbf{x} + \mathbf{y}}$$

and

$$c(\mathbf{x} + \alpha\mathbf{a}) = c\mathbf{x} + (c\alpha)\mathbf{a} \in \widehat{c\mathbf{x}},$$

for arbitrary real numbers α, β and c.

The set $\mathcal{V} = \{\hat{\mathbf{x}} : \mathbf{x} \in \mathbb{R}^2\}$ with the operations defined above is a real vector space.

Example 1.1.9. Let $A = \{(x, x) : x \in \mathbb{R}\}$ and $B = \{(x, -x) : x \in \mathbb{R}\}$. Show that $A \cup B$ is not a real vector space.

Solution. It suffices to note that, for example, the vector $(2, 0) = (1, 1) + (1, -1)$ is not $A \cup B$. Note that both A and B are vector spaces. $\square$

Example 1.1.10. Let $A = \{(x, y) \in \mathbb{R}^2 : y \geq 0\}$. Show that A is not a real vector space.

Solution. It suffices to observe that, for example, the vector $(1, -4)$ is in A, but the vector $(2, -8) = -2(1, -4))$ is not in A. Note that, if $(x_1, y_1), (x_2, y_2) \in A$, then $(x_1 + x_2, y_1 + y_2) = (x_1, y_1) + (x_2, y_2)$ is in A. $\square$

Example 1.1.11. Let $A = \{(x, y, z) \in \mathbb{R}^3 : x + y + z = 1\}$. Show that A is not a real vector space.

Solution 1. It suffices to note that the vectors $(1, 0, 0)$ and $(0, 1, 0)$ are in A, but the vector $(1, 1, 0) = (1, 0, 0) + (0, 1, 0)$ is not in A. $\square$

Solution 2. It suffices to note that the vector $(1,0,0)$ is in A, but the vector $(3,0,0) = 3(1,0,0)$ is not in A.   □

Note that the set $V = \{(x,y,z) \in \mathbb{R}^3 : x+y+z = 0\}$ is a real vector space.

Example 1.1.12. Let V_1 and V_2 be arbitrary vector spaces over $\mathbb{K}$. The set of all pairs $(\mathbf{v}_1, \mathbf{v}_2)$ such that $\mathbf{v}_1 \in V_1$ and $\mathbf{v}_2 \in V_2$ is denoted by $V_1 \times V_2$ and called the *Cartesian product* of spaces V_1 and V_2. In symbols,

$$V_1 \times V_2 = \{(\mathbf{v}_1, \mathbf{v}_2) : \mathbf{v}_1 \in V_1, \mathbf{v}_2 \in V_2\}.$$

It is easy to verify that $V_1 \times V_2$ becomes a vector space if we define

$$(\mathbf{v}_1, \mathbf{v}_2) + (\mathbf{w}_1, \mathbf{w}_2) = (\mathbf{v}_1 + \mathbf{w}_1, \mathbf{v}_2 + \mathbf{w}_2) \quad \text{and} \quad c(\mathbf{v}_1, \mathbf{v}_2) = (c\mathbf{v}_1, c\mathbf{v}_2).$$

More generally, for arbitrary vector spaces $V_1, \ldots, V_n$ over $\mathbb{K}$ we define

$$V_1 \times \cdots \times V_n = \{(\mathbf{v}_1, \ldots, \mathbf{v}_n) : \mathbf{v}_1 \in V_1, \ldots, \mathbf{v}_n \in V_n\}.$$

The set $V_1 \times \cdots \times V_n$, called the Cartesian product of spaces $V_1, \ldots, V_n$, is a vector space with the addition

$$(\mathbf{v}_1, \ldots, \mathbf{v}_n) + (\mathbf{w}_1, \ldots, \mathbf{w}_n) = (\mathbf{v}_1 + \mathbf{w}_1, \ldots, \mathbf{v}_n + \mathbf{w}_n)$$

and scalar multiplication

$$c(\mathbf{v}_1, \ldots, \mathbf{v}_n) = (c\mathbf{v}_1, \ldots, c\mathbf{v}_n).$$

If $V_1 = \cdots = V_n = V$, then we write

$$V_1 \times \cdots \times V_n = \underbrace{V \times \cdots \times V}_{n \text{ times}} = V^n.$$

Example 1.1.13. Let V be a real vector space. Show that $V \times V$ is a complex vector space if we define addition as for the real vector space $V \times V$ and multiplication by complex numbers as follows

$$(a + bi)(\mathbf{x}, \mathbf{y}) = (a\mathbf{x} - b\mathbf{y}, a\mathbf{y} + b\mathbf{x}),$$

where a and b are real numbers and $\mathbf{x}$ and $\mathbf{y}$ are vectors from V.

Solution. We will only verify that

$$(c + di)((a + bi)\mathbf{x}, \mathbf{y}) = ((c + di)(a + bi))(\mathbf{x}, \mathbf{y})$$

because the verification of the other axioms is similar and easier.

$$(c + di)((a + bi)\mathbf{x}, \mathbf{y})) = (c + di)(a\mathbf{x} - b\mathbf{y}, a\mathbf{y} + b\mathbf{x})$$
$$= (c(a\mathbf{x} - b\mathbf{y}) - d(a\mathbf{y} + b\mathbf{x}), c(a\mathbf{y} + b\mathbf{x}) + d(a\mathbf{x} - b\mathbf{y}))$$
$$= ((ac - db)\mathbf{x} - (ad + bc)\mathbf{y}, (ac - bd)\mathbf{y} + (ad + bc)\mathbf{x})$$
$$= ((ac - bd) + (ad + bc)i)(\mathbf{x}, \mathbf{y})$$
$$= ((c + di)(a + bi))(\mathbf{x}, \mathbf{y}).$$

$\square$

Elements of a vector space are called *vectors*. The use of the word "vector" does not imply that we are talking about the familiar vectors in $\mathbb{R}^n$ that we picture as arrows. In the above examples we considered vectors that were functions, matrices, sequences, or even sets (Example 1.1.8). The expressions "$\mathbf{v}$ is an element of a vector space V" and "$\mathbf{v}$ is a vector in a vector space V" are completely equivalent.

We close this section with a theorem that collects some simple but useful properties of addition and scaling in vector spaces.

Theorem 1.1.14. *Let V be a vector space. Then*

(a) *The element $\mathbf{0}$ such that $\mathbf{0} + \mathbf{v} = \mathbf{v}$ for every $\mathbf{v} \in V$ is unique;*

(b) *$0\mathbf{v} = \mathbf{0}$ for every $\mathbf{v} \in V$;*

(c) *If $\mathbf{v} + \mathbf{u} = \mathbf{0}$, then $\mathbf{u} = (-1)\mathbf{v}$;*

(d) *$c\mathbf{0} = \mathbf{0}$ for every $c \in \mathbb{K}$;*

(e) *If $\mathbf{u} + \mathbf{w} = \mathbf{v} + \mathbf{w}$, then $\mathbf{u} = \mathbf{v}$.*

Proof. To prove (a) assume that there are $\mathbf{0}_1, \mathbf{0}_2 \in V$ such that $\mathbf{0}_1 + \mathbf{v} = \mathbf{v}$ and $\mathbf{0}_2 + \mathbf{v} = \mathbf{v}$ for every $\mathbf{v} \in V$. Then we have

$$\mathbf{0}_1 = \mathbf{0}_2 + \mathbf{0}_1 = \mathbf{0}_1 + \mathbf{0}_2 = \mathbf{0}_2.$$

For every $\mathbf{v} \in V$ we have

$$0\mathbf{v} + \mathbf{v} = 0\mathbf{v} + 1\mathbf{v} = (0 + 1)\mathbf{v} = 1\mathbf{v} = \mathbf{v}.$$

Now let $\mathbf{u} \in V$ be such that $\mathbf{v} + \mathbf{u} = \mathbf{0}$. Then $0\mathbf{v} + \mathbf{v} = \mathbf{v}$ implies

$$\mathbf{0} = \mathbf{v} + \mathbf{u} = (0\mathbf{v} + \mathbf{v}) + \mathbf{u} = 0\mathbf{v} + (\mathbf{v} + \mathbf{u}) = 0\mathbf{v} + \mathbf{0} = 0\mathbf{v}.$$

Thus, $0\mathbf{v} = \mathbf{0}$ for every $\mathbf{v} \in V$, proving (b).

To prove (c) we first observe that for every $\mathbf{v} \in V$ we have

$$\mathbf{v} + (-1)\mathbf{v} = 1\mathbf{v} + (-1)\mathbf{v} = (1 - 1)\mathbf{v} = 0\mathbf{v} = \mathbf{0}.$$

It remains to show that $(-1)\mathbf{v}$ is the only element with that property. Indeed, if $\mathbf{u}$ is any element such that $\mathbf{v} + \mathbf{u} = \mathbf{0}$, then

$$\mathbf{u} = \mathbf{u} + \mathbf{0} = \mathbf{u} + (\mathbf{v} + (-1)\mathbf{v}) = (\mathbf{u} + \mathbf{v}) + (-1)\mathbf{v} = \mathbf{0} + (-1)\mathbf{v} = (-1)\mathbf{v}.$$

For any $c \in \mathbb{K}$ we have

$$c\mathbf{0} = c(\mathbf{v} + (-1)\mathbf{v}) = c\mathbf{v} + (-c)\mathbf{v} = (c - c)\mathbf{v} = 0\mathbf{v} = \mathbf{0},$$

where $\mathbf{v}$ is an arbitrary element in $\mathcal{V}$, proving (d).

Finally, if $\mathbf{u} + \mathbf{w} = \mathbf{v} + \mathbf{w}$, then

$$\begin{aligned} \mathbf{u} = \mathbf{u} + \mathbf{0} &= \mathbf{u} + (\mathbf{w} + (-1)\mathbf{w}) = (\mathbf{u} + \mathbf{w}) + (-1)\mathbf{w} \\ &= (\mathbf{v} + \mathbf{w}) + (-1)\mathbf{w} = \mathbf{v} + (\mathbf{w} + (-1)\mathbf{w}) = \mathbf{v} + \mathbf{0} = \mathbf{v}, \end{aligned}$$

proving (e). $\square$

The element $(-1)\mathbf{v}$ is denoted simply by $-\mathbf{v}$. With this notation we have

$$\mathbf{v} - \mathbf{v} = -\mathbf{v} + \mathbf{v} = \mathbf{0}.$$

Note that $-(-\mathbf{v}) = \mathbf{v}$.

1.2   Subspaces

The concept of a subspace of a vector space is one of the fundamental ideas of linear algebra. We begin the discussion of subspaces by considering an example.

Example 1.2.1. Let $\mathcal{V}$ be a vector space and let $\mathbf{x}$ be a vector in $\mathcal{V}$. The set of all vectors of the form $c\mathbf{x}$ where $c \in \mathbb{K}$ is a vector space with the addition and the multiplication by scalars inherited from the vector space $\mathcal{V}$, that is,

$$c_1\mathbf{x} + c_2\mathbf{x} = (c_1 + c_2)\mathbf{x} \quad \text{and} \quad c_1(c_2\mathbf{x}) = (c_1 c_2)\mathbf{x}.$$

This vector space is denoted by $\mathbb{K}\mathbf{x}$ or $\mathrm{Span}\{\mathbf{x}\}$:

$$\mathbb{K}\mathbf{x} = \mathrm{Span}\{\mathbf{x}\} = \{c\mathbf{x} : c \in \mathbb{K}\}.$$

Without being too precise, one can say that the vector space $\mathrm{Span}\{\mathbf{x}\}$ is a "small vector space in a large vector space". It is important that every element of the small space is an element of the large space and that the operations of addition and scaling in the small space are the same as in the large space. These ideas are captured in the following precise definition.

Definition 1.2.2. A nonempty subset $\mathcal{U}$ of a vector space $\mathcal{V}$ is called a *subspace* of $\mathcal{V}$ if the following two conditions are satisfied:

(a) If $\mathbf{v} \in \mathcal{U}$ and $c \in \mathbb{K}$, then $c\mathbf{v} \in \mathcal{U}$;

(b) If $\mathbf{v}, \mathbf{w} \in \mathcal{U}$, then $\mathbf{v} + \mathbf{w} \in \mathcal{U}$.

The set Span$\{\mathbf{x}\}$ defined in Example 1.2.1 is a subspace of the vector space $\mathcal{V}$. Note that, if we take $\mathbf{x} = \mathbf{0}$, then Span$\{\mathbf{x}\} = \{\mathbf{0}\}$. Since $\mathbf{0}$ is an element in every vector space, $\{\mathbf{0}\}$ is a subspace of every vector space.

From the definition of subspaces it follows that every vector space is a subspace of itself. To exclude these special cases, we add the word "proper": $\mathcal{U}$ is a *proper subspace* of $\mathcal{V}$ if $\mathcal{U}$ is a subspace of $\mathcal{V}$ such that $\mathcal{U} \neq \mathcal{V}$ and $\mathcal{U} \neq \{\mathbf{0}\}$.

Now we present several examples of subspaces that often appear in applications.

Example 1.2.3. A polynomial is a function $p : \mathbb{K} \to \mathbb{K}$ defined by

$$p(t) = a_0 + a_1 t + \cdots + a_n t^n$$

where $a_0, a_1, \ldots, a_n \in \mathbb{K}$. If $a_n \neq 0$, then n is called the degree of the polynomial $p(t) = a_0 + a_1 t + \cdots + a_n t^n$ (see Appendix C).

The set of all polynomials, denoted by $\mathcal{P}(\mathbb{K})$, is a subspace of $\mathcal{F}_{\mathbb{K}}(\mathbb{K})$. The set $\mathcal{P}_n(\mathbb{K})$ of all polynomials of the form $a_0 + a_1 t + \cdots + a_n t^n$ is a subspace of $\mathcal{P}(\mathbb{K})$. If $k \leq n$, then $\mathcal{P}_k(\mathbb{K})$ is a subspace of $\mathcal{P}_n(\mathbb{K})$.

Example 1.2.4. For every integer $n \geq 1$ we denote by $\mathcal{D}^n_{\mathbb{R}}(\mathbb{R})$ the set of all real-valued n-times differentiable functions defined on $\mathbb{R}$.

We also denote by $\mathcal{D}_{\mathbb{R}}(\mathbb{R})$ the set of all real-valued functions defined on $\mathbb{R}$ which are n-times differentiable for every integer $n \geq 1$.

Show that $\mathcal{D}^n_{\mathbb{R}}(\mathbb{R})$ is a subspace of $\mathcal{F}_{\mathbb{R}}(\mathbb{R})$ for every integer $n \geq 1$ and that $\mathcal{D}_{\mathbb{R}}(\mathbb{R})$ is a subspace of $\mathcal{D}^n_{\mathbb{R}}(\mathbb{R})$ for every integer $n \geq 1$.

Solution. This is an immediate consequence of the fact that the sum of differentiable functions is differentiable and that a constant multiple of a differentiable function is differentiable. $\square$

Example 1.2.5. Show that the set $\mathcal{S}_{n \times n}$ of all $n \times n$ symmetric matrices (matrices such that $A^T = A$) is a subspace of the vector space of the square matrices $\mathcal{M}_{n \times n}(\mathbb{K})$.

Solution. If A and B are symmetric matrices and $c \in \mathbb{K}$, then we have

$$(A + B)^T = A^T + B^T = A + B \text{ and } (cA)^T = cA^T = cA.$$

□

Example 1.2.6. Show that the set $\mathcal{A}_{n \times n}(\mathbb{K})$ of all antisymmetric matrices (matrices such that $A^T = -A$) is a subspace of the vector space of square matrices $\mathcal{M}_{n \times n}(\mathbb{K})$.

Solution. If A and B are antisymmetric matrices and $c \in \mathbb{K}$, then we have

$$(A + B)^T = A^T + B^T = -A - B = -(A + B)$$

and

$$(cA)^T = cA^T = c(-A) = -cA.$$

□

Example 1.2.7. Let $A \in \mathcal{M}_{m \times n}(\mathbb{K})$. Show that the set

$$\mathbf{C}(A) = \{A\mathbf{x} : \mathbf{x} \in \mathbb{K}^n\}$$

is a subspace of the vector space $\mathbb{K}^m$.

Solution. If $\mathbf{y}_1 = A\mathbf{x}_1$, $\mathbf{y}_2 = A\mathbf{x}_2$, and $c \in \mathbb{K}$, then

$$\mathbf{y}_1 + \mathbf{y}_2 = A(\mathbf{x}_1 + \mathbf{x}_2) \text{ and } c\mathbf{y} = cA\mathbf{x} = A(c\mathbf{x}).$$

□

Example 1.2.8. Let $A \in \mathcal{M}_{m \times n}(\mathbb{K})$. Show that the set

$$\mathbf{N}(A) = \{\mathbf{x} \in \mathbb{K}^n : A\mathbf{x} = \mathbf{0}\}$$

is a subspace of the vector space $\mathbb{K}^n$. Determine

$$\mathbf{N}\left(\begin{bmatrix} 2+i & 1-i & 1 \\ i & 2+i & i \\ 2+2i & 3 & 1+i \end{bmatrix}\right).$$

Solution. If $A\mathbf{x}_1 = A\mathbf{x}_2 = A\mathbf{x} = \mathbf{0}$ and $c \in \mathbb{K}$, then

$$A(\mathbf{x}_1 + \mathbf{x}_2) = A\mathbf{x}_1 + A\mathbf{x}_2 = \mathbf{0} \quad \text{and} \quad A(c\mathbf{x}) = c(A\mathbf{x}) = \mathbf{0},$$

which proves that $\mathbf{N}(A)$ is a subspace of $\mathcal{M}_{m \times n}(\mathbb{K})$.
Since the reduced row echelon form of the matrix

$$\begin{bmatrix} 2+i & 1-i & 1 \\ i & 2+i & i \\ 2+2i & 3 & 1+i \end{bmatrix}$$

is

$$\begin{bmatrix} 1 & 0 & \frac{2}{13} - \frac{3}{13}i \\ 0 & 1 & \frac{1}{13} + \frac{5}{13}i \\ 0 & 0 & 0 \end{bmatrix},$$

the solution of the equation

$$\begin{bmatrix} 2+i & 1-i & 1 \\ i & 2+i & i \\ 2+2i & 3 & 1+i \end{bmatrix} \begin{bmatrix} z_1 \\ z_2 \\ z_3 \end{bmatrix} = \begin{bmatrix} 0 \\ 0 \\ 0 \end{bmatrix}$$

is

$$\begin{bmatrix} z_1 \\ z_2 \\ z_3 \end{bmatrix} = c \begin{bmatrix} -\frac{2}{13} + \frac{3}{13}i \\ -\frac{1}{13} - \frac{5}{13}i \\ 1 \end{bmatrix},$$

where c is an arbitrary complex number. This means that

$$\mathbf{N}\left(\begin{bmatrix} 2+i & 1-i & 1 \\ i & 2+i & i \\ 2+2i & 3 & 1+i \end{bmatrix} \right) = \text{Span}\left\{ \begin{bmatrix} -\frac{2}{13} + \frac{3}{13}i \\ -\frac{1}{13} - \frac{5}{13}i \\ 1 \end{bmatrix} \right\}.$$

$\square$

Example 1.2.9. Let a, b, and c be distinct numbers from $\mathbb{K}$ and let $\mathcal{U}$ be the set of polynomials from $\mathcal{P}_n(\mathbb{K})$ such that $p(a) = p(b) = p(c)$, that is,

$$\mathcal{U} = \{p \in \mathcal{P}_n(\mathbb{K}) : p(a) = p(b) = p(c)\}.$$

Show that $\mathcal{U}$ is a subspace of $\mathcal{P}_n(\mathbb{K})$.

Solution. If $p_1, p_2 \in \mathcal{U}$, then

$$p_1(a) = p_1(b) = p_1(c) \quad \text{and} \quad p_2(a) = p_2(b) = p_2(c),$$

and consequently

$$p_1(a) + p_2(a) = p_1(b) + p_2(b) = p_1(c) + p_2(c),$$

which means that $p_1 + p_2 \in \mathcal{U}$. Similarly, if $p \in \mathcal{U}$ and $k \in \mathbb{K}$, then

$$p(a) = p(b) = p(c)$$

and consequently
$$kp(a) = kp(b) = kp(c),$$

which means that $kp \in \mathcal{U}$.

Note that in our argument we are not using the fact that elements of $\mathcal{U}$ are polynomials. A similar argument can be used to show that, if $\mathcal{V}$ is any subspace of $\mathcal{F}_{\mathbb{K}}(S)$ and $s_1, s_2, \ldots, s_n \in S$, then

$$\mathcal{U} = \{f \in \mathcal{V} : f(s_1) = f(s_2) = \cdots = f(s_n)\}$$

is a subspace of $\mathcal{V}$.   □

Now we are going to consider some general properties of subspaces. We begin with the following simple observation that will help us show that every subspace of a vector space is a vector space itself.

Lemma 1.2.10. *If $\mathcal{U}$ is a subspace of a vector space $\mathcal{V}$, then*

(a) *If $\mathbf{v} \in \mathcal{U}$, then $-\mathbf{v} \in \mathcal{U}$;*

(b) $\mathbf{0} \in \mathcal{U}$.

Proof. To show that (a) holds we note that for any $\mathbf{v} \in \mathcal{V}$ we have

$$-\mathbf{v} = (-1)\mathbf{v},$$

so if $\mathbf{v} \in \mathcal{U}$, then $-\mathbf{v} = (-1)\mathbf{v} \in \mathcal{U}$.

To show that (b) holds we take an arbitrary $\mathbf{u} \in \mathcal{U}$ and note that

$$\mathbf{0} = \mathbf{u} - \mathbf{u} = \mathbf{u} + (-1)\mathbf{u} \in \mathcal{U}.$$

□

Theorem 1.2.11. *A subspace $\mathcal{U}$ of a vector space $\mathcal{V}$ is a vector space itself.*

Proof. First we note that, by the definition of a subspace, for every $\mathbf{v}, \mathbf{w} \in \mathcal{U}$ the element $\mathbf{v} + \mathbf{w}$ is in $\mathcal{U}$ and for every $\mathbf{v} \in \mathcal{U}$ and every number $c \in \mathbb{K}$ the element $c\mathbf{v}$ is in $\mathcal{U}$. Moreover, by Lemma 1.2.10, $\mathbf{0} \in \mathcal{U}$ and $-\mathbf{v} \in \mathcal{U}$ for every $\mathbf{v} \in \mathcal{U}$. Finally, the eight conditions in the definition of a vector space are satisfied for elements of $\mathcal{U}$ because they are satisfied for all elements of $\mathcal{V}$. $\square$

Definition 1.2.12. Let $\mathcal{V}_1, \ldots, \mathcal{V}_n$ be subspaces of a vector space $\mathcal{V}$. The set of all possible sums of the form

$$\mathbf{v}_1 + \cdots + \mathbf{v}_n,$$

where $\mathbf{v}_1 \in \mathcal{V}_1, \ldots, \mathbf{v}_n \in \mathcal{V}_n$, is denoted by

$$\mathcal{V}_1 + \cdots + \mathcal{V}_n,$$

that is,

$$\mathcal{V}_1 + \cdots + \mathcal{V}_n = \{\mathbf{v}_1 + \cdots + \mathbf{v}_n : \mathbf{v}_1 \in \mathcal{V}_1, \ldots, \mathbf{v}_n \in \mathcal{V}_n\}.$$

This operation of "addition of subspaces" has properties similar to ordinary addition.

Theorem 1.2.13. *Let $\mathcal{U}$, $\mathcal{V}$, and $\mathcal{W}$ be subspaces of a vector space $\mathcal{X}$. Then*

$$\mathcal{U} + \mathcal{V} = \mathcal{V} + \mathcal{U}$$

and

$$(\mathcal{U} + \mathcal{V}) + \mathcal{W} = \mathcal{V} + (\mathcal{U} + \mathcal{W}).$$

Proof. The properties follow immediately from the fact that

$$\mathbf{u} + \mathbf{v} = \mathbf{v} + \mathbf{u} \quad \text{and} \quad \mathbf{u} + (\mathbf{v} + \mathbf{w}) = (\mathbf{u} + \mathbf{w}) + \mathbf{v}$$

for any $\mathbf{u}, \mathbf{v}, \mathbf{w} \in \mathcal{X}$. $\square$

While there are some similarities, there are also some differences. For example, in general $\mathcal{U} + \mathcal{W} = \mathcal{V} + \mathcal{W}$ does not imply $\mathcal{U} = \mathcal{V}$.

Theorem 1.2.14. *If $\mathcal{V}_1, \ldots, \mathcal{V}_n$ are subspaces of a vector space $\mathcal{V}$, then the set $\mathcal{V}_1 + \cdots + \mathcal{V}_n$ is a subspace of $\mathcal{V}$.*

Proof. If $\mathbf{v} \in \mathcal{V}_1 + \cdots + \mathcal{V}_n$, then

$$\mathbf{v} = \mathbf{v}_1 + \cdots + \mathbf{v}_n,$$

for some $\mathbf{v}_j \in \mathcal{V}_j$. For any number $c \in \mathbb{K}$ we have

$$c\mathbf{v} = c(\mathbf{v}_1 + \cdots + \mathbf{v}_n) = c\mathbf{v}_1 + \cdots + c\mathbf{v}_n,$$

which shows that $c\mathbf{v} \in \mathcal{V}_1 + \cdots + \mathcal{V}_n$.

If $\mathbf{v}, \mathbf{w} \in \mathcal{V}_1 + \cdots + \mathcal{V}_n$, then

$$\mathbf{v} = \mathbf{v}_1 + \cdots + \mathbf{v}_n \quad \text{and} \quad \mathbf{w} = \mathbf{w}_1 + \cdots + \mathbf{w}_n,$$

for some vectors $\mathbf{v}_j, \mathbf{w}_j \in \mathcal{V}_j$. Since

$$\mathbf{v} + \mathbf{w} = \mathbf{v}_1 + \cdots + \mathbf{v}_n + \mathbf{w}_1 + \cdots + \mathbf{w}_n = \mathbf{v}_1 + \mathbf{w}_1 + \cdots + \mathbf{v}_n + \mathbf{w}_n,$$

we have $\mathbf{v} + \mathbf{w} \in \mathcal{V}_1 + \cdots + \mathcal{V}_n$. $\qquad\square$

Definition 1.2.15. Let $\mathbf{v}_1, \ldots, \mathbf{v}_n$ be elements of a vector space $\mathcal{V}$. Any vector of $\mathcal{V}$ of the form

$$x_1\mathbf{v}_1 + \cdots + x_n\mathbf{v}_n,$$

where $x_1, \ldots, x_n \in \mathbb{K}$, is called a *linear combination* of $\mathbf{v}_1, \ldots, \mathbf{v}_n$. The set of all linear combinations of the vectors $\mathbf{v}_1, \ldots, \mathbf{v}_n$ is denoted by $\mathrm{Span}\{\mathbf{v}_1, \ldots, \mathbf{v}_n\}$, that is,

$$\mathrm{Span}\{\mathbf{v}_1, \ldots, \mathbf{v}_n\} = \{x_1\mathbf{v}_1 + \cdots + x_n\mathbf{v}_n : x_1, \ldots, x_n \in \mathbb{K}\}.$$

The set $\mathrm{Span}\{\mathbf{v}_1, \ldots, \mathbf{v}_n\}$ is called the *linear span* (or simply *span*) of vectors $\mathbf{v}_1, \ldots, \mathbf{v}_n$.

Linear spans play an important role in defining subspaces in vector spaces.

Theorem 1.2.16. *If $\mathbf{v}_1, \ldots, \mathbf{v}_n$ be elements of a vector space $\mathcal{V}$, then the set $\mathrm{Span}\{\mathbf{v}_1, \ldots, \mathbf{v}_n\}$ is a subspace of $\mathcal{V}$.*

Proof. Since

$$\mathrm{Span}\{\mathbf{v}_1, \ldots, \mathbf{v}_n\} = \mathbb{K}\mathbf{v}_1 + \cdots + \mathbb{K}\mathbf{v}_n,$$

the result is a consequence of Theorem 1.2.14. $\qquad\square$

If $\mathcal{U} = \mathrm{Span}\{\mathbf{v}_1, \ldots, \mathbf{v}_n\}$, we say that $\{\mathbf{v}_1, \ldots, \mathbf{v}_n\}$ is a *spanning set* for $\mathcal{U}$. Note that $\mathrm{Span}\{\mathbf{v}_1, \ldots, \mathbf{v}_n\}$ is the smallest vector space containing vectors $\mathbf{v}_1, \ldots, \mathbf{v}_n$.

Example 1.2.17. $\mathcal{P}_n(\mathbb{K}) = \text{Span}\{1, t, t^2, .., t^n\}$.

Example 1.2.18. The set of all real-valued solutions of the differential equation

$$y' - 3y = 0$$

is a subspace on $\mathcal{D}_{\mathbb{R}}(\mathbb{R})$. This subspace is $\text{Span}\{e^{3t}\}$.

Example 1.2.19. The set of all real-valued solutions of the differential equation

$$y'' + y = 0$$

is a subspace of the vector space $\mathcal{D}_{\mathbb{R}}(\mathbb{R})$. This subspace is $\text{Span}\{\cos t, \sin t\}$.

The choice of a spanning set is not unique. For example, we have

$$\text{Span}\{\cos t, \sin t\} = \text{Span}\{\cos t + \sin t, \cos t - \sin t\}.$$

Indeed, if $f \in \text{Span}\{\cos t, \sin t\}$, then for some $a, b \in \mathbb{K}$ we have

$$f(t) = a\cos t + b\sin t = \frac{a+b}{2}(\cos t + \sin t) + \frac{a-b}{2}(\cos t - \sin t),$$

so $\text{Span}\{\cos t, \sin t\} \subseteq \text{Span}\{\cos t + \sin t, \cos t - \sin t\}$.
Similarly, if $g \in \text{Span}\{\cos t + \sin t, \cos t - \sin t\}$, then for some $c, d \in \mathbb{K}$ we have

$$g(t) = c(\cos t + \sin t) + d(\cos t - \sin t) = (c+d)\cos t + (c-d)\sin t,$$

so $\text{Span}\{\cos t + \sin t, \cos t - \sin t\} \subseteq \text{Span}\{\cos t, \sin t\}$.
In the remainder of this section we investigate what changes to the spanning set leave the spanned subspace unchanged.

Theorem 1.2.20. *Let* $\mathbf{v}_1, \ldots, \mathbf{v}_n$ *and* $\mathbf{v}$ *be elements of a vector space* V. *If* $\mathbf{v} \in \text{Span}\{\mathbf{v}_1, \ldots, \mathbf{v}_n\}$, *then*

$$\text{Span}\{\mathbf{v}_1, \ldots, \mathbf{v}_n, \mathbf{v}\} = \text{Span}\{\mathbf{v}_1, \ldots, \mathbf{v}_n\}.$$

Proof. Clearly, $\text{Span}\{\mathbf{v}_1, \ldots, \mathbf{v}_n\} \subseteq \text{Span}\{\mathbf{v}_1, \ldots, \mathbf{v}_n, \mathbf{v}\}$.
Now consider a $\mathbf{w} \in \text{Span}\{\mathbf{v}_1, \ldots, \mathbf{v}_n, \mathbf{v}\}$. Then there are numbers $a_1, \ldots, a_n, a_{n+1} \in \mathbb{K}$ such that

$$\mathbf{w} = a_1\mathbf{v}_1 + \cdots + a_n\mathbf{v}_n + a_{n+1}\mathbf{v}.$$

Since $\mathbf{v} \in \mathrm{Span}\{\mathbf{v}_1, \ldots, \mathbf{v}_n\}$, there are numbers $b_1, \ldots, b_n \in \mathbb{K}$ such that

$$\mathbf{v} = b_1 \mathbf{v}_1 + \cdots + b_n \mathbf{v}_n.$$

Then

$$\begin{aligned}
\mathbf{w} &= a_1 \mathbf{v}_1 + \cdots + a_n \mathbf{v}_n + a_{n+1} \mathbf{v} \\
&= a_1 \mathbf{v}_1 + \cdots + a_n \mathbf{v}_n + a_{n+1}(b_1 \mathbf{v}_1 + \cdots + b_n \mathbf{v}_n) \\
&= (a_1 + a_{n+1} b_1)\mathbf{v}_1 + \cdots + (a_n + a_{n+1} b_n)\mathbf{v}_n.
\end{aligned}$$

Thus $\mathbf{w} \in \mathrm{Span}\{\mathbf{v}_1, \ldots, \mathbf{v}_n\}$. □

The next theorem gives us a condition that can be use to check whether two sets of vectors span the same subspace.

> **Theorem 1.2.21.** *Let $\mathbf{u}_1, \ldots, \mathbf{u}_k$ and $\mathbf{v}_1, \ldots, \mathbf{v}_m$ be elements of a vector space $\mathcal{V}$. Then*
>
> $$\mathrm{Span}\{\mathbf{u}_1, \ldots, \mathbf{u}_k\} = \mathrm{Span}\{\mathbf{v}_1, \ldots, \mathbf{v}_m\}$$
>
> *if and only if*
>
> $$\mathbf{u}_1, \ldots, \mathbf{u}_k \in \mathrm{Span}\{\mathbf{v}_1, \ldots, \mathbf{v}_m\} \quad and \quad \mathbf{v}_1, \ldots, \mathbf{v}_m \in \mathrm{Span}\{\mathbf{u}_1, \ldots, \mathbf{u}_k\}.$$

Proof. If $\mathrm{Span}\{\mathbf{u}_1, \ldots, \mathbf{u}_k\} = \mathrm{Span}\{\mathbf{v}_1, \ldots, \mathbf{v}_m\}$, then clearly we must have $\mathbf{u}_1, \ldots, \mathbf{u}_k \in \mathrm{Span}\{\mathbf{v}_1, \ldots, \mathbf{v}_m\}$ and $\mathbf{v}_1, \ldots, \mathbf{v}_m \in \mathrm{Span}\{\mathbf{u}_1, \ldots, \mathbf{u}_k\}$.

On the other hand, if $\mathbf{u}_1, \ldots, \mathbf{u}_k \in \mathrm{Span}\{\mathbf{v}_1, \ldots, \mathbf{v}_m\}$ and $\mathbf{v}_1, \ldots, \mathbf{v}_m \in \mathrm{Span}\{\mathbf{u}_1, \ldots, \mathbf{u}_k\}$, then

$$\mathrm{Span}\{\mathbf{u}_1, \ldots, \mathbf{u}_k, \mathbf{v}_1, \ldots, \mathbf{v}_m\} = \mathrm{Span}\{\mathbf{u}_1, \ldots, \mathbf{u}_k\}$$

and

$$\mathrm{Span}\{\mathbf{u}_1, \ldots, \mathbf{u}_k, \mathbf{v}_1, \ldots, \mathbf{v}_m\} = \mathrm{Span}\{\mathbf{v}_1, \ldots, \mathbf{v}_m\},$$

by Theorem 1.2.20. Hence $\mathrm{Span}\{\mathbf{u}_1, \ldots, \mathbf{u}_k\} = \mathrm{Span}\{\mathbf{v}_1, \ldots, \mathbf{v}_m\}$. □

In the next three corollaries we show that "elementary operations" on the vectors $\mathbf{v}_1, \ldots, \mathbf{v}_n$ do not affect the vector space $\mathrm{Span}\{\mathbf{v}_1, \ldots, \mathbf{v}_n\}$.

> **Corollary 1.2.22.** *For any $1 \leq i < j \leq n$ we have*
>
> $$\mathrm{Span}\{\mathbf{v}_1, \ldots, \mathbf{v}_j, \ldots, \mathbf{v}_i, \ldots, \mathbf{v}_n\} = \mathrm{Span}\{\mathbf{v}_1, \ldots, \mathbf{v}_i, \ldots, \mathbf{v}_j, \ldots, \mathbf{v}_n\}.$$

Proof. This is a direct consequence of Theorem 1.2.21. □

While the above lemma says that interchanging the position of two vectors in $\{\mathbf{v}_1, \ldots, \mathbf{v}_n\}$ does not affect the span, it follows that writing these vectors in any order will have no affect on the span. More precisely,

$$\text{Span}\{\mathbf{v}_1, \ldots, \mathbf{v}_n\} = \text{Span}\{\mathbf{v}_{\sigma(1)}, \ldots, \mathbf{v}_{\sigma(n)}\}$$

where σ is any permutation on $\{1, \ldots, n\}$.

Corollary 1.2.23. *For any $j \in \{1, \ldots, n\}$ and any scalar $c \neq 0$ we have*

$$\text{Span}\{\mathbf{v}_1, \ldots, c\mathbf{v}_j, \ldots, \mathbf{v}_n\} = \text{Span}\{\mathbf{v}_1, \ldots, \mathbf{v}_j, \ldots, \mathbf{v}_n\}.$$

Proof. This is a direct consequence of Theorem 1.2.21 and the equality $\mathbf{v}_j = \frac{1}{c}(c\mathbf{v}_j)$. $\qquad\square$

Corollary 1.2.24. *For any $i, j \in \{1, \ldots, n\}$ and any scalar c we have*

$$\text{Span}\{\mathbf{v}_1, \ldots, \mathbf{v}_i + c\mathbf{v}_j, \ldots, \mathbf{v}_n\} = \text{Span}\{\mathbf{v}_1, \ldots, \mathbf{v}_i \ldots, \mathbf{v}_n\}.$$

Proof. This is a direct consequence of Theorem 1.2.21 and the equality $\mathbf{v}_i = (\mathbf{v}_i + c\mathbf{v}_j) - c\mathbf{v}_j$. $\qquad\square$

The operations on the spanning set that do not affect the subspace, described in this section, either do not change or increase the number of spanning vectors. Clearly, if we can describe a subspace using fewer vectors it would make sense to use that smaller spanning set. How can we tell whether it is possible to find a smaller spanning set? This question will be addressed in the next section.

1.3 Linearly independent vectors and bases

In this section we introduce two concepts that are of basic importance in every aspect of linear algebra. In the first definition we describe a property that will provide an answer to the question asked at the end of last section, that is, a property of a set of vectors $\mathbf{v}_1, \ldots, \mathbf{v}_n$ that implies that there is a smaller set of vectors spanning the same vector space.

Definition 1.3.1. Vectors $\mathbf{v}_1, \ldots, \mathbf{v}_n$ in a vector space $\mathcal{V}$ are called *linearly dependent* if the equation

$$x_1 \mathbf{v}_1 + \cdots + x_n \mathbf{v}_n = \mathbf{0}$$

has a nontrivial solution, that is, a solution such that at least one of the numbers $x_1, \ldots, x_n$ is different from 0.

Example 1.3.2. In the complex vector space $\mathbb{C}^3$ the vectors $\begin{bmatrix} i \\ -i \\ 0 \end{bmatrix}$, $\begin{bmatrix} i \\ i \\ 2i \end{bmatrix}$ and $\begin{bmatrix} 0 \\ 1 \\ 1 \end{bmatrix}$, are linearity dependent because

$$\frac{i}{2} \begin{bmatrix} i \\ -i \\ 0 \end{bmatrix} - \frac{i}{2} \begin{bmatrix} i \\ i \\ 2i \end{bmatrix} + \begin{bmatrix} 0 \\ 1 \\ 1 \end{bmatrix} = \begin{bmatrix} 0 \\ 0 \\ 0 \end{bmatrix}.$$

Linear dependence of vectors does not depend on the vector space. More precisely, if vectors $\mathbf{v}_1, \ldots, \mathbf{v}_n$ are linearly dependent in the space $\mathrm{Span}\{\mathbf{v}_1, \ldots, \mathbf{v}_n\}$, the smallest vector space containing vectors $\mathbf{v}_1, \ldots, \mathbf{v}_n$, then they are linearly dependent in every vector space that contains $\mathrm{Span}\{\mathbf{v}_1, \ldots, \mathbf{v}_n\}$ as a subspace. On the other hand, linear dependence can depend on the scalar field $\mathbb{K}$. For example, the vectors 1 and i are linearly dependent in the complex vector space $\mathbb{C}$, because $i \cdot 1 - i = 0$, but they are not linearly dependent in the real vector space $\mathbb{C}$.

Theorem 1.3.3. *If one of the vectors* $\mathbf{v}_1, \ldots, \mathbf{v}_n$ *in a vector space* $\mathcal{V}$ *is* $\mathbf{0}$*, then the vectors* $\mathbf{v}_1, \ldots, \mathbf{v}_n$ *are linearly dependent.*

Proof. If $\mathbf{v}_k = \mathbf{0}$ for some $k \in \{1, \ldots, n\}$, then we can take $x_k = 1$ and $x_m = 0$ for $m \neq k$. With this choice we have

$$x_1 \mathbf{v}_1 + \cdots + x_n \mathbf{v}_n = \mathbf{0},$$

which shows that the vectors are linearly dependent. $\qquad\qquad\square$

The next theorem gives a practical criterion for linear dependence of a set vectors.

Theorem 1.3.4. *A set of n vectors, with $n \geq 2$, is linearly dependent if and only if at least one of the vectors can be expressed as a linear combination of the remaining vectors. In other words, vectors $\mathbf{v}_1, \ldots, \mathbf{v}_n$, with $n \geq 2$, are linearly dependent if and only if there is a $k \in \{1, \ldots, n\}$ such that*

$$\mathbf{v}_k \in \mathrm{Span}\{\mathbf{v}_1, \ldots, \mathbf{v}_{k-1}, \mathbf{v}_{k+1}, \ldots, \mathbf{v}_n\}.$$

Proof. If the vectors $\mathbf{v}_1, \ldots, \mathbf{v}_n$ are linearly dependent, then

$$x_1 \mathbf{v}_1 + \cdots + x_n \mathbf{v}_n = \mathbf{0}$$

where $x_1, \ldots, x_n \in \mathbb{K}$ and $x_k \neq 0$ for some $0 \leq k \leq n$. Then

$$\frac{x_1}{x_k} \mathbf{v}_1 + \cdots + 1\mathbf{v}_k + \cdots + \frac{x_n}{x_k} \mathbf{v}_n = \mathbf{0}$$

and thus

$$\mathbf{v}_k = -\frac{x_1}{x_k} \mathbf{v}_1 - \cdots - \frac{x_{k-1}}{x_k} \mathbf{v}_{k-1} - \frac{x_{k+1}}{x_k} \mathbf{v}_{k+1} - \cdots - \frac{x_n}{x_k} \mathbf{v}_n.$$

But this means that $\mathbf{v}_k \in \mathrm{Span}\{\mathbf{v}_1, \ldots, \mathbf{v}_{k-1}, \mathbf{v}_{k+1}, \ldots, \mathbf{v}_n\}$.

Conversely, if $\mathbf{v}_k \in \mathrm{Span}\{\mathbf{v}_1, \ldots, \mathbf{v}_{k-1}, \mathbf{v}_{k+1}, \ldots, \mathbf{v}_n\}$ for some $0 \leq k \leq n$, then there are $a_1, \ldots, a_{k-1}, a_{k+1}, \ldots, a_n \in \mathbb{K}$ such that

$$\mathbf{v}_k = a_1 \mathbf{v}_1 + \cdots + a_{k-1} \mathbf{v}_{k-1} + a_{k+1} \mathbf{v}_{k+1} + \cdots + a_n \mathbf{v}_n$$

and thus

$$x_1 \mathbf{v}_1 + \cdots + x_n \mathbf{v}_n = \mathbf{0},$$

where $x_m = a_m$ for $m \in \{1, \ldots, k-1, k+1, \ldots, n\}$ and $x_k = -1$, which means that the vectors $\mathbf{v}_1, \ldots, \mathbf{v}_n$ are linearly dependent. $\square$

Definition 1.3.5. Vectors $\mathbf{v}_1, \ldots, \mathbf{v}_n$ in a vector space $\mathcal{V}$ are called *linearly independent* if the only solution of the equation

$$x_1 \mathbf{v}_1 + \cdots + x_n \mathbf{v}_n = \mathbf{0}$$

is the trivial solution, that is $x_1 = \cdots = x_n = 0$.

In other words, vectors are linearly independent if they are not linearly dependent.

Example 1.3.6. Show that the polynomials $1, t, t^2, \ldots, t^n$ are linearly independent in $\mathcal{P}_n(\mathbb{K})$.

Solution. Assume

$$x_0 + x_1 t + \cdots + x_n t^n = 0$$

holds for some numbers $x_1, \ldots, x_n \in \mathbb{K}$.

It result from Appendix C, Theorem 5.15, that

$$x_0 = \cdots = x_n = 0.$$

Consequently the polynomials $1, t, t^2, \ldots, t^n$ are linearly independent. □

The result from the next example gives a characterization of linear independence of the columns of a matrix and is usually proved for matrices with real entries in an introductory Matrix Linear Algebra course. This result will be used later in this chapter.

Example 1.3.7. Let $A \in \mathcal{M}_{n \times n}(\mathbb{K})$ with columns $\mathbf{a}_1, \ldots, \mathbf{a}_n \in \mathbb{K}^n$, that is, $A = [\mathbf{a}_1 \ \ldots \ \mathbf{a}_n]$. Show that the following conditions are equivalent:

(a) The vectors $\mathbf{a}_1, \ldots, \mathbf{a}_n$ are linearly independent;

(b) The matrix A is invertible;

(c) The equation $A \begin{bmatrix} x_1 \\ \vdots \\ x_n \end{bmatrix} = \mathbf{0}$ has only the trivial solution, that is, the

solution $x_1 = \cdots = x_n = 0$.

Solution. The equivalence of (a) and (c) follows easily from the definition of linearly independent vectors. It is also easy to show that (b) implies (c). We only show that (c) implies (b).

Let E_{jk} be the matrix with the (j, k) entry being 1 and all other entries being 0 and let $\alpha \in \mathbb{K}$. For $j \neq k$, we denote

$$R_{jk}(\alpha) = I_n + \alpha E_{jk} \quad \text{and} \quad S_j(\alpha) = I_n + (\alpha - 1)E_{jj},$$

where I_n is the identity $n \times n$ matrix. Note that these matrices have the following properties:

(i) $R_{jk}(\alpha)A$ is the matrix obtained from A by adding the kth row multiplied by α to the jth row;

(ii) $S_j(\alpha)A$ is the matrix obtained from A by multiplying the jth row by α;

(iii) $R_{jk}(\alpha)$ is invertible and $R_{jk}(\alpha)^{-1} = R_{jk}(-\alpha)$;

(iv) If $\alpha \neq 0$, then $S_j(\alpha)$ is invertible and $S_j(\alpha)^{-1} = S_j(\alpha^{-1})$.

Now we assume that the equation $A \begin{bmatrix} x_1 \\ \vdots \\ x_n \end{bmatrix} = \mathbf{0}$ has only the trivial solution.

Suppose that

$$A = \begin{bmatrix} 1 & 0 & \cdots & 0 & a_{1p} & \cdots & a_{1n} \\ 0 & 1 & \cdots & 0 & a_{2p} & \cdots & a_{2,n} \\ \vdots & \vdots & & \vdots & \vdots & & \vdots \\ 0 & 0 & \cdots & 1 & a_{p-1,p} & \cdots & a_{p-1,n} \\ 0 & 0 & \cdots & 0 & a_{pp} & \cdots & a_{pn} \\ 0 & 0 & \cdots & 0 & a_{p+1,p} & \cdots & a_{p+1,n} \\ \vdots & \vdots & & \vdots & \vdots & & \vdots \\ 0 & 0 & \cdots & 0 & a_{n-1,p} & \cdots & a_{n-1,n} \\ 0 & 0 & \cdots & 0 & a_{np} & \cdots & a_{nn} \end{bmatrix},$$

where $p \in \{1, \ldots, n\}$.

If $a_{p,p} = a_{p+1,p} = \cdots = a_{n,p} = 0$, then

$$\begin{bmatrix} -a_{1,p} \\ -a_{2,p} \\ \vdots \\ -a_{p-1,p} \\ 1 \\ 0 \\ \vdots \\ 0 \end{bmatrix}$$

is a nontrivial solution, which contradicts our assumption.

Suppose that $a_{p,p} \neq 0$. We multiply A by $S_p(a_{pp}^{-1})$, to get 1 as the (p, p) entry, and then by $R_{r,p}(-a_{rp})$ for every $r \neq p$. The result is a matrix of the form

$$A' = \begin{bmatrix} 1 & 0 & \cdots & 0 & a'_{1,p+1} & \cdots & a'_{1n} \\ 0 & 1 & \cdots & 0 & a'_{2,p+1} & \cdots & a'_{2n} \\ \vdots & \vdots & & \vdots & \vdots & & \vdots \\ 0 & 0 & \cdots & 1 & a'_{p,p+1} & \cdots & a'_{pn} \\ 0 & 0 & \cdots & 0 & a'_{p+1,p+1} & \cdots & a'_{p+1,n} \\ 0 & 0 & \cdots & 0 & a'_{p+2,p+1} & \cdots & a'_{p+2,n} \\ \vdots & \vdots & & \vdots & \vdots & & \vdots \\ 0 & 0 & \cdots & 0 & a'_{n-1,p+1} & \cdots & a'_{n-1,n} \\ 0 & 0 & \cdots & 0 & a'_{n,p+1} & \cdots & a'_{nn} \end{bmatrix}.$$

Note that, since every matrix of the form $S_j(\alpha)$, with $\alpha \neq 0$, or $R_{jk}(\alpha)$ is invertible, a solution of the equation $A' \begin{bmatrix} x_1 \\ \vdots \\ x_n \end{bmatrix} = \mathbf{0}$ is also a solution of the equation $A \begin{bmatrix} x_1 \\ \vdots \\ x_n \end{bmatrix} = \mathbf{0}$ and consequently the equation $A' \begin{bmatrix} x_1 \\ \vdots \\ x_n \end{bmatrix} = \mathbf{0}$ has only the trivial solution.

If $a_{p,p} = 0$ but $a_{q,p} \neq 0$ for some $q > p$, we first multiply A by $R_{pq}(1)$ to get $a_{q,p}$ as the (p,p) entry, and then we continue as in the case $a_{p,p} \neq 0$. Using induction on p we can show that there are matrices $E_1, \ldots, E_m$ such that

$$E_1 \ldots E_m A = I_n,$$

where each matrix E_k is of the form $S_j(\alpha)$, with $\alpha \neq 0$, or $R_{jk}(\alpha)$. This gives us

$$A = E_m^{-1} \ldots E_1^{-1},$$

proving the result, because a product of invertible matrices is invertible. □

Example 1.3.8. Show that the functions $1, \cos t, \cos 2t$ are linearly independent in $\mathcal{F}_{\mathbb{K}}(\mathbb{R})$.

Solution. Assume

$$a + b \cos t + c \cos 2t = 0$$

for some $a, b, c \in \mathbb{K}$. Substituting $t = 0, \frac{\pi}{2}, \pi$ gives us the system of equations

$$\begin{cases} a & + & b & + & c & = 0, \\ a & & & - & c & = 0, \\ a & - & b & + & c & = 0. \end{cases}$$

Since the only solution of this system is $a = b = c = 0$, the functions $1, \cos t, \cos 2t$ are linearly independent. $\qquad\square$

From Theorem 1.3.3 we get the following obvious corollary.

Corollary 1.3.9. *Linearly independent vectors are nonzero.*

Another easy but useful fact about linearly independent vectors is that if we remove some vectors from a linearly independent set of vectors, then the new smaller set is still linearly independent.

Theorem 1.3.10. *If $\mathbf{v}_1, \ldots, \mathbf{v}_n$ are linearly independent vectors in a vector space $\mathcal{V}$, then for any distinct indices $i_1, \ldots, i_m \in \{1, \ldots, n\}$ the vectors $\mathbf{v}_{i_1}, \ldots, \mathbf{v}_{i_m}$ are linearly independent.*

Proof. Since

$$x_{i_1} \mathbf{v}_{i_1} + \cdots + x_{i_m} \mathbf{v}_{i_m} = x_1 \mathbf{v}_1 + \cdots + x_n \mathbf{v}_n$$

where $x_j = 0$ for every j not in the set $\{i_1, \ldots, i_m\}$, linear independence of the vectors $\mathbf{v}_{i_1}, \ldots, \mathbf{v}_{i_m}$ is an immediate consequence of linear independence of the vectors $\mathbf{v}_1, \ldots, \mathbf{v}_n$. $\qquad\square$

The converse of the statement in Theorem 1.3.10 is not true, in general. More precisely, if every proper subset of $\{\mathbf{v}_1, \ldots, \mathbf{v}_k\}$ is linearly independent, it is not necessarily true that the vectors $\mathbf{v}_1, \ldots, \mathbf{v}_k$ are linearly independent. For example, the vectors $(1, 0), (0, 1), (1, 1)$ are linearly dependent even though any two of them are linearly independent.

If $\mathbf{v} \in \mathrm{Span}\{\mathbf{v}_1, \ldots, \mathbf{v}_k\}$, then $\mathbf{v}$ can be written as a linear combination of vectors $\mathbf{v}_1, \ldots, \mathbf{v}_k$, but this representation is not necessarily unique. For example,

$$(1, -1) \in \mathrm{Span}\{(1, 0), (0, 1), (1, 1)\}$$

and we have

$$(1, -1) = (1, 0) - (0, 1) \quad \text{and} \quad (1, -1) = (1, 1) - 2(0, 1).$$

Uniqueness of representation of a vector in terms of the spanning set is a desirable property that is essential in many arguments in linear algebra and in applications. It turns out that linear independence of the spanning set guarantees uniqueness of the representation. This is one of the main reasons why linear independence is so important in linear algebra.

Theorem 1.3.11. *If* $\mathbf{v}_1, \ldots, \mathbf{v}_n$ *are linearly independent vectors in a vector space* $\mathcal{V}$ *and*

$$c_1\mathbf{v}_1 + \cdots + c_n\mathbf{v}_n = d_1\mathbf{v}_1 + \cdots + d_n\mathbf{v}_n, \qquad (1.1)$$

for some $c_1, \ldots, c_n, d_1, \ldots, d_n \in \mathbb{K}$, *then*

$$c_1 = d_1, c_2 = d_2, \ldots, c_n = d_n.$$

Proof. From (1.1) we get

$$(c_1 - d_1)\mathbf{v}_1 + \cdots + (c_n - d_n)\mathbf{v}_n = \mathbf{0},$$

and thus

$$c_1 - d_1 = \cdots = c_n - d_n = 0,$$

by linear independence of the vectors $\mathbf{v}_1, \ldots, \mathbf{v}_n$. □

The converse of the above theorem is also true:

If for every $\mathbf{v} \in \mathrm{Span}\{\mathbf{v}_1, \ldots, \mathbf{v}_n\}$ *there are unique numbers* $c_1, \ldots, c_n \in \mathbb{K}$ *such that* $\mathbf{v} = c_1\mathbf{v}_1 + \cdots + c_n\mathbf{v}_n$, *then the vectors* $\mathbf{v}_1, \ldots, \mathbf{v}_n$ *are linearly independent.*

Indeed, if $c_1\mathbf{v}_1 + \cdots + c_n\mathbf{v}_n = \mathbf{0}$, then since we also have $0\mathbf{v}_1 + \cdots + 0\mathbf{v}_n = \mathbf{0}$, we must have $c_1 = \cdots = c_n = 0$, by the uniqueness.

Definition 1.3.12. A collection of vectors $\{\mathbf{v}_1, \ldots, \mathbf{v}_n\}$ in a vector space $\mathcal{V}$ is called a *basis* of $\mathcal{V}$ if the following two conditions are satisfied:

(a) The vectors $\mathbf{v}_1, \ldots, \mathbf{v}_n$ are linearly independent;

(b) $\mathcal{V} = \mathrm{Span}\{\mathbf{v}_1, \ldots, \mathbf{v}_n\}$.

If $\{\mathbf{v}_1, \ldots, \mathbf{v}_n\}$ is a basis of a vector space $\mathcal{V}$, then every vector in $\mathcal{V}$ has a unique representation in the form $\mathbf{v} = c_1\mathbf{v}_1 + \cdots + c_n\mathbf{v}_n$, by Theorem 1.3.11. Note that Theorem 1.3.4 implies that a basis in a vector space $\mathcal{V}$ is minimal in the following sense: If you remove even one vector from a basis in $\mathcal{V}$, it will no longer span $\mathcal{V}$.

Example 1.3.13. Show that the set $\{1, t, t^2, \ldots, t^n\}$ is a basis in $\mathcal{P}_n(\mathbb{K})$.

Solution. This result is a consequence of the Examples 1.2.17 and 1.3.6. $\square$

Example 1.3.14. The set of all real solutions of the differential equation

$$y''' - y' = 0$$

is a real vector space. The set of functions $\{1, e^t, e^{-t}\}$ is a basis of this vector space.

Example 1.3.15. The set of matrices

$$\left\{ \begin{bmatrix} 0 & 1 \\ 0 & 0 \end{bmatrix}, \begin{bmatrix} 0 & 0 \\ 1 & 0 \end{bmatrix}, \begin{bmatrix} 0 & 0 \\ 0 & 1 \end{bmatrix} \right\}$$

is a basis in the vector space of all matrices of the form $\begin{bmatrix} 0 & a \\ b & c \end{bmatrix}$, where a, b and c are arbitrary numbers from $\mathbb{K}$.

Example 1.3.16. Let $\mathbf{v}_1, \ldots, \mathbf{v}_n$ be arbitrary vectors in a vector space $\mathcal{V}$. If $\mathbf{v}_1 \neq \mathbf{0}$, then we say that $\mathbf{v}_1$ is a pivot. For $k \geq 2$, $\mathbf{v}_k$ is called a pivot if $\mathbf{v}_k \notin \mathrm{Span}\{\mathbf{v}_1, \ldots, \mathbf{v}_{k-1}\}$. Show that the set of all pivots in $\{\mathbf{v}_1, \ldots, \mathbf{v}_n\}$ is a basis of $\mathrm{Span}\{\mathbf{v}_1, \ldots, \mathbf{v}_n\}$.

Solution. First we show by induction on k that

$$\mathrm{Span}\{\mathbf{v}_1, \ldots, \mathbf{v}_k\} = \mathrm{Span}\{\mathbf{v}_{j_1}, \ldots, \mathbf{v}_{j_m}\},$$

where $\mathbf{v}_{j_1}, \ldots, \mathbf{v}_{j_m}$ are the pivots such that $1 \leq j_1 < \cdots < j_m \leq k$. For $k = 1$ the result is trivial.

Assume now that

$$\mathrm{Span}\{\mathbf{v}_1, \ldots, \mathbf{v}_k\} = \mathrm{Span}\{\mathbf{v}_{j_1}, \ldots, \mathbf{v}_{j_m}\},$$

where $\mathbf{v}_{j_1}, \ldots, \mathbf{v}_{j_m}$ are the pivots such that $1 \leq j_1 < \cdots < j_m \leq k$.

If $\mathbf{v}_{k+1}$ is a pivot, then

$$\mathrm{Span}\{\mathbf{v}_1, \ldots, \mathbf{v}_k, \mathbf{v}_{k+1}\} = \mathrm{Span}\{\mathbf{v}_{j_1}, \ldots, \mathbf{v}_{j_m}, \mathbf{v}_{k+1}\}.$$

If we let $j_{m+1} = k+1$, then the vectors $\mathbf{v}_{j_1}, \ldots, \mathbf{v}_{j_m}, \mathbf{v}_{j_{m+1}}$ are the pivots such that $1 \leq j_1 < \cdots < j_m < j_{m+1} \leq k+1$.

If $\mathbf{v}_{k+1}$ is not a pivot, then $\mathbf{v}_{k+1} \in \mathrm{Span}\{\mathbf{v}_1, \ldots, \mathbf{v}_k\} = \mathrm{Span}\{\mathbf{v}_{j_1}, \ldots, \mathbf{v}_{j_m}\}$. Consequently,

$$\mathrm{Span}\{\mathbf{v}_1, \ldots, \mathbf{v}_k, \mathbf{v}_{k+1}\} = \mathrm{Span}\{\mathbf{v}_{j_1}, \ldots, \mathbf{v}_{j_m}\}$$

and the vectors $\mathbf{v}_{j_1}, \ldots, \mathbf{v}_{j_m}$ are the pivots such that $1 \leq j_1 < \cdots < j_m \leq k+1$.

Now we show by induction on k that, if $\mathbf{v}_{j_1}, \ldots, \mathbf{v}_{j_m}$ are the pivots such that $1 \leq j_1 < \cdots < j_m \leq k$, then the vectors $\mathbf{v}_{j_1}, \ldots, \mathbf{v}_{j_m}$ are linearly independent. The statement is trivially true for $k = 1$. Now assume that it is true for some $k \geq 1$ and that $\mathbf{v}_{j_1}, \ldots, \mathbf{v}_{j_m}$ are the pivots such that $1 \leq j_1 < \cdots < j_m \leq k+1$. Let

$$x_1 \mathbf{v}_{j_1} + \cdots + x_m \mathbf{v}_{j_m} = \mathbf{0}$$

for some $x_1, \ldots, x_m \in \mathbb{K}$. If $x_m \neq 0$, then $\mathbf{v}_{j_m} \in \mathrm{Span}\{\mathbf{v}_{j_1}, \ldots, \mathbf{v}_{j_{m-1}}\}$, contradicting the assumption that $\mathbf{v}_{j_m}$ is a pivot. Consequently, $x_m = 0$ and, since $1 \leq j_1 < \cdots < j_{m-1} \leq k$, we must have

$$x_1 = \cdots = x_{m-1} = 0,$$

by our inductive assumption. This shows that the vectors $\mathbf{v}_{j_1}, \ldots, \mathbf{v}_{j_m}$ are linearly independent.

We have shown that the set of the pivots in $\{\mathbf{v}_1, \ldots, \mathbf{v}_k\}$ is a basis of $\mathrm{Span}\{\mathbf{v}_1, \ldots, \mathbf{v}_k\}$, for any $k \leq n$. If we take $k = n$, we get the desired result. $\qquad\square$

Example 1.3.17. Let E_{jk} be the $m \times n$ matrix such that the (j, k) entry is 1 and the all other entries are 0. The set of matrices $\{E_{jk}, 1 \leq j \leq m, 1 \leq k \leq n\}$ is a basis of the vector space $\mathcal{M}_{m \times n}(\mathbb{K})$.

Example 1.3.18. The standard Gaussian elimination method can be applied to a matrix with complex entries in the same way as in the case of a matrix with real entries. For any $A \in \mathcal{M}_{m \times n}(\mathbb{K})$ the set of pivot columns is a basis of $\mathbf{C}(A)$, that is, the vector subspace of $\mathbb{K}^m$ spanned by the columns of A.

Definition 1.3.19. Let $\{\mathbf{v}_1, \ldots, \mathbf{v}_n\}$ be a basis of a vector space $\mathcal{V}$ and let $\mathbf{x} \in \mathcal{V}$. The unique numbers $c_1, \ldots, c_n$ such that $\mathbf{x} = c_1 \mathbf{v}_1 + \cdots + c_n \mathbf{v}_n$ are called the coordinates of $\mathbf{x}$ in the basis $\{\mathbf{v}_1, \ldots, \mathbf{v}_n\}$.

Example 1.3.20. Let a be an arbitrary number in $\mathbb{K}$. Show that

$$\{1, t - a, \ldots, (t-a)^n\}$$

is a basis of $\mathcal{P}_n(\mathbb{K})$ and determine the coordinates of an arbitrary polynomial in this basis.

Solution. We first show that the polynomials $1, t - a, \ldots, (t-a)^n$ are linearly independent. Suppose that we have

$$x_0 + x_1(t-a) + \cdots + x_n(t-a)^n = 0,$$

where $x_0, \ldots, x_n$ are some numbers from $\mathbb{K}$.

If we let $t = a$ we get $x_0 = 0$. Next we differentiate the above equality and get

$$x_1 + 2x_2(t-a) + \cdots + nx_n(t-a)^{n-1} = 0.$$

Again we let $t = a$ and get $x_1 = 0$. We continue in this way to show that $x_2 = \cdots = x_n = 0$. Consequently the polynomials $1, t - a, \ldots, (t-a)^n$ are linearly independent.

To finish the proof we will show that for any polynomial $p \in \mathcal{P}_n(\mathbb{K})$ we have

$$p(t) = p(a) + p'(a)(t-a) + \cdots + \frac{1}{n!}p^{(n)}(a)(t-a)^n. \qquad (1.2)$$

Let $p \in \mathcal{P}_n(\mathbb{K})$ be such that

$$p(t) = a_0 + a_1 t + \cdots + a_n t^n,$$

for some $a_0, a_1, \ldots, a_n \in \mathbb{K}$. Since we can write

$$p(t) = a_0 + a_1(t - a + a) + \cdots + a_n(t - a + a)^n,$$

it is easy to verify, using the binomial expansion, that there are $b_0, b_1, \ldots, b_n \in \mathbb{K}$ such that

$$p(t) = b_0 + b_1(t-a) + \cdots + b_n(t-a)^n.$$

Thus $\mathrm{Span}\{1, t - a, \ldots, (t-a)^n\} = \mathcal{P}_n(\mathbb{K})$.

Since $p(a) = b_0$ and, for $k = 1, 2, \ldots, n$, $p^{(k)}(a) = b_k k!$, we have $b_k = \frac{1}{k!}p^{(k)}(a)$. Hence

$$p(t) = p(a) + p'(a)(t-a) + \cdots + \frac{1}{n!}p^{(n)}(a)(t-a)^n.$$

Note that in this example we use the derivative from Appendix C. $\qquad \square$

Example 1.3.21. Let a be an arbitrary number in $\mathbb{K}$ and let

$$\mathcal{U} = \{p \in \mathcal{P}_n(\mathbb{K}) : p(a) = p'(a) = p''(a) = 0\}.$$

(a) Show that $\mathcal{U}$ is a vector subspace of $\mathcal{P}_n(\mathbb{K})$.

(b) Find a basis $\mathcal{U}$.

(c) Determine the coordinates of an arbitrary polynomial from $\mathcal{U}$ in that basis.

Solution. We are going to use the result from Example 1.3.20. If $p \in \mathcal{P}_n(\mathbb{K})$ and $p(a) = p'(a) = p''(a) = 0$, then

$$p(t) = \frac{1}{3!}p'''(a)(t-a)^3 + \cdots + \frac{1}{n!}p^{(n)}(a)(t-a)^n, \qquad (1.3)$$

by (1.2). Consequently,

$$\mathcal{U} = \mathrm{Span}\{(t-a)^3, \ldots, (t-a)^n\}$$

and, since the functions $(t-a)^3, \ldots, (t-a)^n$ are linearly independent, the set $\{(t-a)^3, \ldots, (t-a)^n\}$ is a basis of $\mathcal{U}$. Finally, the coordinates of an arbitrary polynomial from $\mathcal{U}$ in that basis are already in (1.3). □

The following theorem can often be used to solve problems involving linear independence of vectors using methods from matrix algebra.

Theorem 1.3.22. *Let $\{\mathbf{v}_1, \ldots, \mathbf{v}_n\}$ be a basis of a vector space $\mathcal{V}$ and let*

$$\mathbf{w}_k = a_{1k}\mathbf{v}_1 + \cdots + a_{nk}\mathbf{v}_n$$

for some $a_{jk} \in \mathbb{K}$, $1 \le j, k \le n$. Then the vectors $\mathbf{w}_1, \ldots, \mathbf{w}_n$ are linearly independent if and only if the columns of the matrix

$$\begin{bmatrix} a_{11} & \cdots & a_{1n} \\ \vdots & & \vdots \\ a_{n1} & \cdots & a_{nn} \end{bmatrix}$$

are linearly independent.

Proof. First note that we can write

$$x_1\mathbf{w}_1 + \cdots + x_n\mathbf{w}_n = \left(\begin{bmatrix} a_{11} & \cdots & a_{1n} \end{bmatrix} \begin{bmatrix} x_1 \\ \vdots \\ x_n \end{bmatrix} \right) \mathbf{v}_1 + \cdots + \left(\begin{bmatrix} a_{n1} & \cdots & a_{nn} \end{bmatrix} \begin{bmatrix} x_1 \\ \vdots \\ x_n \end{bmatrix} \right) \mathbf{v}_n.$$

Now, because the vectors $\mathbf{v}_1, \ldots, \mathbf{v}_n$ are linearly independent,

$$x_1\mathbf{w}_1 + \cdots + x_n\mathbf{w}_n = \mathbf{0}$$

is equivalent to

$$\begin{bmatrix} a_{11} & \cdots & a_{1n} \end{bmatrix} \begin{bmatrix} x_1 \\ \vdots \\ x_n \end{bmatrix} = \cdots = \begin{bmatrix} a_{n1} & \cdots & a_{nn} \end{bmatrix} \begin{bmatrix} x_1 \\ \vdots \\ x_n \end{bmatrix} = 0$$

which can be written as

$$\begin{bmatrix} a_{11} & \cdots & a_{1n} \\ \vdots & & \vdots \\ a_{n1} & \cdots & a_{nn} \end{bmatrix} \begin{bmatrix} x_1 \\ \vdots \\ x_n \end{bmatrix} = \begin{bmatrix} 0 \\ \vdots \\ 0 \end{bmatrix}$$

or, equivalently,

$$x_1 \begin{bmatrix} a_{11} \\ \vdots \\ a_{n1} \end{bmatrix} + \cdots + x_n \begin{bmatrix} a_{1n} \\ \vdots \\ a_{nn} \end{bmatrix} = \begin{bmatrix} 0 \\ \vdots \\ 0 \end{bmatrix}.$$

Therefore the vectors $\mathbf{w}_1, \ldots, \mathbf{w}_n$ are linearly independent if and only if the

vectors $\begin{bmatrix} a_{11} \\ \vdots \\ a_{n1} \end{bmatrix}, \ldots, \begin{bmatrix} a_{1n} \\ \vdots \\ a_{nn} \end{bmatrix}$ are linearly independent, proving the theorem. $\square$

Example 1.3.23. Let $\mathbf{v}_1, \mathbf{v}_2, \mathbf{v}_3$ be arbitrary vectors in a vector space. Show that the vectors

$$2\mathbf{v}_1 + 3\mathbf{v}_2 + \mathbf{v}_3, \; 5\mathbf{v}_1 + 2\mathbf{v}_2 + 8\mathbf{v}_3, \; 2\mathbf{v}_1 + 7\mathbf{v}_2 - 3\mathbf{v}_3$$

are linearly dependent and express one of them as a linear combination of the remaining two.

Solution. Because the reduced row echelon form of the matrix

$$\begin{bmatrix} 2 & 5 & 2 \\ 3 & 2 & 7 \\ 1 & 8 & -3 \end{bmatrix}$$

is

$$\begin{bmatrix} 1 & 0 & \frac{31}{11} \\ 0 & 1 & -\frac{8}{11} \\ 0 & 0 & 0 \end{bmatrix}$$

the vectors are linearly dependent and we have

$$2\mathbf{v}_1 + 7\mathbf{v}_2 - 3\mathbf{v}_3 = \frac{31}{11}(2\mathbf{v}_1 + 3\mathbf{v}_2 + \mathbf{v}_3) - \frac{8}{11}(5\mathbf{v}_1 + 2\mathbf{v}_2 + 8\mathbf{v}_3).$$

□

Example 1.3.24. Show that the polynomials

$$t(t-1)^2, \ t^3 + 2t, \ (t+1)^3, \ (t+1)^2$$

are linearly independent in $\mathcal{P}_3(\mathbb{K})$.

Solution. If we use $\{1, t, t^2, t^3\}$ as a basis in $\mathcal{P}_3(\mathbb{K})$, then the matrix in Theorem 1.3.22 is

$$\begin{bmatrix} 0 & 0 & 1 & 1 \\ 1 & 1 & 3 & 2 \\ -2 & 0 & 3 & 1 \\ 1 & 2 & 1 & 0 \end{bmatrix}.$$

Since the reduced row echelon form of the above matrix is

$$\begin{bmatrix} 1 & 0 & 0 & 0 \\ 0 & 1 & 0 & 0 \\ 0 & 0 & 1 & 0 \\ 0 & 0 & 0 & 1 \end{bmatrix},$$

the polynomials $t(t-1)^2, t^3 + 2t, (t+1)^3, (t+1)^2$ are linearly independent. □

Example 1.3.25. Find a basis of the vector space

$$V = \text{Span} \left\{ \begin{bmatrix} 1-i & 2 \\ i & 1+i \end{bmatrix}, \begin{bmatrix} -3+5i & -2 \\ 6-3i & -3-5i \end{bmatrix}, \begin{bmatrix} i & 2 \\ 3 & -i \end{bmatrix} \right\}.$$

Solution. We can solve this problem by finding a basis of the vector space

$$\text{Span} \left\{ \begin{bmatrix} 1-i \\ 2 \\ i \\ 1+i \end{bmatrix}, \begin{bmatrix} -3+5i \\ -2 \\ 6-3i \\ -3-5i \end{bmatrix}, \begin{bmatrix} i \\ 2 \\ 3 \\ -i \end{bmatrix} \right\}.$$

The reduced row echelon form of the matrix

$$\begin{bmatrix} 1-i & -3+5\,i & i \\ 2 & -2 & 2 \\ i & 6-3\,i & 3 \\ 1+i & -3-5\,i & -i \end{bmatrix}$$

is

$$\begin{bmatrix} 1 & 0 & \frac{3}{2} \\ 0 & 1 & \frac{1}{2} \\ 0 & 0 & 0 \\ 0 & 0 & 0 \end{bmatrix},$$

and thus the set

$$\left\{ \begin{bmatrix} 1-i & 2 \\ i & 1+i \end{bmatrix}, \begin{bmatrix} -3+5\,i & -2 \\ 6-3\,i & -3-5\,i \end{bmatrix} \right\} \tag{1.4}$$

is linearly independent. Since

$$\begin{bmatrix} i & 2 \\ 3 & -i \end{bmatrix} = \frac{3}{2} \begin{bmatrix} 1-i & 2 \\ i & 1+i \end{bmatrix} + \frac{1}{2} \begin{bmatrix} -3+5\,i & -2 \\ 6-3\,i & -3-5\,i \end{bmatrix},$$

we have

$$\mathcal{V} = \mathrm{Span} \left\{ \begin{bmatrix} 1-i & 2 \\ i & 1+i \end{bmatrix}, \begin{bmatrix} -3+5\,i & -2 \\ 6-3\,i & -3-5\,i \end{bmatrix} \right\},$$

so the set (1.4) is a basis of $\mathcal{V}$. □

Example 1.3.26. Let a, b, and c be distinct numbers from $\mathbb{K}$. Find a basis in the space

$$\mathcal{U} = \{p \in \mathcal{P}_n(\mathbb{K}) : p(a) = p(b) = p(c)\},$$

where $n \geq 3$.

Solution. Let $\kappa = p(a) = p(b) = p(c)$ and $q(t) = (t-a)(t-b)(t-c)$. If $p \in \mathcal{U}$, then

$$p(t) = s(t)q(t) + \kappa,$$

where $s(t)$ is 0 or a polynomial of degree at most $n-3$. In other words,

$$\mathcal{U} = \{sq + \kappa : s \in \mathcal{P}_{n-3}(\mathbb{K}), k \in \mathbb{K}\}.$$

Consequently,

$$\{1, q(t), tq(t), \ldots, t^{n-3}q(t)\}$$

is a basis of $\mathcal{U}$.

□

1.4   Direct sums

In Section 1.2 we introduced sums of subspaces. Now we are going to refine that idea by introducing direct sums which play a much more important role in linear algebra.

Definition 1.4.1. Let $\mathcal{V}_1, \ldots, \mathcal{V}_n$ be subspaces of a vector space $\mathcal{V}$. The sum

$$\mathcal{V}_1 + \cdots + \mathcal{V}_n$$

is called a *direct sum* if every $\mathbf{v} \in \mathcal{V}_1 + \cdots + \mathcal{V}_n$ can be written in a unique way as a sum $\mathbf{v}_1 + \cdots + \mathbf{v}_n$ with $\mathbf{v}_j \in \mathcal{V}_j$ for every $j \in \{1, \ldots, n\}$. To indicate that the sum $\mathcal{V}_1 + \cdots + \mathcal{V}_n$ is direct we write

$$\mathcal{V}_1 \oplus \cdots \oplus \mathcal{V}_n.$$

The condition of uniqueness of the representation $\mathbf{v}_1 + \cdots + \mathbf{v}_n$ with $\mathbf{v}_j \in \mathcal{V}_j$ for every $j \in \{1, \ldots, n\}$ is similar to a condition that characterizes linear independence of vectors. This is not a coincidence. As we will soon see, direct sums and linear independence are closely related. The following theorem gives a simple condition that characterizes direct sums. It should not be a surprise that it is similar to the condition that characterizes linear independence of vectors.

Theorem 1.4.2. *Let $\mathcal{V}_1, \ldots, \mathcal{V}_n$ be subspaces of a vector space $\mathcal{V}$. The sum $\mathcal{V}_1 + \cdots + \mathcal{V}_n$ is direct if and only if the equality*

$$\mathbf{v}_1 + \cdots + \mathbf{v}_n = \mathbf{0}$$

with $\mathbf{v}_j \in \mathcal{V}_j$ for every $j \in \{1, \ldots, n\}$, implies $\mathbf{v}_1 = \cdots = \mathbf{v}_n = \mathbf{0}$.

Proof. Assume that the sum $\mathcal{V}_1 + \cdots + \mathcal{V}_n$ is direct and that

$$\mathbf{v}_1 + \cdots + \mathbf{v}_n = \mathbf{0}$$

with $\mathbf{v}_j \in \mathcal{V}_j$ for every $j \in \{1, \ldots, n\}$. Since we also have

$$\mathbf{0} + \cdots + \mathbf{0} = \mathbf{0}$$

and $\mathbf{0} \in \mathcal{V}_j$ for every $j \in \{1, \ldots, n\}$, we must have $\mathbf{v}_1 = \cdots = \mathbf{v}_n = \mathbf{0}$ by the uniqueness requirement in the definition of direct sums.

Now assume that for any vectors $\mathbf{v}_j \in \mathcal{V}_j$ for $j \in \{1, \ldots, n\}$, the equality $\mathbf{v}_1 + \cdots + \mathbf{v}_n = \mathbf{0}$ implies $\mathbf{v}_1 = \cdots = \mathbf{v}_n = \mathbf{0}$. We need to show the uniqueness property. If

$$\mathbf{u}_1 + \cdots + \mathbf{u}_n = \mathbf{w}_1 + \cdots + \mathbf{w}_n$$

with $\mathbf{u}_j, \mathbf{w}_j \in V_j$ for every $j \in \{1, \ldots, n\}$, then

$$\mathbf{u}_1 - \mathbf{w}_1 + \cdots + \mathbf{u}_n - \mathbf{w}_n = \mathbf{0}.$$

Since $\mathbf{u}_j - \mathbf{w}_j \in V_j$ for all $j \in \{1, \ldots, n\}$, we can conclude that $\mathbf{u}_j = \mathbf{w}_j$ for all $j \in \{1, \ldots, n\}$. □

Note that, if a sum $V_1 + \cdots + V_n$ is a direct sum and $k_1, k_2, \ldots, k_m \in \{1, 2, \ldots, n\}$ are distinct indices, then the sum $V_{k_1} + \cdots + V_{k_m}$ is also direct.

Example 1.4.3. Suppose that $\mathcal{U}$, $\mathcal{U}'$, $\mathcal{V}$, and $\mathcal{V}'$ are subspaces of a vector space $\mathcal{W}$. If

$$\mathcal{U} = \mathcal{U}' \oplus (\mathcal{U} \cap \mathcal{V}) \quad \text{and} \quad \mathcal{V} = \mathcal{V}' \oplus (\mathcal{U} \cap \mathcal{V}),$$

show that

$$\mathcal{U} + \mathcal{V} = \mathcal{U}' \oplus (\mathcal{U} \cap \mathcal{V}) \oplus \mathcal{V}'.$$

Solution. Because it is easy to see that

$$\mathcal{U} + \mathcal{V} = \mathcal{U}' + (\mathcal{U} \cap \mathcal{V}) + \mathcal{V}'$$

we only have to show that the sum is direct. Suppose that

$$\mathbf{u} + \mathbf{w} + \mathbf{v} = \mathbf{0},$$

where $\mathbf{u} \in \mathcal{U}'$, $\mathbf{w} \in \mathcal{U} \cap \mathcal{V}$, and $\mathbf{v} \in \mathcal{V}'$. From

$$\mathbf{v} = -\mathbf{u} - \mathbf{w} \in \mathcal{U}' \oplus (\mathcal{U} \cap \mathcal{V}) = \mathcal{U}$$

we get

$$\mathbf{v} \in \mathcal{V}' \cap \mathcal{U} \subseteq \mathcal{V} \cap \mathcal{U},$$

which means that

$$\mathbf{v} \in \mathcal{V}' \cap (\mathcal{U} \cap \mathcal{V}) = \{\mathbf{0}\}.$$

Since $\mathbf{v} = \mathbf{0}$, we have $\mathbf{u} + \mathbf{w} = \mathbf{0}$ and thus $\mathbf{u} = \mathbf{w} = \mathbf{0}$. □

Theorem 1.4.4. *Suppose V and W are subspaces of a vector space X. The sum $V + W$ is direct if and only if $V \cap W = \{\mathbf{0}\}$.*

Proof. If the sum $V + W$ is direct and $\mathbf{v} = \mathbf{w}$, with $\mathbf{v} \in V$ and $\mathbf{w} \in W$, then the equality $\mathbf{0} = \mathbf{v} - \mathbf{w}$ implies $\mathbf{v} = \mathbf{w} = \mathbf{0}$.

Now, if $V \cap W = \{\mathbf{0}\}$ and $\mathbf{v} + \mathbf{w} = \mathbf{0}$ with $\mathbf{v} \in V$ and $\mathbf{w} \in W$, then $\mathbf{v} = -\mathbf{w}$. Hence $\mathbf{v} = \mathbf{w} = \mathbf{0}$ because $-\mathbf{w} \in W$. □

Example 1.4.5. Show that the set

$$\mathcal{U} = \{p \in \mathcal{P}_n(\mathbb{K}) : t^3 + 1 \text{ divides } p\},$$

where $n \geq 3$, is a subspace of $\mathcal{P}_n(\mathbb{K})$ and that we have

$$\mathcal{U} \oplus \mathcal{P}_2(\mathbb{K}) = \mathcal{P}_n(\mathbb{K}).$$

Solution. It is easy to show, as a consequence of the definitions, that $\mathcal{U}$ is a subspace of the vector space $\mathcal{P}_n(\mathbb{K})$.

If $q \in \mathcal{P}_n(\mathbb{K})$, then there are polynomials r and s such that

$$q = s(t^3 + 1) + r,$$

where $r = 0$ or $\deg r \leq 2$. Since $s(t^3 + 1) \in \mathcal{U}$, we have shown that

$$\mathcal{U} + \mathcal{P}_2(\mathbb{K}) = \mathcal{P}_n(\mathbb{K}).$$

Clearly the sum $\mathcal{U} + \mathcal{P}_2(\mathbb{K})$ is direct.   □

The property in Theorem 1.4.4 does not extend to three or more subspaces. It is possible that $\mathcal{U} \cap \mathcal{V} \cap \mathcal{W} = \{\mathbf{0}\}$, but the sum $\mathcal{U} + \mathcal{V} + \mathcal{W}$ is not direct. We leave showing this as an exercise.

The next theorem makes the following statement precise: The direct sum of direct sums is direct.

Theorem 1.4.6. *Let $n_1, n_2, \ldots, n_m$ be arbitrary positive integers and let $\mathcal{V}_{j,k}$ be subspaces of a vector space $\mathcal{V}$ for $1 \leq k \leq m$ and $1 \leq j \leq n_k$. If the sum*

$$\mathcal{U}_k = \mathcal{V}_{1,k} + \mathcal{V}_{2,k} + \cdots + \mathcal{V}_{n_k,k}$$

is direct for every $1 \leq k \leq m$ and the sum

$$\mathcal{U}_1 + \cdots + \mathcal{U}_m$$

is direct, then the sum

$$\mathcal{V}_{1,1} + \cdots + \mathcal{V}_{n_1,1} + \cdots + \mathcal{V}_{1,m} + \cdots + \mathcal{V}_{n_m,m}$$

is direct and we have

$$\mathcal{V}_{1,1} \oplus \cdots \oplus \mathcal{V}_{n_1,1} \oplus \cdots \oplus \mathcal{V}_{1,m} \oplus \cdots \oplus \mathcal{V}_{n_m,m} = \mathcal{U}_1 \oplus \cdots \oplus \mathcal{U}_m.$$

Proof. Suppose

$$\mathbf{v}_{1,1} + \cdots + \mathbf{v}_{n_1,1} + \cdots + \mathbf{v}_{1,m} + \cdots + \mathbf{v}_{n_m,m} = \mathbf{0}.$$

Since

$$\mathbf{v}_{1,1} + \cdots + \mathbf{v}_{n_1,1} + \cdots + \mathbf{v}_{1,m} + \cdots + \mathbf{v}_{n_m,m}$$
$$= (\mathbf{v}_{1,1} + \cdots + \mathbf{v}_{n_1,1}) + \cdots + (\mathbf{v}_{1,m} + \cdots + \mathbf{v}_{n_m,m}),$$

we have $\mathbf{v}_{1,k} + \cdots + \mathbf{v}_{n_k,k} = \mathbf{0}$ for every $1 \le k \le m$ and thus

$$\mathbf{v}_{1,1} = \cdots = \mathbf{v}_{n_1,1} = \cdots = \mathbf{v}_{1,k} = \cdots = \mathbf{v}_{n_k,k} = \mathbf{0}.$$

This shows that the sum $\mathcal{V}_{1,1} + \cdots + \mathcal{V}_{n_1,1} + \cdots + \mathcal{V}_{1,k} + \cdots + \mathcal{V}_{n_k,k}$ is direct. The last equality in the theorem is an immediate consequence of Theorem 1.2.13. $\square$

Now we formulate and prove a theorem that connects direct sums with linear independence.

Theorem 1.4.7. *If* $\mathbf{v}_1, \ldots, \mathbf{v}_n$ *are linearly independent vectors in a vector space* $\mathcal{V}$*, then the sum*

$$\mathbb{K}\mathbf{v}_1 + \cdots + \mathbb{K}\mathbf{v}_n$$

is direct.

Proof. Assume the vectors $\mathbf{v}_1, \ldots, \mathbf{v}_n$ are linearly independent. If

$$\mathbf{u}_1 + \cdots + \mathbf{u}_n = \mathbf{0}$$

with $\mathbf{u}_j \in \mathbb{K}\mathbf{v}_j$ for $j \in \{1, \ldots, n\}$, then $\mathbf{u}_j = x_j \mathbf{v}_j$ for some $x_1, \ldots, x_n \in \mathbb{K}$. Then

$$x_1 \mathbf{v}_1 + \cdots + x_n \mathbf{v}_n = \mathbf{0}$$

and, since the vectors $\mathbf{v}_1, \ldots, \mathbf{v}_n$ are linearly independent, $x_1 = \cdots = x_n = 0$. But this means that $\mathbf{u}_1 = \cdots = \mathbf{u}_n = \mathbf{0}$, proving that the sum $\mathbb{K}\mathbf{v}_1 + \cdots + \mathbb{K}\mathbf{v}_n$ is direct. $\square$

The conditions in the above theorems are not equivalent. We need an additional assumption to get the implication in the other direction.

Theorem 1.4.8. *If* $\mathbf{v}_1, \ldots, \mathbf{v}_n$ *are nonzero vectors in a vector space* $\mathcal{V}$ *and the sum*

$$\mathbb{K}\mathbf{v}_1 + \cdots + \mathbb{K}\mathbf{v}_n$$

is direct, then the vectors $\mathbf{v}_1, \ldots, \mathbf{v}_n$ *are linearly independent.*

Proof. Assume the vectors $\mathbf{v}_1, \ldots, \mathbf{v}_n$ are nonzero and the sum $\mathbb{K}\mathbf{v}_1 + \cdots + \mathbb{K}\mathbf{v}_n$ is direct. If

$$x_1\mathbf{v}_1 + \cdots + x_n\mathbf{v}_n = \mathbf{0},$$

then we get $x_1\mathbf{v}_1 = \cdots = x_n\mathbf{v}_n = \mathbf{0}$. Since the vectors $\mathbf{v}_1, \ldots, \mathbf{v}_n$ are nonzero, we must have $x_1 = \cdots = x_n = 0$, proving that the vectors $\mathbf{v}_1, \ldots, \mathbf{v}_n$ are linearly independent. $\qquad\square$

Example 1.4.9. Show that the sum

$$\text{Span}\{t^2 + 1, t^2 + 3\} + \text{Span}\{t^3 + t + 2, t^3 + 3t^2\}$$

is a direct sum of subspaces of $\mathcal{P}_3(\mathbb{K})$.

Proof. It suffices to check that the functions $t^2 + 1, t^2 + 3, t^3 + t + 2, t^3 + 3t^2$ are linearly independent and then use Theorems 1.4.6 and 1.4.7. $\qquad\square$

The next theorem describes another desirable property of direct sums. In general, if $\{\mathbf{u}_1, \ldots, \mathbf{u}_k\}$ is a basis of $\mathcal{U}$ and $\{\mathbf{v}_1, \ldots, \mathbf{v}_m\}$ is a basis of $\mathcal{V}$, then $\{\mathbf{u}_1, \ldots, \mathbf{u}_k, \mathbf{v}_1, \ldots, \mathbf{v}_m\}$ need not be a basis of $\mathcal{U} + \mathcal{V}$. It turns out that we don't have this problem if the sum is direct.

Theorem 1.4.10. *Let $\mathcal{V}_1, \ldots, \mathcal{V}_n$ be subspaces of a vector space $\mathcal{V}$ such that the sum*

$$\mathcal{V}_1 + \cdots + \mathcal{V}_n$$

is direct. If $\{\mathbf{v}_{1,j}, \ldots, \mathbf{v}_{k_j,j}\}$ is a basis of $\mathcal{V}_j$ for each $1 \leq j \leq n$, then

$$\{\mathbf{v}_{1,1}, \ldots, \mathbf{v}_{k_1,1}, \ldots, \mathbf{v}_{1,n}, \ldots, \mathbf{v}_{k_n,n}\}$$

is a basis of $\mathcal{V}_1 \oplus \cdots \oplus \mathcal{V}_n$.

Proof. According to Theorem 1.4.6 we have

$$\mathcal{V}_1 \oplus \cdots \oplus \mathcal{V}_n = (\mathbb{K}\mathbf{v}_{1,1} \oplus \cdots \oplus \mathbb{K}\mathbf{v}_{k_1,1}) \oplus \cdots \oplus (\mathbb{K}\mathbf{v}_{1,n} \oplus \cdots \oplus \mathbb{K}\mathbf{v}_{k_n,n})$$

$$= \mathbb{K}\mathbf{v}_{1,1} \oplus \cdots \oplus \mathbb{K}\mathbf{v}_{k_1,1} \oplus \cdots \oplus \mathbb{K}\mathbf{v}_{1,n} \oplus \cdots \oplus \mathbb{K}\mathbf{v}_{k_n,n}$$

and thus

$$\mathcal{V}_1 \oplus \cdots \oplus \mathcal{V}_n = \text{Span}\{\mathbf{v}_{1,1}, \ldots, \mathbf{v}_{k_1,1}, \ldots, \mathbf{v}_{1,n}, \ldots, \mathbf{v}_{k_n,n}\}.$$

Now the result follows by Theorem 1.4.8. $\qquad\square$

Example 1.4.11. Show that the sum

$$\text{Span}\{1, \cos t, \cos 2t\} + \text{Span}\{\sin t, \sin 2t\} \qquad (1.5)$$

is a direct sum of subspaces of $\mathcal{F}_{\mathbb{K}}(\mathbb{R})$ and that

$$\{1, \cos t, \cos 2t, \sin t, \sin 2t\}$$

is a basis of $\text{Span}\{1, \cos t, \cos 2t\} \oplus \text{Span}\{\sin t, \sin 2t\}$.

Solution. If $a_0, a_1, a_2, b_1, b_2 \in \mathbb{K}$ are such that

$$a_0 + a_1 \cos t + a_2 \cos 2t = b_1 \sin t + b_2 \sin 2t,$$

then also

$$a_0 + a_1 \cos(-t) + a_2 \cos(-2t) = b_1 \sin(-t) + b_2 \sin(-2t),$$

which simplifies to

$$a_0 + a_1 \cos t + a_2 \cos 2t = -(b_1 \sin t + b_2 \sin 2t).$$

Consequently,
$$a_0 + a_1 \cos t + a_2 \cos 2t = 0$$

and
$$b_1 \sin t + b_2 \sin 2t = 0.$$

We have shown in Example 1.3.8 that the functions $1, \cos t, \cos 2t$ are linearly independent. It can be shown, using the same approach, that the functions $\sin t, \sin 2t$ are linearly independent. Linear independence of the functions $1, \cos t, \cos 2t$ gives us $a_0 = a_1 = a_2 = 0$ and linear independence of the functions $\sin t, \sin 2t$ gives us $b_1 = b_2 = 0$. Hence, by Theorem 1.4.4, the sum (1.5) is direct and the set $\{1, \cos t, \cos 2t, \sin t, \sin 2t\}$ is its basis, by Theorem 1.4.10. $\qquad\qquad\square$

Example 1.4.12. Show that the sum

$$\mathcal{S}_{n\times n}(\mathbb{K}) + \mathcal{A}_{n\times n}(\mathbb{K})$$

is direct and we have

$$\mathcal{S}_{n\times n}(\mathbb{K}) \oplus \mathcal{A}_{n\times n}(\mathbb{K}) = \mathcal{M}_{n\times n}(\mathbb{K}),$$

where

$$\mathcal{S}_{n\times n}(\mathbb{K}) = \{A \in \mathcal{M}_{n\times n}(\mathbb{K}) : A^T = A\}$$

and
$$\mathcal{A}_{n\times n}(\mathbb{K}) = \left\{ A \in \mathcal{M}_{n\times n}(\mathbb{K}) : A^T = -A \right\}.$$

Solution. If $A \in \mathcal{S}_{n\times n}(\mathbb{K}) \cap \mathcal{A}_{n\times n}(\mathbb{K})$, then $A = A^T = -A^T$ and thus $A = -A$. Consequently, the entries of the matrix A are all 0. This shows that the sum $\mathcal{S}_{n\times n}(\mathbb{K}) + \mathcal{A}_{n\times n}(\mathbb{K})$ is direct, by Theorem 1.4.4.

Now we observe that any matrix $A \in \mathcal{M}_{n\times n}(\mathbb{K})$ can be written in the form

$$A = \frac{1}{2}(A + A^T) + \frac{1}{2}(A - A^T).$$

Since $A + A^T \in \mathcal{S}_{n\times n}(\mathbb{K})$ and $A - A^T \in \mathcal{A}_{n\times n}(\mathbb{K})$, we have $A \in \mathcal{S}_{n\times n}(\mathbb{K}) \oplus \mathcal{A}_{n\times n}(\mathbb{K})$. $\qquad\square$

So far in this section we were concerned with checking whether a given sum of subspaces was direct. Now we are going to investigate a different problem. We would like to be able to decompose a given vector space into subspaces that give us the original vector space as the direct sum of those subspaces. In practice, we often want those subspaces to have some special properties. We begin by proving a simple lemma that gives us the basic ingredient of the construction.

Lemma 1.4.13. *Let $\mathcal{U}$ be a subspace of a vector space $\mathcal{V}$ and let $\mathbf{v} \in \mathcal{V}$. If $\mathbf{v} \notin \mathcal{U}$, then the sum $\mathcal{U} + \mathbb{K}\mathbf{v}$ is direct.*

Proof. Suppose
$$\mathbf{u} + \alpha\mathbf{v} = \mathbf{0}$$

for some $\mathbf{u} \in \mathcal{U}$ and $\alpha \in \mathbb{K}$. If $\alpha \neq 0$, then $\mathbf{v} = -\frac{1}{\alpha}\mathbf{u} \in \mathcal{U}$, which is not possible, because of our assumption. Consequently, $\alpha = 0$ and thus also $\mathbf{u} = \mathbf{0}$. $\qquad\square$

Theorem 1.4.14. *Let $\mathcal{U}$ be a subspace of a vector space $\mathcal{V}$. If $\mathcal{V} = \mathrm{Span}\{\mathbf{v}_1, \ldots, \mathbf{v}_n\}$, then there are linearly independent vectors $\mathbf{v}_{i_1}, \ldots, \mathbf{v}_{i_m} \in \{\mathbf{v}_1, \ldots, \mathbf{v}_n\}$ such that*

$$\mathcal{V} = \mathcal{U} \oplus \mathbb{K}\mathbf{v}_{i_1} \oplus \cdots \oplus \mathbb{K}\mathbf{v}_{i_m}.$$

Proof. If $\mathcal{U} = \mathcal{V}$, then we have nothing to prove.

Now, if $\mathcal{U} \neq \mathcal{V}$, then there is a vector $\mathbf{v}_{i_1} \in \{\mathbf{v}_1, \ldots, \mathbf{v}_n\}$ such that $\mathbf{v}_{i_1} \notin \mathcal{U}$. According to Lemma 1.4.13 the sum $\mathcal{U} + \mathbb{K}\mathbf{v}_{i_1}$ is direct. If $\mathcal{U} \oplus \mathbb{K}\mathbf{v}_{i_1} = \mathcal{V}$, then we are done. If $\mathcal{U} \oplus \mathbb{K}\mathbf{v}_{i_1} \neq \mathcal{V}$, then we continue using mathematical induction. Suppose that we have proven that the the sum

$$\mathcal{U} + \mathbb{K}\mathbf{v}_{i_1} + \cdots + \mathbb{K}\mathbf{v}_{i_k}$$

is direct. If $\mathcal{U} \oplus \mathbb{K}\mathbf{v}_{i_1} \oplus \cdots \oplus \mathbb{K}\mathbf{v}_{i_k} = \mathcal{V}$, then we are done. If $\mathcal{U} \oplus \mathbb{K}\mathbf{v}_{i_1} \oplus \cdots \oplus \mathbb{K}\mathbf{v}_{i_k} \neq \mathcal{V}$, then there is a vector $\mathbf{v}_{i_{k+1}} \in \{\mathbf{v}_1, \ldots, \mathbf{v}_n\}$ which is not in $\mathcal{U} + \mathbb{K}\mathbf{v}_{i_1} + \cdots + \mathbb{K}\mathbf{v}_{i_k}$. But then the sum $\mathcal{U} + \mathbb{K}\mathbf{v}_{i_1} + \cdots + \mathbb{K}\mathbf{v}_{i_k} + \mathbb{K}\mathbf{v}_{i_{k+1}}$ is direct by Lemma 1.4.13 and Theorem 1.4.6.

Since the set $\{\mathbf{v}_1, \ldots, \mathbf{v}_n\}$ has a finite number of elements, we will reach a point when $\mathcal{U} \oplus \mathbb{K}\mathbf{v}_{i_1} \oplus \cdots \oplus \mathbb{K}\mathbf{v}_{i_k} = \mathcal{V}$. $\square$

Corollary 1.4.15. *Let $\mathcal{U}$ be a subspace of a vector space $\mathcal{V}$. If $\mathcal{V} = \mathrm{Span}\{\mathbf{v}_1, \ldots, \mathbf{v}_n\}$, then there is a subspace $\mathcal{W}$ of the vector space $\mathcal{V}$ such that*

$$\mathcal{V} = \mathcal{U} \oplus \mathcal{W}.$$

Proof. $\mathcal{W} = \mathbb{K}\mathbf{v}_{i_1} \oplus \cdots \oplus \mathbb{K}\mathbf{v}_{i_m}$, where $\mathbf{v}_{i_1}, \ldots, \mathbf{v}_{i_m} \in \{\mathbf{v}_1, \ldots, \mathbf{v}_n\}$ are the vectors in Theorem 1.4.14. $\square$

A subspace $\mathcal{W}$ of a vector space $\mathcal{V}$ such that $\mathcal{V} = \mathcal{U} \oplus \mathcal{W}$ will be called a *complement of $\mathcal{U}$ in $\mathcal{V}$*. It is important to remember that a complement is not unique. For example, both $\mathcal{W} = \{(t,t) : t \in \mathbb{R}\}$ and $\mathcal{W} = \{(0,t) : t \in \mathbb{R}\}$ are complements of $\mathcal{U} = \{(t,0) : t \in \mathbb{R}\}$ in the vector space $\mathcal{V} = \mathbb{R}^2 = \{(s,t) : s, t \in \mathbb{R}\}$. In Chapter 3 we introduce a different notion of complements, namely orthogonal complements, and prove that such complements are unique.

Corollary 1.4.16. *Let $\mathcal{V}$ be a vector space. If $\mathcal{V} = \mathrm{Span}\{\mathbf{v}_1, \ldots, \mathbf{v}_n\}$ and the vectors $\mathbf{w}_1, \ldots \mathbf{w}_k \in \mathcal{V}$ are linearly independent, then there are linearly independent vectors $\mathbf{v}_{i_1}, \ldots, \mathbf{v}_{i_m} \in \{\mathbf{v}_1, \ldots, \mathbf{v}_n\}$ such that the set $\{\mathbf{w}_1, \ldots \mathbf{w}_k, \mathbf{v}_{i_1}, \ldots, \mathbf{v}_{i_m}\}$ is a basis of $\mathcal{V}$.*

Proof. We apply Theorem 1.4.14 to the subspace

$$\mathcal{W} = \mathbb{K}\mathbf{w}_1 \oplus \cdots \oplus \mathbb{K}\mathbf{w}_k$$

and we get linearly independent vectors $\mathbf{v}_{i_1}, \ldots, \mathbf{v}_{i_m}$ such that

$$\mathcal{V} = \mathcal{W} \oplus \mathbb{K}\mathbf{v}_{i_1} \oplus \cdots \oplus \mathbb{K}\mathbf{v}_{i_m} = \mathbb{K}\mathbf{w}_1 \oplus \cdots \oplus \mathbb{K}\mathbf{w}_k \oplus \mathbb{K}\mathbf{v}_{i_1} \oplus \cdots \oplus \mathbb{K}\mathbf{v}_{i_m}.$$

This means that the set $\{\mathbf{w}_1, \ldots, \mathbf{w}_k, \mathbf{v}_{i_1}, \ldots, \mathbf{v}_{i_m}\}$ is a basis of $\mathcal{V}$. $\square$

Corollary 1.4.17. *Let $\mathcal{V}$ be a vector space. If $\mathcal{V} = \mathrm{Span}\{\mathbf{v}_1, \ldots, \mathbf{v}_n\}$, then there are linearly independent vectors $\mathbf{v}_{i_1}, \ldots, \mathbf{v}_{i_m} \in \{\mathbf{v}_1, \ldots, \mathbf{v}_n\}$ such that the set $\{\mathbf{v}_{i_1}, \ldots, \mathbf{v}_{i_m}\}$ is a basis of $\mathcal{V}$.*

Proof. We apply Theorem 1.4.14 to the subspace $\mathcal{W} = \{\mathbf{0}\}$.

Note that in Example 1.3.16 we give an alternative proof of the above fact.

□

Theorem 1.4.18. *Let $\mathcal{U}$, $\mathcal{V}$, and $\mathcal{W}$ be subspaces of a vector space $\mathcal{X}$ such that*

$$\mathcal{V} \oplus \mathcal{U} = \mathcal{W} \oplus \mathcal{U}.$$

If $\mathcal{V}$ has a basis with m vectors, then $\mathcal{W}$ has a basis with m vectors.

Proof. If $\{\mathbf{v}_1, \dots, \mathbf{v}_m\}$ is a basis of $\mathcal{V}$, then $\mathcal{V} = \mathbb{K}\mathbf{v}_1 \oplus \cdots \oplus \mathbb{K}\mathbf{v}_m$. Since

$$\mathbb{K}\mathbf{v}_1 \oplus \cdots \oplus \mathbb{K}\mathbf{v}_m \oplus \mathcal{U} = \mathcal{W} \oplus \mathcal{U},$$

for every $1 \leq j \leq m$, we have

$$\mathbf{v}_j = \mathbf{w}_j + \mathbf{u}_j,$$

for some $\mathbf{w}_j \in \mathcal{W}$ and $\mathbf{u}_j \in \mathcal{U}$. We will show that $\{\mathbf{w}_1, \dots, \mathbf{w}_m\}$ is a basis of $\mathcal{W}$.

Indeed, if $x_1\mathbf{w}_1 + \cdots + x_m\mathbf{w}_m = \mathbf{0}$, then

$$x_1\mathbf{v}_1 + \cdots + x_m\mathbf{v}_m = x_1\mathbf{u}_1 + \cdots + x_m\mathbf{u}_m$$

and thus $x_1\mathbf{v}_1 + \cdots + x_m\mathbf{v}_m = \mathbf{0}$, because $\mathbf{u}_j \in \mathcal{U}$ and $\mathrm{Span}\{\mathbf{v}_1, \dots, \mathbf{v}_m\} \cap \mathcal{U} = \mathbf{0}$. Consequently, $x_1 = \cdots = x_m = 0$, proving linear independence of vectors $\mathbf{w}_1, \dots, \mathbf{w}_m$.

Now, if $\mathbf{w}$ is an arbitrary vector in $\mathcal{W}$, then there are numbers $x_1, \dots, x_m \in \mathbb{K}$ and a vector $\mathbf{u} \in \mathcal{U}$ such that

$$\mathbf{w} = x_1\mathbf{v}_1 + \cdots + x_m\mathbf{v}_m + \mathbf{u} = x_1\mathbf{w}_1 + \cdots + x_m\mathbf{w}_m + x_1\mathbf{u}_1 + \cdots + x_m\mathbf{u}_m + \mathbf{u}.$$

Since $\mathcal{W} \cap \mathcal{U} = \{\mathbf{0}\}$, we must have

$$\mathbf{w} = x_1\mathbf{w}_1 + \cdots + x_m\mathbf{w}_m,$$

proving that $\mathcal{W} = \mathrm{Span}\{\mathbf{w}_1, \dots, \mathbf{w}_m\}$.

□

Example 1.4.19. Let $\mathcal{V}_1$ and $\mathcal{V}_2$ be arbitrary vector spaces over $\mathbb{K}$ and let $\mathcal{W} = \mathcal{V}_1 \times \mathcal{V}_2$. Then

$$\mathcal{W}_1 = \mathcal{V}_1 \times \{\mathbf{0}\} \quad \text{and} \quad \mathcal{W}_2 = \{\mathbf{0}\} \times \mathcal{V}_2$$

are subspaces of $\mathcal{W}$. It is easy to see that

$$\mathcal{W} = \mathcal{W}_1 \oplus \mathcal{W}_2.$$

More generally, if $\mathcal{V}_1, \dots, \mathcal{V}_n$ are arbitrary vector spaces over $\mathbb{K}$ and $\mathcal{W} =$

$\mathcal{V}_1 \times \cdots \times \mathcal{V}_n$, then the spaces

$$\mathcal{W}_1 = \mathcal{V}_1 \times \{0\} \times \cdots \times \{0\}$$

$$\vdots$$

$$\mathcal{W}_j = \{0\} \times \cdots \times \{0\} \times \mathcal{V}_j \times \{0\} \times \cdots \times \{0\}$$

$$\vdots$$

$$\mathcal{W}_n = \{0\} \times \cdots \times \{0\} \times \mathcal{V}_n$$

are subspaces of $\mathcal{W}$ and we have

$$\mathcal{W} = \mathcal{W}_1 \oplus \cdots \oplus \mathcal{W}_n.$$

Note that we cannot write $\mathcal{W} = \mathcal{V}_1 \oplus \cdots \oplus \mathcal{V}_n$ because $\mathcal{V}_1, \ldots, \mathcal{V}_n$ are not subspaces of $\mathcal{W}$.

1.5   Dimension of a vector space

Dimensions of $\mathbb{R}^2$ and $\mathbb{R}^3$ have a clear and intuitive meaning for us. The notion of the dimension of an abstract vector space in less intuitive. Before we can define the dimension of a vector space, we first need to establish some additional properties of bases in vector spaces. To motivate an approach to the proof of the first important theorem in this section (Theorem 1.5.2) we consider a special case.

Example 1.5.1. If a vector space $\mathcal{V} = \text{Span}\{\mathbf{v}_1, \mathbf{v}_2, \mathbf{v}_3\}$ contains three linearly independent vectors $\mathbf{w}_1, \mathbf{w}_2, \mathbf{w}_3$, then the vectors $\mathbf{v}_1, \mathbf{v}_2, \mathbf{v}_3$ are linearly independent and

$$\text{Span}\{\mathbf{v}_1, \mathbf{v}_2, \mathbf{v}_3\} = \text{Span}\{\mathbf{w}_1, \mathbf{w}_2, \mathbf{w}_3\}.$$

Solution. Let $\mathbf{w}_1, \mathbf{w}_2, \mathbf{w}_3$ be linearly independent vectors in $\text{Span}\{\mathbf{v}_1, \mathbf{v}_2, \mathbf{v}_3\}$. Then $\text{Span}\{\mathbf{w}_1, \mathbf{w}_2, \mathbf{w}_3\} \subseteq \text{Span}\{\mathbf{v}_1, \mathbf{v}_2, \mathbf{v}_3\}$ and

$$\mathbf{w}_1 = a_{11}\mathbf{v}_1 + a_{21}\mathbf{v}_2 + a_{31}\mathbf{v}_3$$
$$\mathbf{w}_2 = a_{12}\mathbf{v}_1 + a_{22}\mathbf{v}_2 + a_{32}\mathbf{v}_3$$
$$\mathbf{w}_3 = a_{13}\mathbf{v}_1 + a_{23}\mathbf{v}_2 + a_{33}\mathbf{v}_3,$$

for some $a_{jk} \in \mathbb{K}$. Let

$$A = \begin{bmatrix} a_{11} & a_{12} & a_{13} \\ a_{21} & a_{22} & a_{23} \\ a_{31} & a_{32} & a_{33} \end{bmatrix}.$$

It is easy to verify that the equality

$$A \begin{bmatrix} x_1 \\ x_2 \\ x_3 \end{bmatrix} = \begin{bmatrix} 0 \\ 0 \\ 0 \end{bmatrix}$$

implies

$$x_1 \mathbf{w}_1 + x_2 \mathbf{w}_2 + x_3 \mathbf{w}_3 = \mathbf{0}.$$

Since the vectors $\mathbf{w}_1, \mathbf{w}_2, \mathbf{w}_3$ are linearly independent, we must have $x_1 = x_2 = x_3 = 0$. But this means that the matrix A is invertible. Let

$$B = \begin{bmatrix} b_{11} & b_{12} & b_{13} \\ b_{21} & b_{22} & b_{23} \\ b_{31} & b_{32} & b_{33} \end{bmatrix} = A^{-1}.$$

Now we note that

$$\begin{aligned}
b_{11}\mathbf{w}_1 + b_{21}\mathbf{w}_2 + b_{31}\mathbf{w}_3 &= (b_{11}a_{11} + b_{21}a_{12} + b_{31}a_{13})\mathbf{v}_1 \\
&\quad + (b_{11}a_{21} + b_{21}a_{22} + b_{31}a_{23})\mathbf{v}_2 \\
&\quad + (b_{11}a_{31} + b_{21}a_{32} + b_{31}a_{33})\mathbf{v}_3 \\
&= 1\mathbf{v}_1 + 0\mathbf{v}_2 + 0\mathbf{v}_3 = \mathbf{v}_1,
\end{aligned}$$

that is

$$\mathbf{v}_1 = b_{11}\mathbf{w}_1 + b_{21}\mathbf{w}_2 + b_{31}\mathbf{w}_3.$$

Similarly, we get

$$\mathbf{v}_2 = b_{12}\mathbf{w}_1 + b_{22}\mathbf{w}_2 + b_{32}\mathbf{w}_3,$$
$$\mathbf{v}_3 = b_{13}\mathbf{w}_1 + b_{23}\mathbf{w}_2 + b_{33}\mathbf{w}_3.$$

Consequently $\mathrm{Span}\{\mathbf{v}_1, \mathbf{v}_2, \mathbf{v}_3\} = \mathrm{Span}\{\mathbf{w}_1, \mathbf{w}_2, \mathbf{w}_3\}$.

To show that the vectors $\mathbf{v}_1, \mathbf{v}_2, \mathbf{v}_3$ are linearly independent suppose

$$y_1 \mathbf{v}_1 + y_2 \mathbf{v}_2 + y_3 \mathbf{v}_3 = \mathbf{0}$$

for some $y_1, y_2, y_3 \in \mathbb{K}$. Then

$$y_1(b_{11}\mathbf{w}_1 + b_{21}\mathbf{w}_2 + a_{31}\mathbf{w}_3) + y_2(b_{12}\mathbf{w}_1 + b_{22}\mathbf{w}_2 + b_{32}\mathbf{w}_3) + y_3(b_{13}\mathbf{w}_1 + b_{23}\mathbf{w}_2 + b_{33}\mathbf{w}_3) = \mathbf{0}$$

or, equivalently,

$$(y_1 b_{11} + y_2 b_{12} + y_3 b_{13})\mathbf{w}_1 + (y_1 b_{21} + y_2 b_{22} + y_3 b_{23})\mathbf{w}_2 + (y_1 b_{31} + y_2 b_{32} + y_3 b_{33})\mathbf{w}_3 = \mathbf{0}.$$

Since the vectors $\mathbf{w}_1, \mathbf{w}_2, \mathbf{w}_3$ are linearly independent, we must have

$$y_1 b_{11} + y_2 b_{12} + y_3 b_{13} = y_1 b_{21} + y_2 b_{22} + y_3 b_{23} = y_1 b_{31} + y_2 b_{32} + y_3 b_{33} = 0,$$

which can be written as

$$B \begin{bmatrix} y_1 \\ y_2 \\ y_3 \end{bmatrix} = \begin{bmatrix} 0 \\ 0 \\ 0 \end{bmatrix}.$$

But this means that $y_1 = y_2 = y_3 = 0$, because the matrix B is invertible. Thus the vectors $\mathbf{v}_1, \mathbf{v}_2, \mathbf{v}_3$ are linearly independent. □

And now the general theorem.

Theorem 1.5.2. *Let V be a vector space and let $\mathbf{v}_1, \ldots, \mathbf{v}_n \in V$. If vectors $\mathbf{w}_1, \ldots, \mathbf{w}_n \in \mathrm{Span}\{\mathbf{v}_1, \ldots, \mathbf{v}_n\}$ are linearly independent, then the vectors $\mathbf{v}_1, \ldots, \mathbf{v}_n$ are linearly independent and*

$$\mathrm{Span}\{\mathbf{v}_1, \ldots, \mathbf{v}_n\} = \mathrm{Span}\{\mathbf{w}_1, \ldots, \mathbf{w}_n\}.$$

In other words, every collection of n linearly independent vectors $\mathbf{w}_1, \ldots, \mathbf{w}_n$ in $\mathrm{Span}\{\mathbf{v}_1, \ldots, \mathbf{v}_n\}$ is a basis of $\mathrm{Span}\{\mathbf{v}_1, \ldots, \mathbf{v}_n\}$ and the spanning set $\{\mathbf{v}_1, \ldots, \mathbf{v}_n\}$ is also a basis of $\mathrm{Span}\{\mathbf{v}_1, \ldots, \mathbf{v}_n\}$.

Proof. The proof follows the method presented in detail in Example 1.5.1.

Let $\mathbf{w}_1, \ldots, \mathbf{w}_n$ be linearly independent vectors in $\mathrm{Span}\{\mathbf{v}_1, \ldots, \mathbf{v}_n\}$. Then $\mathrm{Span}\{\mathbf{w}_1, \ldots, \mathbf{w}_n\} \subseteq \mathrm{Span}\{\mathbf{v}_1, \ldots, \mathbf{v}_n\}$ and for every $1 \le k \le n$ we have

$$\mathbf{w}_k = a_{1k}\mathbf{v}_1 + \cdots + a_{nk}\mathbf{v}_n,$$

for some $a_{jk} \in \mathbb{K}$, $1 \le k \le n$. Let

$$A = \begin{bmatrix} a_{11} & \cdots & a_{1n} \\ \vdots & \ddots & \vdots \\ a_{n1} & \cdots & a_{nn} \end{bmatrix}.$$

If

$$A \begin{bmatrix} x_1 \\ \vdots \\ x_n \end{bmatrix} = \begin{bmatrix} 0 \\ \vdots \\ 0 \end{bmatrix},$$

then we have

$$x_1\mathbf{w}_1 + \cdots + x_n\mathbf{w}_n = \mathbf{0},$$

and, since the vectors $\mathbf{w}_1, \ldots, \mathbf{w}_n$ are linearly independent, $x_1 = \cdots = x_n = 0$. This means that A is invertible. Let

$$B = \begin{bmatrix} b_{11} & \cdots & b_{1n} \\ \vdots & \ddots & \vdots \\ b_{n1} & \cdots & b_{nn} \end{bmatrix} = A^{-1}.$$

Now, because we have

$$\mathbf{v}_k = b_{1k}\mathbf{w}_1 + \cdots + b_{nk}\mathbf{w}_n,$$

for every $1 \le k \le n$, we have $\text{Span}\{\mathbf{v}_1, \ldots, \mathbf{v}_n\} = \text{Span}\{\mathbf{w}_1, \ldots, \mathbf{w}_n\}$. Moreover, the equality

$$y_1\mathbf{v}_1 + \cdots + y_n\mathbf{v}_n = \mathbf{0},$$

implies

$$B \begin{bmatrix} y_1 \\ \vdots \\ y_n \end{bmatrix} = \begin{bmatrix} 0 \\ \vdots \\ 0 \end{bmatrix}$$

and thus $y_1 = \cdots = y_n = 0$, because B is invertible. But this means that the vectors $\mathbf{v}_1, \ldots, \mathbf{v}_n$ are linearly independent.   □

The above proof depends heavily on properties of invertible matrices. Below we give a proof that does not use matrices at all. It is based on mathematical induction.

Second proof. We leave as exercise the proof for $n = 1$. Now we assume that the theorem holds for $n - 1$ for some $n \ge 2$ and show that it must also hold for n.

Let $\mathbf{w}_1, \ldots, \mathbf{w}_n$ be linearly independent vectors in $\text{Span}\{\mathbf{v}_1, \ldots, \mathbf{v}_n\}$. As in the first proof, for every $k = 1, 2, \ldots, n$ we write

$$\mathbf{w}_k = a_{1k}\mathbf{v}_1 + \cdots + a_{nk}\mathbf{v}_n.$$

Since $\mathbf{w}_1 \ne \mathbf{0}$ and $\mathbf{w}_1 = a_{11}\mathbf{v}_1 + \cdots + a_{n1}\mathbf{v}_n$, one of the numbers $a_{11}, \ldots, a_{n1}$ is different from 0. Without loss of generality, we can assume that $a_{11} \ne 0$. Then

$$\mathbf{v}_1 = \frac{1}{a_{11}}(\mathbf{w}_1 - a_{21}\mathbf{v}_2 - \cdots - a_{n1}\mathbf{v}_n) = \frac{1}{a_{11}}\mathbf{w}_1 - \frac{a_{21}}{a_{11}}\mathbf{v}_2 - \cdots - \frac{a_{n1}}{a_{11}}\mathbf{v}_n. \quad (1.6)$$

For $k \ge 2$ we get

$$\mathbf{w}_k - \frac{a_{1k}}{a_{11}}\mathbf{w}_1 = \left(a_{2k} - \frac{a_{1k}a_{21}}{a_{11}}\right)\mathbf{v}_2 + \cdots + \left(a_{nk} - \frac{a_{1k}a_{n1}}{a_{11}}\right)\mathbf{v}_n.$$

The vectors $\mathbf{w}_2 - \frac{a_{12}}{a_{11}}\mathbf{w}_1, \ldots, \mathbf{w}_n - \frac{a_{1n}}{a_{11}}\mathbf{w}_1$ are linearly independent. Indeed, if

$$x_2\left(\mathbf{w}_2 - \frac{a_{12}}{a_{11}}\mathbf{w}_1\right) + \cdots + x_n\left(\mathbf{w}_n - \frac{a_{1n}}{a_{11}}\mathbf{w}_1\right) = \mathbf{0},$$

then

$$-\left(x_2\frac{a_{12}}{a_{11}} + \cdots + x_n\frac{a_{1n}}{a_{11}}\right)\mathbf{w}_1 + x_2\mathbf{w}_2 + \cdots + x_n\mathbf{w}_n = \mathbf{0}.$$

This implies $x_2 = \cdots = x_n = 0$, because the vectors $\mathbf{w}_1, \ldots, \mathbf{w}_n$ are linearly independent.

Now, since $\mathbf{w}_2 - \frac{a_{12}}{a_{11}}\mathbf{w}_1, \ldots, \mathbf{w}_n - \frac{a_{1n}}{a_{11}}\mathbf{w}_1$ are linearly independent vectors in $\text{Span}\{\mathbf{v}_2, \ldots, \mathbf{v}_n\}$, by the inductive assumption we have

$$\text{Span}\left\{\mathbf{w}_2 - \frac{a_{12}}{a_{11}}\mathbf{w}_1, \ldots, \mathbf{w}_n - \frac{a_{1n}}{a_{11}}\mathbf{w}_1\right\} = \text{Span}\{\mathbf{v}_2, \ldots, \mathbf{v}_n\}$$

and the vectors $\mathbf{v}_2, \ldots, \mathbf{v}_n$ are linearly independent. Consequently,

$$\text{Span}\{\mathbf{w}_1, \ldots, \mathbf{w}_n\} = \text{Span}\left\{\mathbf{w}_1, \mathbf{w}_2 - \frac{a_{12}}{a_{11}}\mathbf{w}_1, \ldots, \mathbf{w}_n - \frac{a_{1n}}{a_{11}}\mathbf{w}_1\right\}$$
$$= \text{Span}\{\mathbf{w}_1, \mathbf{v}_2 \ldots, \mathbf{v}_n\}$$
$$= \text{Span}\{\mathbf{v}_1, \ldots, \mathbf{v}_n\},$$

where the last equality follows by (1.6). It remains to be shown that the vectors $\mathbf{v}_1, \ldots, \mathbf{v}_n$ are linearly independent. We argue by contradiction.

We already know that the vectors $\mathbf{v}_2, \ldots, \mathbf{v}_n$ are linearly independent. Suppose $\mathbf{v}_1 \in \text{Span}\{\mathbf{v}_2, \ldots, \mathbf{v}_n\}$. Then

$$\text{Span}\{\mathbf{v}_2, \ldots, \mathbf{v}_n\} = \text{Span}\{\mathbf{v}_1, \mathbf{v}_2, \ldots, \mathbf{v}_n\}.$$

Since $\mathbf{w}_2, \ldots, \mathbf{w}_n$ are linearly independent vectors in $\text{Span}\{\mathbf{v}_2, \ldots, \mathbf{v}_n\}$, we have $\text{Span}\{\mathbf{v}_2, \ldots, \mathbf{v}_n\} = \text{Span}\{\mathbf{w}_2, \ldots, \mathbf{w}_n\}$, by our inductive assumption. Consequently, we have $\text{Span}\{\mathbf{w}_2, \ldots, \mathbf{w}_n\} = \text{Span}\{\mathbf{v}_1, \mathbf{v}_2, \ldots, \mathbf{v}_n\}$ and thus also $\mathbf{w}_1 \in \text{Span}\{\mathbf{w}_2, \ldots, \mathbf{w}_n\}$, which contradicts linear independence of the vectors $\mathbf{w}_1, \mathbf{w}_2, \ldots, \mathbf{w}_n$. This proves that $\mathbf{v}_1 \notin \text{Span}\{\mathbf{v}_2, \ldots, \mathbf{v}_n\}$ and therefore the vectors $\mathbf{v}_1, \mathbf{v}_2, \ldots, \mathbf{v}_n$ are linearly independent. $\qquad\square$

Because Theorem 1.5.2 is of central importance in the discussion of the dimension of a vector space, we give still another proof. It is an induction proof that uses Corollaries 1.4.16 and 1.4.17 and Theorem 1.4.18. It is instructive to study and compare these three different arguments.

Third proof. We leave as exercise the proof for $n = 1$. Now we assume that the theorem holds for all integers $p < n$ and show that it must also hold for n.

Let $\mathbf{w}_1, \ldots, \mathbf{w}_n$ be linearly independent vectors in $\text{Span}\{\mathbf{v}_1, \ldots, \mathbf{v}_n\}$.

First we show that the set $\{\mathbf{w}_1, \ldots, \mathbf{w}_n\}$ is a basis of $\text{Span}\{\mathbf{v}_1, \mathbf{v}_2, \ldots, \mathbf{v}_n\}$. We argue by contradiction. If the set $\{\mathbf{w}_1, \ldots, \mathbf{w}_n\}$ is not a basis of $\text{Span}\{\mathbf{v}_1, \mathbf{v}_2, \ldots, \mathbf{v}_n\}$, then, by Corollary 1.4.16, there is an integer $r \geq 1$ and linearly independent vectors $\mathbf{v}_{i_1}, \ldots, \mathbf{v}_{i_r} \in \{\mathbf{v}_1, \ldots, \mathbf{v}_n\}$ such that the set $\{\mathbf{w}_1, \ldots, \mathbf{w}_n, \mathbf{v}_{i_1}, \ldots, \mathbf{v}_{i_r}\}$ is a basis of $\text{Span}\{\mathbf{v}_1, \mathbf{v}_2, \ldots, \mathbf{v}_n\}$. Now, again by Corollary 1.4.16, there are vectors $\mathbf{v}_{j_1}, \ldots, \mathbf{v}_{j_p} \in \{\mathbf{v}_1, \ldots, \mathbf{v}_n\}$ such that

$$\{\mathbf{v}_{j_1}, \ldots, \mathbf{v}_{j_p}, \mathbf{v}_{i_1}, \ldots, \mathbf{v}_{i_r}\}$$

is a basis of $\text{Span}\{\mathbf{v}_1, \mathbf{v}_2, \ldots, \mathbf{v}_n\}$. Note that we must have $p < n$, because if $p = n$ then

$$\text{Span}\{\mathbf{v}_{j_1}, \ldots, \mathbf{v}_{j_p}\} = \text{Span}\{\mathbf{v}_1, \ldots, \mathbf{v}_n\},$$

which is not possible because $r \geq 1$.

Since the subspaces $\mathrm{Span}\{\mathbf{w}_1, \dots, \mathbf{w}_n\}$ and $\mathrm{Span}\{\mathbf{v}_{j_1}, \dots, \mathbf{v}_{j_p}\}$ are complements of the same subspace $\mathrm{Span}\{\mathbf{v}_{i_1}, \dots, \mathbf{v}_{i_r}\}$, the subspace $\mathrm{Span}\{\mathbf{w}_1, \dots, \mathbf{w}_n\}$ has a basis with p elements $\{\mathbf{u}_1, \dots, \mathbf{u}_p\}$, by Theorem 1.4.18. But then the linearly independent vectors $\mathbf{w}_1, \dots, \mathbf{w}_p$ are in the subspace

$$\mathrm{Span}\{\mathbf{u}_1, \dots, \mathbf{u}_p\} = \mathrm{Span}\{\mathbf{w}_1, \dots, \mathbf{w}_n\}.$$

Consequently, by our inductive assumption, $\{\mathbf{w}_1, \dots, \mathbf{w}_p\}$ is a basis of $\mathrm{Span}\{\mathbf{w}_1, \dots, \mathbf{w}_n\}$. But this contradicts independence of the vectors $\mathbf{w}_1, \dots, \mathbf{w}_n$, because $p < n$. This completes the proof of the fact that $\{\mathbf{w}_1, \dots, \mathbf{w}_n\}$ is a basis of $\mathrm{Span}\{\mathbf{v}_1, \mathbf{v}_2, \dots, \mathbf{v}_n\}$.

Now we show that the vectors $\mathbf{v}_1, \dots, \mathbf{v}_n$ are linearly independent. Again we use a proof by contradiction. If $\mathbf{v}_1, \dots, \mathbf{v}_n$ are not linearly independent, then, by Corollary 1.4.17, there is an integer $r < n$ and a set $\{\mathbf{v}_{i_1}, \dots, \mathbf{v}_{i_r}\} \subset \{\mathbf{v}_1, \dots, \mathbf{v}_n\}$ of linearly independent vectors such that

$$\mathrm{Span}\{\mathbf{v}_{i_1}, \dots, \mathbf{v}_{i_r}\} = \mathrm{Span}\{\mathbf{v}_1, \dots, \mathbf{v}_n\}.$$

But then the linearly independent vectors $\mathbf{w}_1, \dots, \mathbf{w}_r$ are in $\mathrm{Span}\{\mathbf{v}_{i_1}, \dots, \mathbf{v}_{i_r}\}$ $= \mathrm{Span}\{\mathbf{v}_1, \dots, \mathbf{v}_n\}$. Consequently, by our inductive assumption, $\{\mathbf{w}_1, \dots, \mathbf{w}_r\}$ is a basis of $\mathrm{Span}\{\mathbf{v}_1, \mathbf{v}_2, \dots, \mathbf{v}_n\}$, contradicting linear independence of the vectors $\mathbf{w}_1, \dots \mathbf{w}_n$, because $r < n$. This contradiction proves that the vectors $\mathbf{v}_1, \dots, \mathbf{v}_n$ are linearly independent. $\qquad\square$

The next theorem is an easy consequence of Theorem 1.5.2.

Theorem 1.5.3. *If $\{\mathbf{v}_1, \dots, \mathbf{v}_n\}$ is a basis of the vector space $\mathcal{V}$, then any $n + 1$ vectors in $\mathcal{V}$ are linearly dependent.*

Proof. Let $\mathbf{w}_1, \dots, \mathbf{w}_n, \mathbf{w}_{n+1} \in \mathcal{V}$. If the vectors $\mathbf{w}_1, \dots, \mathbf{w}_n$ are linearly dependent, we are done. If the vectors $\mathbf{w}_1, \dots, \mathbf{w}_n$ are linearly independent, then the set $\{\mathbf{w}_1, \dots, \mathbf{w}_n\}$ is a basis of $\mathcal{V}$, by Theorem 1.5.2. Consequently $\mathbf{w}_{n+1}$ can be written as a linear combination of the vectors $\mathbf{w}_1, \dots, \mathbf{w}_n$, which implies that the vectors $\mathbf{w}_1, \dots, \mathbf{w}_n, \mathbf{w}_{n+1}$ are linearly dependent. $\qquad\square$

Corollary 1.5.4. *If both $\{\mathbf{v}_1, \dots, \mathbf{v}_m\}$ and $\{\mathbf{w}_1, \dots, \mathbf{w}_n\}$ are bases of a vector space $\mathcal{V}$, then $m = n$.*

Proof. This is an immediate consequence of Theorem 1.5.3. $\qquad\square$

Now we are ready to define the dimension of a vector space spanned by a finite number of vectors.

Definition 1.5.5. If a vector space V has a basis with n vectors, then the number n is called the *dimension* of V and is denoted by $\dim V$. Additionally, we define $\dim\{\mathbf{0}\} = 0$.

If a vector space V has a basis with a finite number of vectors, we say that V is *finite dimensional* and we write $\dim V < \infty$. Not every vector space is finite dimensional.

Example 1.5.6. The space of all polynomials $\mathcal{P}(\mathbb{K})$ is not a finite dimensional space. Suppose, on the contrary, that $\dim \mathcal{P}(\mathbb{K}) = n$ for some positive integer n. Then $\mathcal{P}(\mathbb{K}) = \mathrm{Span}\{p_1, \ldots, p_n\}$ for some nonzero polynomials $p_1, \ldots, p_n$. Let m be the maximum degree of the polynomials $p_1, \ldots, p_n$. Since the degree of every linear combination of the polynomials $p_1, \ldots, p_n$ is at most m, the polynomial $p(t) = t^{m+1}$ is not in $\mathcal{P}(\mathbb{K}) = \mathrm{Span}\{p_1, \ldots, p_n\}$, a contradiction.

The following useful result can be easily obtained from Theorem 1.5.2.

Theorem 1.5.7. *Let V be a vector space such that $\dim V = n$.*

(a) *Any set of n linearly independent vectors from V is basis of V;*

(b) *If $V = \mathrm{Span}\{\mathbf{w}_1, \ldots, \mathbf{w}_n\}$, then $\{\mathbf{w}_1, \ldots, \mathbf{w}_n\}$ is a basis of V.*

Proof. Let $\dim V = n$ and let $\{\mathbf{v}_1, \ldots, \mathbf{v}_n\}$ be a basis of V.

If the vectors $\mathbf{w}_1, \ldots, \mathbf{w}_n \in V$ are linearly independent, then

$$\mathrm{Span}\{\mathbf{w}_1, \ldots, \mathbf{w}_n\} = \mathrm{Span}\{\mathbf{v}_1, \ldots, \mathbf{v}_n\} = V,$$

by Theorem 1.5.2. Thus $\{\mathbf{w}_1, \ldots, \mathbf{w}_n\}$ is a basis of V.

If $V = \mathrm{Span}\{\mathbf{w}_1, \ldots, \mathbf{w}_n\}$, then the linearly independent vectors $\mathbf{v}_1, \ldots, \mathbf{v}_n$ are in $\mathrm{Span}\{\mathbf{w}_1, \ldots, \mathbf{w}_n\}$ and, again by Theorem 1.5.2, the vectors $\mathbf{w}_1, \ldots, \mathbf{w}_n$ are linearly independent. This means that the set $\{\mathbf{w}_1, \ldots, \mathbf{w}_n\}$ is a basis of V. $\square$

Example 1.5.8. The solutions of the differential equation

$$y'' + y = 0$$

is a vector space of dimension 2 over $\mathbb{C}$. Both $\{e^{it}, e^{-it}\}$ and $\{\cos t, \sin t\}$ are bases in this vector space.

Example 1.5.9. Let a, b, and c be distinct numbers from $\mathbb{K}$ and let

$$\mathcal{U} = \{p \in \mathcal{P}_n(\mathbb{K}) : p(a) = p(b) = p(c)\}.$$

The dimension of the vector space $\mathcal{U}$ is $n - 2$ (see Example 1.3.26).

Example 1.5.10. Let a be an arbitrary number from $\mathbb{K}$ and let

$$\mathcal{U} = \{p \in \mathcal{P}_n(\mathbb{K}) : p(a) = p'(a) = p''(a) = 0\}.$$

Then $\dim \mathcal{U} = n - 3$ (see Example 1.3.21).

Example 1.5.11. Let n be an integer greater than 1.

(a) $\dim \mathcal{S}_{n \times n}(\mathbb{K}) = \frac{1}{2}n(n+1)$ (see Example 1.2.5)

(b) $\dim \mathcal{A}_{n \times n}(\mathbb{K}) = \frac{1}{2}n(n-1)$ (see Example 1.2.5)

Example 1.5.12. Show that

$$\dim \mathrm{Span} \left\{ \begin{bmatrix} 1 \\ i \\ -i \\ 0 \end{bmatrix}, \begin{bmatrix} 1-2i \\ -i \\ -2-i \\ -2i \end{bmatrix}, \begin{bmatrix} i \\ i \\ 1 \\ i \end{bmatrix}, \begin{bmatrix} 0 \\ 1 \\ i \\ 1 \end{bmatrix} \right\} = 3$$

and find all bases of the above subspace, in the set

$$\left\{ \begin{bmatrix} 1 \\ i \\ -i \\ 0 \end{bmatrix}, \begin{bmatrix} 1-2i \\ -i \\ -2-i \\ -2i \end{bmatrix}, \begin{bmatrix} i \\ i \\ 1 \\ i \end{bmatrix}, \begin{bmatrix} 0 \\ 1 \\ i \\ 1 \end{bmatrix} \right\}.$$

Solution. Since the reduced row echelon form of the matrix

$$\begin{bmatrix} 1 & 1-2i & i & 0 \\ i & -i & i & 1 \\ -i & -2-i & 1 & i \\ 0 & -2i & i & 1 \end{bmatrix}$$

is

$$\begin{bmatrix} 1 & 0 & \frac{1}{2} & 0 \\ 0 & 1 & -\frac{1}{2} & 0 \\ 0 & 0 & 0 & 1 \\ 0 & 0 & 0 & 0 \end{bmatrix}$$

and the sets

$$\left\{ \begin{bmatrix} 1 \\ 0 \\ 0 \\ 0 \end{bmatrix}, \begin{bmatrix} 0 \\ 1 \\ 0 \\ 0 \end{bmatrix}, \begin{bmatrix} 0 \\ 0 \\ 1 \\ 0 \end{bmatrix} \right\}, \quad \left\{ \begin{bmatrix} 1 \\ 0 \\ 0 \\ 0 \end{bmatrix}, \begin{bmatrix} \frac{1}{2} \\ -\frac{1}{2} \\ 0 \\ 0 \end{bmatrix}, \begin{bmatrix} 0 \\ 0 \\ 1 \\ 0 \end{bmatrix} \right\}, \quad \left\{ \begin{bmatrix} 0 \\ 1 \\ 0 \\ 0 \end{bmatrix}, \begin{bmatrix} \frac{1}{2} \\ -\frac{1}{2} \\ 0 \\ 0 \end{bmatrix}, \begin{bmatrix} 0 \\ 0 \\ 1 \\ 0 \end{bmatrix} \right\},$$

are linearly independent, the following sets are bases:

$$\left\{ \begin{bmatrix} 1 \\ i \\ -i \\ 0 \end{bmatrix}, \begin{bmatrix} 1-2i \\ -i \\ -2-i \\ -2i \end{bmatrix}, \begin{bmatrix} 0 \\ 1 \\ i \\ 1 \end{bmatrix} \right\}, \quad \left\{ \begin{bmatrix} 1 \\ i \\ -i \\ 0 \end{bmatrix}, \begin{bmatrix} i \\ i \\ 1 \\ i \end{bmatrix}, \begin{bmatrix} 0 \\ 1 \\ i \\ 1 \end{bmatrix} \right\}, \quad \left\{ \begin{bmatrix} 1-2i \\ -i \\ -2-i \\ -2i \end{bmatrix}, \begin{bmatrix} i \\ i \\ 1 \\ i \end{bmatrix}, \begin{bmatrix} 0 \\ 1 \\ i \\ 1 \end{bmatrix} \right\}.$$

$\square$

In the next example we use Theorem 1.5.7 to obtain the characterization of invertible matrices in Example 1.3.7.

Example 1.5.13. Show that a matrix $A \in \mathcal{M}_{n \times n}(\mathbb{K})$ is invertible if and only if the only solution of the equation $A\mathbf{x} = \mathbf{0}$ is $\mathbf{x} = \mathbf{0}$.

Solution. If A is invertible and $A\mathbf{x} = \mathbf{0}$, then

$$\mathbf{x} = A^{-1}A\mathbf{x} = A^{-1}\mathbf{0} = \mathbf{0}.$$

Suppose now that the only solution of the equation $A\mathbf{x} = \mathbf{0}$ is $\mathbf{x} = \mathbf{0}$. We write $A = \begin{bmatrix} \mathbf{a}_1 & \cdots & \mathbf{a}_n \end{bmatrix}$, where $\mathbf{a}_1, \ldots, \mathbf{a}_n$ are the columns of A, and $\mathbf{x} = \begin{bmatrix} x_1 \\ \vdots \\ x_n \end{bmatrix}$.

The assumption that the only solution of the equation

$$\begin{bmatrix} \mathbf{a}_1 & \cdots & \mathbf{a}_n \end{bmatrix} \begin{bmatrix} x_1 \\ \vdots \\ x_n \end{bmatrix} = \begin{bmatrix} 0 \\ \vdots \\ 0 \end{bmatrix}$$

is $\begin{bmatrix} 0 \\ \vdots \\ 0 \end{bmatrix}$ implies that the vectors $\mathbf{a}_1, \ldots, \mathbf{a}_n$ are linearly independent. Conse-

quently, by Theorem 1.5.7, the set $\{\mathbf{a}_1, \ldots, \mathbf{a}_n\}$ is a basis of $\mathbb{K}^n$. If $\mathbf{e}_1, \ldots, \mathbf{e}_n$ are the columns of the unit matrix I_n, then for any $j \in \{1, \ldots, n\}$ we have

$$\mathbf{e}_j = b_{1j}\mathbf{a}_1 + \cdots + b_{nj}\mathbf{a}_n,$$

for some $b_{jk} \in \mathbb{K}$, that is,

$$I_n = AB$$

where B is the matrix with entries b_{jk}.

It is easy to see that the only solution of the equation $B\mathbf{x} = \mathbf{0}$ is $\mathbf{x} = \mathbf{0}$. Arguing as before, we can find a matrix C such that

$$I_n = BC.$$

Since

$$A = AI_n = ABC = I_nC = C,$$

we get

$$AB = BA = I_n.$$

$\square$

Example 1.5.14. Show that the dimension of the vector space $\mathcal{U}$ in Example 1.1.8 is 1.

Solution. Let $\mathbf{x}$ be a vector in $\mathbb{R}^2$ which is not on the vector line $\mathbb{R}\mathbf{a}$. Consequently, the vectors $\mathbf{x}$ and $\mathbf{a}$ are linearly independent. If $\mathbf{y}$ is in $\mathbb{R}^2$ then there are real numbers α and β such that

$$\mathbf{y} = \alpha\mathbf{a} + \beta\mathbf{x}.$$

Hence

$$\widehat{\mathbf{y}} = (\alpha\mathbf{a} + \beta\mathbf{x})^\wedge = \widehat{\alpha\mathbf{a}} + \widehat{\beta\mathbf{x}} = \widehat{\beta\mathbf{x}} = \beta\widehat{\mathbf{x}}.$$

This means that $\{\widehat{\mathbf{x}}\}$ is a basis in $\mathcal{U}$ and consequently $\dim \mathcal{U} = 1$. $\square$

In Section 1.3 we remark that linear independence of vectors may depend on the field of scalars $\mathbb{K}$. Since the dimension of a vector space is closely related to linear independence of vectors (being the maximum number of linearly independent vectors in the space), it should be expected that the dimension of a vector space also depends on whether the space is treated as a real vector space or a complex vector space.

Example 1.5.15. Let V be a complex vector of dimension n. Show that the dimension of the real vector space V is $2n$.

Solution. Let V be a complex vector of dimension n and let $\{\mathbf{v}_1, \ldots, \mathbf{v}_n\}$ be a basis of V. This means that every $\mathbf{v} \in V$ has a unique representation of the form

$$\mathbf{v} = z_1 \mathbf{v}_1 + \cdots + z_n \mathbf{v}_n$$

with $z_1, \ldots, z_n \in \mathbb{C}$.

We will show that $\{\mathbf{v}_1, \ldots, \mathbf{v}_n, i\mathbf{v}_1, \ldots, i\mathbf{v}_n\}$ is a basis of the real vector space V. If

$$a_1 \mathbf{v}_1 + \cdots + a_n \mathbf{v}_n + b_1 i\mathbf{v}_1 + \cdots + b_n i\mathbf{v}_n = 0$$

with $a_1, \ldots, a_n, b_1, \ldots, b_n \in \mathbb{R}$, then

$$(a_1 + b_1 i)\mathbf{v}_1 + \cdots + (a_n + b_n i)\mathbf{v}_n = \mathbf{0}.$$

Since the vectors $\mathbf{v}_1, \ldots, \mathbf{v}_n$ are linearly independent in the complex vector space V, we must have

$$a_1 + b_1 i = \cdots = a_n + b_n i = 0,$$

and thus $a_1 = \cdots = a_n = b_1 = \cdots = b_n = 0$. This proves that the vectors $\mathbf{v}_1, \ldots, \mathbf{v}_n, i\mathbf{v}_1, \ldots, i\mathbf{v}_n$ are linearly independent in the real vector space V.

Now we need to show that every vector $\mathbf{v} \in V$ can be written in the form

$$x_1 \mathbf{v}_1 + \cdots + x_n \mathbf{v}_n + y_1 i\mathbf{v}_1 + \cdots + y_n i\mathbf{v}_n$$

with $x_1, \ldots, x_n, y_n, \ldots, y_n \in \mathbb{R}$. Indeed, if $\mathbf{v} \in V$, then

$$\mathbf{v} = z_1 \mathbf{v}_1 + \cdots + z_n \mathbf{v}_n$$

for some $z_1, \ldots, z_n \in \mathbb{C}$. If we write $z_j = x_j + y_j i$, then

$$z_1 \mathbf{v}_1 + \cdots + z_n \mathbf{v}_n = (x_1 + y_1 i)\mathbf{v}_1 + \cdots + (x_n + y_n i)\mathbf{v}_n$$
$$= x_1 \mathbf{v}_1 + \cdots + x_n \mathbf{v}_n + y_1 i\mathbf{v}_1 + \cdots + y_n i\mathbf{v}_n,$$

which is the desired representation. $\qquad\Box$

Theorem 1.5.16. *Let $\mathcal{U}$ be a subspace of a finite dimensional vector space V. Then*

(a) *$\mathcal{U}$ is finite dimensional and $\dim \mathcal{U} \le \dim V$;*

(b) *If $\dim \mathcal{U} = \dim V$, then $\mathcal{U} = V$.*

Proof. Both properties are trivially true if $\mathcal{U} = \{\mathbf{0}\}$.

If $\mathcal{U}$ is a nontrivial subspace of $\mathcal{V}$ and $\{\mathbf{v}_1, \ldots, \mathbf{v}_n\}$ is a basis of $\mathcal{V}$, then there exist an integer $m \geq 1$ and linearly independent vectors $\mathbf{v}_{i_1}, \ldots, \mathbf{v}_{i_m} \in \{\mathbf{v}_1, \ldots, \mathbf{v}_n\}$ such that

$$\mathcal{U} \oplus \mathbb{K}\mathbf{v}_{i_1} \oplus \cdots \oplus \mathbb{K}\mathbf{v}_{i_m} = \mathcal{V},$$

by Theorem 1.4.14. Since we also have

$$\mathbb{K}\mathbf{v}_{j_1} \oplus \cdots \oplus \mathbb{K}\mathbf{v}_{j_{n-m}} \oplus \mathbb{K}\mathbf{v}_{i_1} \oplus \cdots \oplus \mathbb{K}\mathbf{v}_{i_m} = \mathcal{V},$$

where $\{j_1, \ldots, j_{n-m}, i_1, \ldots, i_m\} = \{1, \ldots, n\}$, the vector subspace $\mathcal{U}$ has a basis with $n - m$ vectors, by Theorem 1.4.18. Consequently,

$$\dim \mathcal{U} = n - m < n = \dim \mathcal{V}.$$

If $\mathcal{U} = \mathcal{V}$, then obviously $\dim \mathcal{U} = \dim \mathcal{V}$. In either case, since $\dim \mathcal{U} \leq n$, the space $\mathcal{U}$ is finite dimensional. This completes the proof of part (a).

Part (b) is an immediate consequence of Theorem 1.5.7.   □

Example 1.5.17. Let $\mathcal{U}$ and $\mathcal{V}$ be finite dimensional vector subspaces of a vector space $\mathcal{W}$. Show that

$$\dim(\mathcal{U} + \mathcal{V}) = \dim \mathcal{U} + \dim \mathcal{V} - \dim(\mathcal{U} \cap \mathcal{V}).$$

Solution. First we note that $\mathcal{U} + \mathcal{V}$ is a finite dimensional vector space such that $\mathcal{U} \subseteq \mathcal{U} + \mathcal{V}$ and $\mathcal{V} \subseteq \mathcal{U} + \mathcal{V}$.

Let $\{\mathbf{x}_1, \ldots, \mathbf{x}_p\}$ be a basis of $\mathcal{U} \cap \mathcal{V}$, $\{\mathbf{y}_1, \ldots, \mathbf{y}_q\}$ be a basis of a complement $\mathcal{U}'$ of $\mathcal{U} \cap \mathcal{V}$ in $\mathcal{U}$, and let $\{\mathbf{z}_1, \ldots, \mathbf{z}_r\}$ a basis of a complement $\mathcal{V}'$ of $\mathcal{U} \cap \mathcal{V}$ in $\mathcal{V}$. According to Example 1.4.3 we have

$$\mathcal{U} + \mathcal{V} = \mathcal{U}' \oplus (\mathcal{U} \cap \mathcal{V}) \oplus \mathcal{V}'$$

Consequently, by Theorem 1.4.10,

$$\{\mathbf{x}_1, \ldots, \mathbf{x}_p\, \mathbf{y}_1, \ldots, \mathbf{y}_q, \mathbf{z}_1, \ldots, \mathbf{z}_r\}$$

is a basis of $\mathcal{U} + \mathcal{V}$. Since $\dim(\mathcal{U} + \mathcal{V}) = p + q + r$, $\dim \mathcal{U} = p + q$, $\dim \mathcal{V} = p + r$, and $\dim(\mathcal{U} \cap \mathcal{V}) = p$, the desired equality follows.   □

We close this section with a useful observation that is an immediate consequence of Corollary 1.4.16. Note that the theorem implies that every collection of linearly independent vectors in a vector space $\mathcal{V}$ can be extended to a basis $\mathcal{V}$.

Theorem 1.5.18. *Let $\{\mathbf{v}_1, \ldots, \mathbf{v}_n\}$ be a basis of a vector space $\mathcal{V}$. If the vectors $\mathbf{w}_1, \ldots, \mathbf{w}_k \in \mathcal{V}$ are linearly independent and $k < n$, then there are vectors $\mathbf{w}_{k+1}, \ldots, \mathbf{w}_n \in \{\mathbf{v}_1, \ldots, \mathbf{v}_n\}$ such that*

$$\{\mathbf{w}_1, \ldots, \mathbf{w}_k, \mathbf{w}_{k+1}, \ldots, \mathbf{w}_n\}$$

is a basis of $\mathcal{V}$.

Example 1.5.19. In Example 1.3.26 we use the fact that the polynomials

$$1, q(t), tq(t), \ldots, t^{n-3}q(t),$$

where $q(t) = (t-a)(t-b)(t-c)$, are linearly independent vectors in $\mathcal{P}_n(\mathbb{K})$. Extend this collection of vectors to a basis of $\mathcal{P}_n(\mathbb{K})$.

Solution. We can use two vectors from the standard basis of $\mathcal{P}_n(\mathbb{K})$, that is,

$$\{1, t, t^2, \ldots, t^n\},$$

namely t and t^2. It is easy to check that the set

$$\{1, t, t^2, q(t), tq(t), \ldots, t^{n-3}q(t)\}$$

is a basis in $\mathcal{P}_n(\mathbb{K})$. $\square$

1.6   Change of basis

Problems in linear algebra and its applications often require working with different bases in the same vector space. If coordinates of a vector in one basis are known, we need to be able to efficiently find its coordinates in a different basis. The following theorem describes this process in terms of matrix multiplication.

Theorem 1.6.1. *Let* $\{\mathbf{v}_1, \ldots, \mathbf{v}_n\}$ *and* $\{\mathbf{w}_1, \ldots, \mathbf{w}_n\}$ *be bases of a vector space* V *and let* $a_{jk} \in \mathbb{K}$, *for* $1 \le j, k \le n$, *be the unique numbers such that*

$$\mathbf{v}_k = a_{1k}\mathbf{w}_1 + \cdots + a_{nk}\mathbf{w}_n,$$

for every $1 \le k \le n$. *For every* $\mathbf{v} \in V$, *if*

$$\mathbf{v} = x_1\mathbf{v}_1 + \cdots + x_n\mathbf{v}_n,$$

then

$$\mathbf{v} = y_1\mathbf{w}_1 + \cdots + y_n\mathbf{w}_n, \tag{1.7}$$

where the numbers $y_1, \ldots, y_n$ *are given by*

$$\begin{bmatrix} y_1 \\ \vdots \\ y_n \end{bmatrix} = \begin{bmatrix} a_{11} & \cdots & a_{1n} \\ \vdots & & \vdots \\ a_{n1} & \cdots & a_{nn} \end{bmatrix} \begin{bmatrix} x_1 \\ \vdots \\ x_n \end{bmatrix}.$$

Proof. Let $\mathbf{v}$ be an arbitrary vector in V. If $\mathbf{v} = x_1\mathbf{v}_1 + \cdots + x_n\mathbf{v}_n$, then

$$\mathbf{v} = x_1\mathbf{v}_1 + \cdots + x_n\mathbf{v}_m$$
$$= x_1(a_{11}\mathbf{w}_1 + \cdots + a_{n1}\mathbf{w}_n) + \cdots + x_m(a_{1n}\mathbf{w}_1 + \cdots + a_{nn}\mathbf{w}_n)$$
$$= \left(\begin{bmatrix} a_{11} & \cdots & a_{1n} \end{bmatrix} \begin{bmatrix} x_1 \\ \vdots \\ x_n \end{bmatrix} \right) \mathbf{w}_1 + \cdots + \left(\begin{bmatrix} a_{n1} & \cdots & a_{nn} \end{bmatrix} \begin{bmatrix} x_1 \\ \vdots \\ x_n \end{bmatrix} \right) \mathbf{w}_n.$$

Thus the coordinates of $\mathbf{v}$ in basis $\{\mathbf{w}_1, \ldots, \mathbf{w}_n\}$ are

$$y_k = \begin{bmatrix} a_{k1} & \cdots & a_{kn} \end{bmatrix} \begin{bmatrix} x_1 \\ \vdots \\ x_n \end{bmatrix},$$

for every $k \in \{1, \ldots, n\}$, which can be written as

$$\begin{bmatrix} y_1 \\ \vdots \\ y_n \end{bmatrix} = \begin{bmatrix} a_{11} & \cdots & a_{1n} \\ \vdots & & \vdots \\ a_{n1} & \cdots & a_{nn} \end{bmatrix} \begin{bmatrix} x_1 \\ \vdots \\ x_n \end{bmatrix}.$$

$\square$

Definition 1.6.2. Let $\mathcal{B} = \{\mathbf{v}_1, \ldots, \mathbf{v}_n\}$ and $\mathcal{C} = \{\mathbf{w}_1, \ldots, \mathbf{w}_n\}$ be bases of a vector space V. The $n \times n$ matrix in Theorem 1.6.1 is called the *change of coordinates matrix* from the basis $\mathcal{B}$ to the basis $\mathcal{C}$ and is denoted by $\mathrm{Id}_{\mathcal{B} \to \mathcal{C}}$.

We use Id in the above definition because in Theorem 1.6.1 we have

$$\text{Id}(\mathbf{v}_k) = \mathbf{v}_k = a_{1k}\mathbf{w}_1 + \cdots + a_{nk}\mathbf{w}_n,$$

where the function $\text{Id} : \mathcal{V} \to \mathcal{V}$ is defined by $\text{Id}(\mathbf{v}) = \mathbf{v}$ for every $\mathbf{v} \in \mathcal{V}$. This notation will be explained better in Chapter 2.

Example 1.6.3. Let $\{\mathbf{v}_1, \mathbf{v}_2, \mathbf{v}_3\}$ and $\{\mathbf{w}_1, \mathbf{w}_2, \mathbf{w}_3\}$ be bases for a vector space $\mathcal{V}$. If

$$\mathbf{v}_1 = i\mathbf{w}_1 + i\mathbf{w}_2 + \mathbf{w}_3$$
$$\mathbf{v}_2 = -i\mathbf{w}_1 + i\mathbf{w}_2 + \mathbf{w}_3$$
$$\mathbf{v}_3 = \mathbf{w}_1 + i\mathbf{w}_2 + i\mathbf{w}_3,$$

write the vector $\mathbf{v} = (2-i)\mathbf{v}_1 + 3\mathbf{v}_2 + i\mathbf{v}_3$ as a linear combination of the vectors $\mathbf{w}_1$, $\mathbf{w}_2$, and $\mathbf{w}_3$.

Solution. Since the change of coordinates matrix from the basis $\{\mathbf{v}_1, \mathbf{v}_2, \mathbf{v}_3\}$ to the basis $\{\mathbf{w}_1, \mathbf{w}_2, \mathbf{w}_3\}$ is

$$\begin{bmatrix} i & -i & 1 \\ i & i & i \\ 1 & 1 & i \end{bmatrix}$$

and

$$\begin{bmatrix} i & -i & 1 \\ i & i & i \\ 1 & 1 & i \end{bmatrix} \begin{bmatrix} 2-i \\ 3 \\ i \end{bmatrix} = \begin{bmatrix} 1 \\ 5i \\ 4-i \end{bmatrix},$$

we have

$$\mathbf{v} = \mathbf{w}_1 + 5i\mathbf{w}_2 + (4-i)\mathbf{w}_3.$$

$\square$

Example 1.6.4. Let $a \in \mathbb{K}$. We consider the vector space $\mathcal{P}_3(\mathbb{K})$. Determine the change of coordinates matrix from the basis

$$\{1, t, t^2, t^3\}$$

to the basis

$$\{1, t - a, (t - a)^2, (t - a)^3\}$$

and write an arbitrary polynomial from $\mathcal{P}_3(\mathbb{K})$ in the basis $\{1, t - a, (t - a)^2, (t - a)^3\}$.

Solution. Since

$$1 = 1$$
$$t = a + 1 \cdot (t - a)$$
$$t^2 = a^2 + 2a(t - a) + (t - a)^2$$
$$t^3 = (a + t - a)^3 = a^3 + 3a^2(t - a) + 3a(t - a)^2 + (t - a)^3,$$

the change of coordinates matrix is

$$\begin{bmatrix} 1 & a & a^2 & a^3 \\ 0 & 1 & 2a & 3a^2 \\ 0 & 0 & 1 & 3a \\ 0 & 0 & 0 & 1 \end{bmatrix}.$$

For an arbitrary polynomial $p(t) = b_0 + b_1 t + b_2 t^2 + b_3 t^3$ in $\mathcal{P}_3(\mathbb{K})$ we have

$$\begin{bmatrix} 1 & a & a^2 & a^3 \\ 0 & 1 & 2a & 3a^2 \\ 0 & 0 & 1 & 3a \\ 0 & 0 & 0 & 1 \end{bmatrix} \begin{bmatrix} b_0 \\ b_1 \\ b_2 \\ b_3 \end{bmatrix} = \begin{bmatrix} a^3 b_3 + a^2 b_1 + ab_2 + b_0 \\ 3a^2 b_3 + 2 ab_2 + b_1 \\ 3 ab_3 + b_2 \\ b_3 \end{bmatrix},$$

and thus $p(t)$ becomes

$$a^3 b_3 + a^2 b_1 + ab_2 + b_0 + (3a^2 b_3 + 2 ab_2 + b_1)(t-a) + (3 ab_3 + b_2)(t-a)^2 + b_3(t-a)^3,$$

when written in the basis $\{1, t - a, (t - a)^2, (t - a)^3\}$. Note that we could get this result by direct calculations. $\qquad\square$

Theorem 1.6.5. *Let $\mathcal{B} = \{\mathbf{v}_1, \ldots, \mathbf{v}_n\}$ and $\mathcal{C} = \{\mathbf{w}_1, \ldots, \mathbf{w}_n\}$ be bases of a vector space $\mathcal{V}$. The change of coordinates matrix $\mathrm{Id}_{\mathcal{B} \to \mathcal{C}}$ is invertible and*

$$\mathrm{Id}_{\mathcal{B} \to \mathcal{C}}^{-1} = \mathrm{Id}_{\mathcal{C} \to \mathcal{B}}.$$

Proof. If

$$x_1 \mathbf{v}_1 + \cdots + x_n \mathbf{v}_n = y_1 \mathbf{w}_1 + \cdots + y_n \mathbf{w}_n,$$

then

$$\begin{bmatrix} y_1 \\ \vdots \\ y_n \end{bmatrix} = \mathrm{Id}_{\mathcal{B} \to \mathcal{C}} \begin{bmatrix} x_1 \\ \vdots \\ x_n \end{bmatrix} \quad \text{and} \quad \begin{bmatrix} x_1 \\ \vdots \\ x_n \end{bmatrix} = \mathrm{Id}_{\mathcal{C} \to \mathcal{B}} \begin{bmatrix} y_1 \\ \vdots \\ y_n \end{bmatrix},$$

and consequently

$$
\begin{bmatrix} x_1 \\ \vdots \\ x_n \end{bmatrix} = \mathrm{Id}_{\mathcal{C} \to \mathcal{B}} \begin{bmatrix} y_1 \\ \vdots \\ y_n \end{bmatrix} = \mathrm{Id}_{\mathcal{C} \to \mathcal{B}} \, \mathrm{Id}_{\mathcal{B} \to \mathcal{C}} \begin{bmatrix} x_1 \\ \vdots \\ x_n \end{bmatrix}
$$

and

$$
\begin{bmatrix} y_1 \\ \vdots \\ y_n \end{bmatrix} = \mathrm{Id}_{\mathcal{B} \to \mathcal{C}} \begin{bmatrix} x_1 \\ \vdots \\ x_n \end{bmatrix} = \mathrm{Id}_{\mathcal{B} \to \mathcal{C}} \, \mathrm{Id}_{\mathcal{C} \to \mathcal{B}} \begin{bmatrix} y_1 \\ \vdots \\ y_n \end{bmatrix},
$$

proving that $\mathrm{Id}_{\mathcal{B} \to \mathcal{C}}^{-1} = \mathrm{Id}_{\mathcal{C} \to \mathcal{B}}$. $\qquad\qquad\square$

Example 1.6.6. Let $\{\mathbf{v}_1, \mathbf{v}_2, \mathbf{v}_3\}$ and $\{\mathbf{w}_1, \mathbf{w}_2, \mathbf{w}_3\}$ be bases of the vector space $\mathcal{V}$. If

$$
\begin{aligned}
\mathbf{v}_1 &= i\mathbf{w}_1 + i\mathbf{w}_2 + \mathbf{w}_3 \\
\mathbf{v}_2 &= -i\mathbf{w}_1 + i\mathbf{w}_2 + \mathbf{w}_3 \\
\mathbf{v}_3 &= \mathbf{w}_1 + i\mathbf{w}_2 + i\mathbf{w}_3,
\end{aligned}
$$

write $\mathbf{w}_1$, $\mathbf{w}_2$, and $\mathbf{w}_3$ as linear combinations of the vectors $\mathbf{v}_1$, $\mathbf{v}_2$, and $\mathbf{v}_3$.

Solution. The change of coordinates matrix from the basis $\{\mathbf{v}_1, \mathbf{v}_2, \mathbf{v}_3\}$ to the basis $\{\mathbf{w}_1, \mathbf{w}_2, \mathbf{w}_3\}$ is

$$
\begin{bmatrix} i & -i & 1 \\ i & i & i \\ 1 & 1 & i \end{bmatrix}.
$$

The inverse of this matrix is

$$
\begin{bmatrix} -\frac{i}{2} & 0 & \frac{1}{2} \\ \frac{i}{2} & -\frac{1}{2} - \frac{i}{2} & \frac{i}{2} \\ 0 & \frac{1}{2} - \frac{i}{2} & -\frac{1}{2} - \frac{i}{2} \end{bmatrix}.
$$

Consequently, we can write

$$
\mathbf{w}_1 = -\frac{i}{2}\mathbf{v}_1 + \frac{i}{2}\mathbf{v}_2,
$$

$$
\mathbf{w}_2 = -\left(\frac{1}{2} + \frac{i}{2}\right)\mathbf{v}_2 + \left(\frac{1}{2} - \frac{i}{2}\right)\mathbf{v}_3,
$$

$$
\mathbf{w}_3 = \frac{1}{2}\mathbf{v}_1 + \frac{i}{2}\mathbf{v}_2 - \left(\frac{1}{2} + \frac{i}{2}\right)\mathbf{v}_3.
$$

$\qquad\qquad\square$

Example 1.6.7. It is easy to calculate that the change of coordinates matrix from the basis

$$\{1, t - a, (t-a)^2, (t-a)^3\}$$

to the basis

$$\{1, t, t^2, t^3\}$$

is

$$\begin{bmatrix} 1 & -a & a^2 & -a^3 \\ 0 & 1 & -2\,a & 3\,a^2 \\ 0 & 0 & 1 & -3\,a \\ 0 & 0 & 0 & 1 \end{bmatrix}.$$

Consequently, the change of coordinates matrix from the basis

$$\{1, t, t^2, t^3\}$$

to the basis

$$\{1, t - a, (t-a)^2, (t-a)^3\}$$

is

$$\begin{bmatrix} 1 & -a & a^2 & -a^3 \\ 0 & 1 & -2\,a & 3\,a^2 \\ 0 & 0 & 1 & -3\,a \\ 0 & 0 & 0 & 1 \end{bmatrix}^{-1} = \begin{bmatrix} 1 & a & a^2 & a^3 \\ 0 & 1 & 2\,a & 3\,a^2 \\ 0 & 0 & 1 & 3\,a \\ 0 & 0 & 0 & 1 \end{bmatrix},$$

which is the matrix we found in Example 1.6.4.

1.7   Exercises

1.7.1   Definitions and examples

Exercise 1.1. Let $A = \{(x, y, z) : x \in \mathbb{R}^3, x \geq 0, z \leq 0\}$. Show that A is not a vector space.

Exercise 1.2. Let $A = \{(x, y, z) \in \mathbb{R}^3 : x + y + z = 0\}$ and $B = \{(x, y, z) \in \mathbb{R}^3 : x - y + z = 0\}$. Show that $A \cup B$ is not a vector space.

Exercise 1.3. We define a vector space over $\mathbb{R}$ whose elements are lines in $\mathbb{R}^3$ parallel to a given line. We recall that a line is a set of the form $\mathbf{x} + \mathbb{R}\mathbf{a} = \{\mathbf{x} + c\mathbf{a} : c \in \mathbb{R}\}$ where $\mathbf{x}$ and $\mathbf{a}$ are vectors in $\mathbb{R}^3$ and $\mathbf{a} \neq \mathbf{0}$.

   Let $\mathbf{a}$ be a fixed nonzero vector in $\mathbb{R}^3$. We define

$$\widehat{\mathbf{x}} = \mathbf{x} + \mathbb{R}\mathbf{a}$$

and

$$\widehat{\mathbf{x}} + \widehat{\mathbf{y}} = \widehat{\mathbf{x} + \mathbf{y}} \quad \text{and} \quad c\widehat{\mathbf{x}} = \widehat{c\mathbf{x}}.$$

Show that the set $V = \{\hat{\mathbf{x}} : \mathbf{x} \in \mathbb{R}^3\}$ with the operations defined above is a real vector space.

Exercise 1.4. We define a vector space over $\mathbb{R}$ whose elements are planes in $\mathbb{R}^3$ parallel to a given plane. We recall that a plane is a set of the form $\mathbf{x} + \mathbb{R}\mathbf{a} + \mathbb{R}\mathbf{b} = \{\mathbf{x} + c\mathbf{a} + d\mathbf{b} : c, d \in \mathbb{R}\}$ where $\mathbf{x}, \mathbf{a}$ and $\mathbf{b}$ are vectors in $\mathbb{R}^3$ with $\mathbf{a} \neq \mathbf{0}$ and $\mathbf{b} \neq \mathbf{0}$.

Let $\mathbf{a}$ and $\mathbf{b}$ be fixed nonzero vectors in $\mathbb{R}^3$. We define

$$\hat{\mathbf{x}} = \mathbf{x} + \mathbb{R}\mathbf{a} + \mathbb{R}\mathbf{b}$$

and

$$\hat{\mathbf{x}} + \hat{\mathbf{y}} = \widehat{\mathbf{x} + \mathbf{y}} \quad \text{and} \quad c\hat{\mathbf{x}} = \widehat{c\mathbf{x}}.$$

Show that the set $V = \{\hat{\mathbf{x}} : \mathbf{x} \in \mathbb{R}^3\}$ with the operations defined above is a real vector space.

Exercise 1.5. Let V and W be vector spaces. Show that the set of all functions $f : V \to W$ with the operations of addition and scalar multiplication defined as

$$(f + g)(\mathbf{x}) = f(\mathbf{x}) + g(\mathbf{x}) \quad \text{and} \quad (\alpha f)(\mathbf{x}) = \alpha f(\mathbf{x}),$$

is a vector space.

1.7.2 Subspaces

Exercise 1.6. Let $a, b \in \mathbb{K}$. Write the polynomial $(t + a)^3$ as a linear combination of the polynomials $1, t + b, (t + b)^2, (t + b)^3$.

Exercise 1.7. Let $\mathcal{U}, \mathcal{V}$, and $\mathcal{W}$ be subspaces of a vector space $\mathcal{X}$. If $\mathcal{U} \subset \mathcal{V}$, show that $\mathcal{V} \cap (\mathcal{U} + \mathcal{W}) = \mathcal{U} + (\mathcal{V} \cap \mathcal{W})$.

Exercise 1.8. Let $\mathbf{v}_1, \mathbf{v}_2, \mathbf{v}_3$ be vectors in a vectors space $\mathcal{V}$. If

$$\mathbf{w}_1 = \mathbf{v}_1 + 3\mathbf{v}_2 + 5\mathbf{v}_3$$
$$\mathbf{w}_2 = \mathbf{v}_1 + \mathbf{v}_2 + 3\mathbf{v}_3$$
$$\mathbf{w}_3 = \mathbf{v}_1 + 2\mathbf{v}_2 + 4\mathbf{v}_3$$
$$\mathbf{w}_4 = \mathbf{v}_1 - \mathbf{v}_2 + \mathbf{v}_3$$
$$\mathbf{w}_5 = 3\mathbf{v}_1 - \mathbf{v}_2 + 5\mathbf{v}_3,$$

show that $\text{Span}\{\mathbf{w}_1, \mathbf{w}_2\} = \text{Span}\{\mathbf{w}_3, \mathbf{w}_4, \mathbf{w}_5\}$.

Exercise 1.9. Show that the set $\{p \in \mathcal{P}(\mathbb{R}) : \int_0^1 p(t)t^2 dt = 0\}$ is a vector subspace of $\mathcal{P}(\mathbb{R})$.

Exercise 1.10. Let $\mathcal{V}$ be a subspace of $\mathcal{M}_{n \times n}(\mathbb{K})$. Show that $\mathcal{V}^T = \{A^T : A \in \mathcal{V}\}$ is a subspace of $\mathcal{M}_{n \times n}(\mathbb{K})$.

Exercise 1.11. Let $\mathcal{V}_1, \ldots, \mathcal{V}_m$ be subspaces of a vector space $\mathcal{V}$. If $\mathcal{W}$ is a vector space such that $\mathcal{V}_j \subseteq \mathcal{W}$ for every $j \in \{1, \ldots, m\}$, show that $\mathcal{V}_1 + \cdots + \mathcal{V}_m \subseteq \mathcal{W}$.

Exercise 1.12. If $\mathcal{U}$ is a proper subspace of a vector space $\mathcal{V}$, show that the following conditions are equivalent:

(a) If $\mathcal{W} \neq \mathcal{U}$ is a subspace of $\mathcal{V}$ and $\mathcal{U} \subseteq \mathcal{W} \subseteq \mathcal{V}$, then $\mathcal{W} = \mathcal{V}$;

(b) For every $\mathbf{x} \in \mathcal{V}$ which is not in $\mathcal{U}$ we have $\mathcal{U} + \mathbb{K}\mathbf{x} = \mathcal{V}$.

Exercise 1.13. If $\mathcal{U}$ and $\mathcal{W}$ are subspaces of a vector space $\mathcal{V}$, show that the following conditions are equivalent:

(a) The set $\mathcal{U} \cup \mathcal{W}$ is a subspace of $\mathcal{V}$;

(b) $\mathcal{U} \subseteq \mathcal{W}$ or $\mathcal{W} \subseteq \mathcal{U}$.

Exercise 1.14. Let $\mathcal{U}$ and $\mathcal{W}$ are subspaces of a vector space $\mathcal{V}$ and $\mathcal{A}$ and $\mathcal{B}$ finite subsets of $\mathcal{V}$ such that $\mathrm{Span}\mathcal{A} = \mathcal{U}$ and $\mathrm{Span}\mathcal{B} = \mathcal{W}$. Show that $\mathrm{Span}(\mathcal{A} \cup \mathcal{B}) = \mathcal{U} + \mathcal{W}$.

Exercise 1.15. Let $\mathcal{U}$ be a subspace of the vector space $\mathbb{K}$. Show that $\mathcal{U} = \{\mathbf{0}\}$ or $\mathcal{U} = \mathbb{K}$.

Exercise 1.16. Show that the set of matrices from $\mathcal{M}_{n \times n}(\mathbb{K})$ which satisfy the equation $2A + A^T = \mathbf{0}$ is a subspace of $\mathcal{M}_{n \times n}(\mathbb{K})$. Describe this subspace.

Exercise 1.17. Give an example of subspaces $\mathcal{U}$, $\mathcal{V}$, and $\mathcal{W}$ of $\mathcal{P}_3(\mathbb{K})$ such that $\mathcal{U} \neq \mathcal{V}$ and $\mathcal{U} + \mathcal{W} = \mathcal{V} + \mathcal{W} = \mathcal{P}_3(\mathbb{K})$.

Exercise 1.18. Show that the set

$$\mathcal{UD}_{n \times n}(\mathbb{K}) = \{A = [a_{jk}]_{1 \leq j \leq n, 1 \leq k \leq n} \in \mathcal{M}_{n \times n}(\mathbb{K}) : a_{jk} = 0 \text{ for all } j < k\}$$

is a subspace of $\mathcal{M}_{n \times n}(\mathbb{K})$.

Exercise 1.19. Verify that $\mathcal{M}_{n \times n}(\mathbb{K}) = \mathcal{UD}_{n \times n}(\mathbb{K}) + \mathcal{S}_{n \times n}(\mathbb{K})$, where $\mathcal{UD}_{n \times n}(\mathbb{K})$ is defined in Exercise 1.18.

Exercise 1.20. If $\mathbf{v}_1, \mathbf{v}_2, \mathbf{v}_3$ are vectors in a vector space $\mathcal{V}$, show that

$$\mathrm{Span}\{\mathbf{v}_1, \mathbf{v}_2, \mathbf{v}_3\} = \mathrm{Span}\{7\mathbf{v}_1 + 3\mathbf{v}_2 - 4\mathbf{v}_3, \mathbf{v}_2 + 2\mathbf{v}_3, 5\mathbf{v}_3 - \mathbf{v}_2\}.$$

Exercise 1.21. Let $a \in \mathbb{K}$. Show that a polynomial $p \in \mathcal{P}_n(\mathbb{K})$ is a linear combination of the polynomials $(t-a)^2, \ldots, (t-a)^n$ if and only if $p(a) = p'(a) = 0$.

Exercise 1.22. Let $\mathcal{U}$ be a subspace of a vector space $\mathcal{V}$ and let $\mathbf{x}$ and $\mathbf{y}$ be vectors in $\mathcal{V}$ such that $\mathbf{x} \notin \mathcal{U}$. Show that if $\mathbf{x} \in \mathcal{U} + \mathbb{K}\mathbf{y}$ then $\mathbf{y} \in \mathcal{U} + \mathbb{K}\mathbf{x}$ and $\mathbf{y} \notin \mathcal{U}$.

Exercise 1.23. Let $f, g \in \mathcal{F}_{\mathbb{K}}(S)$. Write the function f as a linear combination of the functions $5f - 3g$ and $4f + 7g$.

Exercise 1.24. Show that the function $|t|$ is not a linear combination of the function $|t - c_1|, \ldots, |t - c_n|$ where $c_1, \ldots, c_n$ are nonzero real numbers.

Exercise 1.25. Write the function $\sin(t + \alpha)$ as a linear combination of the functions $\sin t$ and $\cos t$.

1.7.3   Linearly independent vectors and bases

Exercise 1.26. Let $\mathcal{U}$ be the set of all infinite sequences $(x_1, x_2, \dots)$ of real numbers such that $x_{n+2} = x_n$ for every integer $n \geq 1$. Show that $\mathcal{U}$ is a finite-dimensional subspace of the vector space of all infinite sequences $(x_1, x_2, \dots)$ of real numbers and determine a basis in $\mathcal{U}$.

Exercise 1.27. Let $\mathcal{U}$ and $\mathcal{W}$ be subspaces of a vector space $\mathcal{V}$ such that $\mathcal{U} \cap \mathcal{W} = \{\mathbf{0}\}$. Show that, if the vectors $\mathbf{u}_1, \dots, \mathbf{u}_m \in \mathcal{U}$ are linearly independent and the vectors $\mathbf{w}_1, \dots, \mathbf{w}_n \in \mathcal{W}$ are linearly independent, then the vectors $\mathbf{u}_1, \dots, \mathbf{u}_m, \mathbf{w}_1, \dots, \mathbf{w}_n$ are linearly independent.

Exercise 1.28. If $\mathbf{v}_1, \mathbf{v}_2, \mathbf{v}_3$ are linearly independent vectors in a vector space $\mathcal{V}$, show that the vectors $\mathbf{v}_1 + \mathbf{v}_2, \mathbf{v}_2 + \mathbf{v}_3, \mathbf{v}_3 + \mathbf{v}_1$ are linearly independent.

Exercise 1.29. If $\mathbf{v}_1, \mathbf{v}_2, \mathbf{v}_3, \mathbf{v}_4$ are linearly independent vectors in a vector space $\mathcal{V}$, show that the vectors $\mathbf{v}_1 + \mathbf{v}_2, \mathbf{v}_2 + \mathbf{v}_3, \mathbf{v}_3 + \mathbf{v}_4, \mathbf{v}_4 + \mathbf{v}_1$ are linearly dependent.

Exercise 1.30. Show that the functions $1, \cos^2 t, \cos 2t$ are linearly dependent elements in $\mathcal{D}_\mathbb{R}(\mathbb{R})$.

Exercise 1.31. Show that the matrices

$$\begin{bmatrix} -i & 3 \\ 2 & i \end{bmatrix}, \begin{bmatrix} 1-i & i \\ 3 & i \end{bmatrix}, \begin{bmatrix} -i-2 & 9-2i \\ 0 & i \end{bmatrix}$$

are linearly dependent elements of $\mathcal{M}_{2 \times 2}(\mathbb{C})$.

Exercise 1.32. Let $\{\mathbf{v}_1, \mathbf{v}_2, \mathbf{v}_3\}$ be a basis of a complex vector space $\mathcal{V}$. Show that the vectors

$$\mathbf{w}_1 = i\mathbf{v}_1 + \mathbf{v}_2 + \mathbf{v}_3$$
$$\mathbf{w}_2 = i\mathbf{v}_1 + i\mathbf{v}_2 + \mathbf{v}_3$$
$$\mathbf{w}_3 = \mathbf{v}_1 + \mathbf{v}_2 + i\mathbf{v}_3$$

are linearly independent.

Exercise 1.33. If $\mathbf{v}_1, \dots, \mathbf{v}_n$ are linearly independent vectors in a vector space $\mathcal{V}$, show that the vectors $\mathbf{v}_3 + \mathbf{v}_1 + \mathbf{v}_2, \dots, \mathbf{v}_n + \mathbf{v}_1 + \mathbf{v}_2$ are linearly independent.

Exercise 1.34. Find a basis of $\mathbb{C}^2$ as a vector space over $\mathbb{R}$.

Exercise 1.35. Show that $\{\cos t, \sin t\}$ and $\{\cos t + i \sin t, \cos t - i \sin t\}$ are bases of the same complex vector subspace of $\mathcal{F}_\mathbb{C}(\mathbb{R})$.

Exercise 1.36. Let $\mathbf{v}_1, \dots, \mathbf{v}_k, \mathbf{v}, \mathbf{w}_1, \dots, \mathbf{w}_m$ be elements of a vector space $\mathcal{V}$. If the vectors $\mathbf{v}_1, \dots, \mathbf{v}_k, \mathbf{v}$ are linearly independent and $\mathbf{v} \in \mathrm{Span}\{\mathbf{v}_1, \dots, \mathbf{v}_k, \mathbf{w}_1, \dots, \mathbf{w}_m\}$, then there are integers $j_1, \dots j_{m-1} \in \{1, \dots, m\}$ such that

$$\mathrm{Span}\{\mathbf{v}_1, \dots, \mathbf{v}_k, \mathbf{v}, \mathbf{w}_{j_1}, \dots, \mathbf{w}_{j_{m-1}}\} = \mathrm{Span}\{\mathbf{v}_1, \dots, \mathbf{v}_k, \mathbf{w}_1, \dots, \mathbf{w}_m\}.$$

Exercise 1.37. Let $\mathcal{W}$ be a vector space and let $\mathbf{v}_1, \ldots, \mathbf{v}_k, \mathbf{w}_1, \ldots, \mathbf{w}_m \in \mathcal{W}$. Use Exercise 1.36 to show that, if the vectors $\mathbf{v}_1, \ldots, \mathbf{v}_k$ are linearly independent in $\mathrm{Span}\{\mathbf{w}_1, \ldots, \mathbf{w}_m\}$, then $k \le m$.

Exercise 1.38. Let $\mathbf{v}_1, \ldots, \mathbf{v}_n$ be vectors in a vector space $\mathcal{V}$ such that $\mathcal{V} = \mathrm{Span}\{\mathbf{v}_1, \ldots, \mathbf{v}_n\}$. Let k be an integer such that $1 \le k < n$. We suppose that the vectors $\mathbf{v}_1, \ldots, \mathbf{v}_k$ are linearly independent and that the vectors $\mathbf{v}_1, \ldots, \mathbf{v}_k, \mathbf{v}_m$ are linearly dependent for all $m \in \{k+1, \ldots, n\}$. Show that $\{\mathbf{v}_1, \ldots, \mathbf{v}_k\}$ is a basis of $\mathcal{V}$.

Exercise 1.39. Let $\mathcal{V}$ be a subspace of a vector space $\mathcal{W}$ and let $\mathbf{v}_1, \ldots, \mathbf{v}_k$ be linearly independent vectors in $\mathcal{V}$. Show that either $\{\mathbf{v}_1, \ldots, \mathbf{v}_k\}$ is a basis of $\mathcal{V}$ or there is a vector $\mathbf{v} \in \mathcal{V}$ such that the vectors $\mathbf{v}_1, \ldots, \mathbf{v}_k, \mathbf{v}$ are linearly independent.

Exercise 1.40. Let $\mathcal{V}$ be a subspace of an m-dimensional vector space $\mathcal{W}$. Using Exercises 1.37 and 1.39, show that $\mathcal{V}$ is a k-dimensional vector space for some $k \le m$.

Exercise 1.41. Let $\{\mathbf{v}_1, \mathbf{v}_2, \mathbf{v}_3, \mathbf{v}_4, \mathbf{v}_5\}$ be a basis of a vector space $\mathcal{V}$ and let

$$\mathbf{w}_1 = \mathbf{v}_1 + 3\mathbf{v}_2 + 3\mathbf{v}_3 + 2\mathbf{v}_4 + \mathbf{v}_5$$
$$\mathbf{w}_2 = 2\mathbf{v}_1 + \mathbf{v}_2 + \mathbf{v}_3 + 4\mathbf{v}_4 + \mathbf{v}_5$$
$$\mathbf{w}_3 = \mathbf{v}_1 + \mathbf{v}_2 + \mathbf{v}_3 + 2\mathbf{v}_4 + \mathbf{v}_5.$$

Determine all two-element sets $\{\mathbf{v}_j, \mathbf{v}_k\} \subset \{\mathbf{v}_1, \mathbf{v}_2, \mathbf{v}_3, \mathbf{v}_4, \mathbf{v}_5\}$ such that the set $\{\mathbf{w}_1, \mathbf{w}_2, \mathbf{w}_3, \mathbf{v}_j, \mathbf{v}_k\}$ is a basis of $\mathcal{V}$.

Exercise 1.42. Let $\{\mathbf{v}_1, \mathbf{v}_2, \mathbf{v}_3, \mathbf{v}_4\}$ be a basis of a vector space $\mathcal{V}$. If

$$\mathbf{w}_1 = 3\mathbf{v}_1 + 4\mathbf{v}_2 + 5\mathbf{v}_3 + 2\mathbf{v}_4$$
$$\mathbf{w}_2 = 2\mathbf{v}_1 + 5\mathbf{v}_2 + 3\mathbf{v}_3 + 4\mathbf{v}_4$$
$$\mathbf{w}_3 = \mathbf{v}_1 + a\mathbf{v}_2 + b\mathbf{v}_3$$
$$\mathbf{w}_4 = c\mathbf{v}_2 + d\mathbf{v}_3 + \mathbf{v}_4,$$

find numbers a, b, c, d such that $\{\mathbf{w}_1, \mathbf{w}_2\}$ and $\{\mathbf{w}_3, \mathbf{w}_4\}$ are bases of the same subspace.

Exercise 1.43. Verify that the set

$$\mathcal{S} = \left\{ \begin{bmatrix} x \\ y \\ z \\ t \end{bmatrix} : 3x + y + z + 4t = 0, 2x + y + 3z + t = 0 \right\}$$

is a subspace of $\mathbb{R}^4$ and determine a basis of this subspace.

Exercise 1.44. Show that $\mathcal{V} = \{p \in \mathcal{P}_2(\mathbb{R}) : \int_0^1 p(t)dt = 0, \int_0^2 p(t)dt = 0\}$ is a vector subspace of $\mathcal{P}_2(\mathbb{R})$ and determine a basis in this subspace.

1.7.4 Direct sums

Exercise 1.45. If $U = \{f \in \mathcal{F}_{\mathbb{R}}(\mathbb{R}) : f(-t) = f(t)\}$ and $V = \{f \in \mathcal{F}_{\mathbb{R}}(\mathbb{R}) : f(-t) = -f(t)\}$, show that $\mathcal{F}_{\mathbb{R}}(\mathbb{R}) = U \oplus V$.

Exercise 1.46. Let $\mathcal{D}_{n \times n}(\mathbb{K}) = \{[a_{jk}] \in \mathcal{M}_{n \times n}(\mathbb{K}) : a_{jk} = 0 \text{ if } j \neq k\}$. Find a subspace V of $\mathcal{M}_{n \times n}(\mathbb{K})$ such that

$$\mathcal{A}_{n \times n}(\mathbb{K}) \oplus \mathcal{D}_{n \times n}(\mathbb{K}) \oplus V = \mathcal{M}_{n \times n}(\mathbb{K}).$$

Exercise 1.47. Let $\mathcal{U}$ and $\mathcal{W}$ be vector subspaces of a vector space V. If $\dim \mathcal{U} + \dim \mathcal{W} = \dim V + 1$, show that the sum $\mathcal{U} + \mathcal{W}$ is not direct.

Exercise 1.48. Let $\mathcal{U}_1, \mathcal{U}_2$ and $\mathcal{U}_3$ be subspaces of a vector space V. Show that the sum $\mathcal{U}_1 + \mathcal{U}_2 + \mathcal{U}_3$ is direct if and only if $\mathcal{U}_1 \cap \mathcal{U}_2 = \mathbf{0}$ and $(\mathcal{U}_1 + \mathcal{U}_2) \cap \mathcal{U}_3 = \mathbf{0}$.

Exercise 1.49. Let $\mathcal{U}_1$ and $\mathcal{U}_2$ be subspaces of a vector space V and let $\mathbf{v}_1, \ldots, \mathbf{v}_n \in V$. We assume that the sum

$$\mathcal{U}_1 + \mathcal{U}_2 + \mathbb{K}\mathbf{v}_1 + \cdots + \mathbb{K}\mathbf{v}_n$$

is direct. If $\mathcal{W}_1 = \mathcal{U}_1 + \mathbb{K}\mathbf{v}_1 + \cdots + \mathbb{K}\mathbf{v}_n$ and $\mathcal{W}_2 = \mathcal{U}_2 + \mathbb{K}\mathbf{v}_1 + \cdots + \mathbb{K}\mathbf{v}_n$, show that $\{\mathbf{v}_1, \ldots, \mathbf{v}_n\}$ is a basis of $\mathcal{W}_1 \cap \mathcal{W}_2$.

Exercise 1.50. Let A, B, and S be sets such that A and B are disjoint and $A \cup B = S$. If $V = \{f \in \mathcal{F}_{\mathbb{K}}(S) : f(x) = 0 \text{ for all } x \in A\}$ and $W = \{f \in \mathcal{F}_{\mathbb{K}}(S) : f(x) = 0 \text{ for all } x \in B\}$, show that V and W are subspaces of $\mathcal{F}_{\mathbb{K}}(S)$ and that $\mathcal{F}_{\mathbb{K}}(S) = V \oplus W$.

Exercise 1.51. Show that $\mathcal{M}_{n \times n}(\mathbb{K}) = \mathcal{A}_{n \times n}(\mathbb{K}) \oplus \mathcal{U} \mathcal{D}_{n \times n}(\mathbb{K})$.

Exercise 1.52. Let $x_1, \ldots, x_n$ be distinct elements of a set S. Show that the set $\mathcal{U} = \{f \in \mathcal{F}_{\mathbb{K}}(S) : f(x_1) = \cdots = f(x_n) = 0\}$ is a vector subspace of $\mathcal{F}_{\mathbb{K}}(S)$ and determine a subspace W of $\mathcal{F}_{\mathbb{K}}(S)$ such that

$$\mathcal{U} \oplus W = \mathcal{F}_{\mathbb{K}}(S).$$

1.7.5 Dimension of a vector space

Exercise 1.53. Let $p_0, \ldots, p_n$ be polynomials in $\mathcal{P}_n(\mathbb{K})$ such that $\deg p_j = j$ for every $j \in \{0, \ldots, n\}$. Show that $\{p_0, \ldots, p_n\}$ is a basis of $\mathcal{P}_n(\mathbb{K})$.

Exercise 1.54. Show that the dimension of the vector space V in Exercise 1.4 is 1.

Exercise 1.55. Show that the dimension of the vector space V in Exercise 1.3 is 2.

Exercise 1.56. Show that

$$B = \left\{ \begin{bmatrix} 1 & 1 \\ 1 & 1 \end{bmatrix}, \begin{bmatrix} 1 & 1 \\ 0 & 1 \end{bmatrix}, \begin{bmatrix} 1 & 0 \\ 0 & 1 \end{bmatrix}, \begin{bmatrix} 0 & 0 \\ 0 & 1 \end{bmatrix} \right\}$$

is a basis of the vector space $\mathcal{M}_{2 \times 2}(\mathbb{K})$.

Exercise 1.57. Determine 3 different bases in the vector space $\mathcal{M}_{2\times2}(\mathbb{K})$ which are extensions of the set $\left\{ \begin{bmatrix} 1 & 2 \\ 0 & 0 \end{bmatrix}, \begin{bmatrix} 3 & 0 \\ 0 & 4 \end{bmatrix} \right\}$.

Exercise 1.58. Show that the set of all matrices of the form $\begin{bmatrix} a & b \\ b & b \end{bmatrix}$, where $a, b \in \mathbb{K}$ are arbitrary, is a subspace of $\mathcal{M}_{2\times2}(\mathbb{K})$ and determine the dimension of this subspace.

Exercise 1.59. Show that the set of all matrices of the form $\begin{bmatrix} x & b \\ c & y \end{bmatrix}$, where $a, b, x, y \in \mathbb{K}$ and $2x + 5y = 0$, is a subspace of $\mathcal{M}_{2\times2}(\mathbb{K})$ and determine the dimension of this subspace.

Exercise 1.60. Show that the set

$$\mathcal{U} = \left\{ \begin{bmatrix} a & b \\ c & d \end{bmatrix} \in \mathcal{M}_{2\times2}(\mathbb{K}) : a + b = 0, c + d = 0, a + d = 0 \right\}$$

is a vector subspace of $\mathcal{M}_{2\times2}(\mathbb{K})$ and determine $\dim \mathcal{U}$.

Exercise 1.61. Determine $\dim \mathcal{S}_{3\times3}(\mathbb{K})$.

Exercise 1.62. If $\mathcal{U}_1, \ldots, \mathcal{U}_k$ are subspaces of an n-dimensional vector space $\mathcal{V}$, show that $\dim \mathcal{U}_1 + \cdots + \dim \mathcal{U}_k \leq (k - 1)n + \dim \mathcal{U}_1 \cap \cdots \cap \mathcal{U}_k$.

Exercise 1.63. Let $\mathcal{U}$ and $\mathcal{W}$ be vector subspaces of a vector space $\mathcal{V}$ where $\dim \mathcal{V} = n$. If $\{\mathbf{0}\} \neq \mathcal{U} \not\subseteq \mathcal{W}$, $\dim \mathcal{U} = m$, and $\dim \mathcal{W} = n - 1$, determine $\dim \mathcal{U} \cap \mathcal{W}$.

Exercise 1.64. Let $\{\mathbf{v}_1, \mathbf{v}_2, \mathbf{v}_3, \mathbf{v}_4\}$ be a basis of a vector space $\mathcal{V}$ and let $\mathbf{w}_1, \mathbf{w}_2, \mathbf{w}_3 \in \mathcal{V}$ be linearly independent. Show that at least one the sets

$$\{\mathbf{w}_1, \mathbf{w}_2, \mathbf{w}_3, \mathbf{v}_1\}, \{\mathbf{w}_1, \mathbf{w}_2, \mathbf{w}_3, \mathbf{v}_2\}, \{\mathbf{w}_1, \mathbf{w}_2, \mathbf{w}_3, \mathbf{v}_3\}, \{\mathbf{w}_1, \mathbf{w}_2, \mathbf{w}_3, \mathbf{v}_4\}$$

is a basis of $\mathcal{V}$.

Exercise 1.65. Let $\{\mathbf{v}_1, \mathbf{v}_2, \mathbf{v}_3, \mathbf{v}_4\}$ be a basis of a vector space $\mathcal{V}$. If

$$\mathbf{w}_1 = \mathbf{v}_1 + 3\mathbf{v}_2 + 4\mathbf{v}_3 - \mathbf{v}_4$$
$$\mathbf{w}_2 = 2\mathbf{v}_1 + \mathbf{v}_2 + 3\mathbf{v}_3 + 3\mathbf{v}_4$$
$$\mathbf{w}_3 = \mathbf{v}_1 + \mathbf{v}_2 + 2\mathbf{v}_3 + \mathbf{v}_4,$$

determine the dimension of $\mathrm{Span}\{\mathbf{w}_1, \mathbf{w}_2, \mathbf{w}_3\}$ and find a basis of this subspace.

Exercise 1.66. Show that the set $\mathcal{U} = \{p \in \mathcal{P}_n(\mathbb{K}) : p(1) = p'''(1) = 0\}$ is a subspace of $\mathcal{P}_n(\mathbb{K})$. Determine the dimension of $\mathcal{U}$, find a basis of $\mathcal{U}$, and extend this basis to a basis of $\mathcal{P}_n(\mathbb{K})$.

Exercise 1.67. If $\mathcal{V} = \{p \in \mathcal{P}_n(\mathbb{K}) : p(1) = p(2) = 0\}$ and $\mathcal{W} = \{p \in \mathcal{P}_n(\mathbb{K}) : p(1) = p(3) = 0\}$, describe $\mathcal{V} + \mathcal{W}$ and verify that $\dim(\mathcal{V} + \mathcal{W}) = \dim \mathcal{V} + \dim \mathcal{W} - \dim(\mathcal{V} \cap \mathcal{W})$.

Exercise 1.68. Let $\{\mathbf{v}_1, \mathbf{v}_2, \mathbf{v}_3, \mathbf{v}_4, \mathbf{v}_5\}$ be linearly independent vectors in a vector space V and let $\mathbf{w}_1, \mathbf{w}_2, \mathbf{w}_3 \in \text{Span}\{\mathbf{v}_1, \mathbf{v}_2, \mathbf{v}_3, \mathbf{v}_4, \mathbf{v}_5\}$. We assume that

$$\mathbf{w}_1 = a_1\mathbf{v}_1 + a_2\mathbf{v}_2 + a_3\mathbf{v}_3 + a_4\mathbf{v}_4 + a_5\mathbf{v}_5$$
$$\mathbf{w}_2 = b_1\mathbf{v}_1 + b_2\mathbf{v}_2 + b_3\mathbf{v}_3 + b_4\mathbf{v}_4 + b_5\mathbf{v}_5$$
$$\mathbf{w}_3 = c_1\mathbf{v}_1 + c_2\mathbf{v}_2 + c_3\mathbf{v}_3 + c_4\mathbf{v}_4 + c_5\mathbf{v}_5$$

and that the reduced echelon form of the matrix

$$\begin{bmatrix} a_1 & a_2 & a_3 & a_4 & a_5 \\ b_1 & b_2 & a_3 & b_4 & b_5 \\ c_1 & c_2 & c_3 & c_4 & c_5 \end{bmatrix}$$

is

$$\begin{bmatrix} 1 & t & 0 & 0 & x \\ 0 & 0 & 1 & 0 & y \\ 0 & 0 & 0 & 1 & z \end{bmatrix}.$$

Show that $\{\mathbf{w}_1, \mathbf{w}_2, \mathbf{w}_3, \mathbf{v}_2, \mathbf{v}_5\}$ is a basis of $\text{Span}\{\mathbf{v}_1, \mathbf{v}_2, \mathbf{v}_3, \mathbf{v}_4, \mathbf{v}_5\}$.

Exercise 1.69. Show that $\dim \mathcal{M}_{n \times n}(\mathbb{K}) = \dim \mathcal{UD}_{n \times n} + \dim \mathcal{S}_{n \times n}(\mathbb{K}) - \dim \mathcal{D}_{n \times n}(\mathbb{K})$.

Exercise 1.70. Show that the set $\mathcal{U} = \{p \in \mathcal{P}_3(\mathbb{K}) : p(1) = p(2) = 0\}$ is a subspace of $\mathcal{P}_3(\mathbb{K})$, determine the dimension of $\mathcal{U}$, find a basis of $\mathcal{U}$, and extend that basis to a basis of $\mathcal{P}_3(\mathbb{K})$.

Exercise 1.71. Let $V = \{p \in \mathcal{P}_n(\mathbb{K}) : p(1) = p(-1) = 0\}$ and $W = \{p \in \mathcal{P}_n(\mathbb{K}) : p(i) = p(-i) = 0\}$, where $n \geq 4$. Show that V and W are subspaces of $\mathcal{P}_n(\mathbb{K})$, $V + W = \mathcal{P}_n(\mathbb{K})$, and $\dim(V + W) = \dim V + \dim W - \dim(V \cap W)$.

1.7.6 Change of basis

Exercise 1.72. Find the change of coordinates matrix from $\{\mathbf{w}_1, \mathbf{w}_2\}$ to $\{\mathbf{w}_3, \mathbf{w}_4\}$ defined in Exercise 1.42.

Exercise 1.73. Show that $\{t - a, t(t - a), t^2(t - a)\}$ and $\{t - a, (t - a)^2, (t - a)^3\}$ are bases in the vector subspace $\{p \in \mathcal{P}_4(\mathbb{K}) : p(a) = 0\}$ and find the change of coordinates matrix from $\{t - a, (t - a)^2, (t - a)^3\}$ to $\{t - a, t(t - a), t(t - a)^2\}$.

Exercise 1.74. Let $\{\mathbf{v}_1, \mathbf{v}_2, \mathbf{v}_3, \mathbf{v}_4\}$ be a basis of a vector space V. If

$$\mathbf{w}_1 = \mathbf{v}_1 + 2\mathbf{v}_2 + 4\mathbf{v}_3 + \mathbf{v}_4$$
$$\mathbf{w}_2 = 2\mathbf{v}_1 + 4\mathbf{v}_2 + 7\mathbf{v}_3 + 3\mathbf{v}_4$$
$$\mathbf{w}_3 = \mathbf{v}_1 + 2\mathbf{v}_2 + 5\mathbf{v}_4$$
$$\mathbf{w}_4 = \mathbf{v}_3 - \mathbf{v}_4,$$

show that $\{\mathbf{w}_1, \mathbf{w}_2\}$ and $\{\mathbf{w}_3, \mathbf{w}_4\}$ are bases of the same subspace and find the change of coordinates matrix from $\{\mathbf{w}_3, \mathbf{w}_4\}$ to $\{\mathbf{w}_1, \mathbf{w}_2\}$ and from $\{\mathbf{w}_1, \mathbf{w}_2\}$ to $\{\mathbf{w}_3, \mathbf{w}_4\}$.

Chapter 2

Linear Transformations

Introduction

When limits are introduced in calculus, one of the first properties of limits that we learn is

$$\lim_{x \to a} (f(x) + g(x)) = \lim_{x \to a} f(x) + \lim_{x \to a} +g(x) \quad \text{and} \quad \lim_{x \to a} cf(x) = c \lim_{x \to a} f(x),$$

where c is an arbitrary constant. These properties are then used in a more general version, namely,

$$\lim_{x \to a} (c_1 f_1(x) + \cdots + c_n f_n(x)) = c_1 \lim_{x \to a} f_1(x) + \cdots + c_n \lim_{x \to a} f_n(x)$$

where $c_1, \ldots, c_n$ are arbitrary constants. Then we see a similar property for derivatives

$$\frac{d}{dx}(c_1 f_1(x) + \cdots + c_n f_n(x)) = c_1 \frac{d}{dx} f_1(x) + \cdots + c_n \frac{d}{dx} f_n(x)$$

and integrals

$$\int_a^b (c_1 f_1(x) + \cdots + c_n f_n(x)) \, dx = c_1 \int_a^b f_1(x) \, dx + \cdots + c_n \int_a^b f_n(x) \, dx.$$

This property is referred to as linearity.

When matrix multiplication is introduced in an introductory matrix linear algebra course, again one of the first properties we mention is linearity of matrix multiplication.

Linearity of functions between vector spaces is one of the fundamental ideas of linear algebra. In this chapter we study properties of linear functions in the abstract setting of vector spaces.

2.1   Basic properties

Definition 2.1.1. Let V and W be vector spaces. A function $f : V \to W$ is called *linear* if it satisfies the following two conditions

   (a) $f(\mathbf{x} + \mathbf{y}) = f(\mathbf{x}) + f(\mathbf{y})$ for every $\mathbf{x}, \mathbf{y} \in V$;

   (b) $f(\alpha \mathbf{x}) = \alpha f(\mathbf{x})$ for every $\mathbf{x} \in V$ and every $\alpha \in \mathbb{K}$.

Linear functions from V to W are called *linear transformations*.

Note that, if $f : V \to W$ is a linear transformation, then

$$f(\alpha_1 \mathbf{x}_1 + \cdots + \alpha_j \mathbf{x}_j) = \alpha_1 f(\mathbf{x}_1) + \cdots + \alpha_j f(\mathbf{x}_j)$$

for any vectors $\mathbf{x}_1, \ldots, \mathbf{x}_j \in V$ and any numbers $\alpha_1, \ldots, \alpha_j \in \mathbb{K}$.

Proposition 2.1.2. *If $f : V \to W$ is a linear transformation, then*

   (a) $f(\mathbf{0}) = \mathbf{0}$;

   (b) $f(-\mathbf{v}) = -f(\mathbf{v})$.

Proof. To prove (a) it suffices to note that

$$f(\mathbf{0}) = f(\mathbf{0} + \mathbf{0}) = f(\mathbf{0}) + f(\mathbf{0})$$

and to prove (b) it suffices to note that

$$\mathbf{0} = f(\mathbf{0}) = f(\mathbf{v} - \mathbf{v}) = f(\mathbf{v}) + f(-\mathbf{v}).$$

$\square$

Example 2.1.3. Let V be a vector space. Show that the function $\mathrm{Id} : V \to V$ defined by $\mathrm{Id}(\mathbf{x}) = \mathbf{x}$ is linear.

Solution. We have

$$\mathrm{Id}(\mathbf{x} + \mathbf{y}) = \mathbf{x} + \mathbf{y} = \mathrm{Id}(\mathbf{x}) + \mathrm{Id}(\mathbf{y})$$

for every $\mathbf{x}, \mathbf{y} \in V$, and

$$\mathrm{Id}(\alpha \mathbf{x}) = \alpha \mathbf{x} = \alpha \, \mathrm{Id}(\mathbf{x})$$

for every $\mathbf{x} \in V$ and every $\alpha \in \mathbb{K}$.

$\square$

Example 2.1.4. Let V and W be vector spaces. If $f : V \to W$ and $g : V \to W$ are linear transformations, then the function $f + g : V \to W$ defined by

$$(f + g)(\mathbf{x}) = f(\mathbf{x}) + g(\mathbf{x}),$$

for every $\mathbf{x} \in V$, is a linear transformation.

Solution. For any vectors $\mathbf{x}, \mathbf{y} \in V$ and any number $\alpha \in \mathbb{K}$ we have

$$
\begin{aligned}
(f + g)(\mathbf{x} + \mathbf{y}) &= f(\mathbf{x} + \mathbf{y}) + g(\mathbf{x} + \mathbf{y}) \\
&= f(\mathbf{x}) + f(\mathbf{y}) + g(\mathbf{x}) + g(\mathbf{y}) = f(\mathbf{x}) + g(\mathbf{x}) + f(\mathbf{y}) + g(\mathbf{y}) \\
&= (f + g)(\mathbf{x}) + (f + g)(\mathbf{y})
\end{aligned}
$$

and

$$
\begin{aligned}
(f + g)(\alpha \mathbf{x}) &= f(\alpha \mathbf{x}) + g(\alpha \mathbf{x}) = \alpha f(\mathbf{x}) + \alpha g(\mathbf{x}) \\
&= \alpha(f(\mathbf{x}) + g(\mathbf{x})) = \alpha(f + g)(\mathbf{x}).
\end{aligned}
$$

This means that $f + g$ is a linear transformation. □

Example 2.1.5. Let V and W be vector spaces. If $f : V \to W$ is a linear transformation and $\alpha \in \mathbb{K}$, then the function $\alpha f : V \to W$ defined by

$$(\alpha f)(\mathbf{x}) = \alpha f(\mathbf{x})$$

for every $\mathbf{x} \in V$, is a linear transformation.

Solution. For any vectors $\mathbf{x}, \mathbf{y} \in V$ and any number $\beta \in \mathbb{K}$ we have

$$
\begin{aligned}
(\alpha f)(\mathbf{x} + \mathbf{y}) &= \alpha(f(\mathbf{x} + \mathbf{y})) = \alpha(f(\mathbf{x}) + f(\mathbf{y})) \\
&= \alpha f(\mathbf{x}) + \alpha f(\mathbf{y}) = (\alpha f)(\mathbf{x}) + (\alpha f)(\mathbf{y})
\end{aligned}
$$

and

$$(\alpha f)(\beta \mathbf{x}) = \alpha(f(\beta \mathbf{x})) = \alpha \beta f(\mathbf{x}) = \beta \alpha f(\mathbf{x}) = \beta(\alpha f)(\mathbf{x}).$$

This means that αf is a linear transformation. □

The proof of the following important result is a consequence of the definitions.

Theorem 2.1.6. *Let V and W be vector spaces. The set of all linear transformations from V to W with the operations $f + g$ and αf defined as*

$$(f + g)(\mathbf{x}) = f(\mathbf{x}) + g(\mathbf{x}) \quad and \quad (\alpha f)(\mathbf{x}) = \alpha f(\mathbf{x})$$

is a vector space.

Definition 2.1.7. The vector space of all linear transformations from a vector space V to a vector space W is denoted by $\mathcal{L}(V, W)$.
A linear transformation $f : V \to V$ is also called an *operator* or an *endomorphism*. The vector space $\mathcal{L}(V, V)$ is often denoted by $\mathcal{L}(V)$.

Example 2.1.8. If $A \in \mathcal{M}_{n \times m}(\mathbb{K})$, show that the function $f : \mathbb{K}^m \to \mathbb{K}^n$ defined by $f(\mathbf{x}) = A\mathbf{x}$ is a linear transformation.

Solution. From properties of matrix multiplication we get

$$f(\mathbf{x} + \mathbf{y}) = A(\mathbf{x} + \mathbf{y}) = A\mathbf{x} + A\mathbf{y} = f(\mathbf{x}) + f(\mathbf{y})$$

and

$$f(\alpha \mathbf{x}) = A(\alpha \mathbf{x}) = \alpha A \mathbf{x} = \alpha f(\mathbf{x}).$$

This means that f is a linear transformation.   □

Example 2.1.9. Let $V_1, \ldots, V_n$ be subspaces of a vector space V. Show that the function $f : V_1 \times \cdots \times V_n \to V$ defined by $f(\mathbf{x}_1, \ldots, \mathbf{x}_n) = \mathbf{x}_1 + \cdots + \mathbf{x}_n$ is a linear transformation.

Solution. The proof is an immediate consequence of the definition of a linear transformation.   □

Another important property of linear transformations is that the composition of linear transformations is a linear transformation.

Theorem 2.1.10. *Let V, W, and X be vector spaces. If $f : V \to W$ and $g : W \to X$ are linear transformations, then the function $g \circ f : V \to X$ is a linear transformation.*
In other words, if $f \in \mathcal{L}(V, W)$ and $g \in \mathcal{L}(W, X)$, then $g \circ f \in \mathcal{L}(V, X)$.

Proof. If $\mathbf{x}, \mathbf{y} \in \mathcal{V}$ and $\alpha \in \mathbb{K}$, then

$$g \circ f(\mathbf{x} + \mathbf{y}) = g(f(\mathbf{x} + \mathbf{y})) = g(f(\mathbf{x}) + f(\mathbf{y}))$$
$$= g(f(\mathbf{x})) + g(f(\mathbf{y})) = g \circ f(\mathbf{x}) + g \circ f(\mathbf{y})$$

and

$$g \circ f(\alpha \mathbf{x}) = g(f(\alpha \mathbf{x})) = g(\alpha f(\mathbf{x})) = \alpha g(f(\mathbf{x})) = \alpha g \circ f(\mathbf{x}).$$

$\square$

Example 2.1.11. Let $f : \mathbb{K} \to \mathbb{K}$ and $g : \mathbb{K} \to \mathbb{K}$ be the linear transformations defined by $f(x) = \alpha x$ and $g(x) = \beta x$, where α and β are numbers from $\mathbb{K}$. The linear transformation $g \circ f$ is defined by $g \circ f(x) = (\alpha \beta)x$. Note that in this case $g \circ f = f \circ g$. This equality is not generally true.

The above example can be generalized as follows.

Example 2.1.12. Let $f : \mathbb{K}^m \to \mathbb{K}^n$ and $g : \mathbb{K}^n \to \mathbb{K}^p$ be the linear transformations defined by

$$f(\mathbf{x}) = A\mathbf{x} \quad \text{and} \quad g(\mathbf{y}) = B\mathbf{y},$$

where $A \in \mathcal{M}_{n \times m}(\mathbb{K})$ and $B \in \mathcal{M}_{p \times n}(\mathbb{K})$. Then

$$g \circ f(\mathbf{x}) = (BA)\mathbf{x}$$

for every $\mathbf{x} \in \mathbb{K}^m$.

In linear algebra it is customary to write the composition $f \circ g$ simply as fg and call it the product of f and g. Note that, if f and g are defined in terms of matrices, as in Example 2.1.12, then the product of f and g corresponds to the product of matrices.

Composition of linear transformations has properties similar to multiplication. The main difference is that composition is not commutative, that is, in general fg is different from gf. Moreover, if fg is well-defined, it does not mean that gf makes sense.

Proposition 2.1.13. *Let $\mathcal{V}$ and $\mathcal{W}$ be vector spaces, let $f, f' : \mathcal{V} \to \mathcal{W}$ and $g, g' : \mathcal{W} \to \mathcal{X}$ be linear transformations, and let $\alpha \in \mathbb{K}$. Then*

(a) $g(f + f') = gf + gf'$;

(b) $(g + g')f = gf + g'f$;

(c) $(\alpha g)f = g(\alpha f) = \alpha(gf)$.

Proof. The properties are direct consequences of the definitions. The proof is left as an exercise.   □

The following theorem implies that a linear transformation $f \in \mathcal{L}(\mathcal{V}, \mathcal{W})$ is completely determined by its values at elements of an arbitrary basis of $\mathcal{V}$. This is very different from arbitrary functions from $\mathcal{V}$ to $\mathcal{W}$ and has important consequences.

Theorem 2.1.14. *Let $\mathcal{V}$ and $\mathcal{W}$ be vector spaces and let $\{\mathbf{v}_1, \ldots, \mathbf{v}_n\}$ be a basis of $\mathcal{V}$. For any $\mathbf{w}_1, \ldots, \mathbf{w}_n \in \mathcal{W}$ there is a unique linear transformation $f : \mathcal{V} \to \mathcal{W}$ such that*

$$f(\mathbf{v}_1) = \mathbf{w}_1, \quad \ldots, \quad f(\mathbf{v}_n) = \mathbf{w}_n.$$

Proof. Since $\{\mathbf{v}_1, \ldots, \mathbf{v}_n\}$ is a basis of $\mathcal{V}$, for every $\mathbf{x} \in \mathcal{V}$ there are unique numbers $x_1, \ldots, x_n \in \mathbb{K}$ such that $\mathbf{x} = x_1 \mathbf{v}_1 + \cdots + x_n \mathbf{v}_n$. We define

$$f(\mathbf{x}) = f(x_1 \mathbf{v}_1 + \cdots + x_n \mathbf{v}_n) = x_1 \mathbf{w}_1 + \cdots + x_n \mathbf{w}_n.$$

Since for any $\alpha \in \mathbb{K}$ we have

$$\alpha \mathbf{x} = \alpha(x_1 \mathbf{v}_1 + \cdots + x_n \mathbf{v}_n) = (\alpha x_1)\mathbf{v}_1 + \cdots + (\alpha x_n)\mathbf{v}_n,$$

we also have

$$f(\alpha \mathbf{x}) = (\alpha x_1)\mathbf{w}_1 + \cdots + (\alpha x_n)\mathbf{w}_n = \alpha(x_1 \mathbf{w}_1 + \cdots + x_n \mathbf{w}_n) = \alpha f(\mathbf{x}).$$

Now, if $\mathbf{x}, \mathbf{y} \in \mathcal{V}$, then

$$\mathbf{x} = x_1 \mathbf{v}_1 + \cdots + x_n \mathbf{v}_n \quad \text{and} \quad \mathbf{y} = y_1 \mathbf{v}_1 + \cdots + y_n \mathbf{v}_n,$$

for some numbers $x_1, \ldots, x_n, y_1, \ldots, y_n \in \mathbb{K}$. Since

$$\mathbf{x} + \mathbf{y} = x_1 \mathbf{v}_1 + \cdots + x_n \mathbf{v}_n + y_1 \mathbf{v}_1 + \cdots + y_n \mathbf{v}_n = (x_1 + y_1)\mathbf{v}_1 + \cdots + (x_n + y_n)\mathbf{v}_n,$$

we get

$$\begin{aligned}
f(\mathbf{x} + \mathbf{y}) &= (x_1 + y_1)\mathbf{w}_1 + \cdots + (x_n + y_n)\mathbf{w}_k \\
&= x_1 \mathbf{w}_1 + \cdots + x_k \mathbf{w}_n + y_1 \mathbf{w}_1 + \cdots + y_n \mathbf{w}_n \\
&= f(\mathbf{x}) + f(\mathbf{y}).
\end{aligned}$$

This shows that the defined function f is a linear transformation. Clearly, $f(\mathbf{v}_1) = \mathbf{w}_1, \ldots, f(\mathbf{v}_n) = \mathbf{w}_n$.

Now we need to show that defined function f is a unique linear transformation such that $f(\mathbf{v}_1) = \mathbf{w}_1, \ldots, f(\mathbf{v}_n) = \mathbf{w}_n$. Let g be any linear transformation such that $g(\mathbf{v}_1) = \mathbf{w}_1, \ldots, g(\mathbf{v}_n) = \mathbf{w}_n$ and let $\mathbf{x} \in \mathcal{V}$. Then $\mathbf{x} = x_1 \mathbf{v}_1 + \cdots + x_n \mathbf{v}_n$ for some $x_1, \ldots, x_n \in \mathbb{K}$ and we have

$$\begin{aligned}
g(\mathbf{x}) &= g(x_1 \mathbf{v}_1 + \cdots + x_n \mathbf{v}_n) = x_1 g(\mathbf{v}_1) + \cdots + x_n g(\mathbf{v}_n) \\
&= x_1 \mathbf{w}_1 + \cdots + x_n \mathbf{w}_n = f(x_1 \mathbf{v}_1 + \cdots + x_n \mathbf{v}_n) = f(\mathbf{x}).
\end{aligned}$$

This proves the uniqueness and completes the proof.   □

2.1.1 The kernel and range of a linear transformation

Definition 2.1.15. Let $f : V \to W$ be a linear transformation. The set

$$\ker f = \{\mathbf{x} \in V : f(\mathbf{x}) = \mathbf{0}\}$$

is called the *kernel* of f.

Example 2.1.16. Consider an $m \times n$ matrix A with entries in $\mathbb{K}$ and the linear transformation $f : \mathbb{K}^n \to \mathbb{K}^m$ defined by $f(\mathbf{x}) = A\mathbf{x}$. Then $\ker f = \mathbf{N}(A)$ (see Example 1.2.8).

Theorem 2.1.17. *Let $f : V \to W$ be a linear transformation. Then $\ker f$ is a subspace of V.*

Proof. If $\mathbf{v} \in \ker f$ and $\alpha \in \mathbb{K}$, then

$$f(\alpha\mathbf{v}) = \alpha f(\mathbf{v}) = \mathbf{0},$$

and thus $\alpha\mathbf{v} \in \ker f$. Similarly, if $\mathbf{v}_1, \mathbf{v}_2 \in \ker f$, then

$$f(\mathbf{v}_1 + \mathbf{v}_2) = f(\mathbf{v}_1) + f(\mathbf{v}_2) = \mathbf{0},$$

and thus $\mathbf{v}_1 + \mathbf{v}_2 \in \ker f$. $\qquad\square$

Example 2.1.18. Consider the linear transformation $f : \mathcal{P}_2(\mathbb{R}) \to \mathbb{R}$ defined by $f(p) = \int_0^1 p(t)dt$. Determine $\ker f$ and $\dim \ker f$.

Solution. An arbitrary element of $\mathcal{P}_2(\mathbb{R}))$ is of the form $at^2 + bt + c$ where a, b, and c are real numbers. Since

$$\int_0^1 (at^2 + bt + c)\, dt = \frac{a}{3} + \frac{b}{2} + c,$$

$f(at^2 + bt + c) = 0$ if and only if $\frac{a}{3} + \frac{b}{2} + c = 0$ or, equivalently, $c = -\frac{a}{3} - \frac{b}{2}$. Consequently, $f(at^2 + bt + c) = 0$ if and only if

$$at^2 + bt + c = at^2 + bt - \frac{a}{3} - \frac{b}{2} = a\left(t^2 - \frac{1}{3}\right) + b\left(t - \frac{1}{2}\right).$$

Hence
$$\ker f = \operatorname{Span}\left\{ t^2 - \frac{1}{3},\ t - \frac{1}{2} \right\}$$

and $\dim \ker f = 2$.   □

Example 2.1.19. Let $f : \mathcal{V} \to \mathcal{V}$ be a linear transformation such that

$$(f - \alpha \operatorname{Id})(f - \beta \operatorname{Id}) = 0$$

(where $\operatorname{Id} : \mathcal{V} \to \mathcal{V}$ is the identity linear transformation) and $\alpha, \beta \in \mathbb{K}$ with $\alpha \neq \beta$. Show that

$$\mathcal{V} = \ker(f - \alpha \operatorname{Id}) \oplus \ker(f - \beta \operatorname{Id}).$$

Solution. First we note that

$$f - \alpha \operatorname{Id} - f - \beta \operatorname{Id} = (\beta - \alpha)\operatorname{Id}. \tag{2.1}$$

Consequently, for any $\mathbf{v} \in \mathcal{V}$, we have

$$(f - \alpha \operatorname{Id})\mathbf{v} - (f - \beta \operatorname{Id})\mathbf{v} = (\beta - \alpha)\mathbf{v}$$

and thus
$$\mathbf{v} = \frac{1}{\beta - \alpha}(f - \alpha \operatorname{Id})\mathbf{v} - \frac{1}{\beta - \alpha}(f - \beta \operatorname{Id})\mathbf{v}.$$

Since $(f - \alpha \operatorname{Id})\mathbf{v} \in \ker(f - \beta \operatorname{Id})$ and $(f - \beta \operatorname{Id})\mathbf{v} \in \ker(f - \alpha \operatorname{Id})$, we have

$$\mathcal{V} = \ker(f - \alpha \operatorname{Id}) + \ker(f - \beta \operatorname{Id}).$$

To finish the proof we have to show that the sum is direct. Indeed, if $\mathbf{v} \in \ker(f - \alpha \operatorname{Id})$ and $\mathbf{v} \in \ker(f - \beta \operatorname{Id})$, then $\mathbf{v} = \mathbf{0}$ by (2.1).   □

Definition 2.1.20. Let $f : \mathcal{V} \to \mathcal{W}$ be a linear transformation. The set

$$\operatorname{ran} f = \{f(\mathbf{x}) : \mathbf{x} \in \mathcal{V}\}$$

is called the *range* of f.

In other words, if $f : \mathcal{V} \to \mathcal{W}$, then $\operatorname{ran} f$ is the set of all $\mathbf{y} \in \mathcal{W}$ such that $\mathbf{y} = f(\mathbf{x})$ for some $\mathbf{x} \in \mathcal{V}$. We can also write $\operatorname{ran} f = f(\mathcal{V})$.

Theorem 2.1.21. *Let* $f : V \to W$ *be a linear transformation. Then* ran f *is a subspace of* W.

Proof. If $\mathbf{w} \in \text{ran } f$, then $\mathbf{w} = f(\mathbf{v})$ for some $\mathbf{v} \in V$. Then for any $\alpha \in \mathbb{K}$ we have

$$\alpha \mathbf{w} = \alpha f(\mathbf{v}) = f(\alpha \mathbf{v}),$$

so $\alpha \mathbf{w} \in \text{ran } f$. Similarly, if $\mathbf{w}_1, \mathbf{w}_2 \in \text{ran } f$, then $\mathbf{w}_1 = f(\mathbf{v}_1)$ and $\mathbf{w}_2 = f(\mathbf{v}_2)$ for some $\mathbf{v}_1, \mathbf{v}_2 \in V$ and we have

$$\mathbf{w}_1 + \mathbf{w}_2 = f(\mathbf{v}_1) + f(\mathbf{v}_2) = f(\mathbf{v}_1 + \mathbf{v}_2),$$

so $\mathbf{w}_1 + \mathbf{w}_2 \in \text{ran } f$. $\qquad \square$

Example 2.1.22. Consider an $m \times n$ matrix A with entries in $\mathbb{K}$ and the linear transformation $f : \mathbb{K}^n \to \mathbb{K}^m$ defined by $f(\mathbf{x}) = A\mathbf{x}$. Show that ran $f = \mathbf{C}(A)$ (see Example 1.2.7).

Solution. Let $A = \begin{bmatrix} \mathbf{a}_1 & \cdots & \mathbf{a}_n \end{bmatrix}$ where $\mathbf{a}_1, \dots, \mathbf{a}_n$ are the columns of the matrix A. Then

$$\text{ran } f = \{A\mathbf{x} : \mathbf{x} \in \mathbb{K}^n\} = \{x_1 \mathbf{a}_1 + \cdots + x_n \mathbf{a}_n : \begin{bmatrix} x_1 \\ \vdots \\ x_n \end{bmatrix} \in \mathbb{K}^n\} = \mathbf{C}(A).$$

$\qquad \square$

Theorem 2.1.23. *A linear transformation* $f : V \to W$ *is injective if and only if* ker $f = \{\mathbf{0}\}$.

Proof. Since $f(\mathbf{0}) = \mathbf{0}$, if f is injective, then the only $\mathbf{v} \in V$ such that $f(\mathbf{v}) = \mathbf{0}$ is $\mathbf{v} = \mathbf{0}$. This means that ker $f = \{\mathbf{0}\}$.

Now assume ker $f = \{\mathbf{0}\}$. If $f(\mathbf{v}_1) = f(\mathbf{v}_2)$ for some $\mathbf{v}_1, \mathbf{v}_2 \in V$, then

$$f(\mathbf{v}_1 - \mathbf{v}_2) = f(\mathbf{v}_1) - f(\mathbf{v}_2) = \mathbf{0}.$$

So $\mathbf{v}_1 - \mathbf{v}_2 \in \text{ker } f = \{\mathbf{0}\}$, which means that $\mathbf{v}_1 - \mathbf{v}_2 = \mathbf{0}$ or $\mathbf{v}_1 = \mathbf{v}_2$. This shows that f is injective. $\qquad \square$

2.1.2   Projections

Projections on subspaces play an important role in linear algebra. In this section we discuss projections associated with direct sums. In Chapter 3 we will discuss orthogonal projections that are a special type of projections discussed in this section.

Recall that, if V is a vector space and U and W are subspaces of V such that $V = U \oplus W$, then for every $\mathbf{v} \in V$ there are unique $\mathbf{u} \in U$ and $\mathbf{w} \in W$ such that $\mathbf{v} = \mathbf{u} + \mathbf{w}$. This property is essential for the following definition.

Definition 2.1.24. Let V be a vector space and let U and W be subspaces of V such that $V = U \oplus W$. The function $f : V \to V$ defined by

$$f(\mathbf{u} + \mathbf{w}) = \mathbf{u},$$

where $\mathbf{u} \in U$ and $\mathbf{w} \in W$ is called the *projection on U along W*.

Note that, if f is the projection on U along W, then $f^2 = f$ and $\mathrm{Id} - f$ is the projection on W along U. Indeed, if $\mathbf{u} \in U$ and $\mathbf{w} \in W$, then

$$f^2(\mathbf{u} + \mathbf{w}) = f(f(\mathbf{u} + \mathbf{w})) = f(\mathbf{u}) = f(\mathbf{u} + \mathbf{0}) = \mathbf{u} = f(\mathbf{u} + \mathbf{w})$$

and

$$(\mathrm{Id} - f)(\mathbf{u} + \mathbf{w}) = \mathbf{u} + \mathbf{w} - f(\mathbf{u} + \mathbf{w}) = \mathbf{u} + \mathbf{w} - \mathbf{u} = \mathbf{w}.$$

Example 2.1.25. Let V be a vector space and let $f : V \to V$ be a linear transformation such that $f^2 = f$. Show that

$$V = \mathrm{ran}\, f \oplus \ker f$$

and that f is the projection on $\mathrm{ran}\, f$ along $\ker f$.

Solution. For any $\mathbf{v} \in V$ we have

$$\mathbf{v} = f(\mathbf{v}) + (\mathbf{v} - f(\mathbf{v}))$$

and

$$f(\mathbf{v} - f(\mathbf{v})) = f(\mathbf{v}) - f(f(\mathbf{v})) = f(\mathbf{v}) - f(\mathbf{v}) = \mathbf{0}.$$

Since $f(\mathbf{v}) \in \mathrm{ran}\, f$ and $\mathbf{v} - f(\mathbf{v}) \in \ker f$, this shows that $V = \mathrm{ran}\, f + \ker f$. We need to show that this sum is direct, that is, that the only vector that is in both $\mathrm{ran}\, f$ and $\ker f$ is the zero vector.

Suppose $\mathbf{v} \in \operatorname{ran} f$ and $\mathbf{v} \in \ker f$. Since $\mathbf{v} \in \operatorname{ran} f$, there is a $\mathbf{w} \in \mathcal{V}$ such that $f(\mathbf{w}) = \mathbf{v}$. Then

$$\mathbf{v} = f(\mathbf{w}) = f(f(\mathbf{w})) = f(\mathbf{v}) = \mathbf{0},$$

because $\mathbf{v} \in \ker f$.

Clearly, f is the projection on $\operatorname{ran} f$ along $\ker f$. □

Example 2.1.26. Let $\mathcal{V}$ be a vector space and let $f_1, \ldots, f_n : \mathcal{V} \to \mathcal{V}$ be linear transformations such that $f_j f_k = \mathbf{0}$ for $j \neq k$ and $f_1 + \cdots + f_n = \mathrm{Id}$. Show that

(a) The linear transformations $f_1, \ldots, f_n$ are projections;

(b) $\mathcal{V} = \operatorname{ran} f_1 \oplus \cdots \oplus \operatorname{ran} f_n$;

(c) For every $j \in \{1, \ldots, n\}$ the transformation f_j is the projection on $\operatorname{ran} f_j$ along $\operatorname{ran} f_1 \oplus \cdots \oplus \operatorname{ran} f_{j-1} \oplus \operatorname{ran} f_{j+1} \oplus \cdots \oplus \operatorname{ran} f_n$.

Solution. Let $\mathbf{v}$ be an arbitrary vector in $\mathcal{V}$. Since

$$\mathbf{v} = f_1(\mathbf{v}) + \cdots + f_n(\mathbf{v}),$$

for every $j \in \{1, \ldots, n\}$ we have

$$f_j(\mathbf{v}) = f_j f_1(\mathbf{v}) + \cdots + f_j f_n(\mathbf{v}) = f_j f_j(\mathbf{v}).$$

This shows that $f_j f_j = f_j$ and thus, by Example 2.1.25, f_j is the projection on $\operatorname{ran} f_j$ along $\ker f_j$.

To prove (b) we first note that, since $\mathbf{v} = f_1(\mathbf{v}) + \cdots + f_n(\mathbf{v})$ for every $\mathbf{v} \in \mathcal{V}$, we have

$$\mathcal{V} = f_1(V) + \cdots + f_n(V).$$

We need to show that this sum is direct. If

$$f_1(\mathbf{v}) + \cdots + f_n(\mathbf{v}) = \mathbf{0},$$

then

$$f_j f_1(\mathbf{v}) + \cdots + f_j f_n(\mathbf{v}) = f_j(\mathbf{0}) = \mathbf{0}.$$

On the other hand, since $f_j f_k = \mathbf{0}$ for $j \neq k$, we have

$$f_j f_1(\mathbf{v}) + \cdots + f_j f_n(\mathbf{v}) = f_j f_j(\mathbf{v}) = f_j(\mathbf{v})$$

and consequently $f_j(\mathbf{v}) = \mathbf{0}$ for every $j \in \{1, \ldots, n\}$. This shows that the sum $f_1(V) + \cdots + f_n(V)$ is direct.

Finally, to prove (c) we take a $\mathbf{v} \in \ker f_j$. Then we have

$$\mathbf{v} = f_1(\mathbf{v}) + \cdots + f_n(\mathbf{v})$$
$$= f_1(\mathbf{v}) + \cdots + f_{j-1}(\mathbf{v}) + f_{j+1}(\mathbf{v}) + \cdots + f_n(\mathbf{v})$$

and thus

$$\mathbf{v} \in f_1(V) \oplus \cdots \oplus f_{j-1}(V) \oplus f_{j+1}(V) \oplus \cdots \oplus f_n(V).$$

On the other hand, since $f_j f_k = \mathbf{0}$ for $j \neq k$, every $\mathbf{v} \in f_1(V) \oplus \cdots \oplus f_{j-1}(V) \oplus f_{j+1}(V) \oplus \cdots \oplus f_n(V)$ is in $\ker f_j$. Consequently,

$$\ker f_j = f_1(V) \oplus \cdots \oplus f_{j-1}(V) \oplus f_{j+1}(V) \oplus \cdots \oplus f_n(V),$$

completing the proof, by Example 2.1.25.   □

2.1.3   The Rank-Nullity Theorem

The main result of this section is an important theorem that connects the dimension of the domain of a linear transformation with the dimensions of its range and the subspace on which the transformation is zero, that is, the kernel of the transformation. We start with an example that will motivate the result.

Example 2.1.27. Let $f : V \to W$ be a linear transformation. If $\{\mathbf{v}_1, \mathbf{v}_2, \mathbf{v}_3\}$ is a basis of $\ker f$ and $\{\mathbf{w}_1, \mathbf{w}_2\}$ is a basis of $\operatorname{ran} f$, show that $\dim V = 5$.

Solution. For any $\mathbf{v} \in V$ there are $x_1, x_2 \in \mathbb{K}$ such that

$$f(\mathbf{v}) = x_1 \mathbf{w}_1 + x_2 \mathbf{w}_2.$$

If $\mathbf{u}_1$ and $\mathbf{u}_2$ are vectors in V such that $f(\mathbf{u}_1) = \mathbf{w}_1$ and $f(\mathbf{u}_2) = \mathbf{w}_2$, then

$$f(\mathbf{v}) = x_1 f(\mathbf{u}_1) + x_2 f(\mathbf{u}_2) = f(x_1 \mathbf{u}_1 + x_2 \mathbf{u}_2)$$

and thus

$$f(\mathbf{v} - (x_1 \mathbf{u}_1 + x_2 \mathbf{u}_2)) = \mathbf{0}.$$

This means that $\mathbf{v} - (x_1 \mathbf{u}_1 + x_2 \mathbf{u}_2) \in \ker f$ and consequently there are $y_1, y_2, y_3 \in \mathbb{K}$ such that

$$\mathbf{v} - (x_1 \mathbf{u}_1 + x_2 \mathbf{u}_2) = y_1 \mathbf{v}_1 + y_2 \mathbf{v}_2 + y_3 \mathbf{v}_3$$

or

$$\mathbf{v} = x_1 \mathbf{u}_1 + x_2 \mathbf{u}_2 + y_1 \mathbf{v}_1 + y_2 \mathbf{v}_2 + y_3 \mathbf{v}_3.$$

Since $\mathbf{v}$ is an arbitrary vector in $\mathcal{V}$, this shows that $\mathrm{Span}\{\mathbf{u}_1, \mathbf{u}_2, \mathbf{v}_1, \mathbf{v}_2, \mathbf{v}_3\} = \mathcal{V}$. To finish the proof we have to show that the vectors $\mathbf{u}_1, \mathbf{u}_2, \mathbf{v}_1, \mathbf{v}_2, \mathbf{v}_3$ are linearly independent. To this end suppose that

$$x_1\mathbf{u}_1 + x_2\mathbf{u}_2 + y_1\mathbf{v}_1 + y_2\mathbf{v}_2 + y_3\mathbf{v}_3 = \mathbf{0}. \tag{2.2}$$

Then

$$x_1 f(\mathbf{u}_1) + x_2 f(\mathbf{u}_2) + y_1 f(\mathbf{v}_1) + y_2 f(\mathbf{v}_2) + y_3 f(\mathbf{v}_3) = 0$$

and consequently

$$x_1\mathbf{w}_1 + x_2\mathbf{w}_2 = \mathbf{0},$$

which gives us $x_1 = x_2 = 0$, because the vectors $\mathbf{w}_1$ and $\mathbf{w}_2$ are linearly independent. Now, since $x_1 = x_2 = 0$, equation (2.2) becomes

$$y_1\mathbf{v}_1 + y_2\mathbf{v}_2 + y_3\mathbf{v}_3 = \mathbf{0},$$

which gives us $y_1 = y_2 = y_3 = 0$, because the vectors $\mathbf{v}_1, \mathbf{v}_2, \mathbf{v}_3$ are linearly independent. $\square$

It turns out that the property of the linear transformation in the example above holds for all linear transformations.

Theorem 2.1.28 (Rank-Nullity Theorem). *Let $\mathcal{V}$ be a finite dimensional vector space and let $f : \mathcal{V} \to \mathcal{W}$ be a linear transformation. Then*

$$\dim \ker f + \dim \operatorname{ran} f = \dim \mathcal{V}.$$

Proof. The proof is a generalization of the argument presented in Example 2.1.27. Let $\{\mathbf{v}_1, \ldots, \mathbf{v}_m\}$ be a basis of $\ker f$ and let $\{\mathbf{w}_1, \ldots, \mathbf{w}_n\}$ be a basis of $\operatorname{ran} f$. Then there are $\mathbf{u}_1, \ldots, \mathbf{u}_n \in \mathcal{V}$ such that $f(\mathbf{u}_j) = \mathbf{w}_j$ for $1 \le j \le n$.

If $\mathbf{v} \in \mathcal{V}$, then there are $x_1, \ldots, x_n \in \mathbb{K}$ such that

$$f(\mathbf{v}) = x_1\mathbf{w}_1 + \cdots + x_n\mathbf{w}_n = x_1 f(\mathbf{u}_1) + \cdots + x_n f(\mathbf{u}_n) = f(x_1\mathbf{u}_1 + \cdots + x_n\mathbf{u}_n)$$

and thus

$$f(\mathbf{v} - x_1\mathbf{u}_1 + \cdots + x_n\mathbf{u}_n) = \mathbf{0}.$$

Consequently $\mathbf{v} - x_1\mathbf{u}_1 + \cdots + x_n\mathbf{u}_n \in \ker f$ and

$$\mathbf{v} = x_1\mathbf{u}_1 + \cdots + x_n\mathbf{u}_n + y_1\mathbf{v}_1 + \cdots + y_m\mathbf{v}_m$$

for some $y_1, \ldots, y_m \in \mathbb{K}$. This shows that $\mathrm{Span}\{\mathbf{u}_1, \ldots, \mathbf{u}_n, \mathbf{v}_1, \ldots, \mathbf{v}_m\} = \mathcal{V}$. To finish the proof we have to show that the vectors $\mathbf{u}_1, \ldots, \mathbf{u}_n, \mathbf{v}_1, \ldots, \mathbf{v}_m$ are linearly independent. Suppose that

$$x_1\mathbf{u}_1 + \cdots + x_n\mathbf{u}_n + y_1\mathbf{v}_1 + \cdots + y_m\mathbf{v}_m = \mathbf{0}. \tag{2.3}$$

By applying f to the above equation and using the fact that $\{\mathbf{v}_1, \ldots, \mathbf{v}_m\}$ is a basis of ker f we obtain

$$x_1\mathbf{w}_1 + \cdots + x_n\mathbf{w}_n = \mathbf{0},$$

which gives us $x_1 = \cdots = x_n = 0$, because the vectors $\mathbf{w}_1, \ldots, \mathbf{w}_n$ are linearly independent. Now equation (2.3) reduces to

$$y_1\mathbf{v}_1 + \cdots + y_m\mathbf{v}_m = \mathbf{0},$$

which gives us $y_1 = \cdots = y_m = 0$ in view of linear independence of vectors $\mathbf{v}_1, \ldots, \mathbf{v}_m$. $\qquad\square$

The above theorem is called the Rank-Nullity Theorem because the number $\dim \operatorname{ran} f$ is called the *rank* of f and the number $\dim \ker f$ is called the *nullity* of f.

Example 2.1.29. Let $f : \mathcal{P}_5(\mathbb{R}) \to \mathcal{P}_5(\mathbb{R})$ be the linear transformation defined by $f(p) = p'''$. Determine ker f, dim ker f, ran f, dim ran f, and verify the Rank-Nullity Theorem.

Solution. If $p''' = 0$, then $p(t) = at^2 + bt + c$ for some $a, b, c \in \mathbb{R}$. Consequently ker $f = \operatorname{Span}\{1, t, t^2\}$ and dim ker $f = 3$. On the other hand, since

$$(a_5t^5 + a_4t^4 + a_3t^3 + a_2t^2 + a_1t + a_0)''' = 60a_5t^2 + 24a_4t + 6a_3,$$

ran $f = \operatorname{Span}\{1, t, t^2\}$ and dim ran $f = 3$. As stated in the Rank-Nullity Theorem,

$$\dim \ker f + \dim \operatorname{ran} f = 6 = \dim \mathcal{P}_5(\mathbb{R}).$$

$\qquad\square$

Example 2.1.30. Let $f : \mathcal{P}_5(\mathbb{R}) \to \mathbb{R}^2$ be the linear transformation defined by $f(p) = \begin{bmatrix} p'(5) \\ p(5) \end{bmatrix}$. Determine ker f, dim ker f, ran f, dim ran f, and verify the the Rank-Nullity Theorem.

Solution. If $f \in \mathcal{P}_5(\mathbb{R})$ and $p'(5) = p(5) = 0$, then

$$p(t) = (at^3 + bt^2 + ct + d)(t - 5)^2$$

for some $a, b, c, d \in \mathbb{R}$. Consequently

$$\ker f = \operatorname{Span}\{(t - 5)^2, t(t - 5)^2, t^2(t - 5)^2, t^3(t - 5)^2\}$$

and $\dim \ker f = 4$. On the other hand, since

$$f(p) = \begin{bmatrix} p'(5) \\ p(5) \end{bmatrix} = p'(5) \begin{bmatrix} 1 \\ 0 \end{bmatrix} + p(5) \begin{bmatrix} 0 \\ 1 \end{bmatrix},$$

$\operatorname{ran} f = \operatorname{Span} \left\{ \begin{bmatrix} 1 \\ 0 \end{bmatrix}, \begin{bmatrix} 0 \\ 1 \end{bmatrix} \right\} = \mathbb{R}^2$ and $\dim \operatorname{ran} f = 2$. As stated in the Rank-Nullity Theorem,

$$\dim \ker f + \dim \operatorname{ran} f = 6 = \dim \mathcal{P}_5(\mathbb{R}).$$

□

The following simple consequence of the Rank-Nullity Theorem is often used.

Corollary 2.1.31. *If $f : \mathcal{V} \to \mathbb{K}$ is a nonzero linear transformation, then $\dim \ker f = \dim \mathcal{V} - 1$.*

2.2 Isomorphisms

Consider the vector space $\mathcal{P}_n(\mathbb{K})$ of all functions $p : \mathbb{K} \to \mathbb{K}$ of the form $p(t) = a_0 + a_1 t + \cdots + a_n t^n$ where $a_0, a_1, \ldots, a_n \in \mathbb{K}$. Since the polynomial $p(t) = a_0 + a_1 t + \cdots + a_n t^n$ is completely determined by the numbers $a_0, a_1, \ldots, a_n$, one could say that the space $\mathcal{P}_n(\mathbb{K})$ can be "identified" with the space $\mathbb{K}^{n+1}$. We expect the vector spaces $\mathcal{P}_n(\mathbb{K})$ and $\mathbb{K}^{n+1}$ to have the same algebraic properties. We could say that, from the point of view of linear algebra, $\mathcal{P}_n(\mathbb{K})$ and $\mathbb{K}^{n+1}$ are two "representations" of the same vector space. This point of view is important in linear algebra. In this section we will make this idea precise and examine some of its consequences.

Definition 2.2.1. Let $\mathcal{V}$ and $\mathcal{W}$ be vector spaces. A linear transformation $f : \mathcal{V} \to \mathcal{W}$ that is both injective and surjective is called an *isomorphism of vector spaces* or simply an *isomorphism*. Vector spaces $\mathcal{V}$ and $\mathcal{W}$ are called *isomorphic* if there is an isomorphism $f : \mathcal{V} \to \mathcal{W}$.

Example 2.2.2. The vector spaces $\mathcal{P}_n(\mathbb{K})$ and $\mathbb{K}^{n+1}$ are isomorphic. Indeed, the function

$$f(a_0 + a_1 t + \cdots + a_n t^n) = \begin{bmatrix} a_0 \\ a_1 \\ \vdots \\ a_n \end{bmatrix}$$

is an isomorphism from $\mathcal{P}_n(\mathbb{K})$ onto $\mathbb{K}^{n+1}$.

Theorem 2.2.3. *Let V and W be vector spaces. If $f : V \to W$ is an isomorphism, then its inverse $f^{-1} : W \to V$ is a linear transformation.*

Proof. Let $\mathbf{w} \in W$ and $\alpha \in \mathbb{K}$. Since f is surjective, $\mathbf{w} = f(\mathbf{v})$ for some $\mathbf{v} \in V$. From linearity of f we get

$$f^{-1}(\alpha \mathbf{w}) = f^{-1}(\alpha f(\mathbf{v})) = f^{-1}(f(\alpha \mathbf{v})) = \alpha \mathbf{v} = \alpha f^{-1}(\mathbf{w}).$$

Now, let $\mathbf{w}_1, \mathbf{w}_2 \in W$. Since f is surjective, $\mathbf{w}_1 = f(\mathbf{v}_1)$ and $\mathbf{w}_2 = f(\mathbf{v}_2)$ for some $\mathbf{v}_1, \mathbf{v}_2 \in V$ and, from linearity of f, we get

$$f^{-1}(\mathbf{w}_1 + \mathbf{w}_2) = f^{-1}(f(\mathbf{v}_1) + f(\mathbf{v}_2)) = f^{-1}(f(\mathbf{v}_1 + \mathbf{v}_2))$$
$$= \mathbf{v}_1 + \mathbf{v}_2 = f^{-1}(\mathbf{w}_1) + f^{-1}(\mathbf{w}_2).$$

$\square$

Corollary 2.2.4. *The inverse of an isomorphism is an isomorphism.*

Since the function f in the definition of an isomorphism maps V onto W, it may seem that the role of V in the definition is different from the role of W, but in view of the above corollary we know that it is not the case.

In the next theorem we characterize isomorphisms in terms of bases.

Theorem 2.2.5. *Let V and W be vector spaces and let $\{\mathbf{v}_1, \ldots, \mathbf{v}_n\}$ be a basis of V. A linear transformation $f : V \to W$ is an isomorphism if and only if the set $\{f(\mathbf{v}_1), \ldots, f(\mathbf{v}_n)\}$ is a basis of W.*

Proof. Assume that $\{f(\mathbf{v}_1), \ldots, f(\mathbf{v}_n)\}$ is a basis of W. We first show that f is injective. If $f(x_1 \mathbf{v}_1 + \cdots + x_n \mathbf{v}_n) = \mathbf{0}$ for some $x_1, \ldots, x_n \in \mathbb{K}$, then $x_1 f(\mathbf{v}_1) + \cdots + x_n f(\mathbf{v}_n) = \mathbf{0}$ and thus $x_1 = \cdots = x_n = 0$. This means that $\ker f = \mathbf{0}$ and consequently f is injective, by Theorem 2.1.23.

To show that f is surjective we consider an arbitrary $\mathbf{w} \in \mathcal{W}$. Then $\mathbf{w} = x_1 f(\mathbf{v}_1) + \cdots + x_n f(\mathbf{v}_n)$ for some $x_1, \ldots, x_n \in \mathbb{K}$. Since

$$\mathbf{w} = x_1 f(\mathbf{v}_1) + \cdots + x_n f(\mathbf{v}_n) = f(x_1 \mathbf{v}_1 + \cdots + x_n \mathbf{v}_n),$$

we have $\mathbf{w} \in \operatorname{ran} f$.

Now we assume that f is an isomorphism. If $\mathbf{w} \in \mathcal{W}$, then there is $\mathbf{v} \in \mathcal{V}$ such that $f(\mathbf{v}) = \mathbf{w}$. Since $\mathbf{v} = x_1 \mathbf{v}_1 + \cdots + x_n \mathbf{v}_n$ for some $x_1, \ldots, x_n \in \mathbb{K}$, we have

$$\mathbf{w} = f(\mathbf{v}) = f(x_1 \mathbf{v}_1 + \cdots + x_k \mathbf{v}_n) = x_1 f(\mathbf{v}_1) + \cdots + x_n f(\mathbf{v}_n).$$

This shows that $\operatorname{Span}\{f(\mathbf{v}_1), \ldots, f(\mathbf{v}_n)\} = \mathcal{W}$. To show that the vectors $f(\mathbf{v}_1), \ldots, f(\mathbf{v}_n)$ are linearly independent we suppose that $x_1 f(\mathbf{v}_1) + \cdots + x_n f(\mathbf{v}_n) = \mathbf{0}$ for some $x_1, \ldots, x_n \in \mathbb{K}$. Then $f(x_1 \mathbf{v}_1 + \cdots + x_n \mathbf{v}_n) = \mathbf{0}$ and thus $x_1 \mathbf{v}_1 + \cdots + x_n \mathbf{v}_n = \mathbf{0}$, because $\ker f = \mathbf{0}$. Since $\{\mathbf{v}_1, \ldots, \mathbf{v}_n\}$ is a basis, we conclude that $x_1 = \cdots = x_n = 0$. Consequently, the set $f(\mathbf{v}_1), \ldots, f(\mathbf{v}_n)$ is a basis of the vector space $\mathcal{W}$. □

As an immediate consequence of the above theorem we obtain the following important result.

Corollary 2.2.6. *Finite dimensional vector spaces $\mathcal{V}$ and $\mathcal{W}$ are isomorphic if and only if* $\dim \mathcal{V} = \dim \mathcal{W}$.

Example 2.2.7. Let $\mathcal{V}$ be a finite dimensional vector space and let $\mathcal{U}$, $\mathcal{W}_1$, and $\mathcal{W}_2$ be subspaces of $\mathcal{V}$ such that $\mathcal{V} = \mathcal{U} \oplus \mathcal{W}_1 = \mathcal{U} \oplus \mathcal{W}_2$. Then $\dim \mathcal{W}_1 = \dim \mathcal{W}_2$, by Theorem 1.4.18. Consequently, the vector spaces $\mathcal{W}_1$ and $\mathcal{W}_2$ are isomorphic.

Isomorphic vector spaces have the same algebraic properties and, as we mentioned at the beginning of this section, from the point of view of linear algebra, isomorphic vector spaces can be thought of as different representations of the same vector space. The following corollary says that every vector space over $\mathbb{K}$ of dimension n is basically a version of $\mathbb{K}^n$.

Corollary 2.2.8. *Let $\{\mathbf{v}_1, \ldots, \mathbf{v}_n\}$ be a basis of a vector space $\mathcal{V}$. The function $f : \mathcal{V} \to \mathbb{K}^n$ defined by*

$$f(x_1\mathbf{v}_1 + \cdots + x_n\mathbf{v}_n) = \begin{bmatrix} x_1 \\ \vdots \\ x_n \end{bmatrix}$$

for all $x_1, \ldots, x_n \in \mathbb{K}$, is an isomorphism.

Proof. We have
$$f(\mathbf{v}_1) = \mathbf{e}_1, \quad \ldots, \quad f(\mathbf{v}_n) = \mathbf{e}_n,$$
where $\mathbf{e}_1, \ldots, \mathbf{e}_n$ is the standard basis of $\mathbb{K}^n$.

$\square$

Example 2.2.9. Let $\mathcal{V}_1, \ldots, \mathcal{V}_n$ be subspaces of a vector space $\mathcal{V}$. Show that the function $f : \mathcal{V}_1 \times \cdots \times \mathcal{V}_n \to \mathcal{V}$ defined by

$$f(\mathbf{v}_1, \ldots, \mathbf{v}_n) = \mathbf{v}_1 + \cdots + \mathbf{v}_n$$

is an isomorphism if and only if

$$\mathcal{V}_1 \oplus \cdots \oplus \mathcal{V}_n = \mathcal{V}.$$

Solution. The function f is clearly a linear transformation.

If f is an isomorphism, then $\operatorname{ran} f = \mathcal{V}$ and $\ker f = \{\mathbf{0}\}$. Since $\operatorname{ran} f = \mathcal{V}$, we have $\mathcal{V}_1 + \cdots + \mathcal{V}_n = \mathcal{V}$. Since $\ker f = \{\mathbf{0}\}$, $\mathbf{v}_1 + \cdots + \mathbf{v}_n = \mathbf{0}$ implies $\mathbf{v}_1 = \cdots = \mathbf{v}_n = \mathbf{0}$, which means that the sum $\mathcal{V}_1 + \cdots + \mathcal{V}_n$ is direct.

Now we assume that $\mathcal{V}_1 \oplus \cdots \oplus \mathcal{V}_n = \mathcal{V}$. Then $\operatorname{ran} f = \mathcal{V}$ and $\ker f = \{\mathbf{0}\}$, so f is an isomorphism. $\square$

Example 2.2.10. Let $\mathcal{V}$ be an arbitrary vector space. Show that $\mathcal{V}$ and $\mathcal{L}(\mathbb{K}, \mathcal{V})$ (the vector space of all linear transformations $f : \mathbb{K} \to \mathcal{V}$) are isomorphic.

Solution. For $\mathbf{v} \in \mathcal{V}$ let $t_{\mathbf{v}} : \mathbb{K} \to \mathcal{V}$ be the function defined by

$$t_{\mathbf{v}}(x) = x\mathbf{v}.$$

Note that $t_{\mathbf{v}} \in \mathcal{L}(\mathbb{K}, \mathcal{V})$. We will show that $f : \mathcal{V} \to \mathcal{L}(\mathbb{K}, \mathcal{V})$ defined by

$$f(\mathbf{v}) = t_{\mathbf{v}}$$

is an isomorphism.

It is easy to verify that f is a linear injection. We need to show that f is a surjection. Consider an arbitrary $s \in \mathcal{L}(\mathbb{K}, \mathcal{V})$. Then

$$s(x) = s(x \cdot 1) = xs(1),$$

If we let $\mathbf{v} = s(1)$, then have $s = t_{\mathbf{v}} = f(\mathbf{v})$. Consequently, ran $f = \mathcal{L}(\mathbb{K}, \mathcal{V})$.

Note that, as a particular case, we get that $\mathbb{K}$ and $\mathcal{L}(\mathbb{K}, \mathbb{K}) = \mathcal{L}(\mathbb{K})$ are isomorphic. $\qquad\square$

We close this section with a theorem that gives several useful characterizations of isomorphisms from a vector space to itself.

Theorem 2.2.11. *Let $\mathcal{V}$ be a finite dimensional vector space and let $f : \mathcal{V} \to \mathcal{V}$ be a linear transformation. The following conditions are equivalent:*

(a) *f is an isomorphism;*

(b) *$\ker f = \{\mathbf{0}\}$;*

(c) *$\operatorname{ran} f = \mathcal{V}$;*

(d) *f is left invertible, that is, there is a function $g : \mathcal{V} \to \mathcal{V}$ such that $gf = \operatorname{Id}$;*

(e) *f is right invertible, that is, there is a function $g : \mathcal{V} \to \mathcal{V}$ such that $fg = \operatorname{Id}$.*

Proof. Clearly (a) implies each of the remaining four conditions. Let $\{\mathbf{v}_1, \ldots, \mathbf{v}_n\}$ be a basis of $\mathcal{V}$.

Assume $\ker f = \{\mathbf{0}\}$. If

$$x_1 f(\mathbf{v}_1) + \cdots + x_n f(\mathbf{v}_n) = \mathbf{0},$$

then

$$f(x_1 \mathbf{v}_1 + \cdots + x_n \mathbf{v}_n) = \mathbf{0}$$

and consequently $x_1 \mathbf{v}_1 + \cdots + x_n \mathbf{v}_n = \mathbf{0}$, since $\ker f = \{\mathbf{0}\}$. Hence $x_1 = \cdots = x_n = 0$, because the vectors $\mathbf{v}_1, \ldots, \mathbf{v}_n$ are linearly independent. This proves that the vectors $f(\mathbf{v}_1), \ldots, f(\mathbf{v}_n)$ are linearly independent. Consequently $\{f(\mathbf{v}_1), \ldots, f(\mathbf{v}_n)\}$ is a basis of $\mathcal{V}$ and we have ran $f = \mathcal{V}$. This shows that (b) implies (c).

If ran $f = \mathcal{V}$, then $\operatorname{Span}\{f(\mathbf{v}_1), \ldots, f(\mathbf{v}_n)\} = \mathcal{V}$ and thus $\{f(\mathbf{v}_1), \ldots, f(\mathbf{v}_n)\}$ is a basis of $\mathcal{V}$. Hence $\ker f = \{\mathbf{0}\}$, by the Rank-Nullity Theorem. This shows that (c) implies (b).

Since (b) and (c) are equivalent and (b) and (c) together are equivalent to (a), all three conditions are equivalent.

Now we show that (d) implies (b). Indeed, if there is function $g : V \to V$ such that $gf = \text{Id}$ and $f(\mathbf{x}) = f(\mathbf{y})$, then

$$\mathbf{x} = g(f(\mathbf{x})) = g(f(\mathbf{y})) = \mathbf{y}.$$

This shows that f is injective and thus $\ker f = \{\mathbf{0}\}$.

Finally we show that (e) implies (c). Assume there is a function $g : V \to V$ such that $fg = \text{Id}$. Then for every $\mathbf{x} \in V$ we have $\mathbf{x} = f(g(\mathbf{x}))$ and thus $\text{ran } f = V$. $\qquad\square$

2.3   Linear transformations and matrices

2.3.1   The matrix of a linear transformation

At the beginning of this chapter we observed that an $n \times m$ matrix with entries in $\mathbb{K}$ defines a linear transformation from $\mathbb{K}^m$ to $\mathbb{K}^n$. It turns out that all linear transformations between finite dimensional spaces can be described in terms of matrix multiplication. We begin with an example.

Example 2.3.1. Let $\{\mathbf{v}_1, \mathbf{v}_2, \mathbf{v}_3\}$ and $\{\mathbf{w}_1, \mathbf{w}_2\}$ be bases of V and W, respectively and let $f : V \to W$ be the linear transformation defined by

$$f(\mathbf{v}_1) = a_{11}\mathbf{w}_1 + a_{21}\mathbf{w}_2,$$
$$f(\mathbf{v}_2) = a_{12}\mathbf{w}_1 + a_{22}\mathbf{w}_2,$$
$$f(\mathbf{v}_3) = a_{13}\mathbf{w}_1 + a_{22}\mathbf{w}_2.$$

Show that for every $\mathbf{v} = x_1\mathbf{v}_1 + x_2\mathbf{v}_2 + x_3\mathbf{v}_3 \in V$ we have

$$f(\mathbf{v}) = y_1\mathbf{w}_1 + y_2\mathbf{w}_2,$$

where the numbers y_1 and y_2 are given by the equality

$$\begin{bmatrix} y_1 \\ y_2 \end{bmatrix} = \begin{bmatrix} a_{11} & a_{12} & a_{13} \\ a_{21} & a_{22} & a_{23} \end{bmatrix} \begin{bmatrix} x_1 \\ x_2 \\ x_3 \end{bmatrix}.$$

Solution. Since

$$
\begin{aligned}
f(\mathbf{v}) &= f(x_1\mathbf{v}_1 + x_2\mathbf{v}_2 + x_3\mathbf{v}_3) \\
&= x_1 f(\mathbf{v}_1) + x_2 f(\mathbf{v}_2) + x_3 f(\mathbf{v}_3) \\
&= x_1(a_{11}\mathbf{w}_1 + a_{21}\mathbf{w}_2) + x_2(a_{12}\mathbf{w}_1 + a_{22}\mathbf{w}_2) + x_3(a_{13}\mathbf{w}_1 + a_{22}\mathbf{w}_2) \\
&= \left(\begin{bmatrix} a_{11} & a_{12} & a_{13} \end{bmatrix} \begin{bmatrix} x_1 \\ x_2 \\ x_3 \end{bmatrix} \right) \mathbf{w}_1 + \left(\begin{bmatrix} a_{21} & a_{22} & a_{23} \end{bmatrix} \begin{bmatrix} x_1 \\ x_2 \\ x_3 \end{bmatrix} \right) \mathbf{w}_2,
\end{aligned}
$$

we have

$$
\begin{bmatrix} y_1 \\ y_2 \end{bmatrix} = \begin{bmatrix} \begin{bmatrix} a_{11} & a_{12} & a_{13} \end{bmatrix} \begin{bmatrix} x_1 \\ x_2 \\ x_3 \end{bmatrix} \\ \begin{bmatrix} a_{21} & a_{22} & a_{23} \end{bmatrix} \begin{bmatrix} x_1 \\ x_2 \\ x_3 \end{bmatrix} \end{bmatrix} = \begin{bmatrix} a_{11} & a_{12} & a_{13} \\ a_{21} & a_{22} & a_{23} \end{bmatrix} \begin{bmatrix} x_1 \\ x_2 \\ x_3 \end{bmatrix}.
$$

□

The observation in the above example can be generalized to an arbitrary linear transformation between finite dimensional spaces.

Theorem 2.3.2. *Let* $\{\mathbf{v}_1,\ldots,\mathbf{v}_m\}$ *and* $\{\mathbf{w}_1,\ldots,\mathbf{w}_n\}$ *be bases of vector spaces* V *and* W, *respectively. For every linear transformation* $f : V \rightarrow W$ *there is a unique* $n \times m$ *matrix*

$$
\begin{bmatrix} a_{11} & \cdots & a_{1m} \\ \vdots & & \vdots \\ a_{n1} & \cdots & a_{nm} \end{bmatrix}
$$

such that for every $\mathbf{v} = x_1\mathbf{v}_1 + \cdots + x_m\mathbf{v}_m \in V$ *we have*

$$
f(x_1\mathbf{v}_1 + \cdots + x_m\mathbf{v}_m) = y_1\mathbf{w}_1 + \cdots + y_n\mathbf{w}_n
$$

where the numbers $y_1,\ldots,y_n \in \mathbb{K}$ *are given by*

$$
\begin{bmatrix} y_1 \\ \vdots \\ y_n \end{bmatrix} = \begin{bmatrix} a_{11} & \cdots & a_{1m} \\ \vdots & & \vdots \\ a_{n1} & \cdots & a_{nm} \end{bmatrix} \begin{bmatrix} x_1 \\ \vdots \\ x_m \end{bmatrix}.
$$

Proof. For every $1 \leq j \leq m$ there are unique $a_{1j},\ldots,a_{nj} \in \mathbb{K}$ such that

$$
f(\mathbf{v}_j) = a_{1j}\mathbf{w}_1 + \cdots + a_{nj}\mathbf{w}_n.
$$

If $\mathbf{v} = x_1\mathbf{v}_1 + \cdots + x_n\mathbf{v}_n$ is an arbitrary vector in $\mathcal{V}$, then

$$
\begin{aligned}
f(\mathbf{v}) &= f(x_1\mathbf{v}_1 + \cdots + x_m\mathbf{v}_m) \\
&= x_1 f(\mathbf{v}_1) + \cdots + x_m f(\mathbf{v}_m) \\
&= x_1(a_{11}\mathbf{w}_1 + \cdots + a_{n1}\mathbf{w}_n) + \cdots + x_m(a_{1m}\mathbf{w}_1 + \cdots + a_{nm}\mathbf{w}_n) \\
&= \left(\begin{bmatrix} a_{11} & \cdots & a_{1m} \end{bmatrix} \begin{bmatrix} x_1 \\ \vdots \\ x_m \end{bmatrix} \right) \mathbf{w}_1 + \cdots + \left(\begin{bmatrix} a_{n1} & \cdots & a_{nm} \end{bmatrix} \begin{bmatrix} x_1 \\ \vdots \\ x_m \end{bmatrix} \right) \mathbf{w}_n.
\end{aligned}
$$

Consequently, if $f(x_1\mathbf{v}_1 + \cdots + x_m\mathbf{v}_m) = y_1\mathbf{w}_1 + \cdots + y_n\mathbf{w}_n$, then

$$
y_k = \begin{bmatrix} a_{k1} & \cdots & a_{km} \end{bmatrix} \begin{bmatrix} x_1 \\ \vdots \\ x_m \end{bmatrix},
$$

for all $1 \leq k \leq n$, which is equivalent to

$$
\begin{bmatrix} y_1 \\ \vdots \\ y_n \end{bmatrix} = \begin{bmatrix} a_{11} & \cdots & a_{1m} \\ \vdots & & \vdots \\ a_{n1} & \cdots & a_{nm} \end{bmatrix} \begin{bmatrix} x_1 \\ \vdots \\ x_m \end{bmatrix}.
$$

$\square$

Definition 2.3.3. Let $f : \mathcal{V} \to \mathcal{W}$ be a linear transformation and let $\mathcal{B} = \{\mathbf{v}_1, \ldots, \mathbf{v}_m\}$ and $\mathcal{C} = \{\mathbf{w}_1, \ldots, \mathbf{w}_n\}$ be bases of $\mathcal{V}$ and $\mathcal{W}$, respectively. The matrix

$$
\begin{bmatrix} a_{11} & \cdots & a_{1m} \\ \vdots & & \vdots \\ a_{n1} & \cdots & a_{nm} \end{bmatrix}
$$

in Theorem 2.3.2 is called the *matrix of f relative to the bases* $\{\mathbf{v}_1, \ldots, \mathbf{v}_m\}$ *and* $\{\mathbf{w}_1, \ldots, \mathbf{w}_n\}$ and is denoted by $f_{\mathcal{B} \to \mathcal{C}}$.

Example 2.3.4. Let $f : \mathcal{V} \to \mathcal{W}$ be a linear transformation and let $\mathcal{B} = \{\mathbf{v}_1, \ldots, \mathbf{v}_m\}$ and $\mathcal{C} = \{\mathbf{w}_1, \ldots, \mathbf{w}_n\}$ be bases of $\mathcal{V}$ and $\mathcal{W}$, respectively. If

$$
A = f_{\mathcal{B} \to \mathcal{C}} = \begin{bmatrix} a_{11} & \cdots & a_{1m} \\ \vdots & & \vdots \\ a_{n1} & \cdots & a_{nm} \end{bmatrix}
$$

is the matrix of f relative to the bases $\mathcal{B}$ and $\mathcal{C}$, show that there is an isomorphism $g : \ker f \to \mathbf{N}(A)$ such that

$$
g(x_1\mathbf{v}_1 + \cdots + x_m\mathbf{v}_m) = \begin{bmatrix} x_1 \\ \vdots \\ x_m \end{bmatrix}
$$

whenever $x_1\mathbf{v}_1 + \cdots + x_m\mathbf{v}_m \in \ker f$.

Solution. It suffices to observe that $x_1\mathbf{v}_1 + \cdots + x_m\mathbf{v}_m \in \ker f$ is equivalent to

$$
\begin{bmatrix} 0 \\ \vdots \\ 0 \end{bmatrix} = \begin{bmatrix} a_{11} & \cdots & a_{1m} \\ \vdots & & \vdots \\ a_{n1} & \cdots & a_{nm} \end{bmatrix} \begin{bmatrix} x_1 \\ \vdots \\ x_m \end{bmatrix},
$$

by Theorem 2.3.2, and that the function $h : \ker f \to \mathbb{K}^m$ defined by

$$
h(x_1\mathbf{v}_1 + \cdots + x_m\mathbf{v}_m) = \begin{bmatrix} x_1 \\ \vdots \\ x_m \end{bmatrix}
$$

is an isomorphism. Consequently, $g : \ker f \to \mathbf{N}(A)$ is an isomorphism.

□

Example 2.3.5. Let $\mathcal{V}$ be a vector space with a basis $\{\mathbf{v}_1, \mathbf{v}_2, \mathbf{v}_3, \mathbf{v}_4\}$ and let $\mathcal{W}$ be a vector space with a basis $\{\mathbf{w}_1, \mathbf{w}_2, \mathbf{w}_3\}$. Let $f : \mathcal{V} \to \mathcal{W}$ be a linear transformation such that the matrix of f relative to the bases $\{\mathbf{v}_1, \mathbf{v}_2, \mathbf{v}_3, \mathbf{v}_4\}$ and $\{\mathbf{w}_1, \mathbf{w}_2, \mathbf{w}_3\}$ is

$$
\begin{bmatrix} 1 & 2 & 2 & 1 \\ 2 & 1 & 3 & 5 \\ 4 & 5 & 7 & 7 \end{bmatrix}.
$$

Find a basis of $\ker f$.

Solution. It is easy to verify that

$$
\left\{ \begin{bmatrix} 4 \\ 1 \\ -3 \\ 0 \end{bmatrix}, \begin{bmatrix} -7 \\ 0 \\ 3 \\ 1 \end{bmatrix} \right\}
$$

is a basis of

$$\mathbf{N}\left(\begin{bmatrix} 1 & 2 & 2 & 1 \\ 2 & 1 & 3 & 5 \\ 4 & 5 & 7 & 7 \end{bmatrix}\right).$$

Consequently,

$$\{4\mathbf{v}_1 + \mathbf{v}_2 - 3\mathbf{v}_4, -7\mathbf{v}_1 + 3\mathbf{v}_3 + \mathbf{v}_4\}$$

is a basis of ker f.   □

Example 2.3.6. Let $\{\mathbf{v}_1, \ldots, \mathbf{v}_m\}$ and $\{\mathbf{w}_1, \ldots, \mathbf{w}_n\}$ be bases of vector spaces V and W, respectively. If $f : V \to W$ is the linear transformation such that the matrix of f relative to the bases $\{\mathbf{v}_1, \ldots, \mathbf{v}_m\}$ and $\{\mathbf{w}_1, \ldots, \mathbf{w}_n\}$ is

$$A = \begin{bmatrix} a_{11} & \cdots & a_{1m} \\ \vdots & & \vdots \\ a_{n1} & \cdots & a_{nm} \end{bmatrix},$$

show that there is an isomorphism $g : \operatorname{ran} f \to \mathbf{C}(A)$ such that

$$g(f(\mathbf{v}_j)) = \begin{bmatrix} a_{1j} \\ \vdots \\ a_{nj} \end{bmatrix},$$

for all $1 \le j \le m$.

Solution. We define $g : \operatorname{ran} f \to \mathbf{C}(A)$ by

$$g(f(x_1\mathbf{v}_1 + \cdots + x_m\mathbf{v}_n)) = x_1 \begin{bmatrix} a_{11} \\ \vdots \\ a_{n1} \end{bmatrix} + \cdots + x_m \begin{bmatrix} a_{1m} \\ \vdots \\ a_{nm} \end{bmatrix}.$$

To show that g is well-defined assume that $f(x_1\mathbf{v}_1 + \cdots + x_m\mathbf{v}_n) = f(x_1'\mathbf{v}_1 + \cdots + x_m'\mathbf{v}_n)$. Then

$$f(x_1\mathbf{v}_1 + \cdots + x_m\mathbf{v}_n) = y_1\mathbf{w}_1 + \cdots + y_n\mathbf{w}_n = f(x_1'\mathbf{v}_1 + \cdots + x_m'\mathbf{v}_n),$$

for some $y_1, \ldots, y_n \in \mathbb{K}$. By Theorem 2.3.2, this is equivalent to

$$\begin{bmatrix} a_{11} & \cdots & a_{1m} \\ \vdots & & \vdots \\ a_{n1} & \cdots & a_{nm} \end{bmatrix} \begin{bmatrix} x_1 \\ \vdots \\ x_m \end{bmatrix} = \begin{bmatrix} y_1 \\ \vdots \\ y_n \end{bmatrix} = \begin{bmatrix} a_{11} & \cdots & a_{1m} \\ \vdots & & \vdots \\ a_{n1} & \cdots & a_{nm} \end{bmatrix} \begin{bmatrix} x_1' \\ \vdots \\ x_m' \end{bmatrix}.$$

Consequently,

$$x_1 \begin{bmatrix} a_{11} \\ \vdots \\ a_{n1} \end{bmatrix} + \cdots + x_m \begin{bmatrix} a_{1m} \\ \vdots \\ a_{nm} \end{bmatrix} = x_1' \begin{bmatrix} a_{11} \\ \vdots \\ a_{n1} \end{bmatrix} + \cdots + x_m' \begin{bmatrix} a_{1m} \\ \vdots \\ a_{nm} \end{bmatrix},$$

proving that the function g is well-defined.

Clearly, g is a linear transformation. If $g(f(x_1\mathbf{v}_1 + \cdots + x_m\mathbf{v}_n)) = \mathbf{0}$, then

$$\mathbf{0} = g(f(x_1\mathbf{v}_1 + \cdots + x_m\mathbf{v}_n))$$

$$= x_1 \begin{bmatrix} a_{11} \\ \vdots \\ a_{n1} \end{bmatrix} + \cdots + x_m \begin{bmatrix} a_{1m} \\ \vdots \\ a_{nm} \end{bmatrix}$$

$$= \begin{bmatrix} a_{11} & \cdots & a_{1m} \\ \vdots & & \vdots \\ a_{n1} & \cdots & a_{nm} \end{bmatrix} \begin{bmatrix} x_1 \\ \vdots \\ x_m \end{bmatrix} = \begin{bmatrix} y_1 \\ \vdots \\ y_n \end{bmatrix}.$$

Hence
$$f(x_1\mathbf{v}_1 + \cdots + x_m\mathbf{v}_n) = y_1\mathbf{w}_1 + \cdots + y_n\mathbf{w}_n = \mathbf{0},$$

proving that g is injective.

Finally, if $\begin{bmatrix} y_1 \\ \vdots \\ y_n \end{bmatrix} \in \mathbf{C}(A)$, then

$$\begin{bmatrix} y_1 \\ \vdots \\ y_n \end{bmatrix} = x_1 \begin{bmatrix} a_{11} \\ \vdots \\ a_{n1} \end{bmatrix} + \cdots + x_m \begin{bmatrix} a_{1m} \\ \vdots \\ a_{nm} \end{bmatrix}$$

for some $x_1, \ldots, x_m \in \mathbb{K}$ and, consequently,

$$\begin{bmatrix} y_1 \\ \vdots \\ y_n \end{bmatrix} = g(f(x_1\mathbf{v}_1 + \cdots + x_m\mathbf{v}_n)),$$

proving that g is surjective. $\qquad\square$

Example 2.3.7. Let $\mathcal{V}$ be vector space with a basis $\{\mathbf{v}_1, \mathbf{v}_2, \mathbf{v}_3, \mathbf{v}_4\}$ and let $\mathcal{W}$ be vector space with a basis $\{\mathbf{w}_1, \mathbf{w}_2, \mathbf{w}_3\}$. If $f : \mathcal{V} \to \mathcal{W}$ is a linear transformation such that the matrix of f relative to the bases $\{\mathbf{v}_1, \mathbf{v}_2, \mathbf{v}_3, \mathbf{v}_4\}$ and $\{\mathbf{w}_1, \mathbf{w}_2, \mathbf{w}_3\}$ is

$$\begin{bmatrix} 1 & 2 & 2 & 1 \\ 2 & 1 & 3 & 5 \\ 4 & 5 & 7 & 7 \end{bmatrix},$$

find a basis of ran f.

Solution. Since the reduced row echelon form of the matrix $\begin{bmatrix} 1 & 2 & 2 & 1 \\ 2 & 1 & 3 & 5 \\ 4 & 5 & 7 & 7 \end{bmatrix}$

is $\begin{bmatrix} 1 & 0 & 4/3 & 3 \\ 0 & 1 & 1/3 & -1 \\ 0 & 0 & 0 & 0 \end{bmatrix}$, the set

$$\{\mathbf{w}_1 + 2\mathbf{w}_2 + 4\mathbf{w}_3, 2\mathbf{w}_1 + \mathbf{w}_2 + 5\mathbf{w}_3\}$$

is a basis of ran f.   □

Theorem 2.3.8. *Let $\mathcal{B}$, $\mathcal{C}$, and $\mathcal{D}$ be bases of vector spaces $\mathcal{V}$, $\mathcal{W}$, and $\mathcal{X}$, respectively, and let $f : \mathcal{V} \to \mathcal{W}$ and $g : \mathcal{W} \to \mathcal{X}$ be linear transformations. If A is the matrix of f relative to the bases $\mathcal{B}$ and $\mathcal{C}$ and B is the matrix of g relative to the bases $\mathcal{C}$ and $\mathcal{D}$, then the matrix BA is the matrix of gf relative to the bases $\mathcal{B}$ and $\mathcal{D}$. In other words*

$$(gf)_{\mathcal{B} \to \mathcal{D}} = g_{\mathcal{C} \to \mathcal{D}} f_{\mathcal{B} \to \mathcal{C}}.$$

Proof. If $\mathcal{B} = \{\mathbf{v}_1, \ldots, \mathbf{v}_m\}$, $\mathcal{C} = \{\mathbf{w}_1, \ldots, \mathbf{w}_n\}$, $\mathcal{D} = \{\mathbf{x}_1, \ldots, \mathbf{x}_p\}$, and

$$f_{\mathcal{B} \to \mathcal{C}} = A = \begin{bmatrix} a_{11} & \cdots & a_{1m} \\ \vdots & & \vdots \\ a_{n1} & \cdots & a_{nm} \end{bmatrix} \quad \text{and} \quad g_{\mathcal{C} \to \mathcal{D}} = B = \begin{bmatrix} b_{11} & \cdots & b_{1n} \\ \vdots & & \vdots \\ b_{p1} & \cdots & b_{pn} \end{bmatrix},$$

where $A \in \mathcal{M}_{n \times m}(\mathbb{K})$ and $B \in \mathcal{M}_{p \times n}(\mathbb{K})$, then

$$f(\mathbf{v}_j) = a_{1j}\mathbf{w}_1 + \cdots + a_{nj}\mathbf{w}_n,$$

for all $1 \leq j \leq m$, and

$$g(\mathbf{w}_k) = b_{1k}\mathbf{x}_1 + \cdots + b_{pk}\mathbf{x}_p,$$

for all $1 \leq k \leq n$. Since

$$g(f(\mathbf{v}_j)) = a_{1j}g(\mathbf{w}_1) + \cdots + a_{nj}g(\mathbf{w}_n)$$

$$= a_{1j}(b_{11}\mathbf{x}_1 + \cdots + b_{p1}\mathbf{x}_p) + \cdots + a_{nj}(b_{1n}\mathbf{x}_1 + \cdots + b_{pn}\mathbf{x}_p)$$

$$= \left(\begin{bmatrix} b_{11} & \cdots & b_{1n} \end{bmatrix} \begin{bmatrix} a_{1j} \\ \vdots \\ a_{nj} \end{bmatrix} \right) \mathbf{x}_1 + \cdots + \left(\begin{bmatrix} b_{p1} & \cdots & b_{pn} \end{bmatrix} \begin{bmatrix} a_{1j} \\ \vdots \\ a_{nj} \end{bmatrix} \right) \mathbf{x}_p$$

and the j-th column of the matrix

$$\begin{bmatrix} b_{11} & \cdots & b_{1n} \\ \vdots & & \vdots \\ b_{p1} & \cdots & b_{pn} \end{bmatrix} \begin{bmatrix} a_{11} & \cdots & a_{1m} \\ \vdots & & \vdots \\ a_{n1} & \cdots & a_{nm} \end{bmatrix}$$

is

$$\begin{bmatrix} \begin{bmatrix} b_{11} & \cdots & b_{1n} \end{bmatrix} \begin{bmatrix} a_{1j} \\ \vdots \\ a_{nj} \end{bmatrix} \\ \vdots \\ \begin{bmatrix} b_{p1} & \cdots & b_{pn} \end{bmatrix} \begin{bmatrix} a_{1j} \\ \vdots \\ a_{nj} \end{bmatrix} \end{bmatrix} = \begin{bmatrix} b_{11} & \cdots & b_{1n} \\ \vdots & & \vdots \\ b_{p1} & \cdots & b_{pn} \end{bmatrix} \begin{bmatrix} a_{1j} \\ \vdots \\ a_{nj} \end{bmatrix},$$

the matrix of gf relative to the bases $\mathcal{B}$ and $\mathcal{D}$ is BA. □

Corollary 2.3.9. *Let $\mathcal{B}$ and $\mathcal{C}$ be bases of a vector space V. Then the matrix of the identity function $\mathrm{Id} : V \to V$ relative to the bases $\mathcal{B}$ and $\mathcal{C}$ is invertible and its inverse is the matrix of the identity function relative to the bases $\mathcal{C}$ and $\mathcal{B}$.*

Proof. This is an immediate consequence of Theorem 2.3.8 because

$$\mathrm{Id}_{\mathcal{C} \to \mathcal{B}} \, \mathrm{Id}_{\mathcal{B} \to \mathcal{C}} = \mathrm{Id}_{\mathcal{B} \to \mathcal{B}} \quad \text{and} \quad \mathrm{Id}_{\mathcal{B} \to \mathcal{C}} \, \mathrm{Id}_{\mathcal{C} \to \mathcal{B}} = \mathrm{Id}_{\mathcal{C} \to \mathcal{C}} .$$

□

Example 2.3.10. We consider the linear vector space $\mathcal{M}_{2 \times 2}(\mathbb{K})$ and the bases

$$\mathcal{B} = \left\{ \begin{bmatrix} 1 & 0 \\ 0 & 0 \end{bmatrix}, \begin{bmatrix} 0 & 1 \\ 0 & 0 \end{bmatrix}, \begin{bmatrix} 0 & 0 \\ 1 & 0 \end{bmatrix}, \begin{bmatrix} 0 & 0 \\ 0 & 1 \end{bmatrix} \right\}$$

and
$$C = \left\{ \begin{bmatrix} 1 & 1 \\ 1 & 1 \end{bmatrix}, \begin{bmatrix} 1 & 1 \\ 0 & 1 \end{bmatrix}, \begin{bmatrix} 1 & 0 \\ 0 & 1 \end{bmatrix}, \begin{bmatrix} 0 & 0 \\ 0 & 1 \end{bmatrix} \right\}.$$

Determine the matrix of Id relative to the bases $\mathcal{B}$ and $\mathcal{C}$.

Solution. Since the matrix of Id relative to the bases $\mathcal{C}$ and $\mathcal{B}$ is

$$\begin{bmatrix} 1 & 1 & 1 & 0 \\ 1 & 1 & 0 & 0 \\ 1 & 0 & 0 & 0 \\ 1 & 1 & 1 & 1 \end{bmatrix},$$

the matrix of Id relative to the bases $\mathcal{B}$ and $\mathcal{C}$ is

$$\begin{bmatrix} 1 & 1 & 1 & 0 \\ 1 & 1 & 0 & 0 \\ 1 & 0 & 0 & 0 \\ 1 & 1 & 1 & 1 \end{bmatrix}^{-1} = \begin{bmatrix} 0 & 0 & 1 & 0 \\ 0 & 1 & -1 & 0 \\ 1 & -1 & 0 & 0 \\ -1 & 0 & 0 & 1 \end{bmatrix}.$$

$\square$

If $\{\mathbf{v}_1, \ldots, \mathbf{v}_n\}$ is a basis of a vector space $\mathcal{V}$ and $f : \mathcal{V} \to \mathcal{V}$ is a linear transformation, then there is a unique $n \times n$ matrix A such that for every $\mathbf{v} = x_1\mathbf{v}_1 + \cdots + x_n\mathbf{v}_n \in \mathcal{V}$ we have

$$f(x_1\mathbf{v}_1 + \cdots + x_n\mathbf{v}_n) = y_1\mathbf{v}_1 + \cdots + y_n\mathbf{v}_n$$

where the numbers $y_1, \ldots, y_n \in \mathbb{K}$ are given by

$$\begin{bmatrix} y_1 \\ \vdots \\ y_n \end{bmatrix} = A \begin{bmatrix} x_1 \\ \vdots \\ x_n \end{bmatrix}.$$

This is simply a special case of Theorem 2.3.2. We will say that A is the *matrix of the linear transformation* $f : \mathcal{V} \to \mathcal{V}$ *relative to the basis* $\{\mathbf{v}_1, \ldots, \mathbf{v}_n\}$.

Theorem 2.3.11. *Let $\mathcal{B}$ and $\mathcal{C}$ be bases of a vector space $\mathcal{V}$ and let $f : \mathcal{V} \to \mathcal{V}$ be a linear transformation. Then the matrix of f relative to the basis $\mathcal{C}$ is*
$$M = P^{-1}NP,$$
where N is the matrix of the linear transformation f relative to the basis $\mathcal{B}$ and P is the matrix of the identity function $\mathrm{Id} : \mathcal{V} \to \mathcal{V}$ relative to the bases $\mathcal{C}$ and $\mathcal{B}$.

Proof. Since $f = \mathrm{Id}\, f\, \mathrm{Id}$ and consequently

$$\mathrm{Id}_{\mathcal{C}\to\mathcal{B}}\, f_{\mathcal{B}\to\mathcal{B}}\, \mathrm{Id}_{\mathcal{B}\to\mathcal{C}} = (\mathrm{Id}_{\mathcal{B}\to\mathcal{C}})^{-1} f_{\mathcal{B}\to\mathcal{B}}\, \mathrm{Id}_{\mathcal{B}\to\mathcal{C}} = f_{\mathcal{C}\to\mathcal{C}},$$

the result follows from Theorem 2.3.8 because P^{-1} is the matrix of the identity function $\mathrm{Id} : \mathcal{V} \to \mathcal{V}$ relative to the bases $\mathcal{B}$ and $\mathcal{C}$. $\qquad\square$

2.3.2 The isomorphism between $\mathcal{M}_{n\times m}(\mathbb{K})$ and $\mathcal{L}(\mathcal{V}, \mathcal{W})$

The main result of this section is the fact that, if $\mathcal{V}$ and $\mathcal{W}$ are vector spaces such that $\dim \mathcal{V} = m$ and $\dim \mathcal{W} = n$, then the space of all linear transformations from $\mathcal{V}$ to $\mathcal{W}$ can be identified with the space of all $n \times m$ matrices. The following theorem formalizes this claim.

Theorem 2.3.12. *Let $\{\mathbf{v}_1, \ldots, \mathbf{v}_m\}$ and $\{\mathbf{w}_1, \ldots, \mathbf{w}_n\}$ be bases of vector spaces $\mathcal{V}$ and $\mathcal{W}$, respectively. For every $n \times m$ matrix*

$$A = \begin{bmatrix} a_{11} & \cdots & a_{1m} \\ \vdots & & \vdots \\ a_{n1} & \cdots & a_{nm} \end{bmatrix}$$

we define the linear transformation $f_A : \mathcal{V} \to \mathcal{W}$ via

$$f_A(\mathbf{v}_j) = a_{1j}\mathbf{w}_1 + \cdots + a_{nj}\mathbf{w}_n$$

for every $j \in \{1, \ldots, m\}$. Then the function $\Delta : \mathcal{M}_{n\times m}(\mathbb{K}) \to \mathcal{L}(\mathcal{V}, \mathcal{W})$ defined by

$$\Delta(A) = f_A$$

is an isomorphism.

Proof. We first show the Δ is a linear transformation. Let

$$A = \begin{bmatrix} a_{11} & \cdots & a_{1m} \\ \vdots & & \vdots \\ a_{n1} & \cdots & a_{nm} \end{bmatrix} \quad \text{and} \quad B = \begin{bmatrix} b_{11} & \cdots & b_{1m} \\ \vdots & & \vdots \\ b_{n1} & \cdots & b_{nm} \end{bmatrix}.$$

Since

$$\begin{aligned}
\Delta(A + B)(\mathbf{v}_j) &= (a_{1j} + b_{1j})\mathbf{w}_1 + \cdots + (a_{nj} + b_{nj})\mathbf{w}_n \\
&= a_{1j}\mathbf{w}_1 + \cdots + a_{nj}\mathbf{w}_n + b_{1j}\mathbf{w}_1 + \cdots + b_{nj}\mathbf{w}_n \\
&= \Delta(A)(\mathbf{v}_j) + \Delta(B)(\mathbf{v}_j),
\end{aligned}$$

we have $\Delta(A + B) = \Delta(A) + \Delta(B)$. If $\alpha \in \mathbb{K}$, then

$$\Delta(\alpha A)(\mathbf{v}_j) = (\alpha a_{1j})\mathbf{w}_1 + \cdots + (\alpha a_{nj})\mathbf{w}_n$$
$$= \alpha(a_{1j}\mathbf{w}_1 + \cdots + a_{nj}\mathbf{w}_n)$$
$$= (\alpha\Delta(A))(\mathbf{v}_j),$$

so $\Delta(\alpha A) = \alpha\Delta(A)$. Consequently Δ is a linear transformation.
If $\Delta(A) = \Delta(B)$, then

$$a_{1j}\mathbf{w}_1 + \cdots + a_{nj}\mathbf{w}_n = b_{1j}\mathbf{w}_1 + \cdots + b_{nj}\mathbf{w}_n$$

for all $1 \leq j \leq m$. Consequently $A = B$, proving that Δ is injective.
Finally, if $g : \mathcal{V} \to \mathcal{W}$ is an arbitrary linear transformation and

$$g(\mathbf{v}_j) = a_{1j}\mathbf{w}_1 + \cdots + a_{nj}\mathbf{w}_n$$

for all $1 \leq j \leq m$, then $g = \Delta(A)$ where

$$A = \begin{bmatrix} a_{11} & \cdots & a_{1m} \\ \vdots & & \vdots \\ a_{n1} & \cdots & a_{nm} \end{bmatrix}.$$

This shows that Δ is surjective.   □

Example 2.3.13. The function that assigns to the matrix

$$A = \begin{bmatrix} a_1 & \cdots & a_m \end{bmatrix}$$

the linear transformation $f_A : \mathbb{K}^m \to \mathbb{K}$ defined by

$$f_A\left(\begin{bmatrix} x_1 \\ \vdots \\ x_m \end{bmatrix}\right) = a_1 x_1 + \cdots + a_m x_m$$

is an isomorphism from the vector space $\mathcal{M}_{1 \times m}(\mathbb{K})$ to the vector space $\mathcal{L}(\mathbb{K}^m, \mathbb{K})$.

2.4   Duality

In this section we study the vector space $\mathcal{L}(\mathcal{V}, \mathbb{K})$, that is the space of all linear transformations from a vector space to the number field $\mathbb{K}$. While $\mathcal{L}(\mathcal{V}, \mathbb{K})$ is a special case of the vector space of all linear transformations between vector spaces, it has some distinct properties.

2.4.1 The dual space

> **Definition 2.4.1.** Let V be vector space. A linear transformation $f : V \to \mathbb{K}$ is called a *functional* or a *linear form*. The vector space $\mathcal{L}(V, \mathbb{K})$ is called the *dual space* of the vector space V and is denoted by V'.

Example 2.4.2. Let V be an n-dimensional vector space and let $f : V \to \mathbb{K}$ be a nonzero linear form. If $\mathbf{a} \in V$ is such that $f(\mathbf{a}) \neq 0$, show that

$$V = \mathbb{K}\mathbf{a} \oplus \ker f$$

and determine the projection of V on $\mathbb{K}\mathbf{a}$ along $\ker f$.

Solution. For every $\mathbf{v} \in V$ we have

$$f\left(\mathbf{v} - \frac{f(\mathbf{v})}{f(\mathbf{a})}\mathbf{a}\right) = 0$$

and thus $\mathbf{b} = \mathbf{v} - \frac{f(\mathbf{v})}{f(\mathbf{a}}\mathbf{a} \in \ker f$. Since we can write

$$\mathbf{v} = \frac{f(\mathbf{v})}{f(\mathbf{a})}\mathbf{a} + \mathbf{b},$$

we have

$$V = \mathbb{K}\mathbf{a} + \ker f.$$

Now we show that this sum is direct. If $\mathbf{v} \in \mathbb{K}\mathbf{a} \cap \ker f$, then $\mathbf{v} = \alpha\mathbf{a}$ for some $\alpha \in \mathbb{K}$ and $f(\mathbf{v}) = 0$. This yields $f(\alpha\mathbf{a}) = \alpha f(\mathbf{a}) = 0$. Since $f(\mathbf{a}) \neq 0$, we have $\alpha = 0$ and consequently $\mathbf{v} = \mathbf{0}$. Therefore

$$V = \mathbb{K}\mathbf{a} \oplus \ker f$$

and the projection of the vector $\mathbf{v} \in V$ on $\mathbb{K}\mathbf{a}$ along $\ker f$ is $\dfrac{f(\mathbf{v})}{f(\mathbf{a})}\mathbf{a}$. □

Note that, since $\dim \mathbb{K}\mathbf{a} = 1$ and $\dim \mathbb{K}\mathbf{a} + \dim \ker f = \dim V = n$, we have $\dim \ker f = n - 1$, as shown in Corollary 2.1.31.

Example 2.4.3. Let V be an n-dimensional vector space and let $f, g \in V'$. If f is nonzero and $\ker f \subseteq \ker g$, show that $g \in \text{Span}\{f\}$.

Solution. Let $\mathbf{v} \in \ker f$. With the notation from Example 2.4.2, we have

$$g(\alpha \mathbf{a} + \mathbf{v}) = \alpha g(\mathbf{a}) = \frac{\alpha g(\mathbf{a})}{f(\mathbf{a})} f(\mathbf{a}) = \frac{g(\mathbf{a})}{f(\mathbf{a})} f(\alpha \mathbf{a}) = \frac{g(\mathbf{a})}{f(\mathbf{a})} f(\alpha \mathbf{a} + \mathbf{v}).$$

Thus $g = \dfrac{g(\mathbf{a})}{f(\mathbf{a})} f$, because $V = \mathbb{K}\mathbf{a} + \ker f$. $\square$

Definition 2.4.4. Let $\{\mathbf{v}_1, \ldots, \mathbf{v}_n\}$ be a basis of a vector space V. For every $j \in \{1, \ldots, n\}$ by $l_{\mathbf{v}_j}$ we mean the unique linear form $l_{\mathbf{v}_j} : V \to \mathbb{K}$ such that $l_{\mathbf{v}_j}(\mathbf{v}_j) = 1$ and $l_{\mathbf{v}_j}(\mathbf{v}_k) = 0$ for every $k \neq j$. In other words,

$$l_{\mathbf{v}_j}(x_1 \mathbf{v}_1 + \cdots + x_n \mathbf{v}_n) = x_j.$$

Theorem 2.4.5. *If $\{\mathbf{v}_1, \ldots, \mathbf{v}_n\}$ is a basis of V, then $\{l_{\mathbf{v}_1}, \ldots, l_{\mathbf{v}_n}\}$ is a basis of V'.*

Proof. Assume $x_1 l_{\mathbf{v}_1} + \cdots + x_n l_{\mathbf{v}_n} = \mathbf{0}$. Since, for every $1 \leq k \leq n$ we have

$$0 = x_1 l_{\mathbf{v}_1}(\mathbf{v}_k) + \cdots + x_n l_{\mathbf{v}_n}(\mathbf{v}_k) = x_k l_{\mathbf{v}_k}(\mathbf{v}_k) = x_k,$$

the functions $l_{\mathbf{v}_1}, \ldots, l_{\mathbf{v}_n}$ are linearly independent.

If $f : V \to \mathbb{K}$ is the linear transformation such that $f(\mathbf{v}_j) = a_j$ for every $1 \leq j \leq n$, then it is easy to verify that

$$f = a_1 l_{\mathbf{v}_1} + \cdots + a_n l_{\mathbf{v}_n}.$$

This shows that

$$\mathrm{Span}\{l_{\mathbf{v}_j}, 1 \leq j \leq n\} = V',$$

completing the proof. $\square$

Definition 2.4.6. Let $\{\mathbf{v}_1, \ldots, \mathbf{v}_n\}$ be a basis of the vector space V. The basis $\{l_{\mathbf{v}_1}, \ldots, l_{\mathbf{v}_n}\}$ of V' is called the *dual basis* of the basis $\{\mathbf{v}_1, \ldots, \mathbf{v}_n\}$.

Example 2.4.7. Find the dual basis of the basis $\{1, t, \ldots, t^n\}$ in the space $\mathcal{P}_n(\mathbb{K})$.

Solution. According to the definition of the dual basis we have

$$l_1(a_0 + a_1t + \cdots + a_nt^n) = a_0,$$
$$l_t(a_0 + a_1t + \cdots + a_nt^n) = a_1,$$

$$\vdots$$

$$l_{t^n}(a_0 + a_1t + \cdots + a_nt^n) = a_n.$$

□

Note that for any $p \in \mathcal{P}_n(\mathbb{K})$ we could write

$$l_1(p) = p(0), \; l_t(p) = p'(0), \; l_{t^2}(p) = \frac{1}{2}p''(0), \; \ldots, \; l_{t^n}(p) = \frac{1}{n!}p^{(n)}(0).$$

This formulation has the advantage that we don't have to write p in the form $a_0 + a_1t + \cdots + a_nt^n$. For example, if $p(t) = ((t^2 + t + 1)^3 + t + 3)^7$, it would be quite time consuming to calculate $l_t(p)$ using the first formula. Calculating it using $l_t(p) = p'(0)$ is much simpler.

Theorem 2.4.8. *Let $\mathcal{V}$ be a vector space such that* $\dim \mathcal{V} = n$. *If* $\mathbf{v}_1, \ldots, \mathbf{v}_j$ *are linearly independent vectors in $\mathcal{V}$, then the set*

$$\mathcal{Q} = \{f \in \mathcal{V}' : f(\mathbf{v}_1) = \cdots = f(\mathbf{v}_j) = 0\}$$

is a vector subspace of $\mathcal{V}'$ and $\dim \mathcal{Q} = n - j$.

Proof. First we extend $\{\mathbf{v}_1, \ldots, \mathbf{v}_j\}$ to a basis $\{\mathbf{v}_1, \ldots, \mathbf{v}_n\}$ of $\mathcal{V}$. Let $\{l_{\mathbf{v}_1}, \ldots, l_{\mathbf{v}_n}\}$ be its dual basis. It is easy to see that $\mathcal{Q}$ is a subspace of $\mathcal{V}'$.

Now we show that $\{l_{\mathbf{v}_{j+1}}, \ldots, l_{\mathbf{v}_n}\}$ is a basis of $\mathcal{Q}$. The linear functionals $l_{\mathbf{v}_{j+1}}, \ldots, l_{\mathbf{v}_n}$ are in $\mathcal{Q}$ and are linearly independent, so we only have to show that $\mathrm{Span}\{l_{\mathbf{v}_{j+1}}, \ldots, l_{\mathbf{v}_n}\} = \mathcal{Q}$. If $f \in \mathcal{Q}$, then we can write

$$f = x_1 l_{\mathbf{v}_1} + \cdots + x_n l_{\mathbf{v}_n}$$

where $x_1, \ldots, x_n$ are numbers from $\mathbb{K}$. Since, for every $1 \le k \le j$, we have $0 = f(\mathbf{v}_k) = x_k$, we conclude that

$$f = x_{j+1} l_{\mathbf{v}_{j+1}} + \cdots + x_n l_{\mathbf{v}_n}.$$

□

2.4.2   The bidual

For any vector space $\mathcal{V}$ the dual space $\mathcal{V}'$ is a vector space, so it makes sense to consider its dual, that is, $(\mathcal{V}')'$.

Definition 2.4.9. Let $\mathcal{V}$ be vector space. The vector space $(\mathcal{V}')'$ is called the *bidual* of $\mathcal{V}$.

Example 2.4.10. Let $\mathcal{V}$ be a vector space and let $\mathbf{v} \in \mathcal{V}$. It's easy to verify that the function $g_{\mathbf{v}} : \mathcal{V}' \to \mathbb{K}$ defined by $g_{\mathbf{v}}(l) = l(\mathbf{v})$ is an element of the bidual, that is, a linear form on $\mathcal{V}'$.

In this section we will show that, if $\mathcal{V}$ is finite dimensional, then the spaces $\mathcal{V}$ and $(\mathcal{V}')'$ are isomorphic (see Theorem 2.4.12). First we need to prove an auxiliary result.

Lemma 2.4.11. *Let $\mathcal{V}$ be a vector space of finite dimension and let $\mathbf{v} \in \mathcal{V}$. If $l(\mathbf{v}) = 0$ for every $l \in \mathcal{V}'$, then $\mathbf{v} = \mathbf{0}$.*

Proof. Let $\{\mathbf{v}_1, \ldots, \mathbf{v}_n\}$ be a basis of $\mathcal{V}$ and let $\{l_{\mathbf{v}_1}, \ldots, l_{\mathbf{v}_n}\}$ be the dual basis. If $\mathbf{v} \in \mathcal{V}$, then $\mathbf{v} = x_1 \mathbf{v}_1 + \cdots + x_n \mathbf{v}_n$ for some $x_1, \ldots, x_n \in \mathbb{K}$. Since, for every $1 \leq j \leq n$, we have $0 = l_{\mathbf{v}_j}(\mathbf{v}) = x_j$, which means that $\mathbf{v} = \mathbf{0}$. $\qquad\square$

Now we prove the main result of this section.

Theorem 2.4.12. *Let $\mathcal{V}$ be a finite dimensional vector space. The function $\Gamma : \mathcal{V} \to (\mathcal{V}')'$, which associates with every vector $\mathbf{v} \in \mathcal{V}$ the linear form $g_{\mathbf{v}} : \mathcal{V}' \to \mathbb{K}$ defined by $g_{\mathbf{v}}(l) = l(\mathbf{v})$, is an isomorphism from $\mathcal{V}$ to $(\mathcal{V}')'$.*

Proof. For any $\mathbf{v}, \mathbf{v}_1, \mathbf{v}_2 \in \mathcal{V}$, $\alpha \in \mathbb{K}$, and $l \in \mathcal{V}'$, we have

$$g_{\mathbf{v}_1 + \mathbf{v}_2}(l) = l(\mathbf{v}_1 + \mathbf{v}_2) = l(\mathbf{v}_1) + l(\mathbf{v}_2) = g_{\mathbf{v}_1}(l) + g_{\mathbf{v}_2}(l)$$

and

$$g_{\alpha \mathbf{v}}(l) = l(\alpha \mathbf{v}) = \alpha l(\mathbf{v}) = \alpha g_{\mathbf{v}}(l).$$

This shows that Γ is a linear transformation. If $g_{\mathbf{v}}(l) = l(\mathbf{v}) = 0$ for every $l \in \mathcal{V}'$, then $\mathbf{v} = \mathbf{0}$ by Lemma 2.4.11. Consequently, Γ is injective and thus $\dim \operatorname{ran} \Gamma = \dim \mathcal{V}$.

Finally, since $\dim \mathcal{V} = \dim \mathcal{V}' = \dim(\mathcal{V}')'$, we have $\operatorname{ran} \Gamma = (\mathcal{V}')'$ because $\operatorname{ran} \Gamma$ is a subspace of $(\mathcal{V}')'$ such that $\dim \operatorname{ran} \Gamma = \dim(\mathcal{V}')'$. $\qquad\square$

The linear form $g_{\mathbf{v}} : \mathcal{V}' \to \mathbb{K}$ defined by $g_{\mathbf{v}}(l) = l(\mathbf{v})$ is called the *canonical isomorphism* from $\mathcal{V}$ to $(\mathcal{V}')'$.

From Theorem 2.4.12 it follows that every basis of $\mathcal{V}'$ is the dual basis of some basis of $\mathcal{V}$.

Theorem 2.4.13. *Let $\mathcal{V}$ be vector space and let $\{f_1, \ldots, f_n\}$ be a basis of $\mathcal{V}'$. Then there is a basis $\{\mathbf{v}_1, \ldots, \mathbf{v}_n\}$ of $\mathcal{V}$ such that $\{f_1, \ldots, f_n\}$ is its dual basis.*

Proof. Let $\mathcal{V}$ be an n-dimensional vector space. Let $\{f_1, \ldots, f_n\}$ be a basis of $\mathcal{V}'$ and let $\{l_{f_1}, \ldots, l_{f_n}\}$ be its dual basis in $(\mathcal{V}')'$. By Theorem 2.4.12, there exist vectors $\mathbf{v}_1, \ldots, \mathbf{v}_n \in \mathcal{V}$ such that $\Gamma(\mathbf{v}_j) = l_{f_j}$ for $j \in \{1, \ldots, n\}$. Since $\{l_{f_1}, \ldots, l_{f_n}\}$ is a basis of $(\mathcal{V}')'$ and Γ is an isomorphism, $\{\mathbf{v}_1, \ldots, \mathbf{v}_n\}$ is a basis of $\mathcal{V}$. The set $\{f_1, \ldots, f_n\}$ is the dual basis of $\{\mathbf{v}_1, \ldots, \mathbf{v}_n\}$ because for every $j \in \{1, \ldots, n\}$ we have

$$f_j(\mathbf{v}_j) = \Gamma(\mathbf{v}_j)(f_j) = l_{f_j}(f_j) = 1$$

and

$$f_j(\mathbf{v}_k) = \Gamma(\mathbf{v}_k)(f_j) = l_{f_k}(f_j) = 0$$

whenever $j \neq k$. $\qquad\qquad\square$

Note the similarity between the next theorem and Theorem 2.4.8. We could say that Theorem 2.4.14 is a "dual version" of Theorem 2.4.8. The proofs of these two theorems are also similar.

Theorem 2.4.14. *Let $\mathcal{V}$ be a vector space such that $\dim \mathcal{V} = n$. If $f_1, \ldots, f_j \in \mathcal{V}'$ are linearly independent, then the set*

$$\mathcal{U} = \{\mathbf{v} \in \mathcal{V} : f_1(\mathbf{v}) = \cdots = f_j(\mathbf{v}) = 0\}$$

is a vector subspace of $\mathcal{V}$ and $\dim \mathcal{U} = n - j$.

Proof. First we extend $\{f_1, \ldots, f_j\}$ to a basis $\{f_1, \ldots, f_n\}$ of $\mathcal{V}'$. Let $\{\mathbf{v}_1, \ldots, \mathbf{v}_n\}$ be a basis of $\mathcal{V}$ such that $\{f_1, \ldots, f_n\}$ is its dual basis. It is easy to see that $\mathcal{U}$ is a subspace of $\mathcal{V}$.

Now we show that $\{\mathbf{v}_{j+1}, \ldots, \mathbf{v}_n\}$ is a basis of $\mathcal{U}$. The vectors $\mathbf{v}_{j+1}, \ldots, \mathbf{v}_n$ are linearly independent, so we only have to show that $\mathrm{Span}\{\mathbf{v}_{j+1}, \ldots, \mathbf{v}_n\} = \mathcal{U}$. If $\mathbf{u} \in \mathcal{U}$, then

$$\mathbf{u} = x_1 \mathbf{v}_1 + \cdots + x_n \mathbf{v}_n$$

where $x_1, \ldots, x_n$ are numbers from $\mathbb{K}$.

Since, for every $1 \leq k \leq j$, we have $0 = f_k(\mathbf{u}) = x_k$, we conclude that

$$\mathbf{u} = x_{j+1}\mathbf{v}_{j+1} + \cdots + x_n\mathbf{v}_n.$$

$\square$

2.5 Quotient spaces

For a vector space $\mathcal{V}$ and its subspace $\mathcal{U}$ there is a subspace $\mathcal{W}$ such that $\mathcal{V} = \mathcal{U} \oplus \mathcal{W}$. While the space $\mathcal{W}$ is not unique, we can show that, if $\mathcal{U} \oplus \mathcal{W}_1 = \mathcal{U} \oplus \mathcal{W}_2$, then the spaces $\mathcal{W}_1$ and $\mathcal{W}_2$ are isomorphic. In this section we present a canonical way of constructing, for a given vector space $\mathcal{V}$ and its subspace $\mathcal{U}$, a space that is isomorphic to every space $\mathcal{W}$ such that $\mathcal{V} = \mathcal{U} \oplus \mathcal{W}$.

If $\mathcal{U}$ is a subspace of a vector space $\mathcal{V}$, then we define

$$\mathbf{x} + \mathcal{U} = \{\mathbf{x} + \mathbf{u} : \mathbf{u} \in \mathcal{U}\}.$$

In this section we will use the following notation

$$\widehat{\mathbf{x}} = \mathbf{x} + \mathcal{U},$$

which is a generalization of what was introduced in Example 1.1.8. This notation makes sense only if it is clear what the subspace $\mathcal{U}$ is. Note that, while $\mathbf{x}$ is a vector, $\widehat{\mathbf{x}}$ is a set of vectors. In particular, we have $\mathbf{x} = \mathbf{x} + \mathbf{0} \in \widehat{\mathbf{x}}$.

Lemma 2.5.1. *Let $\mathbf{x}$ and $\mathbf{y}$ be vectors in a vector space $\mathcal{V}$ and let $\mathcal{U}$ be a subspace of $\mathcal{V}$. Then $\widehat{\mathbf{x}} = \widehat{\mathbf{y}}$ if and only if $\mathbf{y} = \mathbf{x} + \mathbf{u}$ for some $\mathbf{u} \in \mathcal{U}$.*

Proof. First assume that $\widehat{\mathbf{y}} = \widehat{\mathbf{x}}$. Then

$$\mathbf{y} \in \widehat{\mathbf{y}} = \widehat{\mathbf{x}} = \mathbf{x} + \mathcal{U}.$$

Consequently, $\mathbf{y} = \mathbf{x} + \mathbf{u}$ for some $\mathbf{u} \in \mathcal{U}$.

Now assume that $\mathbf{y} = \mathbf{x} + \mathbf{u}$ for some $\mathbf{u} \in \mathcal{U}$. Then

$$\widehat{\mathbf{y}} = \mathbf{y} + \mathcal{U} = (\mathbf{x} + \mathbf{u}) + \mathcal{U} = \mathbf{x} + (\mathbf{u} + \mathcal{U}) = \mathbf{x} + \mathcal{U} = \widehat{\mathbf{x}}.$$

$\square$

Corollary 2.5.2. *Let $\mathcal{V}$ be a vector space and let $\mathcal{U}$ be a subspace of $\mathcal{V}$. If $\mathbf{x} \in \mathcal{V}$ and $\mathbf{u} \in \mathcal{U}$, then*
$$\widehat{\mathbf{x}} = \widehat{\mathbf{x} + \mathbf{u}}.$$

Theorem 2.5.3. *Let* V *be a vector space and let* U *be a subspace of* V. *If* $\widehat{\mathbf{x}} \cap \widehat{\mathbf{y}} \neq \emptyset$, *then*

$$\widehat{\mathbf{x}} = \widehat{\mathbf{y}}.$$

Proof. If $\widehat{\mathbf{x}} \cap \widehat{\mathbf{y}} \neq \emptyset$, then there are vectors $\mathbf{u}_1, \mathbf{u}_2 \in U$ such that $\mathbf{x} + \mathbf{u}_1 = \mathbf{y} + \mathbf{u}_2$ and we have

$$\widehat{\mathbf{x}} = \widehat{\mathbf{x} + \mathbf{u}_1} = \widehat{\mathbf{y} + \mathbf{u}_2} = \widehat{\mathbf{y}}.$$

$\square$

Note that the above implies that $\widehat{\mathbf{x}} = \widehat{\mathbf{0}}$ if and only if $\mathbf{x} \in U$.

Definition 2.5.4. Let V be a vector space and let U be a subspace of V. The set

$$V/U = \{\mathbf{x} + U : \mathbf{x} \in V\}$$

is called the *quotient space* of V by U.

In other words, but less precisely, V/U is the set of all $\widehat{\mathbf{x}}$'s with $\mathbf{x} \in V$. The quotient space V/U becomes a vector space if we define

$$\widehat{\mathbf{x}} + \widehat{\mathbf{y}} = \widehat{\mathbf{x} + \mathbf{y}} \quad \text{and} \quad \alpha\widehat{\mathbf{x}} = \widehat{\alpha\mathbf{x}}$$

for any $\mathbf{x}, \mathbf{y} \in V$ and $\alpha \in \mathbb{K}$. It is easy to verify that these operations are well-defined. Note that $\widehat{\mathbf{0}} = U$ is the zero vector in V/U. It is important that the operations in V/U are defined in such a way that the function $q : V \to V/U$ defined by $q(\mathbf{x}) = \widehat{\mathbf{x}}$ is a linear transformation.

Definition 2.5.5. Let U be a subspace of a vector space V. The function $q : V \to V/U$ defined by

$$q(\mathbf{x}) = \widehat{\mathbf{x}}$$

is called the *quotient linear transformation*.

The following example will be generalized in Exercise 2.75.

Example 2.5.6. Show that $\dim V/U = 1$ if and only if $U \oplus \mathbb{K}\mathbf{v} = V$ for some $\mathbf{v} \in V$.

Solution. Let $q : V \to V/U$ be the quotient linear transformation and let $\{q(\mathbf{v})\} = \{\widehat{\mathbf{v}}\}$ be a basis for V/U. Note that because $\widehat{\mathbf{v}} \neq \widehat{\mathbf{0}} = U$ the vector $\mathbf{v}$ is not in U. Then for every $\mathbf{x} \in V$ there is $\alpha \in \mathbb{K}$ such that $\widehat{\mathbf{x}} = \alpha\widehat{\mathbf{v}} = \widehat{\alpha\mathbf{v}}$, and

thus
$$\mathbf{x} = \alpha \mathbf{v} + \mathbf{u},$$

for some $\mathbf{u} \in \mathcal{U}$. Consequently, $\mathcal{V} = \mathcal{U} + \mathbb{K}\mathbf{v}$ and the sum $\mathcal{U} + \mathbb{K}\mathbf{v}$ is direct, because $\mathbf{v} \notin \mathcal{U}$.

Conversely, if $\mathcal{U} \oplus \mathbb{K}\mathbf{v} = \mathcal{V}$, then every vector $\mathbf{x} \in \mathcal{V}$ can be written as $\mathbf{x} = \mathbf{u} + \alpha \mathbf{v}$ for some $\alpha \in \mathbb{K}$. Consequently, $\widehat{\mathbf{x}} = \widehat{\alpha \mathbf{v}} = \alpha \widehat{\mathbf{v}}$ and thus $\{\widehat{\mathbf{v}}\}$ is a basis for $\mathcal{V}/\mathcal{U}$.   □

Example 2.5.7. Let $\mathbf{v}, \mathbf{u}, \mathbf{w}$ be linearly independent vectors in $\mathbb{R}^3$. The set $\{\mathbf{u} + \mathbb{R}\mathbf{v} + \mathbb{R}\mathbf{w}\}$ is a basis of $\mathbb{R}^3/(\mathbb{R}\mathbf{v} + \mathbb{R}\mathbf{w})$.

Theorem 2.5.8. *Let $\mathcal{U}$ be a subspace of a finite dimensional vector space $\mathcal{V}$. Then*
$$\dim \mathcal{V} = \dim \mathcal{U} + \dim \mathcal{V}/\mathcal{U}.$$

Proof. This result can be obtained from the Rank-Nullity Theorem 2.1.28. Indeed, $\mathcal{U} = \ker q$ and $\mathcal{V}/\mathcal{U} = \operatorname{ran} q$ where $q : \mathcal{V} \to \mathcal{V}/\mathcal{U}$ is the quotient linear transformation (see Definition 2.5.5).   □

Theorem 2.5.9. *Let $\mathcal{V}$ and $\mathcal{W}$ be vector spaces and let $f : \mathcal{V} \to \mathcal{W}$ be a linear transformation. There is an isomorphism $g : \mathcal{V}/\ker f \to \operatorname{ran} f$ such that*
$$f(\mathbf{x}) = gq(\mathbf{x}) = g(\widehat{\mathbf{x}}),$$

where $\mathbf{x}$ is a vector from $\mathcal{V}$ and $q : \mathcal{V} \to \mathcal{V}/\ker f$ is the quotient linear transformation.

Proof. For $\mathbf{x} \in \mathcal{V}$ we define $g(\widehat{\mathbf{x}}) = f(\mathbf{x})$. Note that g is a well-defined function $g : \mathcal{V}/\ker f \to \operatorname{ran} f$. Indeed, if $\widehat{\mathbf{x}} = \widehat{\mathbf{x} + \mathbf{y}}$ for some $\mathbf{y} \in \ker f$, then we have $g(\widehat{\mathbf{x} + \mathbf{y}}) = f(\mathbf{x} + \mathbf{y}) = f(\mathbf{x})$ because $f(\mathbf{y}) = \mathbf{0}$.

Since for every $\mathbf{x}_1, \mathbf{x}_2 \in \mathcal{V}$ we have
$$g(\widehat{\mathbf{x}}_1 + \widehat{\mathbf{x}}_2) = g(\widehat{\mathbf{x}_1 + \mathbf{x}_2}) = f(\mathbf{x}_1 + \mathbf{x}_2) = f(\mathbf{x}_1) + f(\mathbf{x}_2) = g(\widehat{\mathbf{x}_1}) + g(\widehat{\mathbf{x}_2})$$

and for every $\mathbf{x} \in \mathcal{V}$ and $\alpha \in \mathbb{K}$ we have
$$g(\alpha \widehat{\mathbf{x}}) = g(\widehat{\alpha \mathbf{x}}) = f(\alpha \mathbf{x}) = \alpha f(\mathbf{x}) = \alpha g(\widehat{\mathbf{x}}),$$

g is a linear transformation.

If $g(\widehat{\mathbf{x}}) = f(\mathbf{x}) = \mathbf{0}$, then $\mathbf{x} \in \ker f$ and thus $\widehat{\mathbf{x}} = \widehat{\mathbf{0}}$. Consequently, g is injective.

Finally, if $\mathbf{y} \in \operatorname{ran} f$, then the there is $\mathbf{x} \in V$ such that $f(\mathbf{x}) = \mathbf{y}$. Consequently, $g(\widehat{\mathbf{x}}) = \mathbf{y}$ and thus g is surjective. $\square$

Corollary 2.5.10. *If U and W are subspaces of a vector space V such that $V = U \oplus W$, then the spaces V/U and W are isomorphic.*

Proof. If $f : U \oplus W \to W$ is the projection on W along U, then $\ker f = U$ and $\operatorname{ran} f = W$. Therefore the result is a consequence of Theorem 2.5.9. $\square$

Note that if the vector space V is finite dimensional we can get Theorem 2.5.8 as a consequence of Corollary 2.5.10.

Example 2.5.11. Let $\mathbb{R}^3 = \mathbb{R}\mathbf{u} \oplus \mathbb{R}\mathbf{v} \oplus \mathbb{R}\mathbf{w}$. If $f : \mathbb{R}^3 \to \mathbb{R}\mathbf{v} \oplus \mathbb{R}\mathbf{w}$ is the projection on $\mathbb{R}\mathbf{v} \oplus \mathbb{R}\mathbf{w}$ along $\mathbb{R}\mathbf{u}$, then the function $g : \mathbb{R}^3/\mathbb{R}\mathbf{u} \to \mathbb{R}\mathbf{v} \oplus \mathbb{R}\mathbf{w}$ defined by

$$g(q(\alpha\mathbf{v} + \beta\mathbf{w} + \gamma\mathbf{u})) = \alpha\mathbf{v} + \beta\mathbf{w},$$

where $q : \mathbb{R}^3 \to \mathbb{R}/\mathbb{R}\mathbf{u}$ is the quotient linear transformation, is an isomorphism.

2.6 Exercises

2.6.1 Basic properties

Exercise 2.1. Let V be a vector space and let U be a subspace of V. If $f : V \to V$ and $f(U) \subseteq U$, then we say that U is *f-invariant*.

Let $f, g : V \to V$ be linear transformations. If U is an f-invariant and a g-invariant subspace, show that U is a (gf)-invariant subspace.

Exercise 2.2. Let V be a vector space and let $f : V \to V$ be a linear transformation. If U is an f-invariant subspace (see Exercise 2.1), show that the restriction of f to U is a linear transformation $f_U : U \to U$.

Exercise 2.3. Let V be a vector space and $f : V \to V$ be a linear transformation. We suppose that U and W are f-invariant subspaces (see Exercise 2.1). Show that $U \cap W$ is an f-invariant subspace.

Exercise 2.4. Let V be a vector space and $f : V \to V$ be a linear transformation. We suppose that U and W are f-invariant subspaces (see Exercise 2.1). Show that $U + W$ is an f-invariant subspace.

Exercise 2.5. Let V and W be vector spaces, let U be a subspace of V, and let $g : U \to W$ be a linear transformation. If V is a finite dimensional space, show that there is a linear transformation $f : V \to W$ such that the restriction of f to U is g.

Exercise 2.6. Let $f : V \to W$ be a linear transformation and let $\{\mathbf{w}_1, \ldots, \mathbf{w}_n\}$ be a basis of ran f. If $\mathbf{w}_j = f(\mathbf{u}_j)$ for every $j \in \{1, \ldots, n\}$ and some $\mathbf{u}_1, \ldots, \mathbf{u}_n \in V$, then $V = \mathrm{Span}\{\mathbf{u}_1, \ldots, \mathbf{u}_n\} \oplus \ker f$.

Exercise 2.7. If $f : V \to \mathbb{K}$ is a nonzero linear transformation, then there is a vector $\mathbf{u} \in V$ such that $f(\mathbf{u}) = 1$ and $V = \mathrm{Span}\{\mathbf{u}\} \oplus \ker f$. Determine the projection on $\ker f$ along $\mathrm{Span}\{\mathbf{u}\}$.

Exercise 2.8. Let V and W be vector spaces. For arbitrary $\mathbf{w} \in W$ and a linear transformation $f : V \to \mathbb{K}$ we define a function $\mathbf{w} \otimes f : V \to W$ by $(\mathbf{w} \otimes f)(\mathbf{v}) = f(\mathbf{v})\mathbf{w}$. Show that, if $g : V \to W$ is a linear transformation such that $\dim \mathrm{ran}\, g = 1$, then there exist $\mathbf{w} \in W$ and a linear transformation $f : V \to \mathbb{K}$ such that $g = \mathbf{w} \otimes f$.

Exercise 2.9. Let V be a vector space and let $\mathbf{v}_1, \ldots, \mathbf{v}_n \in V$. We define a function $f : \mathbb{K}^n \to V$ by

$$f\left(\begin{bmatrix} x_1 \\ \vdots \\ x_n \end{bmatrix}\right) = x_1 \mathbf{v}_1 + \cdots + x_n \mathbf{v}_n$$

Show that $\ker f \neq \mathbf{0}$ if and only if the vectors $\mathbf{v}_1, \ldots, \mathbf{v}_n$ are linearly dependent.

Exercise 2.10. Let V be a vector space and let $\mathbf{v}$ be a nonzero vector in V. If there is a vector subspace $U \subseteq V$ such that $U \oplus \mathbb{K}\mathbf{v} = V$, show that there is a linear transformation $f : V \to \mathbb{K}$ such that $\ker f = U$.

Exercise 2.11. Let V be a vector space and $f : V \to \mathbb{K}$ be a nonzero linear transformation. Show that there is a nonzero vector $\mathbf{v} \in V$ such that $\ker f \oplus \mathbb{K}\mathbf{v} = V$.

Exercise 2.12. Let V be a vector space and let $f, g : V \to \mathbb{K}$ be nonzero linear transformations such that $\ker f = \ker g$. Show that there is a nonzero number $\alpha \in \mathbb{K}$ such that $g = \alpha f$.

Exercise 2.13. Let V and W be vector spaces and let $U_1, \ldots, U_n$ be subspaces of V such that $V = U_1 \oplus \cdots \oplus U_n$. If $f_1 : U_1 \to W, \ldots, f_n : U_n \to W$ are linear transformations, show that there is an unique linear transformation $g : V \to W$ such that $g(\mathbf{u}_j) = f_j(\mathbf{u}_j)$ for every $j \in \{1, \ldots, n\}$ and $\mathbf{u}_j \in U_j$.

Exercise 2.14. Let V, W, and U be vector spaces and let $V_1 \subseteq V$ and $W_1 \subseteq W$ be subspaces. If $V = U \oplus V_1$ and $\dim V_1 = \dim W_1 = n$ for some $n \geq 1$, show that there is a linear transformation $f \in \mathcal{L}(V, W)$ such that $\ker f = U$ and $f(V_1) = W_1$.

Exercise 2.15. Let $\mathcal{V}$ and $\mathcal{W}$ be vector spaces and let $f : \mathcal{V} \to \mathcal{W}$ be a linear transformation. Show that the set $\{(\mathbf{v}, f(\mathbf{v})) | \mathbf{v} \in \mathcal{V}\}$ is a vector subspace of $\mathcal{V} \times \mathcal{W}$.

Exercise 2.16. If $\mathcal{V}$ and $\mathcal{W}$ are finite dimensional vector spaces, show that there is an injective linear transformation $f : \mathcal{V} \to \mathcal{W}$ if and only if $\dim \mathcal{V} \leq \dim \mathcal{W}$.

Exercise 2.17. Show that the function $f : \mathcal{D}_{\mathbb{R}}^1(\mathbb{R}) \to \mathcal{F}_{\mathbb{R}}(\mathbb{R})$ defined by $f(\varphi) = \varphi + 2\varphi'$ is linear and determine $\ker f$.

Exercise 2.18. Show that the function $f : \mathcal{D}_{\mathbb{R}}^2(\mathbb{R}) \to \mathcal{F}_{\mathbb{R}}(\mathbb{R})$ defined by $f(\varphi) = \varphi + \varphi''$ is linear and determine $\ker f$

Exercise 2.19. Find a basis of $\ker f$ for the linear transformation $f : M_{2 \times 2}(\mathbb{K}) \to \mathbb{K}$ defined by

$$f\left(\begin{bmatrix} a & b \\ c & d \end{bmatrix}\right) = a + b + c + d.$$

Exercise 2.20. Let $\mathcal{U}$ be the subspace of $\mathbb{R}^3$ defined by

$$\mathcal{U} = \left\{ \begin{bmatrix} x \\ y \\ z \end{bmatrix} \in \mathbb{R}^3 : x + y + z = 0 \right\}$$

and let $f : \mathcal{U} \to \mathbb{R}^2$ be the linear transformation such that

$$f\left(\begin{bmatrix} 1 \\ -1 \\ 0 \end{bmatrix}\right) = \begin{bmatrix} 2 \\ 3 \end{bmatrix} \quad \text{and} \quad f\left(\begin{bmatrix} -1 \\ 0 \\ 1 \end{bmatrix}\right) = \begin{bmatrix} 1 \\ 4 \end{bmatrix}.$$

Determine f.

Exercise 2.21. Let $\mathcal{V}$ and $\mathcal{W}$ be finite dimensional vector spaces and let $f : \mathcal{V} \to \mathcal{W}$ be a linear transformation. If $\dim \mathcal{W} = n$, show that for $j \in \{1, \dots, n\}$ there are vectors $\mathbf{w}_j \in \mathcal{W}$ and linear transformations $f_j : \mathcal{V} \to \mathbb{K}$ such that $f = \mathbf{w}_1 \otimes f_1 + \dots + \mathbf{w}_n \otimes f_n$, where $\mathbf{w}_j \otimes f_j$ is defined as in Exercise 2.8.

Exercise 2.22. Let $\mathcal{V}$ and $\mathcal{W}$ be vector spaces and let $f : \mathcal{V} \to \mathcal{W}$ be a linear transformation. Show that, if $\mathbf{u}_1, \dots, \mathbf{u}_n \in \mathcal{V}$ are linearly independent vectors such that $\mathcal{V} = \mathrm{Span}\{\mathbf{u}_1, \dots, \mathbf{u}_n\} \oplus \ker f$, then $\{f(\mathbf{u}_1), \dots, f(\mathbf{u}_n)\}$ is a basis of $\mathrm{ran}\, f$.

Exercise 2.23. Let $\mathcal{V}$ and $\mathcal{W}$ be finite dimensional vector spaces and let $f \in \mathcal{L}(\mathcal{V}, \mathcal{W})$. Use the rank-nullity theorem to show that, if f is surjective, then $\dim \mathcal{V} \geq \dim \mathcal{W}$.

Exercise 2.24. Let $\mathcal{V}$ and $\mathcal{W}$ be finite dimensional vector spaces. Explain the meaning of the rank-nullity theorem for the function $f : \mathcal{V} \times \mathcal{W} \to \mathcal{V}$ defined by $f(\mathbf{v}, \mathbf{w}) = \mathbf{v}$.

2.6.2   Isomorphisms

Exercise 2.25. Let $f : V \to W$ be a linear transformation and let $\{\mathbf{w}_1, \ldots, \mathbf{w}_n\}$ be a basis of ran f. Then there are $\mathbf{u}_1, \ldots, \mathbf{u}_n \in V$ such that $f(\mathbf{u}_j) = \mathbf{w}_j$ for every $j \in \{1, \ldots, n\}$. Consider the linear transformation $g : \mathrm{Span}\{\mathbf{u}_1, \ldots, \mathbf{u}_n\} \to$ ran f defined by $g(\mathbf{u}) = f(\mathbf{u})$ for every $\mathbf{u} \in \mathrm{Span}\{\mathbf{u}_1, \ldots, \mathbf{u}_n\}$. Using Exercise 2.6, show that g is an isomorphism such that for every $\mathbf{v} \in V$ we have $g(h(\mathbf{v})) = f(\mathbf{v})$, where $h : V \to V$ is the projection on $\mathrm{Span}\{\mathbf{u}_1, \ldots, \mathbf{u}_n\}$ along $\ker f$.

Exercise 2.26. Let $f : \mathcal{P}_4(\mathbb{R}) \to \mathcal{P}_4(\mathbb{R})$ be the linear transformation defined by $f(p) = p''$. Find n, $\mathbf{w}_1, \ldots, \mathbf{w}_n, \mathbf{u}_1, \ldots, \mathbf{u}_n$, and h that satisfy the conditions in Exercise 2.25.

Exercise 2.27. Let V, W_1, W_2 be arbitrary vector spaces. Show that the vector space $\mathcal{L}(V, W_1 \times W_2)$ is isomorphic to the vector space $\mathcal{L}(V, W_1) \times \mathcal{L}(V, W_2)$.

Exercise 2.28. Let V be a vector space and let $\mathcal{U}, W_1$, and W_2 be subspaces of V. If $V = \mathcal{U} \oplus W_1 = \mathcal{U} \oplus W_2$, show that there is an isomorphism $g : W_1 \to W_2$.

Exercise 2.29. Show that there is an isomorphism $f : \mathbb{K}^4 \to \mathbb{K}^4$ such that $f^2(\mathbf{x}) = -\mathbf{x}$.

Exercise 2.30. Let V be a finite dimensional vector space and let $f, g \in \mathcal{L}(V)$. Show that the operator fg is invertible if and only if both f and g are invertible.

Exercise 2.31. Let V and W be vector spaces and let $\{\mathbf{v}_1, \ldots, \mathbf{v}_n\}$ be a basis of V. Show that the function $g : \mathcal{L}(V, W) \to W^n$ defined by $g(f) = (f(\mathbf{v}_1), \ldots, f(\mathbf{v}_n))$ is an isomorphism.

Exercise 2.32. Let V and W be vector spaces and let $\{\mathbf{v}_1, \ldots, \mathbf{v}_n\}$ be a basis of V. If W is finite dimensional, show that the set $\mathcal{S} = \{f \in \mathcal{L}(V, W) : f(\mathbf{v}_1) = f(\mathbf{v}_2) = \mathbf{0}\}$ is a vector subspace of $\mathcal{L}(V, W)$ and find $\dim \mathcal{S}$.

Exercise 2.33. Let V and W be vector spaces. Show that there is an isomorphism between $V \times W$ and $W \times V$.

Exercise 2.34. Let V and W be finite dimensional vector spaces and let $f : V \to W$ be a linear transformation. Show that the spaces V and $\ker f \times \mathrm{ran} f$ are isomorphic.

Exercise 2.35. Let $\mathcal{U}, V$, and W be vector spaces. Show that, if $\mathcal{U}$ and V are isomorphic and V and W are isomorphic, then $\mathcal{U}$ and W are isomorphic.

Exercise 2.36. Let V be a vector space and let $f : V \to V$ be a linear operator such that $f^3 = \mathbf{0}$. Show that $\mathrm{Id} - f$ is an isomorphism.

Exercise 2.37. Let V and W be vector spaces and let $f : V \to W$ be an isomorphism. Show that $\varphi : \mathcal{L}(V) \to \mathcal{L}(W)$ defined by $\varphi(g) = fgf^{-1}$ is an isomorphism.

Exercise 2.38. Show that the function $f : \mathcal{M}_{m \times n}(\mathbb{K}) \to \mathcal{M}_{n \times m}(\mathbb{K})$ defined by $f(A) = A^T$ is an isomorphism.

2.6.3 Linear transformations and matrices

Exercise 2.39. Let V be a vector space such that $\dim V = 4$ and let $f : V \to V$ be an operator such that $\dim \operatorname{ran} f = 2$. Show that there are bases $\mathcal{B}$ and $\mathcal{C}$ of V such that

$$f_{\mathcal{B} \to \mathcal{C}} = \begin{bmatrix} 1 & 0 & 0 & 0 \\ 0 & 0 & 0 & 0 \\ 0 & 0 & 1 & 0 \\ 0 & 0 & 0 & 0 \end{bmatrix}.$$

Exercise 2.40. We consider the linear transformation $f : M_{2 \times 2}(\mathbb{K}) \to M_{2 \times 2}(\mathbb{K})$ defined by $f(X) = \frac{1}{2}(X + X^T)$. Determine the matrix of f relative to the basis

$$\mathcal{B} = \left\{ \begin{bmatrix} 1 & 0 \\ 0 & 0 \end{bmatrix}, \begin{bmatrix} 0 & 1 \\ 0 & 0 \end{bmatrix}, \begin{bmatrix} 0 & 0 \\ 1 & 0 \end{bmatrix}, \begin{bmatrix} 0 & 0 \\ 0 & 1 \end{bmatrix} \right\}.$$

Exercise 2.41. Let $\mathcal{U} = \operatorname{Span}\{\cos t, t \cos t, \sin t, t \sin t\}$ and let $f : \mathcal{U} \to \mathcal{U}$ be the operator $f(\varphi) = \varphi'$. Show that $\mathcal{B} = \{\cos t, t \cos t, \sin t, t \sin t\}$ is a basis of $\mathcal{U}$ and determine the $\mathcal{B}$-matrix of f.

Exercise 2.42. Let $\mathcal{U} = \operatorname{Span}\{\cos t, t \cos t, \sin t, t \sin t\}$ and let $f : \mathcal{U} \to \mathcal{U}$ be the operator $f(\varphi) = \varphi''$. Determine the matrix of f relative to the basis $\{\cos t, t \cos t, \sin t, t \sin t\}$.

Exercise 2.43. Let $V = \operatorname{Span}\{\mathbf{v}_1, \mathbf{v}_2\} = \operatorname{Span}\{\mathbf{w}_1, \mathbf{w}_2\}$ where $\{\mathbf{v}_1, \mathbf{v}_2\}$ and $\{\mathbf{w}_1, \mathbf{w}_2\}$ are bases. Let $f : V \to V$ be defined by $f(\mathbf{v}_1) = 5\mathbf{v}_1 + 7\mathbf{v}_2$ and $f(\mathbf{v}_2) = 2\mathbf{v}_1 + 3\mathbf{v}_2$. If $\mathbf{v}_1 = 2\mathbf{w}_1 - \mathbf{w}_2$ and $\mathbf{v}_2 = 5\mathbf{w}_1 + 4\mathbf{w}_2$, determine the matrix of f relative to the basis $\{\mathbf{w}_1, \mathbf{w}_2\}$.

2.6.4 Duality

Exercise 2.44. We define $f_1, f_2, f_3 \in (\mathbb{R}^3)'$ by

$$f_1 \left(\begin{bmatrix} x \\ y \\ z \end{bmatrix} \right) = 2x + y + z, \quad f_2 \left(\begin{bmatrix} x \\ y \\ z \end{bmatrix} \right) = x + 2y + z, \quad \text{and } f_3 \left(\begin{bmatrix} x \\ y \\ z \end{bmatrix} \right) = x + y + 2z.$$

Show that f_1, f_2, and f_3 are linearly independent.

Exercise 2.45. Let V be a vector space and let $\{\mathbf{v}_1, \ldots, \mathbf{v}_n\}$ be a basis of V. Show that $f = f(\mathbf{v}_1)l_{\mathbf{v}_1} + \cdots + f(\mathbf{v}_n)l_{\mathbf{v}_n}$ for every $f \in V'$.

Exercise 2.46. Let V be an n-dimensional vector space and let $\{f_1, \ldots, f_n\}$ be a basis of V'. Show that the function $g : V \to \mathbb{K}^n$ defined by $g(\mathbf{v}) = (f_1(\mathbf{v}), \ldots, f_n(\mathbf{v}))$ is an isomorphism.

Exercise 2.47. Let V be a finite dimensional vector space, let $\mathcal{U}$ be a subspace of V, and let $\mathbf{x} \in V$ be such that $\mathbf{x} \notin \mathcal{U}$. Show that there is a linear form $f \in V'$ such that $f(\mathbf{x}) \neq 0$ and $f(\mathbf{u}) = 0$ for every $\mathbf{u} \in \mathcal{U}$.

Exercise 2.48. Let V be a finite dimensional vector space and let $f, g_1 \ldots, g_n \in V'$. If $f(\mathbf{x}) = 0$ for every $\mathbf{x} \in \ker g_1 \cap \cdots \cap \ker g_n$, show that f is a linear combination of $g_1, \ldots, g_n$.

Exercise 2.49. Let $f : V \to W$ be a linear transformation. We define the function $f^T : W' \to V'$ by $f^T(l)(\mathbf{v}) = l(f(\mathbf{v}))$ for $l \in W'$ and $\mathbf{v} \in V$. Show that the function f^T is a linear transformation.

Exercise 2.50. Let V and W be vector spaces and let $\mathcal{B}_V = \{\mathbf{v}_1, \ldots, \mathbf{v}_n\}$ and $\mathcal{B}_W = \{\mathbf{w}_1, \ldots, \mathbf{w}_m\}$ be a bases of V and W, respectively. Let $f : V \to W$ be a linear transformation and let A be the matrix of f relative to the bases $\mathcal{B}_V$ and $\mathcal{B}_W$. Show that, if $f^T : W' \to V'$ is the linear transformation defined in Exercise 2.49, then the matrix of f^T relative to the dual bases $\{l_{\mathbf{w}_1}, \ldots, l_{\mathbf{w}_m}\}$ and $\{l_{\mathbf{v}_1}, \ldots, l_{\mathbf{v}_n}\}$ is A^T.

Exercise 2.51. Let $f : V \to W$ and $g : W \to X$ be linear transformations. Show that $(gf)^T = f^T g^T$.

Exercise 2.52. Let V and W be vector spaces and let $f \in \mathcal{L}(V, W)$. Show that, if f is an isomorphism, then f^T is an isomorphism.

Exercise 2.53. Let V and W be finite dimensional vector spaces and let $f \in \mathcal{L}(V, W)$. Let $G : (V')' \to (W')'$ be defined by $G(F)(l) = F(lf)$ for $F \in (V')'$ and $l \in W'$. If $S : V \to (V')'$ and $T : W \to (W')'$ be the canonical isomorphisms, show that, if $S(\mathbf{v}) = F$ for some $\mathbf{v} \in V$, then $T(f(\mathbf{v})) = G(F)$.

Exercise 2.54. Let V and W be finite dimensional vector spaces and let $f \in \mathcal{L}(V, W)$. Show that there is a unique linear transformation $g \in \mathcal{L}(V, W)$ such that $l(g(\mathbf{v})) = f^T(l)(\mathbf{v})$ for every $l \in W'$ and $\mathbf{v} \in V$.

Exercise 2.55. Let $\mathcal{U}$ be a subspace of a finite dimensional vector space V. Show that the set $\mathcal{U}^0 = \{l \in V' : l(\mathbf{u}) = 0 \text{ for every } \mathbf{u} \in \mathcal{U}\}$ is a subspace of V' and that $\dim \mathcal{U}^0 + \dim \mathcal{U} = \dim V$.

Exercise 2.56. Let V and W be finite dimensional vector spaces. Show that the function $f : \mathcal{L}(V, W) \to \mathcal{L}(W', V')$ defined by $f(g) = g^T$ is an isomorphism.

Exercise 2.57. Let V_1, V_2, and W be vector spaces. Show that the vector space $\mathcal{L}(V_1 \times V_2, W)$ is isomorphic to the vector space $\mathcal{L}(V_1, W) \times \mathcal{L}(V_2, W)$.

Exercise 2.58. Let V and W be vector spaces and let $f : V \to W$ be a linear transformation. Show that $(\operatorname{ran} f)^0 = \ker f^T$, where $(\operatorname{ran} f)^0$ is defined in Exercise 2.55 and f^T is as in Exercise 2.49.

Exercise 2.59. Let V and W be finite dimensional vector spaces and let $f : V \to W$ be a linear transformation. Using Exercise 2.58 show that f is surjective if and only if f^T is injective.

Exercise 2.60. Let $\mathcal{U}$ be a subspace of a vector space V and let $\mathcal{U}^0$ be as defined in Exercise 2.55. If $\mathcal{U}^{00} = \{\mathbf{x} \in V : l(\mathbf{x}) = 0 \text{ for every } l \in \mathcal{U}^0\}$, show that $\mathcal{U}^{00} = \mathcal{U}$.

Exercise 2.61. Let $\mathcal{V}$ and $\mathcal{W}$ be finite dimensional vector spaces, $f \in \mathcal{L}(\mathcal{V}, \mathcal{W})$, and $l \in \mathcal{V}'$. If $l \in (\ker f)^0$, show that there is $m \in \mathcal{W}'$ such that $l = mf$.

Exercise 2.62. Let $\mathcal{V}$ and $\mathcal{W}$ be finite dimensional vector spaces and let $f \in \mathcal{L}(\mathcal{V}, \mathcal{W})$. Show that $(\ker f)^0 = \operatorname{ran} f^T$, where $(\ker f)^0$ is defined in Exercise 2.55 and f^T is as in Exercise 2.49.

Exercise 2.63. Let $\mathcal{V}$ and $\mathcal{W}$ be finite dimensional vector spaces and let $f : \mathcal{V} \to \mathcal{W}$ be a linear transformation. Using Exercises 2.55 and 2.58 show that $\dim \operatorname{ran} f = \dim \operatorname{ran} f^T$.

Exercise 2.64. Use Exercise 2.63 to show that f is injective if and only if f^T is surjective.

Exercise 2.65 (Rank Theorem). Use Exercises 2.50 and 2.63 to show that if $A \in \mathcal{M}_{n \times m}(\mathbb{K})$, then $\dim \mathbf{C}(A) = \dim \mathbf{C}(A^T)$.

Exercise 2.66. Let $\mathcal{V}$ and $\mathcal{W}$ be vector spaces with bases $\{\mathbf{v}_1, \ldots, \mathbf{v}_m\}$ and $\{\mathbf{w}_1, \ldots, \mathbf{w}_n\}$, respectively. Show that the set of all linear transformations $\mathbf{w}_k \otimes l_{\mathbf{v}_j}$, where $1 \le j \le m$ and $1 \le k \le n$, is a basis of $\mathcal{L}(\mathcal{V}, \mathcal{W})$. (See Exercise 2.8 for the definition of $\mathbf{w}_k \otimes l_{\mathbf{v}_j}$.)

Exercise 2.67. Let $\mathcal{V}$ and $\mathcal{W}$ be vector spaces with bases $\{\mathbf{v}_1, \ldots, \mathbf{v}_m\}$ and $\{\mathbf{w}_1, \ldots, \mathbf{w}_n\}$, respectively. For every $(j, k) \in \{1, \ldots, m\} \times \{1, \ldots, n\}$ we define the linear transformation $f_{jk} : \mathcal{V} \to \mathcal{W}$ by

$$f_{jk}(\mathbf{v}_i) = \begin{cases} \mathbf{w}_k & \text{if } i = j, \\ \mathbf{0} & \text{if } i \ne j. \end{cases}$$

Show that the set $\{f_{jk} : (j, k) \in \{1, \ldots, m\} \times \{1, \ldots, n\}\}$ is a basis of $\mathcal{L}(\mathcal{V}, \mathcal{W})$.

Exercise 2.68. Let $\mathcal{V}$ be a finite dimensional vector space and let $f, f_1, \ldots, f_j \in \mathcal{V}'$. If $f \notin \operatorname{Span}\{f_1, \ldots, f_j\}$, then there is $\mathbf{w} \in \mathcal{V}$ such that $f(\mathbf{w}) \ne 0$ and $f_k(\mathbf{w}) = 0$ for every $k \in \{1, \ldots, j\}$.

2.6.5 Quotient spaces

Exercise 2.69. Let $\mathcal{V}$ and $\mathcal{W}$ be vector spaces and let $f : \mathcal{V} \to \mathcal{W}$ be a linear transformation. Show that the function $\hat{f} : \mathcal{V} / \ker f \to \operatorname{ran} f$ defined by $\hat{f}(\mathbf{v} + \ker f) = f(\mathbf{v})$ is an isomorphism.

Exercise 2.70. Let $\mathbb{R}^3 = \mathbb{R}\mathbf{u} \oplus \mathbb{R}\mathbf{v} \oplus \mathbb{R}\mathbf{w}$. If $f : \mathbb{R}^3 \to \mathbb{R}\mathbf{v} \oplus \mathbb{R}\mathbf{w}$ is the projection on $\mathbb{R}\mathbf{v} \oplus \mathbb{R}\mathbf{w}$ along $\mathbb{R}\mathbf{u}$, then the function $g : \mathbb{R}^3 / \mathbb{R}\mathbf{u} \to \mathbb{R}\mathbf{v} \oplus \mathbb{R}\mathbf{w}$ defined by

$$g((\alpha\mathbf{v} + \beta\mathbf{w} + \gamma\mathbf{u})^\wedge) = \alpha\mathbf{v} + \beta\mathbf{w},$$

is an isomorphism.

Exercise 2.71. Let $\mathcal{V}$ be a vector space and let $f : \mathcal{V} \to \mathbb{K}$ be a nonzero linear transformation. Show that there is an isomorphism between $\mathcal{V} / \ker f$ and $\mathbb{K}$.

Exercise 2.72. Let V and W be vector spaces and let $f : V \to W$ be a linear transformation. Show that the spaces V and $\ker f \times V / \ker f$ are isomorphic.

Exercise 2.73. Let V and W be finite dimensional vector spaces. If $V_1 \subseteq V$ and $W_1 \subseteq W$ are subspaces, show that the spaces $(V \times W)/(V_1 \times W_1)$ and $V/V_1 \times W/W_1$ are isomorphic.

Exercise 2.74. Let V and W be vector spaces and let U a subspace of V. If $f : V \to W$ is a linear transformation such that $U \subseteq \ker f$, show that the function $g : V/U \to W$ defined by $g(q(\mathbf{x})) = f(\mathbf{x})$, where $q : V \to V/U$ is the quotient linear transformation, is a well-defined linear transformation.

Exercise 2.75. Let U be a subspace of a vector space V. Show that $\dim V/U = n$ if and only if there are linearly independent vectors $\mathbf{v}_1, \ldots, \mathbf{v}_n \in V$ such that $V = U \oplus \mathbb{K}\mathbf{v}_1 \cdots \oplus \mathbb{K}\mathbf{v}_n$.

Exercise 2.76. Let V be a vector space and let U be a subspace of V. If $f : V \to V$ is a linear transformation and U is f-invariant, then there is a unique linear transformation $g : V/U \to V/U$ such that $qf = gq$, where $q : V \to V/U$ is the quotient linear transformation.

Exercise 2.77. If U and W are subspaces of a vector space V such that $V = U \oplus W$, show that the linear transformation $h : W \to V/U$ defined by $h(\mathbf{w}) = \widehat{\mathbf{w}}$ is an isomorphism, without using Theorem 2.5.9.

Exercise 2.78. Let $\mathbf{v}$, $\mathbf{u}$, and $\mathbf{w}$ be linearly independent vectors in $\mathbb{R}^3$. Show that the set $\{\mathbf{u} + \mathbb{R}\mathbf{v}, \mathbf{w} + \mathbb{R}\mathbf{v}\}$ is a basis of $\mathbb{R}^3/\mathbb{R}\mathbf{v}$.

Exercise 2.79. Let U and W be subspaces of a vector space V such that $V = U \oplus W$. If $\{\mathbf{w}_1, \ldots, \mathbf{w}_n\}$ is a basis of W, show that $\{\mathbf{w}_1 + U, \ldots, \mathbf{w}_p + U\}$ is a basis of V/U.

Chapter 3

Inner Product Spaces

Introduction

The dot product is an important tool in the linear algebra of Euclidean spaces as well as many applications. In this chapter we investigate properties of vector spaces where an abstract form of the dot product is available. In the context of general vector spaces the name *inner product* is used instead of *dot product*.

In some examples and exercises in this chapter we will use determinants. In particular, we will use the fact that a matrix $\begin{bmatrix} \alpha & \beta \\ \gamma & \delta \end{bmatrix} \in \mathcal{M}_{2\times 2}(\mathbb{K})$ is invertible if and only if

$$\det \begin{bmatrix} \alpha & \beta \\ \gamma & \delta \end{bmatrix} = \alpha\delta - \beta\gamma \neq 0.$$

The use of determinants in those examples and exercises is not essential, but it is convenient and it leads to simplifications. Unlike some other textbooks at the same level, we do not consider determinants a forbidden tool.

3.1 Definitions and examples

In Chapter 2 we used the name *linear form* to mean a linear function $f : \mathcal{V} \to \mathbb{K}$. In this chapter we consider functions $f : \mathcal{V} \times \mathcal{V} \to \mathbb{K}$. Since $\mathcal{V} \times \mathcal{V}$ is a vector space, we can talk about linearity of $f : \mathcal{V} \times \mathcal{V} \to \mathbb{K}$:

$$f(a_1(\mathbf{x}_1, \mathbf{y}_1) + a_2(\mathbf{x}_2, \mathbf{y}_2)) = a_1 f(\mathbf{x}_1, \mathbf{y}_1) + a_2 f(\mathbf{x}_2, \mathbf{y}_2).$$

In the context of inner product spaces it is natural to consider a different property of functions $f : \mathcal{V} \times \mathcal{V} \to \mathbb{K}$ related to linearity, namely *bilinearity*.

Definition 3.1.1. By a *bilinear form* on a vector space V we mean a function $f : V \times V \to \mathbb{K}$ such that

(a) $f(\mathbf{x}_1 + \mathbf{x}_2, \mathbf{y}) = f(\mathbf{x}_1, \mathbf{y}) + f(\mathbf{x}_2, \mathbf{y})$,

(b) $f(\mathbf{x}, \mathbf{y}_1 + \mathbf{y}_2) = f(\mathbf{x}, \mathbf{y}_1) + f(\mathbf{x}, \mathbf{y}_2)$,

(c) $f(\alpha \mathbf{x}, \mathbf{y}) = \alpha f(\mathbf{x}, \mathbf{y})$,

(d) $f(\mathbf{x}, \alpha \mathbf{y}) = \alpha f(\mathbf{x}, \mathbf{y})$,

for all vectors $\mathbf{x}, \mathbf{x}_1, \mathbf{x}_2, \mathbf{y}, \mathbf{y}_1, \mathbf{y}_2 \in V$ and all numbers $\alpha \in \mathbb{K}$.

Note that the conditions (a)-(d) in the above definition can be expressed as a single equality:

$$f(a_1\mathbf{x}_1 + a_2\mathbf{x}_2, b_1\mathbf{y}_1 + b_2\mathbf{y}_2)$$
$$= a_1 b_1 f(\mathbf{x}_1, \mathbf{y}_1) + a_2 b_1 f(\mathbf{x}_2, \mathbf{y}_1) + a_1 b_2 f(\mathbf{x}_1, \mathbf{y}_2) + a_2 b_2 f(\mathbf{x}_2, \mathbf{y}_2).$$

Clearly, this condition implies

$$f\left(\sum_{j=1}^{m} a_j \mathbf{x}_j, \sum_{k=1}^{n} b_k \mathbf{y}_k \right) = \sum_{j=1}^{m} \sum_{k=1}^{n} a_j b_k f(\mathbf{x}_j, \mathbf{y}_k).$$

The conditions for linearity and bilinearity of a function $f : V \times V \to \mathbb{K}$ are not equivalent. Both linear and bilinear functions satisfy conditions (a) and (b), but for a linear f we have $f(a\mathbf{x}, a\mathbf{y}) = af(\mathbf{x}, \mathbf{y})$ and for a bilinear f we have $f(a\mathbf{x}, a\mathbf{y}) = a^2 f(\mathbf{x}, \mathbf{y})$.

Example 3.1.2. If f and g are linear forms on a vector space V, then the function
$$h(\mathbf{x}, \mathbf{y}) = f(\mathbf{x}) + g(\mathbf{y})$$
is a linear form on $V \times V$ and the function
$$k(\mathbf{x}, \mathbf{y}) = f(\mathbf{x})g(\mathbf{y})$$
is a bilinear form on V.

If $\mathbb{K} = \mathbb{C}$, the field of complex numbers, then there are reasons to replace condition (d) in the definition of bilinearity with the condition $f(\mathbf{x}, a\mathbf{y}) = \bar{a} f(\mathbf{x}, \mathbf{y})$, where $\bar{a}$ denotes the complex conjugate of a.

Definition 3.1.3. By a *sesquilinear form* on a vector space V we mean a function $s : V \times V \to \mathbb{K}$ such that

(a) $s(\mathbf{x}_1 + \mathbf{x}_2, \mathbf{y}) = s(\mathbf{x}_1, \mathbf{y}) + s(\mathbf{x}_2, \mathbf{y})$,

(b) $s(\mathbf{x}, \mathbf{y}_1 + \mathbf{y}_2) = s(\mathbf{x}, \mathbf{y}_1) + s(\mathbf{x}, \mathbf{y}_2)$,

(c) $s(\alpha\mathbf{x}, \mathbf{y}) = \alpha s(\mathbf{x}, \mathbf{y})$,

(d) $s(\mathbf{x}, \alpha\mathbf{y}) = \overline{\alpha} s(\mathbf{x}, \mathbf{y})$,

for all vectors $\mathbf{x}, \mathbf{x}_1, \mathbf{x}_2, \mathbf{y}, \mathbf{y}_1, \mathbf{y}_2 \in V$ and all numbers $\alpha \in \mathbb{K}$.

As in the case of bilinear form, the conditions (a)-(d) in the above definition can be expressed as a single equality:

$$f(a_1\mathbf{x}_1 + a_2\mathbf{x}_2, b_1\mathbf{y}_1 + b_2\mathbf{y}_2)$$
$$= a_1\overline{b_1} f(\mathbf{x}_1, \mathbf{y}_1) + a_2\overline{b_1} f(\mathbf{x}_2, \mathbf{y}_1) + a_1\overline{b_2} f(\mathbf{x}_1, \mathbf{y}_2) + a_2\overline{b_2} f(\mathbf{x}_2, \mathbf{y}_2).$$

In general, we have

$$f\left(\sum_{j=1}^{m} a_j\mathbf{x}_j, \sum_{k=1}^{n} b_k\mathbf{y}_k \right) = \sum_{j=1}^{m}\sum_{k=1}^{n} a_j\overline{b_k} f(\mathbf{x}_j, \mathbf{y}_k).$$

Note that for a function $f : V \times V \to \mathbb{R}$ the conditions for bilinearity and sesquilinearity are equivalent.

Example 3.1.4. The function

$$s\left(\begin{bmatrix} u_1 \\ u_2 \\ u_3 \end{bmatrix}, \begin{bmatrix} v_1 \\ v_2 \\ v_3 \end{bmatrix} \right) = 3u_1\overline{v_1} + \sqrt{2}u_2\overline{v_2} + \frac{1}{5}u_3\overline{v_3}$$

is a sesquilinear form on the the vector space $\mathbb{C}^3$ over $\mathbb{C}$.

Definition 3.1.5. A form $f : V \times V \to \mathbb{K}$ is called *symmetric*, if $f(\mathbf{x}, \mathbf{y}) = f(\mathbf{y}, \mathbf{x})$ for all $\mathbf{x}, \mathbf{y} \in V$.

The functions h and k in Example 3.1.2 are symmetric if and only if $f(\mathbf{x}) = g(\mathbf{x})$ for all $\mathbf{x} \in V$. Note that, if $f : V \times V \to \mathbb{K}$ is symmetric, then (b) in Definition 3.1.1 follows from (a). Similarly, (d) follows from (c).

The following theorem is an easy consequence of the definition of symmetric bilinear forms.

Theorem 3.1.6. *Let V be a vector space over $\mathbb{R}$ and let $s : V \times V \to \mathbb{R}$ be a symmetric bilinear form. Then*

$$s(\mathbf{x}, \mathbf{y}) = \frac{1}{4}\left[s(\mathbf{x} + \mathbf{y}, \mathbf{x} + \mathbf{y}) - s(\mathbf{x} - \mathbf{y}, \mathbf{x} - \mathbf{y})\right]$$

for all $\mathbf{x}, \mathbf{y} \in V$.

The identity in the above theorem is often referred to as a *polarization iden- tity*. It implies that, if the values of $s(\mathbf{x}, \mathbf{x})$ are known for all $\mathbf{x} \in V$, then the values of $s(\mathbf{x}, \mathbf{y})$ are known for all $\mathbf{x}, \mathbf{y} \in V$, which is often used in arguments.

Corollary 3.1.7. *If s_1 and s_2 are symmetric bilinear forms on a real vector space V such that*

$$s_1(\mathbf{x}, \mathbf{x}) = s_2(\mathbf{x}, \mathbf{x}) \quad \text{for all } \mathbf{x} \in V,$$

then

$$s_1(\mathbf{x}, \mathbf{y}) = s_2(\mathbf{x}, \mathbf{y}) \quad \text{for all } \mathbf{x}, \mathbf{y} \in V.$$

In particular, if s is a symmetric bilinear form such that $s(\mathbf{x}, \mathbf{x}) = 0$ for every $\mathbf{x} \in V$, then $s = 0$.

Note that the above property is not true for all bilinear forms. Indeed, for the bilinear form $s : \mathbb{R}^2 \times \mathbb{R}^2 \to \mathbb{R}$ defined by

$$s(\mathbf{x}, \mathbf{y}) = s\left(\begin{bmatrix} x_1 \\ x_2 \end{bmatrix}, \begin{bmatrix} y_1 \\ y_2 \end{bmatrix}\right) = x_1 y_2 - x_2 y_1$$

we have $s(\mathbf{x}, \mathbf{x}) = 0$ for every $\mathbf{x} \in \mathbb{R}^2$, but it is not true that $s(\mathbf{x}, \mathbf{y}) = 0$ for every $\mathbf{x}, \mathbf{y} \in \mathbb{R}^2$.

It turns out that for sesquilinear forms a different condition is more natural than symmetry.

Definition 3.1.8. A sesquilinear form $s : V \times V \to \mathbb{K}$ is called a *hermi- tian form* if

$$s(\mathbf{x}, \mathbf{y}) = \overline{s(\mathbf{y}, \mathbf{x})}$$

for all $\mathbf{x}, \mathbf{y} \in V$.

Example 3.1.9. The function

$$
s\left(\begin{bmatrix} u_1 \\ u_2 \\ u_3 \end{bmatrix}, \begin{bmatrix} v_1 \\ v_2 \\ v_3 \end{bmatrix}\right) = 3u_1\overline{v_1} + \sqrt{2}u_2\overline{v_2} + \frac{1}{5}u_3\overline{v_3}
$$

considered in Example 3.1.4 is a hermitian sesquilinear form, but the function

$$
s\left(\begin{bmatrix} u_1 \\ u_2 \\ u_3 \end{bmatrix}, \begin{bmatrix} v_1 \\ v_2 \\ v_3 \end{bmatrix}\right) = 3u_1\overline{v_1} + \sqrt{2}u_2\overline{v_2} + \frac{i}{5}u_3\overline{v_3}
$$

is not.

As in the case of symmetric forms, if s is a hermitian form, then the condition (b) in Definition 3.1.3 follows from (a) and (d) follows from (c).

Theorem 3.1.10. *Let V be a vector space over $\mathbb{C}$ and let $s : V \times V \to \mathbb{C}$ be a sesquilinear form on V. Then s is hermitian if and only if $s(\mathbf{x}, \mathbf{x}) \in \mathbb{R}$ for every $\mathbf{x} \in V$.*

Proof. If s is hermitian, then for every $\mathbf{x} \in V$ we have $s(\mathbf{x}, \mathbf{x}) = \overline{s(\mathbf{x}, \mathbf{x})}$ which means that $s(\mathbf{x}, \mathbf{x}) \in \mathbb{R}$.

Suppose now that $s(\mathbf{v}, \mathbf{v}) \in \mathbb{R}$ for every $\mathbf{v} \in V$. Then

$$
\alpha = s(\mathbf{x}, \mathbf{y}) + s(\mathbf{y}, \mathbf{x}) = s(\mathbf{x} + \mathbf{y}, \mathbf{x} + \mathbf{y}) - s(\mathbf{x}, \mathbf{x}) + s(\mathbf{y}, \mathbf{y}) \in \mathbb{R}
$$

and

$$
\begin{aligned}
\beta = i(-s(\mathbf{x}, \mathbf{y}) + s(\mathbf{y}, \mathbf{x})) &= s(\mathbf{x}, i\mathbf{y}) + s(i\mathbf{y}, \mathbf{x}) \\
&= s(\mathbf{x} + i\mathbf{y}, \mathbf{x} + i\mathbf{y}) - s(\mathbf{x}, \mathbf{x}) + s(i\mathbf{y}, i\mathbf{y}) \in \mathbb{R}.
\end{aligned}
$$

Since

$$
s(\mathbf{x}, \mathbf{y}) = \frac{1}{2}(\alpha + i\beta) \quad \text{and} \quad s(\mathbf{y}, \mathbf{x}) = \frac{1}{2}(\alpha - i\beta),
$$

s is hermitian. $\qquad\qquad\square$

Theorem 3.1.11. *Let V be a vector space over $\mathbb{C}$ and let $s : V \times V \to \mathbb{C}$ be a hermitian form. Then*

$$s(\mathbf{x}, \mathbf{y}) = \frac{1}{4}[s(\mathbf{x}+\mathbf{y}, \mathbf{x}+\mathbf{y}) - s(\mathbf{x}-\mathbf{y}, \mathbf{x}-\mathbf{y})$$
$$+ is(\mathbf{x}+i\mathbf{y}, \mathbf{x}+i\mathbf{y}) - is(\mathbf{x}-i\mathbf{y}, \mathbf{x}-i\mathbf{y})]$$

for all $\mathbf{x}, \mathbf{y} \in V$.

Proof. The result is a consequence of the following equalities:

$$s(\mathbf{x}+\mathbf{y}, \mathbf{x}+\mathbf{y}) - s(\mathbf{x}-\mathbf{y}, \mathbf{x}-\mathbf{y}) = 2s(\mathbf{x}, \mathbf{y}) + 2s(\mathbf{y}, \mathbf{x})$$
$$= 2s(\mathbf{x}, \mathbf{y}) + 2\overline{s(\mathbf{x}, \mathbf{y})}$$
$$= 4\operatorname{Re} s(\mathbf{x}, \mathbf{y})$$

and

$$s(\mathbf{x}+i\mathbf{y}, \mathbf{x}+i\mathbf{y}) - s(\mathbf{x}-i\mathbf{y}, \mathbf{x}-i\mathbf{y}) = 4\operatorname{Re} s(\mathbf{x}, i\mathbf{y})$$
$$= 4\operatorname{Re}(-is(\mathbf{x}, \mathbf{y}))$$
$$= 4\operatorname{Im} s(\mathbf{x}, \mathbf{y}).$$

□

The identity in the above theorem is a complex version of the *polarization identity*. As in the real case it implies the following useful property of hermitian sesquilinear forms.

Corollary 3.1.12. *If s_1 and s_2 are hermitian forms on a complex vector space V such that*

$$s_1(\mathbf{x}, \mathbf{x}) = s_2(\mathbf{x}, \mathbf{x}) \quad \textit{for all } \mathbf{x} \in V,$$

then

$$s_1(\mathbf{x}, \mathbf{y}) = s_2(\mathbf{x}, \mathbf{y}) \quad \textit{for all } \mathbf{x}, \mathbf{y} \in V.$$

In particular, if s is a sesquilinear form such that $s(\mathbf{x}, \mathbf{x}) = 0$ for every $\mathbf{x} \in V$, then $s = 0$.

Definition 3.1.13. A sesquilinear form $s : V \times V \to \mathbb{K}$ is called a *positive form* if

$$s(\mathbf{x}, \mathbf{x}) \geq 0$$

for every $\mathbf{x} \in V$.
A positive form is called *positive definite* if

$$s(\mathbf{x}, \mathbf{x}) > 0$$

whenever $\mathbf{x} \neq \mathbf{0}$.

The condition $s(\mathbf{x}, \mathbf{x}) \geq 0$ implicitly assumes that $s(\mathbf{x}, \mathbf{x}) \in \mathbb{R}$ for every $\mathbf{x} \in V$. Consequently, by Theorem 3.1.10, every positive form is hermitian.

Example 3.1.14. The function $s : \mathcal{M}_{2 \times 2}(\mathbb{C}) \times \mathcal{M}_{2 \times 2}(\mathbb{C}) \to \mathbb{C}$ defined by

$$s\left(\begin{bmatrix} u_1 & u_2 \\ u_3 & u_4 \end{bmatrix}, \begin{bmatrix} v_1 & v_2 \\ v_3 & v_4 \end{bmatrix}\right) = u_1 \overline{v_1} + u_2 \overline{v_2} + u_3 \overline{v_3} + u_4 \overline{v_4}$$

is a positive definite sesquilinear form on $\mathcal{M}_{2 \times 2}(\mathbb{C})$.

Now we are in a position to define the generalization of the dot product to arbitrary vector spaces.

Definition 3.1.15. By an *inner product* on a vector space V we mean a positive definite sesquilinear form on V. A vector space V with an inner product is called an *inner product space.*

The inner product of two vectors $\mathbf{x}$ and $\mathbf{y}$ in V is denoted by $\langle \mathbf{x}, \mathbf{y} \rangle$. Below we list all properties that constitute the definition of an inner product.

A function
$$\langle \cdot, \cdot \rangle : \mathcal{V} \times \mathcal{V} \to \mathbb{K}$$
is an inner product on $\mathcal{V}$ if the following conditions are satisfied.

1. $\langle \cdot, \cdot \rangle$ is sesquilinear:

   (a) $\langle \mathbf{x}_1 + \mathbf{x}_2, \mathbf{y} \rangle = \langle \mathbf{x}_1, \mathbf{y} \rangle + \langle \mathbf{x}_2, \mathbf{y} \rangle$ for all $\mathbf{x}_1, \mathbf{x}_2, \mathbf{y} \in \mathcal{V}$,

   (b) $\langle \mathbf{x}, \mathbf{y}_1 + \mathbf{y}_2 \rangle = \langle \mathbf{x}, \mathbf{y}_1 \rangle + \langle \mathbf{x}, \mathbf{y}_2 \rangle$ for all $\mathbf{x}, \mathbf{y}_1, \mathbf{y}_2 \in \mathcal{V}$,

   (c) $\langle \alpha \mathbf{x}, \mathbf{y} \rangle = \alpha \langle \mathbf{x}, \mathbf{y} \rangle$ for all $\mathbf{x}, \mathbf{y} \in \mathcal{V}$ and $\alpha \in \mathbb{K}$,

   (d) $\langle \mathbf{x}, \alpha \mathbf{y} \rangle = \overline{\alpha} \langle \mathbf{x}, \mathbf{y} \rangle$ for all $\mathbf{x}, \mathbf{y} \in \mathcal{V}$ and $\alpha \in \mathbb{K}$,

2. $\langle \cdot, \cdot \rangle$ is hermitian: $\langle \mathbf{x}, \mathbf{y} \rangle = \overline{\langle \mathbf{y}, \mathbf{x} \rangle}$ for all $\mathbf{x}, \mathbf{y} \in \mathcal{V}$,

3. $\langle \cdot, \cdot \rangle$ is positive definite: $\langle \mathbf{x}, \mathbf{x} \rangle > 0$ for all $\mathbf{0} \neq \mathbf{x} \in \mathcal{V}$.

In view of the previous comments and Theorem 3.1.10, in order to verify that a function $\langle \cdot, \cdot \rangle : \mathcal{V} \times \mathcal{V} \to \mathbb{C}$ is an inner product on $\mathcal{V}$ it suffices to check the following three conditions:

(i) $\langle \alpha_1 \mathbf{x}_1 + \alpha_2 \mathbf{x}_2, \mathbf{y} \rangle = \alpha_1 \langle \mathbf{x}_1, \mathbf{y} \rangle + \alpha_2 \langle \mathbf{x}_2, \mathbf{y} \rangle$ for all $\mathbf{x}_1, \mathbf{x}_2, \mathbf{y} \in \mathcal{V}$ and $\alpha_1, \alpha_2 \in \mathbb{K}$,

(ii) $\langle \mathbf{x}, \mathbf{y} \rangle = \overline{\langle \mathbf{y}, \mathbf{x} \rangle}$ for all $\mathbf{x}, \mathbf{y} \in \mathcal{V}$,

(iii) $\langle \mathbf{x}, \mathbf{x} \rangle > 0$ for all nonzero $\mathbf{x} \in \mathcal{V}$.

Example 3.1.16. The standard inner product in the vector space $\mathbb{C}^n$ is defined by

$$\left\langle \begin{bmatrix} x_1 \\ \vdots \\ x_n \end{bmatrix}, \begin{bmatrix} y_1 \\ \vdots \\ y_n \end{bmatrix} \right\rangle = \sum_{j=1}^{n} x_j \overline{y_j} = x_1 \overline{y_1} + \cdots + x_n \overline{y_n}.$$

Example 3.1.17. The functions defined in Examples 3.1.4 and 3.1.14 are examples of inner products.

The vector space $\mathbb{C}^3$ is an inner product space with the inner product defined by

$$\left\langle \begin{bmatrix} u_1 \\ u_2 \\ u_3 \end{bmatrix}, \begin{bmatrix} v_1 \\ v_2 \\ v_3 \end{bmatrix} \right\rangle = 3u_1 \overline{v_1} + \sqrt{2} u_2 \overline{v_2} + \frac{1}{5} u_3 \overline{v_3}.$$

More generally, for any positive real numbers $\alpha_1, \ldots, \alpha_n$ the form

$$\left\langle \begin{bmatrix} x_1 \\ \vdots \\ x_n \end{bmatrix}, \begin{bmatrix} y_1 \\ \vdots \\ y_n \end{bmatrix} \right\rangle = \sum_{j=1}^{n} \alpha_j x_j \overline{y_j} = \alpha_1 x_1 \overline{y_1} + \cdots + \alpha_n x_n \overline{y_n}$$

is an inner product in $\mathbb{C}^n$.

The vector space $\mathcal{M}_{2 \times 2}(\mathbb{C})$ is an inner product space with the inner product defined by

$$\left\langle \begin{bmatrix} u_1 & u_2 \\ u_3 & u_4 \end{bmatrix}, \begin{bmatrix} v_1 & v_2 \\ v_3 & v_4 \end{bmatrix} \right\rangle = u_1 \overline{v_1} + u_2 \overline{v_2} + u_3 \overline{v_3} + u_4 \overline{v_4}.$$

This example can be easily generalized to $\mathcal{M}_{m \times n}(\mathbb{C})$ for any positive integers m and n.

Example 3.1.18. The vector space $\mathcal{C}_{[a,b]}(\mathbb{C})$ of all continuous complex-valued functions on the interval $[a, b]$ is an inner product space with the inner product defined by

$$\langle f, g \rangle = \int_a^b f(t) \overline{g(t)} dt.$$

More generally, for any $\varphi \in \mathcal{C}_{[a,b]}(\mathbb{C})$ such that $\varphi(t) > 0$ for all $t \in [a, b]$, the form

$$\langle f, g \rangle = \int_a^b f(t) \overline{g(t)} \varphi(t) dt.$$

is an inner product in $\mathcal{C}_{[a,b]}(\mathbb{C})$.

Example 3.1.19. Let $\mathcal{V} = \mathcal{C}_{[0,1]}^2(\mathbb{C})$, the space of complex-valued functions on $[0, 1]$ with continuous second derivatives. Show that

$$\langle f, g \rangle = f(0) \overline{g(0)} + f'(0) \overline{g'(0)} + \int_0^1 f''(t) \overline{g''(t)} dt$$

is an inner product.

Solution. The only nontrivial part is showing that $\langle f, f \rangle = 0$ implies $f = \mathbf{0}$.

Since

$$\langle f, f \rangle = f(0)\overline{f(0)} + f'(0)\overline{f'(0)} + \int_0^1 f''(t)\overline{f''(t)}dt$$

$$= |f(0)|^2 + |f'(0)|^2 + \int_0^1 |f''(t)|^2 dt,$$

if $\langle f, f \rangle = 0$, then $f(0) = 0$, $f'(0) = 0$, and $f''(t) = 0$ for all $t \in [0,1]$. From $f''(t) = 0$ we get $f(t) = at + b$. Then, form $f(0) = 0$ we get $b = 0$ and finally from $f'(0) = 0$ we get $a = 0$. Consequently, $f(t) = 0$ for all $t \in [0,1]$.   □

Example 3.1.20. Let $\mathcal{V} = \mathcal{C}^1_{[0,1]}(\mathbb{C})$ the space of complex-valued functions on $[0,1]$ with continuous derivatives. Show that

$$\langle f, g \rangle = \int_0^1 f'(t)\overline{g'(t)}dt$$

is not an inner product.

Solution. The defined function is not an inner product because it is not positive definite. Indeed, $\langle f, f \rangle = 0$ implies $f' = \mathbf{0}$, but this does not mean that $f = \mathbf{0}$ because f could be any constant, not necessarily 0.

Note that the defined function is a positive sesquilinear form.   □

The inequality in the next theorem, known as *Schwarz's Inequality*, is one of the most important and useful properties of the inner product.

Theorem 3.1.21 (Schwarz's Inequality). *Let $\mathcal{V}$ be an inner product space. Then*

$$|\langle \mathbf{x}, \mathbf{y} \rangle|^2 \leq \langle \mathbf{x}, \mathbf{x} \rangle \langle \mathbf{y}, \mathbf{y} \rangle$$

for all $\mathbf{x}, \mathbf{y} \in \mathcal{V}$.

Proof. If $\mathbf{y} = \mathbf{0}$, then the inequality is trivially true since both sides are equal to zero. If $\mathbf{y} \neq \mathbf{0}$, then

$$0 \leq \langle \mathbf{x} + \alpha\mathbf{y}, \mathbf{x} + \alpha\mathbf{y} \rangle = \langle \mathbf{x}, \mathbf{x} \rangle + \overline{\alpha}\langle \mathbf{x}, \mathbf{y} \rangle + \alpha\langle \mathbf{y}, \mathbf{x} \rangle + |\alpha|^2 \langle \mathbf{y}, \mathbf{y} \rangle,$$

for any $\alpha \in \mathbb{C}$. If we let $\alpha = -\dfrac{\langle \mathbf{x}, \mathbf{y} \rangle}{\langle \mathbf{y}, \mathbf{y} \rangle}$, then the above inequality becomes

$$0 \leq \langle \mathbf{x}, \mathbf{x} \rangle - \frac{\langle \mathbf{y}, \mathbf{x} \rangle}{\langle \mathbf{y}, \mathbf{y} \rangle}\langle \mathbf{x}, \mathbf{y} \rangle - \frac{\langle \mathbf{x}, \mathbf{y} \rangle}{\langle \mathbf{y}, \mathbf{y} \rangle}\langle \mathbf{y}, \mathbf{x} \rangle + \frac{|\langle \mathbf{x}, \mathbf{y} \rangle|^2}{(\langle \mathbf{y}, \mathbf{y} \rangle)^2}\langle \mathbf{y}, \mathbf{y} \rangle.$$

After multiplying the above inequality by $\langle \mathbf{y}, \mathbf{y} \rangle$ and simplifying we get

$$0 \le \langle \mathbf{x}, \mathbf{x} \rangle \langle \mathbf{y}, \mathbf{y} \rangle - |\langle \mathbf{x}, \mathbf{y} \rangle|^2,$$

which is Schwarz's inequality. □

Theorem 3.1.22. *Let $\mathcal{V}$ be an inner product space and let $\mathbf{x}, \mathbf{y} \in \mathcal{V}$. Then*

$$|\langle \mathbf{x}, \mathbf{y} \rangle|^2 = \langle \mathbf{x}, \mathbf{x} \rangle \langle \mathbf{y}, \mathbf{y} \rangle$$

if and only if the vectors $\mathbf{x}$ and $\mathbf{y}$ are linearly dependent.

Proof. Let $\mathbf{x}$ and $\mathbf{y}$ be linearly dependent vectors in $\mathcal{V}$. Without loss of generality we can assume that $\mathbf{x} = \alpha \mathbf{y}$ for some $\alpha \in \mathbb{K}$. Then

$$|\langle \mathbf{x}, \mathbf{y} \rangle|^2 = |\langle \mathbf{x}, \alpha \mathbf{x} \rangle|^2 = |\overline{\alpha}|^2 (\langle \mathbf{x}, \mathbf{x} \rangle)^2$$
$$= \langle \mathbf{x}, \mathbf{x} \rangle \alpha \overline{\alpha} \langle \mathbf{x}, \mathbf{x} \rangle = \langle \mathbf{x}, \mathbf{x} \rangle \langle \alpha \mathbf{x}, \alpha \mathbf{x} \rangle = \langle \mathbf{x}, \mathbf{x} \rangle \langle \mathbf{y}, \mathbf{y} \rangle.$$

Now, we assume that $\mathbf{x}$ and $\mathbf{y}$ are vectors in $\mathcal{V}$ such that $|\langle \mathbf{x}, \mathbf{y} \rangle|^2 = \langle \mathbf{x}, \mathbf{x} \rangle \langle \mathbf{y}, \mathbf{y} \rangle$. Then $\langle \mathbf{x}, \mathbf{y} \rangle \langle \mathbf{y}, \mathbf{x} \rangle = \langle \mathbf{x}, \mathbf{x} \rangle \langle \mathbf{y}, \mathbf{y} \rangle$ and consequently

$$\langle \langle \mathbf{y}, \mathbf{y} \rangle \mathbf{x} - \langle \mathbf{x}, \mathbf{y} \rangle \mathbf{y}, \langle \mathbf{y}, \mathbf{y} \rangle \mathbf{x} - \langle \mathbf{x}, \mathbf{y} \rangle \mathbf{y} \rangle$$
$$= \langle \mathbf{y}, \mathbf{y} \rangle^2 \langle \mathbf{x}, \mathbf{x} \rangle - \langle \mathbf{y}, \mathbf{y} \rangle \langle \mathbf{y}, \mathbf{x} \rangle \langle \mathbf{x}, \mathbf{y} \rangle - \langle \mathbf{x}, \mathbf{y} \rangle \langle \mathbf{y}, \mathbf{y} \rangle \langle \mathbf{y}, \mathbf{x} \rangle + \langle \mathbf{x}, \mathbf{y} \rangle \langle \mathbf{y}, \mathbf{x} \rangle \langle \mathbf{y}, \mathbf{y} \rangle = 0.$$

This shows that $\langle \mathbf{y}, \mathbf{y} \rangle \mathbf{x} - \langle \mathbf{x}, \mathbf{y} \rangle \mathbf{y} = 0$, which implies linear dependence of $\mathbf{x}$ and $\mathbf{y}$. □

The dot product in $\mathbb{R}^n$ has an important geometric meaning. For example, if $\mathbf{x}$ and $\mathbf{y}$ are nonzero vectors in $\mathbb{R}^3$ and $\mathbf{x} \cdot \mathbf{y} = 0$, then the vectors are perpendicular, that is, the angle between them is $90°$. In general vector spaces we do not have that geometric interpretation. For example, what would "the angle between functions $\sin t$ and $\cos t$" even mean? On the other hand, the importance of the dot product in $\mathbb{R}^n$ goes far beyond its connection with the angle between vectors. Many of those properties and applications of the dot product extend to the inner product in general vector spaces.

Definition 3.1.23. Let $\mathcal{V}$ be an inner product space. Vectors $\mathbf{x}, \mathbf{y} \in \mathcal{V}$ are called *orthogonal* if

$$\langle \mathbf{x}, \mathbf{y} \rangle = 0.$$

Example 3.1.24. Show that the vectors

$$\begin{bmatrix} 1+2i \\ 2-i \end{bmatrix} \quad \text{and} \quad \begin{bmatrix} -2-i \\ 1-2i \end{bmatrix}$$

are orthogonal in the inner product space $\mathbb{C}^2$.

Solution.

$$\left\langle \begin{bmatrix} 1+2i \\ 2-i \end{bmatrix}, \begin{bmatrix} -2-i \\ 1-2i \end{bmatrix} \right\rangle = (1+2i)\overline{(-2-i)} + (2-i)\overline{(1-2i)} = 0.$$

$\square$

Example 3.1.25. Consider the vector space of continuous functions defined on the interval $[-1,1]$ with the inner product

$$\langle f, g \rangle = \int_{-1}^{1} f(t)\overline{g(t)}\, dt.$$

Show that an odd function $f(t)$, that is a function such that $f(-t) = -f(t)$ for all $t \in [-1,1]$, and $\cos t$ are orthogonal.

Solution. First we note that the function $f(t)\cos t$ is odd since

$$f(-t)\cos(-t) = -f(t)\cos t$$

for all $t \in [-1,1]$. Hence,

$$\langle f(t), \cos t \rangle = \int_{-1}^{0} f(t)\cos t\, dt + \int_{0}^{1} f(t)\cos t\, dt$$

$$= -\int_{0}^{1} f(t)\cos t\, dt + \int_{0}^{1} f(t)\cos t\, dt = 0.$$

$\square$

Another tool that plays an important role in the linear algebra of $\mathbb{R}^n$ is the norm. The standard norm, called the *Euclidean norm*, is defined as

$$\|(x_1,\ldots,x_n)\| = \sqrt{\sum_{k=1}^{n} |x_k|^2}.$$

A norm in a general vector spaces V is introduced as a function $\|\cdot\| : V \to [0,\infty)$ satisfying certain conditions.

Definition 3.1.26. Let $\mathcal{V}$ be a vector space. By a *norm* in $\mathcal{V}$ we mean a function $\|\cdot\| : \mathcal{V} \to [0, \infty)$ such that

(a) $\|\mathbf{x} + \mathbf{y}\| \leq \|\mathbf{x}\| + \|\mathbf{y}\|$,

(b) $\|\alpha \mathbf{x}\| = |\alpha| \|\mathbf{x}\|$,

(c) $\|\mathbf{x}\| = 0$ if and only if $\mathbf{x} = \mathbf{0}$,

for all vectors $\mathbf{x}, \mathbf{y} \in \mathcal{V}$ and all numbers $\alpha \in \mathbb{K}$. A vector spece with a norm is called a *normed space*.

The inequality $\|\mathbf{x} + \mathbf{y}\| \leq \|\mathbf{x}\| + \|\mathbf{y}\|$ is called the *triangle inequality*.

Example 3.1.27. Here are two examples of norms in $\mathbb{K}^n$:

$$\|(x_1, \ldots, x_n)\| = \sum_{k=1}^{n} |x_k|,$$

$$\|(x_1, \ldots, x_n)\| = \max\{|x_1|, \ldots, |x_n|\}.$$

Example 3.1.28. The function

$$\|f\| = \int_a^b |f(t)| dt$$

is a norm in the vector space $\mathcal{C}_{[a,b]}(\mathbb{C})$ of all continuous functions $f : [a, b] \to \mathbb{C}$.

It turns out the inner product in a vector space defines in a natural way a norm in that space. That norm has the best algebraic and geometric properties.

Theorem 3.1.29. *Let $\mathcal{V}$ be an inner product space. The function*

$$\|\mathbf{x}\| = \sqrt{\langle \mathbf{x}, \mathbf{x} \rangle}$$

is a norm in $\mathcal{V}$.

Proof. First notice that $\|\mathbf{x}\|$ is well-defined because $\langle \mathbf{x}, \mathbf{x} \rangle$ is always a non-negative real number. Since the inner product is a positive definite form, $\|\mathbf{x}\| = 0$ if and only if $\mathbf{x} = \mathbf{0}$. Moreover

$$\|\alpha \mathbf{x}\| = \sqrt{\langle \alpha \mathbf{x}, \alpha \mathbf{x} \rangle} = \sqrt{\alpha \overline{\alpha} \langle \mathbf{x}, \mathbf{x} \rangle} = |\alpha| \|\mathbf{x}\|.$$

The triangle inequality follows from Schwarz's inequality:

$$
\begin{aligned}
\|\mathbf{x} + \mathbf{y}\|^2 &= \langle \mathbf{x} + \mathbf{y}, \mathbf{x} + \mathbf{y} \rangle \\
&= \langle \mathbf{x}, \mathbf{x} \rangle + 2\operatorname{Re}\langle \mathbf{x}, \mathbf{y} \rangle + \langle \mathbf{y}, \mathbf{y} \rangle \\
&\leq \langle \mathbf{x}, \mathbf{x} \rangle + 2|\langle \mathbf{x}, \mathbf{y} \rangle| + \langle \mathbf{y}, \mathbf{y} \rangle \\
&\leq \|\mathbf{x}\|^2 + 2\|\mathbf{x}\|\|\mathbf{y}\| + \|\mathbf{y}\|^2 \quad \text{(by Schwarz's inequality)} \\
&= (\|\mathbf{x}\| + \|\mathbf{y}\|)^2.
\end{aligned}
$$

Hence $\|\mathbf{x} + \mathbf{y}\| \leq \|\mathbf{x}\| + \|\mathbf{y}\|$. $\qquad\qquad\qquad\qquad\qquad\qquad\qquad$ □

When we say *the norm in an inner product space* we always mean the norm $\|\mathbf{x}\| = \sqrt{\langle \mathbf{x}, \mathbf{x} \rangle}$.

Example 3.1.30. The standard norm in the vector space $\mathbb{C}^n$ is defined by the inner product

$$
\left\langle \begin{bmatrix} x_1 \\ \vdots \\ x_n \end{bmatrix}, \begin{bmatrix} y_1 \\ \vdots \\ y_n \end{bmatrix} \right\rangle,
$$

which means that

$$
\left\| \begin{bmatrix} x_1 \\ \vdots \\ x_n \end{bmatrix} \right\| = \sqrt{\left\langle \begin{bmatrix} x_1 \\ \vdots \\ x_n \end{bmatrix}, \begin{bmatrix} x_1 \\ \vdots \\ x_n \end{bmatrix} \right\rangle} = \sqrt{\sum_{j=1}^n x_j \overline{x_j}}.
$$

Example 3.1.31. Since

$$
\left\| \begin{bmatrix} 1+2i \\ 2-3i \\ -2 \end{bmatrix} \right\|^2 = \left\langle \begin{bmatrix} 1+2i \\ 2-3i \\ -2 \end{bmatrix}, \begin{bmatrix} 1+2i \\ 2-3i \\ -2 \end{bmatrix} \right\rangle = 5 + 13 + 4 = 22
$$

the norm of the vector $\begin{bmatrix} 1+2i \\ 2-3i \\ -2 \end{bmatrix} \in \mathbb{C}^3$ is $\sqrt{22}$.

Example 3.1.32. Consider the vector space of continuous functions defined on the interval $[-\pi, \pi]$ with the inner product

$$\langle f, g \rangle = \frac{1}{\pi} \int_{-\pi}^{\pi} f(t)\overline{g(t)} \, dt.$$

Find $\| \sin nt \|$ for any positive integer n.

Solution. Since

$$\| \sin nt \|^2 = \langle \sin nt, \sin nt \rangle = \frac{1}{\pi} \int_{-\pi}^{\pi} (\sin nt)^2 \, dt = \frac{1}{2\pi} \int_{-\pi}^{\pi} (1 - \cos(2nt)) \, dt = 1,$$

we have $\| \sin nt \| = 1$. □

Schwarz's Inequality is often stated and applied in the form stated in the following Corollary.

Corollary 3.1.33. *Let V be an inner product space. Then*

$$|\langle \mathbf{x}, \mathbf{y} \rangle| \leq \|\mathbf{x}\| \|\mathbf{y}\|$$

for all $\mathbf{x}, \mathbf{y} \in V$.

From Theorem 3.1.11 we obtain the polarization identity for the inner product expressed in terms of the norm.

Corollary 3.1.34 (Polarization identity). *Let V be an inner product space over $\mathbb{C}$. Then*

$$\langle \mathbf{x}, \mathbf{y} \rangle = \frac{1}{4} \left(\|\mathbf{x} + \mathbf{y}\|^2 - \|\mathbf{x} - \mathbf{y}\|^2 + i\|\mathbf{x} + i\mathbf{y}\|^2 - i\|\mathbf{x} - i\mathbf{y}\|^2 \right)$$

for all $\mathbf{x}, \mathbf{y} \in V$.

We close this section with a general version of the Pythagorean Theorem that we are all familiar with from geometry.

Theorem 3.1.35 (The Pythagorean Theorem). *If $\mathbf{u}$ and $\mathbf{v}$ are orthogonal vectors in an inner product space, then*

$$\|\mathbf{u} + \mathbf{v}\|^2 = \|\mathbf{u}\|^2 + \|\mathbf{v}\|^2.$$

Proof. If $\mathbf{u}$ and $\mathbf{v}$ are orthogonal, then $\langle \mathbf{u}, \mathbf{v} \rangle = \langle \mathbf{v}, \mathbf{u} \rangle = 0$ and thus

$$\|\mathbf{u} + \mathbf{v}\|^2 = \langle \mathbf{u} + \mathbf{v}, \mathbf{u} + \mathbf{v} \rangle = \langle \mathbf{u}, \mathbf{u} \rangle + \langle \mathbf{u}, \mathbf{v} \rangle + \langle \mathbf{v}, \mathbf{u} \rangle + \langle \mathbf{v}, \mathbf{v} \rangle = \|\mathbf{u}\|^2 + \|\mathbf{v}\|^2.$$

$\square$

3.2   Orthogonal projections

In an elementary calculus course you have probably seen an exercise similar to the following one:

Find a point on the line $2x + 5y = 7$ that is closest to the point $(3, 6)$.

In calculus we use derivatives to find such a point. Linear algebra offers a simpler and more elegant way of solving the above problem. Moreover, the algebraic method generalizes in a natural way to more difficult problems. For example, it would be rather difficult to use methods form calculus to solve the following problem:

Find numbers a, b, and c that minimize the integral $\int_0^1 \left| e^t - a - bt - ct^2 \right|^2 dt$.

We will see in this section that both problems are quite similar from the point of view of linear algebra and that the method uses the inner product in a substantial way.

3.2.1   Orthogonal projections on lines

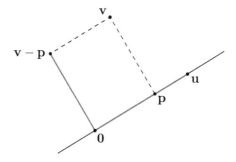

We start by considering the simplest case of an orthogonal projection in a general vector space, namely, projection on a one-dimensional subspace. In $\mathbb{R}^2$ we observe that, if $\mathbf{u}$ is a nonzero vector and $\mathbf{v}$ is a point not on the line through $\mathbf{u}$ and the origin, then the point $\mathbf{p}$ on that line that is the closest to $\mathbf{v}$ is characterized by the fact that $\langle \mathbf{v} - \mathbf{p}, \mathbf{u} \rangle = 0$. This property generalizes to any inner product space.

Theorem 3.2.1. *Let V be an inner product space and let $\mathbf{u}, \mathbf{v} \in V$. If $\mathbf{u} \neq \mathbf{0}$, then there is a unique vector $\mathbf{p} \in \mathrm{Span}\{\mathbf{u}\}$ such that*

$$\langle \mathbf{v} - \mathbf{p}, \mathbf{u} \rangle = 0.$$

Proof. If $\mathbf{p} \in \mathrm{Span}\{\mathbf{u}\}$, then there is a number $\alpha \in \mathbb{K}$ such that $\mathbf{p} = \alpha \mathbf{u}$. If $\langle \mathbf{v} - \mathbf{p}, \mathbf{u} \rangle = 0$, then

$$\langle \mathbf{v} - \mathbf{p}, \mathbf{u} \rangle = \langle \mathbf{v} - \alpha \mathbf{u}, \mathbf{u} \rangle = 0.$$

The equation $\langle \mathbf{v} - \alpha \mathbf{u}, \mathbf{u} \rangle = 0$ has a unique solution $\alpha = \dfrac{\langle \mathbf{v}, \mathbf{u} \rangle}{\langle \mathbf{u}, \mathbf{u} \rangle}$. Consequently, $\mathbf{p} = \dfrac{\langle \mathbf{v}, \mathbf{u} \rangle}{\langle \mathbf{u}, \mathbf{u} \rangle} \mathbf{u}$ is the unique vector in $\mathrm{Span}\{\mathbf{u}\}$ such that $\langle \mathbf{v} - \mathbf{p}, \mathbf{u} \rangle = 0$. $\qquad \square$

Definition 3.2.2. Let V be an inner product space and let $\mathbf{u} \in V$ be a nonzero vector. For $\mathbf{v} \in V$ we define

$$\mathrm{proj}_{\mathrm{Span}\{\mathbf{u}\}}(\mathbf{v}) = \frac{\langle \mathbf{v}, \mathbf{u} \rangle}{\langle \mathbf{u}, \mathbf{u} \rangle} \mathbf{u} = \frac{\langle \mathbf{v}, \mathbf{u} \rangle}{\|\mathbf{u}\|^2} \mathbf{u}.$$

The vector $\mathrm{proj}_{\mathrm{Span}\{\mathbf{u}\}}(\mathbf{v})$ is called the *orthogonal projection* of $\mathbf{v}$ on the subspace $\mathrm{Span}\{\mathbf{u}\}$.

If $\mathbf{u}$ is a unit vector, that is, $\|\mathbf{u}\| = 1$, then the expression for the projection on $\mathrm{Span}\{\mathbf{u}\}$ can be simplified:

$$\mathrm{proj}_{\mathrm{Span}\{\mathbf{u}\}}(\mathbf{v}) = \langle \mathbf{v}, \mathbf{u} \rangle \mathbf{u}.$$

Assuming that $\mathbf{u}$ is a unit vector is not restrictive, since for any $\mathbf{u} \neq \mathbf{0}$ we have

$$\mathrm{Span}\{\mathbf{u}\} = \mathrm{Span}\left\{ \frac{\mathbf{u}}{\|\mathbf{u}\|} \right\}.$$

Note that $\mathrm{proj}_{\mathrm{Span}\{\mathbf{u}\}} : V \to V$ is a linear operator.

Example 3.2.3. In the inner product space $\mathbb{C}^3$ find the projection of the vector $\begin{bmatrix} 1 \\ 1 \\ 1 \end{bmatrix}$ on $\mathrm{Span}\left\{ \begin{bmatrix} i \\ -1 \\ 1 \end{bmatrix} \right\}$.

Solution. Since

$$\left\langle \begin{bmatrix} i \\ -1 \\ 1 \end{bmatrix}, \begin{bmatrix} i \\ -1 \\ 1 \end{bmatrix} \right\rangle = 3 \quad \text{and} \quad \left\langle \begin{bmatrix} 1 \\ 1 \\ 1 \end{bmatrix}, \begin{bmatrix} i \\ -1 \\ 1 \end{bmatrix} \right\rangle = -i,$$

the projection of the vector $\begin{bmatrix} 1 \\ 1 \\ 1 \end{bmatrix}$ on Span $\left\{ \begin{bmatrix} i \\ -1 \\ 1 \end{bmatrix} \right\}$ is

$$\frac{-i}{3} \begin{bmatrix} i \\ -1 \\ 1 \end{bmatrix} = \begin{bmatrix} \frac{1}{3} \\ \frac{1}{3}i \\ -\frac{1}{3}i \end{bmatrix}.$$

Note that, as expected, we have

$$\left\langle \begin{bmatrix} 1 \\ 1 \\ 1 \end{bmatrix} - \begin{bmatrix} \frac{1}{3} \\ \frac{1}{3}i \\ -\frac{1}{3}i \end{bmatrix}, \begin{bmatrix} i \\ -1 \\ 1 \end{bmatrix} \right\rangle = 0.$$

$\square$

Example 3.2.4. Consider the vector space of continuous functions defined on the interval $[-\pi, \pi]$ with the inner product

$$\langle f, g \rangle = \frac{1}{\pi} \int_{-\pi}^{\pi} f(t)\overline{g(t)}\, dt.$$

Find the projection of the function $f(t) = t$ on Span$\{\sin nt\}$ where n is an arbitrary positive integer.

Solution. The projection is

$$\frac{\langle t, \sin nt \rangle}{\langle \sin nt, \sin nt \rangle} \sin nt.$$

Since

$$\int t \sin nt\, dt = t\left(-\frac{\cos nt}{n}\right) + \frac{1}{n}\int \cos nt\, dt = t\left(-\frac{\cos nt}{n}\right) + \frac{\sin nt}{n^2},$$

we have

$$\langle t, \sin nt \rangle = \frac{1}{\pi} \int_{-\pi}^{\pi} t \sin nt\, dt = (-1)^{n+1}\frac{2}{n}.$$

And, since

$$\langle \sin nt, \sin nt \rangle = \| \sin nt \|^2 = \frac{1}{\pi} \int_{-\pi}^{\pi} (\sin nt)^2 \, dt = \frac{1}{2\pi} \int_{-\pi}^{\pi} (1 - \cos 2nt) \, dt = 1,$$

the projection of the function t on $\mathrm{Span}\{\sin(nt)\}$ is

$$(-1)^{n+1} \frac{2}{n} \sin nt.$$

□

Example 3.2.5. In the vector space $\mathbb{C}^n$ with the standard inner product the projection of the vector $\begin{bmatrix} x_1 \\ \vdots \\ x_n \end{bmatrix}$ on $\mathrm{Span}\left\{ \begin{bmatrix} y_1 \\ \vdots \\ y_n \end{bmatrix} \right\}$ is

$$\frac{\left\langle \begin{bmatrix} x_1 \\ \vdots \\ x_n \end{bmatrix}, \begin{bmatrix} y_1 \\ \vdots \\ y_n \end{bmatrix} \right\rangle}{\left\| \begin{bmatrix} y_1 \\ \vdots \\ y_n \end{bmatrix} \right\|^2} \begin{bmatrix} y_1 \\ \vdots \\ y_n \end{bmatrix} = \frac{1}{\left\| \begin{bmatrix} y_1 \\ \vdots \\ y_n \end{bmatrix} \right\|^2} \left(\begin{bmatrix} \overline{y_1} & \cdots & \overline{y_n} \end{bmatrix} \begin{bmatrix} x_1 \\ \vdots \\ x_n \end{bmatrix} \right) \begin{bmatrix} y_1 \\ \vdots \\ y_n \end{bmatrix}$$

$$= \frac{1}{\left\| \begin{bmatrix} y_1 \\ \vdots \\ y_n \end{bmatrix} \right\|^2} \left(\begin{bmatrix} y_1 \\ \vdots \\ y_n \end{bmatrix} \begin{bmatrix} \overline{y_1} & \cdots & \overline{y_n} \end{bmatrix} \right) \begin{bmatrix} x_1 \\ \vdots \\ x_n \end{bmatrix}.$$

The matrix $\begin{bmatrix} y_1 \\ \vdots \\ y_n \end{bmatrix} \begin{bmatrix} \overline{y_1} & \cdots & \overline{y_n} \end{bmatrix}$ is called the *projection matrix* on $\mathrm{Span}\left\{ \begin{bmatrix} y_1 \\ \vdots \\ y_n \end{bmatrix} \right\}$.

The following theorem establishes properties of projections in general vector spaces that should look familiar from our experience with projections on lines in $\mathbb{R}^2$ and $\mathbb{R}^3$. Part (a) says that the projections is the unique vector minimizing the distance. In part (b) we say that the magnitude of the projection cannot exceed the magnitude of the original vector. And in part (c) we show that the projection and the original vector are the same if and only if the original vector is on the line.

Theorem 3.2.6. *Let $\mathcal{V}$ be an inner product space and let $\mathbf{u}$ be a nonzero vector in $\mathcal{V}$. For all $\mathbf{v} \in \mathcal{V}$ we have*

(a) $\|\mathbf{v} - \mathrm{proj}_{\mathrm{Span}\{\mathbf{u}\}}(\mathbf{v})\| < \|\mathbf{v} - \beta\mathbf{u}\|$ *for every $\beta \in \mathbb{K}$ such that* $\beta \neq \frac{\langle \mathbf{v},\mathbf{u}\rangle}{\langle \mathbf{u},\mathbf{u}\rangle}$;

(b) $\|\mathrm{proj}_{\mathrm{Span}\{\mathbf{u}\}}(\mathbf{v})\| \leq \|\mathbf{v}\|$;

(c) $\mathrm{proj}_{\mathrm{Span}\{\mathbf{u}\}}(\mathbf{v}) = \mathbf{v}$ *if and only if* $\mathbf{v} \in \mathrm{Span}\{\mathbf{u}\}$.

Proof. (a) From Theorem 3.2.1 we get

$$\left\langle \mathbf{v} - \mathrm{proj}_{\mathrm{Span}\{\mathbf{u}\}}(\mathbf{v}), \alpha\mathbf{u}\right\rangle = \overline{\alpha}\left\langle \mathbf{v} - \mathrm{proj}_{\mathrm{Span}\{\mathbf{u}\}}(\mathbf{v}), \mathbf{u}\right\rangle = 0$$

for any $\alpha \in \mathbb{C}$. Consequently, for any $\beta \in \mathbb{C}$ we have

$$\left\langle \mathbf{v} - \mathrm{proj}_{\mathrm{Span}\{\mathbf{u}\}}(\mathbf{v}), \mathrm{proj}_{\mathrm{Span}\{\mathbf{u}\}}(\mathbf{v}) - \beta\mathbf{u}\right\rangle = 0, \tag{3.1}$$

because $\mathrm{proj}_{\mathrm{Span}\{\mathbf{u}\}}(\mathbf{v}) - \beta\mathbf{u} \in \mathrm{Span}\{\mathbf{u}\}$. From (3.1) we get

$$\|\mathbf{v} - \mathrm{proj}_{\mathrm{Span}\{\mathbf{u}\}}(\mathbf{v})\|^2 + \|\mathrm{proj}_{\mathrm{Span}\{\mathbf{u}\}}(\mathbf{v}) - \beta\mathbf{u}\|^2 = \|\mathbf{v} - \beta\mathbf{u}\|^2. \tag{3.2}$$

Hence

$$\|\mathbf{v} - \mathrm{proj}_{\mathrm{Span}\{\mathbf{u}\}}(\mathbf{v})\| < \|\mathbf{v} - \beta\mathbf{u}\|$$

whenever $\|\mathrm{proj}_{\mathrm{Span}\{\mathbf{u}\}}(\mathbf{v}) - \beta\mathbf{u}\| \neq 0$, that is, for every $\beta \neq \frac{\langle \mathbf{v},\mathbf{u}\rangle}{\langle \mathbf{u},\mathbf{u}\rangle}$.

(b) If we let $\beta = 0$ in (3.2), then we get

$$\|\mathbf{v} - \mathrm{proj}_{\mathrm{Span}\{\mathbf{u}\}}(\mathbf{v})\|^2 + \|\mathrm{proj}_{\mathrm{Span}\{\mathbf{u}\}}(\mathbf{v})\|^2 = \|\mathbf{v}\|^2.$$

Consequently

$$\|\mathrm{proj}_{\mathrm{Span}\{\mathbf{u}\}}(\mathbf{v})\| \leq \|\mathbf{v}\|.$$

(c) If $\mathrm{proj}_{\mathrm{Span}\{\mathbf{u}\}}(\mathbf{v}) = \mathbf{v}$, then

$$\mathbf{v} = \mathrm{proj}_{\mathrm{Span}\{\mathbf{u}\}}(\mathbf{v}) \in \mathrm{Span}\{\mathbf{u}\}.$$

Now, if $\mathbf{v} \in \mathrm{Span}\{\mathbf{u}\}$, then $\mathbf{v} = \alpha\mathbf{u}$ for some $\alpha \in \mathbb{C}$ and thus

$$\mathrm{proj}_{\mathrm{Span}\{\mathbf{u}\}}(\mathbf{v}) = \frac{\langle \mathbf{v},\mathbf{u}\rangle}{\langle \mathbf{u},\mathbf{u}\rangle}\mathbf{u} = \frac{\langle \alpha\mathbf{u},\mathbf{u}\rangle}{\langle \mathbf{u},\mathbf{u}\rangle}\mathbf{u} = \alpha\mathbf{u} = \mathbf{v}.$$

$\square$

From the properties of projections proved in Theorem 3.2.6 we can obtain the results in Theorems 3.1.21 and 3.1.22 in a different way.

Theorem 3.2.7. *Let V be an inner product space. For all $\mathbf{x}, \mathbf{y} \in V$ we have*

(a) $|\langle \mathbf{x}, \mathbf{y} \rangle| \leq \|\mathbf{x}\| \|\mathbf{y}\|$;

(b) $|\langle \mathbf{x}, \mathbf{y} \rangle| = \|\mathbf{x}\| \|\mathbf{y}\|$ *if and only if* $\mathbf{x}$ *and* $\mathbf{y}$ *are linearly dependent.*

Proof. (a) Since the inequality is trivial when $\mathbf{x} = \mathbf{0}$, we can assume that $\mathbf{x} \neq \mathbf{0}$. By Theorem 3.2.6 we have $\|\text{proj}_{\text{Span}\{\mathbf{x}\}}(\mathbf{y})\| \leq \|\mathbf{y}\|$, which means that

$$\frac{|\langle \mathbf{x}, \mathbf{y} \rangle|}{\|\mathbf{x}\|^2} \|\mathbf{x}\| \leq \|\mathbf{y}\|.$$

Hence $|\langle \mathbf{x}, \mathbf{y} \rangle| \leq \|\mathbf{x}\| \|\mathbf{y}\|$.

(b) Again, without loss of generality, we can assume that $\mathbf{x} \neq \mathbf{0}$. Then the following statements are equivalent:

$\mathbf{x}$ and $\mathbf{y}$ are linearly dependent

$\mathbf{y} \in \text{Span}\{\mathbf{x}\}$

$\|\text{proj}_{\text{Span}\{\mathbf{x}\}}(\mathbf{y})\| = \|\mathbf{y}\|$ (by (c) in Theorem 3.2.6)

$\dfrac{|\langle \mathbf{x}, \mathbf{y} \rangle|}{\|\mathbf{x}\|^2} \|\mathbf{x}\| = \|\mathbf{y}\|$

$|\langle \mathbf{x}, \mathbf{y} \rangle| = \|\mathbf{x}\| \|\mathbf{y}\|$

$\square$

3.2.2 Orthogonal projections on arbitrary subspaces

Now we generalize orthogonal projections to arbitrary subspaces. While some properties remain the same, some aspects of projections become more complicated when the dimension of the subspace is more than one.

We use the property described in Theorem 3.2.1 to define an orthogonal projection on a subspace.

Definition 3.2.8. Let $\mathcal{U}$ be a subspace of an inner product space V and let $\mathbf{v} \in V$. A vector $\mathbf{p} \in \mathcal{U}$ is called an *orthogonal projection* of the vector $\mathbf{v}$ on the subspace $\mathcal{U}$ if

$$\langle \mathbf{v} - \mathbf{p}, \mathbf{u} \rangle = 0$$

for every vector $\mathbf{u} \in \mathcal{U}$.

Note that it is not clear if the projection always exists. This question will be addressed later.

> **Theorem 3.2.9.** *Let $\mathcal{U}$ be a subspace of an inner product space $\mathcal{V}$ and let $\mathbf{v} \in \mathcal{V}$. If an orthogonal projection of $\mathbf{v}$ on the subspace $\mathcal{U}$ exists, then it is unique.*

Proof. Assume that both $\mathbf{p}_1$ and $\mathbf{p}_2$ are orthogonal projections of $\mathbf{v}$ on the subspace $\mathcal{U}$, that is, $\langle \mathbf{v} - \mathbf{p}_1, \mathbf{u} \rangle = \langle \mathbf{v} - \mathbf{p}_2, \mathbf{u} \rangle = 0$ for every $\mathbf{u} \in \mathcal{U}$. Since $\mathbf{p}_1, \mathbf{p}_2 \in \mathcal{U}$, we have

$$0 = \langle \mathbf{v} - \mathbf{p}_1, \mathbf{p}_2 \rangle = \langle \mathbf{v}, \mathbf{p}_2 \rangle - \langle \mathbf{p}_1, \mathbf{p}_2 \rangle$$

and

$$0 = \langle \mathbf{v} - \mathbf{p}_2, \mathbf{p}_2 \rangle = \langle \mathbf{v}, \mathbf{p}_2 \rangle - \langle \mathbf{p}_2, \mathbf{p}_2 \rangle = \langle \mathbf{v}, \mathbf{p}_2 \rangle - \|\mathbf{p}_2\|^2.$$

Consequently, $\langle \mathbf{p}_1, \mathbf{p}_2 \rangle = \|\mathbf{p}_2\|^2$. Similarly, we can show that $\langle \mathbf{p}_2, \mathbf{p}_1 \rangle = \|\mathbf{p}_1\|^2$. Hence

$$\|\mathbf{p}_1 - \mathbf{p}_2\|^2 = \langle \mathbf{p}_1 - \mathbf{p}_2, \mathbf{p}_1 - \mathbf{p}_2 \rangle = \|\mathbf{p}_1\|^2 - \langle \mathbf{p}_1, \mathbf{p}_2 \rangle - \langle \mathbf{p}_2, \mathbf{p}_1 \rangle + \|\mathbf{p}_2\|^2 = 0,$$

proving that $\mathbf{p}_1 = \mathbf{p}_2$. $\qquad\square$

The unique orthogonal projection of a vector $\mathbf{v}$ on a subspace $\mathcal{U}$ is denoted by $\text{proj}_{\mathcal{U}}(\mathbf{v})$.

Example 3.2.10. Consider the vector space $\mathcal{V}$ of continuous functions defined on the interval $[-\pi, \pi]$ with the inner product

$$\langle f, g \rangle = \frac{1}{2\pi} \int_{-\pi}^{\pi} f(t)\overline{g(t)}\, dt$$

and its subspace $\mathcal{U} = \text{Span}\{1, \cos t, \sin t\}$. Show that the function $2\sin t$ is a projection of the function t on the subspace $\mathcal{U}$.

Solution. Since

$$\langle t - 2\sin t, 1 \rangle = \frac{1}{2\pi} \int_{-\pi}^{\pi} (t - 2\sin t)\, dt = 0,$$

$$\langle t - 2\sin t, \cos t \rangle = \frac{1}{2\pi} \int_{-\pi}^{\pi} (t - 2\sin t)\cos t\, dt = 0,$$

and

$$\langle t - 2\sin t, \sin t \rangle = \frac{1}{2\pi} \int_{-\pi}^{\pi} (t - 2\sin t)\sin t\, dt = 0,$$

we have $\text{proj}_{\mathcal{U}}(t) = 2\sin t$. $\qquad\square$

Example 3.2.11. Consider the vector space $\mathcal{V}$ of continuous functions defined on the interval $[-1, 1]$ with the inner product $\langle f, g \rangle = \int_{-1}^{1} f(t)\overline{g(t)}\, dt$. Find a projection of a function $f \in \mathcal{V}$ on the subspace $\mathcal{E}$ of even functions.

Solution. First we note that, if $f \in \mathcal{V}$, then the function $\frac{1}{2}(f(t) + f(-t))$ is even and, for any function $g \in \mathcal{E}$, the function $(f(t) - f(-t))g(t)$ is odd. Since

$$\int_{-1}^{1} \left(f(t) - \frac{1}{2}(f(t) + f(-t)) \right) \overline{g(t)}\, dt = \frac{1}{2} \int_{-1}^{1} (f(t) - f(-t))\overline{g(t)}\, dt = 0$$

for any function $g \in \mathcal{E}$, we conclude that $\operatorname{proj}_{\mathcal{E}}(f(t)) = \frac{1}{2}(f(t) + f(-t))$. $\quad\square$

Definition 3.2.12. Let $\mathcal{U}$ be a subspace of an inner product space $\mathcal{V}$ and let $\mathbf{v}$ be a vector in $\mathcal{V}$. A vector $\mathbf{p} \in \mathcal{U}$ is called the *best approximation* to the vector $\mathbf{v}$ by vectors from the subspace $\mathcal{U}$ if

$$\|\mathbf{v} - \mathbf{p}\| < \|\mathbf{v} - \mathbf{u}\|$$

for every $\mathbf{u} \in \mathcal{U}$ such that $\mathbf{u} \neq \mathbf{p}$.

Note that the definition of the best approximation implies that, if a vector has a best approximation by vectors from the subspace, then that approximation is unique.

A projection of a vector $\mathbf{v} \in \mathcal{V}$ on a subspace $\mathcal{U}$ can be interpreted as the best approximation of $\mathbf{v}$ by vectors from $\mathcal{U}$. This point of view is natural in many applications.

Theorem 3.2.13. *Let $\mathcal{U}$ be a subspace of an inner product space $\mathcal{V}$ and let $\mathbf{v}$ be a vector in $\mathcal{V}$. The following conditions are equivalent:*

(a) $\mathbf{p} \in \mathcal{U}$ *is the orthogonal projection of $\mathbf{v}$ on $\mathcal{U}$;*

(b) $\mathbf{p} \in \mathcal{U}$ *is the best approximation to the vector $\mathbf{v}$ by vectors from $\mathcal{U}$.*

In other words, $\langle \mathbf{v} - \mathbf{p}, \mathbf{u} \rangle = 0$ for every $\mathbf{u} \in \mathcal{U}$ if and only if $\|\mathbf{v} - \mathbf{p}\| < \|\mathbf{v} - \mathbf{u}\|$ for every $\mathbf{u} \in \mathcal{U}$ such that $\mathbf{u} \neq \mathbf{p}$.

Proof. If $\mathbf{p}$ is an orthogonal projection of $\mathbf{v}$ on $\mathcal{U}$, then $\langle \mathbf{v} - \mathbf{p}, \mathbf{u} \rangle = 0$ for every vector $\mathbf{u} \in \mathcal{U}$. Let $\mathbf{q}$ be an arbitrary vector from $\mathcal{U}$. Then

$$\|\mathbf{v} - \mathbf{q}\|^2 = \|\mathbf{v} - \mathbf{p} + \mathbf{p} - \mathbf{q}\|^2.$$

Since $\mathbf{p} - \mathbf{q} \in \mathcal{U}$, we have $\langle \mathbf{v} - \mathbf{p}, \mathbf{p} - \mathbf{q} \rangle = 0$ and thus

$$\|\mathbf{v} - \mathbf{q}\|^2 = \|\mathbf{v} - \mathbf{p}\|^2 + \|\mathbf{p} - \mathbf{q}\|^2,$$

by the Pythagorean Theorem 3.1.35. Hence the vector $\mathbf{q} = \mathbf{p}$ is the best approximation to the vector $\mathbf{v}$ by vectors from $\mathcal{U}$. This shows that (a) implies (b).

Now assume that the vector $\mathbf{p}$ is the best approximation to the vector $\mathbf{v}$ by vectors from $\mathcal{U}$. Let $\mathbf{u}$ be an arbitrary nonzero vector in $\mathcal{U}$ and let z be a number from $\mathbb{K}$. Then

$$\|\mathbf{v} - \mathbf{p} + z\mathbf{u}\|^2 \geq \|\mathbf{v} - \mathbf{p}\|^2$$

and consequently

$$|z|^2 \|\mathbf{u}\|^2 + \overline{z}\langle \mathbf{u}, \mathbf{v} - \mathbf{p} \rangle + z\overline{\langle \mathbf{v} - \mathbf{p}, \mathbf{u} \rangle} \geq 0.$$

If we take $z = t\langle \mathbf{v} - \mathbf{p}, \mathbf{u} \rangle$, where t is a real number, and suppose that $\langle \mathbf{v} - \mathbf{p}, \mathbf{u} \rangle \neq 0$, then we get

$$t^2 \|\mathbf{u}\|^2 + 2t \geq 0$$

for every real numbers t, which is not possible because $\|\mathbf{u}\| \neq 0$, and thus $\langle \mathbf{v} - \mathbf{p}, \mathbf{u} \rangle = 0$. Since $\mathbf{u}$ is an arbitrary nonzero vector in $\mathcal{U}$, $\mathbf{p}$ is an orthogonal projection of $\mathbf{v}$ on $\mathcal{U}$. Therefore (b) implies (a). $\square$

The next theorem can be helpful when calculating the projection of a vector on the subspace spanned by vectors $\mathbf{u}_1, \ldots, \mathbf{u}_k$.

Theorem 3.2.14. *Let* $\mathbf{u}_1, \mathbf{u}_2, \ldots, \mathbf{u}_k$ *be vectors in an inner product space* V *and let* $\mathbf{v} \in V$. *The following conditions are equivalent:*

(a) $\mathrm{proj}_{\mathrm{Span}\{\mathbf{u}_1, \ldots, \mathbf{u}_k\}}(\mathbf{v}) = x_1\mathbf{u}_1 + x_2\mathbf{u}_2 + \cdots + x_k\mathbf{u}_k$;

(b) $\begin{cases} x_1\langle \mathbf{u}_1, \mathbf{u}_1 \rangle + x_2\langle \mathbf{u}_2, \mathbf{u}_1 \rangle + \cdots + x_k\langle \mathbf{u}_k, \mathbf{u}_1 \rangle = \langle \mathbf{v}, \mathbf{u}_1 \rangle \\ x_1\langle \mathbf{u}_1, \mathbf{u}_2 \rangle + x_2\langle \mathbf{u}_2, \mathbf{u}_2 \rangle + \cdots + x_k\langle \mathbf{u}_k, \mathbf{u}_2 \rangle = \langle \mathbf{v}, \mathbf{u}_2 \rangle \\ \quad \vdots \qquad\qquad \vdots \qquad \vdots \qquad\qquad \vdots \\ x_1\langle \mathbf{u}_1, \mathbf{u}_k \rangle + x_2\langle \mathbf{u}_2, \mathbf{u}_k \rangle + \cdots + x_k\langle \mathbf{u}_k, \mathbf{u}_k \rangle = \langle \mathbf{v}, \mathbf{u}_k \rangle \end{cases}$

Proof. $\mathrm{proj}_{\mathrm{Span}\{\mathbf{u}_1, \ldots, \mathbf{u}_k\}}(\mathbf{v}) = x_1\mathbf{u}_1 + \cdots + x_k\mathbf{u}_k$ if and only if $\langle \mathbf{v} - \mathbf{p}, \mathbf{u} \rangle = 0$ for every $\mathbf{u} \in \mathrm{Span}\{\mathbf{u}_1, \ldots, \mathbf{u}_k\}$. Since every $\mathbf{u} \in \mathrm{Span}\{\mathbf{u}_1, \ldots, \mathbf{u}_k\}$ is of the form $\mathbf{u} = y_1\mathbf{u}_1 + \cdots + y_k\mathbf{u}_k$, for some $y_1, \ldots, y_k \in \mathbb{K}$, the equation $\langle \mathbf{v} - \mathbf{p}, \mathbf{u} \rangle = 0$ is equivalent to the equations

$$\langle \mathbf{v} - \mathbf{p}, \mathbf{u}_1 \rangle = 0, \langle \mathbf{v} - \mathbf{p}, \mathbf{u}_2 \rangle = 0, \ldots, \langle \mathbf{v} - \mathbf{p}, \mathbf{u}_k \rangle = 0$$

or

$$\langle \mathbf{v} - (x_1\mathbf{u}_1 + \cdots + x_k\mathbf{u}_k), \mathbf{u}_1 \rangle = 0,$$
$$\langle \mathbf{v} - (x_1\mathbf{u}_1 + \cdots + x_k\mathbf{u}_k), \mathbf{u}_2 \rangle = 0,$$
$$\vdots$$
$$\langle \mathbf{v} - (x_1\mathbf{u}_1 + \cdots + x_k\mathbf{u}_k), \mathbf{u}_k \rangle = 0,$$

which can also be written as

$$x_1\langle \mathbf{u}_1, \mathbf{u}_1 \rangle + \cdots + x_k\langle \mathbf{u}_k, \mathbf{u}_1 \rangle = \langle \mathbf{v}, \mathbf{u}_1 \rangle,$$
$$x_1\langle \mathbf{u}_1, \mathbf{u}_2 \rangle + \cdots + x_k\langle \mathbf{u}_k, \mathbf{u}_2 \rangle = \langle \mathbf{v}, \mathbf{u}_2 \rangle,$$
$$\vdots$$
$$x_1\langle \mathbf{u}_1, \mathbf{u}_k \rangle + \cdots + x_k\langle \mathbf{u}_k, \mathbf{u}_k \rangle = \langle \mathbf{v}, \mathbf{u}_k \rangle.$$

$\square$

Example 3.2.15. Find the projection of the vector $\begin{bmatrix} 0 \\ 1 \\ 0 \end{bmatrix}$ on Span $\left\{ \begin{bmatrix} 1 \\ 1 \\ 1 \end{bmatrix}, \begin{bmatrix} i \\ 0 \\ 0 \end{bmatrix} \right\}$

in the inner product space $\mathbb{C}^3$.

Solution. We have to solve the system

$$\begin{cases} x_1\langle \mathbf{u}_1, \mathbf{u}_1 \rangle & + & x_2\langle \mathbf{u}_2, \mathbf{u}_1 \rangle & = & \langle \mathbf{v}, \mathbf{u}_1 \rangle \\ x_1\langle \mathbf{u}_1, \mathbf{u}_2 \rangle & + & x_2\langle \mathbf{u}_2, \mathbf{u}_2 \rangle & = & \langle \mathbf{v}, \mathbf{u}_2 \rangle \end{cases},$$

where $\mathbf{u}_1 = \begin{bmatrix} 1 \\ 1 \\ 1 \end{bmatrix}$, $\mathbf{u}_2 = \begin{bmatrix} i \\ 0 \\ 0 \end{bmatrix}$, and $\mathbf{v} = \begin{bmatrix} 0 \\ 1 \\ 0 \end{bmatrix}$, that is, the system

$$\begin{cases} 3x_1 & + & ix_2 & = & 1 \\ ix_1 & - & x_2 & = & 0 \end{cases}.$$

Since the solutions are $x_1 = \frac{1}{2}$ and $x_2 = \frac{i}{2}$, the projection is

$$\frac{1}{2}\begin{bmatrix} 1 \\ 1 \\ 1 \end{bmatrix} + \frac{i}{2}\begin{bmatrix} i \\ 0 \\ 0 \end{bmatrix} = \begin{bmatrix} 0 \\ \frac{1}{2} \\ \frac{1}{2} \end{bmatrix}.$$

It is easy to verify that $\left\langle \mathbf{v} - \begin{bmatrix} 0 \\ \frac{1}{2} \\ \frac{1}{2} \end{bmatrix}, \mathbf{u}_1 \right\rangle = 0$ and $\left\langle \mathbf{v} - \begin{bmatrix} 0 \\ \frac{1}{2} \\ \frac{1}{2} \end{bmatrix}, \mathbf{u}_2 \right\rangle = 0.$ □

We are now in a position to prove that the orthogonal projection on a finite dimensional subspace of an inner product space always exists. We show that by showing that the system of equations in Theorem 3.2.14 always has a unique solution.

> **Theorem 3.2.16.** *Let $\mathcal{U}$ be a finite dimensional subspace of an inner product space $\mathcal{V}$. For any vector $\mathbf{v} \in \mathcal{V}$ the orthogonal projection of $\mathbf{v}$ on the subspace $\mathcal{U}$ exists.*

Proof. Let $\mathcal{U} = \text{Span}\{\mathbf{u}_1, \ldots, \mathbf{u}_k\}$. Without loss of generality we can suppose that the vectors $\mathbf{u}_1, \ldots, \mathbf{u}_k$ are linearly independent. In view of Theorem 3.2.14 it suffices to show that the matrix

$$\begin{bmatrix} \langle \mathbf{u}_1, \mathbf{u}_1 \rangle & \cdots & \langle \mathbf{u}_k, \mathbf{u}_1 \rangle \\ \vdots & & \vdots \\ \langle \mathbf{u}_1, \mathbf{u}_k \rangle & \cdots & \langle \mathbf{u}_k, \mathbf{u}_k \rangle \end{bmatrix}$$

is invertible. If

$$\begin{bmatrix} \langle \mathbf{u}_1, \mathbf{u}_1 \rangle & \cdots & \langle \mathbf{u}_k, \mathbf{u}_1 \rangle \\ \vdots & & \vdots \\ \langle \mathbf{u}_1, \mathbf{u}_k \rangle & \cdots & \langle \mathbf{u}_k, \mathbf{u}_k \rangle \end{bmatrix} \begin{bmatrix} x_1 \\ \vdots \\ x_k \end{bmatrix} = \begin{bmatrix} 0 \\ \vdots \\ 0 \end{bmatrix},$$

for some $x_1, \ldots, x_k \in \mathbb{K}$, then

$$\langle x_1 \mathbf{u}_1 + \cdots + x_k \mathbf{u}_k, \mathbf{u}_j \rangle = 0$$

for $1 \leq j \leq k$. Hence

$$\langle x_1 \mathbf{u}_1 + \cdots + x_k \mathbf{u}_k, x_1 \mathbf{u}_1 + \cdots + x_k \mathbf{u}_k \rangle = \|x_1 \mathbf{u}_1 + \cdots + x_k \mathbf{u}_k\|^2 = 0$$

and thus

$$x_1 \mathbf{u}_1 + \cdots + x_k \mathbf{u}_k = \mathbf{0}.$$

Since the vectors $\mathbf{u}_1, \ldots, \mathbf{u}_k$ are linearly independent, we get $x_1 = \cdots = x_k = 0$.
                                                                                              □

The assumption that the subspace $\mathcal{U}$ in the above theorem is finite dimensional is essential. For infinite dimensional subspaces the projection may not exist. For example, consider the space $\mathcal{V}$ of all continuous functions on the interval $[0, 1]$ with the inner product $\langle f, g \rangle = \int_0^1 f(t)\overline{g(t)}\, dt$ and the subspace $\mathcal{U}$ of all polynomials. Since there is no polynomial p such that

$$\int_0^1 \left(e^t - p(t)\right) \overline{q(t)}\, dt = 0$$

for every polynomial q, the function e^t does not have an orthogonal projection on the subspace of all polynomials.

3.2.3 Calculations and applications of orthogonal projections

Theorem 3.2.14 gives us a method for effectively calculating projections on subspaces spanned by arbitrary vectors $\mathbf{u}_1, \ldots, \mathbf{u}_k$. It turns out that the calculations are significantly simplified if the vectors $\mathbf{u}_1, \ldots, \mathbf{u}_k$ are orthogonal.

Definition 3.2.17. Let $\mathbf{v}_1, \ldots, \mathbf{v}_k$ be vectors in an inner product space $\mathcal{V}$. We say that the set $\{\mathbf{v}_1, \ldots, \mathbf{v}_k\}$ is an *orthogonal set* if $\langle \mathbf{v}_i, \mathbf{v}_j \rangle = 0$ for all $i, j = 1, \ldots, k$ such that $i \neq j$.
An orthogonal set $\{\mathbf{v}_1, \ldots, \mathbf{v}_k\}$ is called an *orthonormal set* if $\|\mathbf{v}_i\| = 1$ for all $i = 1, \ldots, k$.

The condition of orthonormality is often expressed in terms of the *Kronecker delta* function:

$$\delta_{ij} = \begin{cases} 1 & \text{if } i = j, \\ 0 & \text{if } i \neq j. \end{cases}$$

Using the Kronecker delta function we can say that a set $\{\mathbf{v}_1, \ldots, \mathbf{v}_k\}$ is orthonormal if $\langle \mathbf{v}_i, \mathbf{v}_j \rangle = \delta_{ij}$ for all $i, j \in \{1, \ldots, k\}$.

Theorem 3.2.18. *Let $\{\mathbf{u}_1, \ldots, \mathbf{u}_k\}$ be an orthogonal set of nonzero vectors in an inner product space $\mathcal{V}$ and let $\mathcal{U} = \text{Span}\{\mathbf{u}_1, \ldots, \mathbf{u}_k\}$. Then*

$$\text{proj}_{\mathcal{U}} \mathbf{v} = \frac{\langle \mathbf{v}, \mathbf{u}_1 \rangle}{\langle \mathbf{u}_1, \mathbf{u}_1 \rangle} \mathbf{u}_1 + \cdots + \frac{\langle \mathbf{v}, \mathbf{u}_k \rangle}{\langle \mathbf{u}_k, \mathbf{u}_k \rangle} \mathbf{u}_k$$

for every vector $\mathbf{v}$ in $\mathcal{V}$.

Proof. Let

$$\mathbf{p} = \frac{\langle \mathbf{v}, \mathbf{u}_1 \rangle}{\langle \mathbf{u}_1, \mathbf{u}_1 \rangle} \mathbf{u}_1 + \cdots + \frac{\langle \mathbf{v}, \mathbf{u}_k \rangle}{\langle \mathbf{u}_k, \mathbf{u}_k \rangle} \mathbf{u}_k.$$

Then, for every $j \in \{1, \dots, k\}$, we have

$$\langle \mathbf{v} - \mathbf{p}, \mathbf{u}_j \rangle = \left\langle \mathbf{v} - \frac{\langle \mathbf{v}, \mathbf{u}_1 \rangle}{\langle \mathbf{u}_1, \mathbf{u}_1 \rangle} \mathbf{u}_1 - \cdots - \frac{\langle \mathbf{v}, \mathbf{u}_k \rangle}{\langle \mathbf{u}_k, \mathbf{u}_k \rangle} \mathbf{u}_k, \mathbf{u}_j \right\rangle$$

$$= \langle \mathbf{v}, \mathbf{u}_j \rangle - \frac{\langle \mathbf{v}, \mathbf{u}_1 \rangle}{\langle \mathbf{u}_1, \mathbf{u}_1 \rangle} \langle \mathbf{u}_1, \mathbf{u}_j \rangle - \cdots - \frac{\langle \mathbf{v}, \mathbf{u}_k \rangle}{\langle \mathbf{u}_k, \mathbf{u}_k \rangle} \langle \mathbf{u}_k, \mathbf{u}_j \rangle$$

$$= \langle \mathbf{v}, \mathbf{u}_j \rangle - \frac{\langle \mathbf{v}, \mathbf{u}_j \rangle}{\langle \mathbf{u}_j, \mathbf{u}_j \rangle} \langle \mathbf{u}_j, \mathbf{u}_j \rangle = 0.$$

Since every $\mathbf{u} \in \mathrm{Span}\{\mathbf{u}_1, \dots, \mathbf{u}_k\}$ is of the form

$$\mathbf{u} = x_1 \mathbf{u}_1 + \cdots + x_k \mathbf{u}_k,$$

for some $x_1, x_2, \dots, x_k \in \mathbb{K}$, it follows that

$$\langle \mathbf{v} - \mathbf{p}, \mathbf{u} \rangle = 0$$

for every $\mathbf{u} \in \mathrm{Span}\{\mathbf{u}_1, \dots, \mathbf{u}_k\}$, which means that $\mathbf{p} = \mathrm{proj}_{\mathcal{U}} \mathbf{v}$. □

Note that Theorem 3.2.18 implies that, if $\{\mathbf{u}_1, \dots, \mathbf{u}_k\}$ is an orthogonal set of nonzero vectors, then the best approximation to the vector $\mathbf{v}$ by vectors from the subspace $\mathrm{Span}\{\mathbf{u}_1, \dots, \mathbf{u}_k\}$ is the vector

$$\frac{\langle \mathbf{v}, \mathbf{u}_1 \rangle}{\langle \mathbf{u}_1, \mathbf{u}_1 \rangle} \mathbf{u}_1 + \cdots + \frac{\langle \mathbf{v}, \mathbf{u}_k \rangle}{\langle \mathbf{u}_k, \mathbf{u}_k \rangle} \mathbf{u}_k.$$

If $\{\mathbf{u}_1, \dots, \mathbf{u}_k\}$ is an orthonormal set, then the formula for the projection becomes even simpler.

Corollary 3.2.19. *Let* $\{\mathbf{u}_1, \dots, \mathbf{u}_k\}$ *be an orthonormal set in an inner product space* $\mathcal{V}$ *and let* $\mathcal{U} = \mathrm{Span}\{\mathbf{u}_1, \dots, \mathbf{u}_k\}$. *Then*

$$\mathrm{proj}_{\mathcal{U}} \mathbf{v} = \langle \mathbf{v}, \mathbf{u}_1 \rangle \mathbf{u}_1 + \cdots + \langle \mathbf{v}, \mathbf{u}_k \rangle \mathbf{u}_k$$

for every vector $\mathbf{v}$ *in* $\mathcal{V}$.

Example 3.2.20. Consider the vector space of continuous functions defined on the interval $[\alpha, \beta]$ with the inner product

$$\langle f, g \rangle = \frac{2}{\beta - \alpha} \int_{\alpha}^{\beta} f(t) \overline{g(t)} \, dt.$$

Show that the set $\left\{ \cos \frac{2\pi t}{\beta - \alpha}, \sin \frac{10\pi t}{\beta - \alpha} \right\}$ is an orthonormal set and thus for every function f continuous on the interval $[\alpha, \beta]$ we have

$$\text{proj}_{\mathcal{U}}(f) = \frac{2}{\beta - \alpha} \int_{\alpha}^{\beta} f(t) \cos \frac{2\pi t}{\beta - \alpha} dt \cos \frac{2\pi t}{\beta - \alpha} + \frac{2}{\beta - \alpha} \int_{\alpha}^{\beta} f(t) \sin \frac{10\pi t}{\beta - \alpha} dt \sin \frac{10\pi t}{\beta - \alpha},$$

where $\mathcal{U} = \text{Span} \left\{ \cos \frac{2\pi t}{\beta - \alpha}, \sin \frac{10\pi t}{\beta - \alpha} \right\}$.

Solution. First we recall that for an arbitrary positive integer k we have

$$\int_{\alpha}^{\beta} \sin \frac{2k\pi t}{\beta - \alpha} dt = \frac{\alpha - \beta}{2k\pi} \left(\cos \frac{2k\pi \alpha}{\beta - \alpha} - \cos \frac{2k\pi \beta}{\beta - \alpha} \right)$$

$$= \frac{\alpha - \beta}{2k\pi} \left(\cos \frac{2k\pi (\alpha - \beta + \beta)}{\beta - \alpha} - \cos \frac{2k\pi \beta}{\beta - \alpha} \right)$$

$$= \frac{\alpha - \beta}{2k\pi} \left(\cos \frac{2k\pi (\alpha - \beta)}{\beta - \alpha} \cos \frac{2k\pi \beta}{\beta - \alpha} - \sin \frac{2k\pi (\alpha - \beta)}{\beta - \alpha} \sin \frac{2k\pi \beta}{\beta - \alpha} - \cos \frac{2k\pi \beta}{\beta - \alpha} \right) = 0.$$

Now we use the trigonometric identity

$$\sin \alpha \cos \beta = \frac{1}{2} (\sin(\alpha + \beta) + \sin(\alpha - \beta))$$

and the above result to calculate the inner product:

$$\left\langle \cos \frac{2\pi t}{\beta - \alpha}, \sin \frac{10\pi t}{\beta - \alpha} \right\rangle = \frac{2}{\beta - \alpha} \int_{\alpha}^{\beta} \cos \frac{2\pi t}{\beta - \alpha} \sin \frac{10\pi t}{\beta - \alpha} dt$$

$$= \frac{1}{\beta - \alpha} \int_{\alpha}^{\beta} \frac{1}{2} \left(\sin \frac{12\pi t}{\beta - \alpha} + \sin \frac{8\pi t}{\beta - \alpha} \right) dt = 0.$$

This proves orthogonality of the set. Using a similar approach we can show that

$$\left\| \cos \frac{2\pi t}{\beta - \alpha} \right\|^2 = \left\| \sin \frac{10\pi t}{\beta - \alpha} \right\|^2 = 1,$$

which completes the proof of orthonormality. Then the formula for the inner product $\langle f, g \rangle$ follows from Corollary 3.2.19. $\qquad \square$

Example 3.2.21. Let $\mathcal{P}_2([-1, 1])$ be the space of complex valued polynomials on the interval $[-1, 1]$ of degree at most 2 with the inner product defined as

$$\langle p(t), q(t) \rangle = \int_{-1}^{1} p(t) q(t) \, dt.$$

Show that $\left\{ \frac{1}{\sqrt{2}}, \sqrt{\frac{3}{2}} t, \sqrt{\frac{5}{2}} \left(\frac{3}{2} t^2 - \frac{1}{2} \right) \right\}$ is an orthonormal set in $\mathcal{P}_2([-1, 1])$.

Solution. We need to calculate the following integrals:

$$\int_{-1}^{1} \frac{1}{\sqrt{2}} \sqrt{\frac{3}{2}} t \, dt = \frac{\sqrt{3}}{2} \int_{-1}^{1} t \, dt = 0,$$

$$\int_{-1}^{1} \frac{1}{\sqrt{2}} \sqrt{\frac{5}{2}} \left(\frac{3}{2} t^2 - \frac{1}{2} \right) dt = \frac{\sqrt{5}}{4} \int_{-1}^{1} (3t^2 - 1) \, dt = 0,$$

$$\int_{-1}^{1} \sqrt{\frac{3}{2}} t \sqrt{\frac{5}{2}} \left(\frac{3}{2} t^2 - \frac{1}{2} \right) dt = \frac{\sqrt{15}}{4} \int_{-1}^{1} t(3t^2 - 1) \, dt = 0,$$

$$\int_{-1}^{1} \left(\frac{1}{\sqrt{2}} \right)^2 dt = \frac{1}{2} \int_{-1}^{1} 1 \, dt = 1,$$

$$\int_{-1}^{1} \left(\sqrt{\frac{3}{2}} t \right)^2 dt = \frac{3}{2} \int_{-1}^{1} t^2 \, dt = 1,$$

$$\int_{-1}^{1} \left(\sqrt{\frac{5}{2}} \left(\frac{3}{2} t^2 - \frac{1}{2} \right) \right)^2 dt = \frac{5}{8} \int_{-1}^{1} (3t^2 - 1)^2 \, dt = 1.$$

□

Example 3.2.22. Let $\mathcal{V}$ be an inner product space and let $\{\mathbf{v}_1, \mathbf{v}_2\}$ be an orthonormal set in $\mathcal{V}$. Show that the set $\left\{ \frac{1}{\sqrt{2}}(\mathbf{v}_2 + \mathbf{v}_1), \frac{1}{\sqrt{2}}(\mathbf{v}_2 - \mathbf{v}_1) \right\}$ is orthonormal.

Solution. We have

$$\left\langle \frac{1}{\sqrt{2}}(\mathbf{v}_2 + \mathbf{v}_1), \frac{1}{\sqrt{2}}(\mathbf{v}_2 - \mathbf{v}_1) \right\rangle = \frac{1}{2}(\langle \mathbf{v}_2, \mathbf{v}_2 \rangle - \langle \mathbf{v}_2, \mathbf{v}_1 \rangle + \langle \mathbf{v}_2, \mathbf{v}_1 \rangle - \langle \mathbf{v}_1, \mathbf{v}_1 \rangle) = 0,$$

$$\left\| \frac{1}{\sqrt{2}}(\mathbf{v}_2 + \mathbf{v}_1) \right\|^2 = \left\langle \frac{1}{\sqrt{2}}(\mathbf{v}_2 + \mathbf{v}_1), \frac{1}{\sqrt{2}}(\mathbf{v}_2 + \mathbf{v}_1) \right\rangle$$

$$= \frac{1}{2}(\langle \mathbf{v}_2, \mathbf{v}_2 \rangle + \langle \mathbf{v}_1, \mathbf{v}_2 \rangle + \langle \mathbf{v}_2, \mathbf{v}_1 \rangle + \langle \mathbf{v}_1, \mathbf{v}_1 \rangle)$$

$$= \frac{1}{2}(\langle \mathbf{v}_2, \mathbf{v}_2 \rangle + \langle \mathbf{v}_1, \mathbf{v}_1 \rangle) = 1,$$

and similarly

$$\left\| \frac{1}{\sqrt{2}}(\mathbf{v}_2 - \mathbf{v}_1) \right\|^2 = \left\langle \frac{1}{\sqrt{2}}(\mathbf{v}_2 - \mathbf{v}_1), \frac{1}{\sqrt{2}}(\mathbf{v}_2 - \mathbf{v}_1) \right\rangle = 1.$$

□

For many results on subspaces $\mathcal{U} = \text{Span}\{\mathbf{u}_1, \ldots, \mathbf{u}_k\}$ it was necessary to assume that the vectors $\mathbf{u}_1, \ldots, \mathbf{u}_k$ were linearly independent. It turns out that, if $\mathbf{u}_1, \ldots, \mathbf{u}_k$ are nonzero orthogonal vectors, then they are always linearly independent. Consequently, any orthonormal set is linearly independent.

Theorem 3.2.23. *If* $\{\mathbf{v}_1, \ldots, \mathbf{v}_k\}$ *is an orthogonal set of nonzero vectors in an inner product space* $\mathcal{V}$, *then the vectors* $\mathbf{v}_1, \ldots, \mathbf{v}_k$ *are linearly independent.*

Proof. If $\{\mathbf{v}_1, \ldots, \mathbf{v}_k\}$ is an orthogonal set of nonzero vectors and

$$x_1\mathbf{v}_1 + \cdots + x_k\mathbf{v}_k = \mathbf{0}$$

for some numbers $x_1, \ldots, x_k \in \mathbb{K}$, then for any $j \in \{1, \ldots, k\}$ we have

$$\langle x_1\mathbf{v}_1 + x_2\mathbf{v}_2 + \cdots + x_k\mathbf{v}_k, \mathbf{v}_j \rangle = x_1\langle \mathbf{v}_1, \mathbf{v}_j \rangle + x_2\langle \mathbf{v}_2, \mathbf{v}_j \rangle + \cdots + x_k\langle \mathbf{v}_k, \mathbf{v}_j \rangle$$
$$= x_j\langle \mathbf{v}_j, \mathbf{v}_j \rangle = x_1\|\mathbf{v}_1\|^2.$$

On the other hand,

$$\langle x_1\mathbf{v}_1 + x_2\mathbf{v}_2 + \cdots + x_k\mathbf{v}_k, \mathbf{v}_j \rangle = \langle \mathbf{0}, \mathbf{v}_j \rangle = 0.$$

Since $x_j\|\mathbf{v}_1\|^2 = 0$ and $\|\mathbf{v}_j\| \neq 0$, we must have $x_j = 0$. Consequently, the vectors $\mathbf{v}_1, \ldots, \mathbf{v}_k$ are linearly independent. $\square$

3.2.4 The annihilator and the orthogonal complement

In Chapter 1 we introduced the notion of complementary subspaces: If $\mathcal{U}$ is a subspace of a vector space $\mathcal{V}$, then a subspace $\mathcal{W}$ is called a complement of $\mathcal{U}$ in $\mathcal{V}$ if $\mathcal{V} = \mathcal{U} \oplus \mathcal{W}$. We pointed out that such space is not unique. In inner product spaces we can define orthogonal complements that have better properties.

Definition 3.2.24. Let $\mathcal{A}$ be a nonempty subset of an inner product space $\mathcal{V}$. The set of all vectors in $\mathcal{V}$ orthogonal to every vector in $\mathcal{A}$ is called the *annihilator* of $\mathcal{A}$ and is denoted by $\mathcal{A}^\perp$:

$$\mathcal{A}^\perp = \{\mathbf{x} \in \mathcal{V} : \langle \mathbf{x}, \mathbf{v} \rangle = 0 \text{ for every } \mathbf{v} \in \mathcal{V}\}.$$

If $\mathcal{U}$ is a subspace of $\mathcal{V}$, then $\mathcal{U}^\perp$ is called the *orthogonal complement* of $\mathcal{U}$.

From the definition of the annihilator and basic properties of the inner product we get the following useful result.

Theorem 3.2.25. *Let $\mathcal{A}$ be a subset of an inner product space $\mathcal{V}$. The annihilator $\mathcal{A}^\perp$ is a subspace of $\mathcal{V}$.*

Example 3.2.26. Show that

$$\langle f, g \rangle = f(0)\overline{g(0)} + f'(0)\overline{g'(0)} + f''(0)\overline{g''(0)}$$

is an inner product in the vector space $\mathcal{P}_2(\mathbb{C})$ and determine $(\operatorname{Span}\{t^2 + 1\})^\perp$.

Solution. First we note that $\langle f, g \rangle$ is an inner product because

$$\langle \alpha_1 t^2 + \beta_1 t + \gamma_1, \alpha_2 t^2 + \beta_2 t + \gamma_3 \rangle = \gamma_1\overline{\gamma_2} + \beta_1\overline{\beta_2} + 4\alpha_1\overline{\alpha_2}.$$

Now we describe $(\operatorname{Span}\{t^2 + 1\})^\perp$. Since

$$\langle \alpha t^2 + \beta t + \gamma, t^2 + 1 \rangle = 4\alpha + \gamma,$$

we have

$$(\operatorname{Span}\{t^2 + 1\})^\perp = \{\alpha t^2 + \beta t + \gamma : 4\alpha + \gamma = 0\} = \operatorname{Span}\{t, t^2 - 4\}.$$

$\square$

If $\mathcal{U}$ is a subspace of an inner product space $\mathcal{V}$, then $\mathcal{U}^\perp$ is a subspace of $\mathcal{V}$, so it makes sense to consider the subspace $(\mathcal{U}^\perp)^\perp$. How is this subspace related to $\mathcal{U}$? If $\mathbf{u} \in \mathcal{U}$, then $\mathbf{u}$ is orthogonal to every vector in $\mathcal{U}^\perp$, so $\mathbf{u} \in (\mathcal{U}^\perp)^\perp$. This means that $\mathcal{U} \subseteq (\mathcal{U}^\perp)^\perp$. In general, $\mathcal{U}$ and $(\mathcal{U}^\perp)^\perp$ need not be equal. For example, consider the space $\mathcal{V}$ of all continuous functions on the interval $[0, 1]$ with the inner product $\langle f, g \rangle = \int_0^1 f(t)\overline{g(t)}\,dt$ and the subspace $\mathcal{U}$ of all polynomials. It can be shown that, for any continuous function f, if

$$\int_0^1 f(t)\overline{q(t)}\,dt = 0$$

for every polynomial q, then $f = 0$. This means that $\mathcal{U}^\perp = \{\mathbf{0}\}$ and thus $(\mathcal{U}^\perp)^\perp = \mathcal{V} \neq \mathcal{U}$.

If we assume that $\mathcal{U}$ is finite dimensional, then we can show that $\mathcal{U}$ and $(\mathcal{U}^\perp)^\perp$ are equal.

Theorem 3.2.27. *If $\mathcal{U}$ is a finite dimensional subspace of an inner product space $\mathcal{V}$, then*

$$(\mathcal{U}^\perp)^\perp = \mathcal{U}.$$

Proof. We need to show that $(\mathcal{U}^\perp)^\perp \subseteq \mathcal{U}$. Let $\mathbf{v} \in (\mathcal{U}^\perp)^\perp$. If $\mathcal{U}$ is finite dimensional, then $\text{proj}_\mathcal{U}(\mathbf{v})$ exists, by Theorem 3.2.16. Since $\langle \mathbf{w}, \mathbf{v} \rangle = 0$ for every $\mathbf{w} \in \mathcal{U}^\perp$ and $\mathbf{v} - \text{proj}_\mathcal{U}(\mathbf{v}) \in \mathcal{U}^\perp$, we have $\langle \mathbf{v} - \text{proj}_\mathcal{U}(\mathbf{v}), \mathbf{v} \rangle = 0$. Consequently,

$$
\begin{aligned}
0 &= \langle \mathbf{v} - \text{proj}_\mathcal{U}(\mathbf{v}), \mathbf{v} \rangle \\
&= \langle \mathbf{v} - \text{proj}_\mathcal{U}(\mathbf{v}), \mathbf{v} - \text{proj}_\mathcal{U}(\mathbf{v}) + \text{proj}_\mathcal{U}(\mathbf{v}) \rangle \\
&= \langle \mathbf{v} - \text{proj}_\mathcal{U}(\mathbf{v}), \mathbf{v} - \text{proj}_\mathcal{U}(\mathbf{v}) \rangle + \langle \mathbf{v} - \text{proj}_\mathcal{U}(\mathbf{v}), \text{proj}_\mathcal{U}(\mathbf{v}) \rangle \\
&= \langle \mathbf{v} - \text{proj}_\mathcal{U}(\mathbf{v}), \mathbf{v} - \text{proj}_\mathcal{U}(\mathbf{v}) \rangle \\
&= \| \mathbf{v} - \text{proj}_\mathcal{U}(\mathbf{v}) \|^2,
\end{aligned}
$$

which means that $\mathbf{v} - \text{proj}_\mathcal{U}(\mathbf{v}) = \mathbf{0}$. Thus $\mathbf{v} = \text{proj}_\mathcal{U}(\mathbf{v})$, which implies $\mathbf{v} \in \mathcal{U}$. $\square$

Theorem 3.2.28. *Let V be an inner product space and let $\mathcal{U}$ be a finite dimensional subspace of V. Then for every $\mathbf{v} \in V$ the projection of $\mathbf{v}$ on $\mathcal{U}^\perp$ exists and*

$$
\text{proj}_{\mathcal{U}^\perp}(\mathbf{v}) = \mathbf{v} - \text{proj}_\mathcal{U}(\mathbf{v}).
$$

Proof. First we note that $\langle \mathbf{v} - \text{proj}_\mathcal{U}(\mathbf{v}), \mathbf{u} \rangle = 0$ for every $\mathbf{u} \in \mathcal{U}$, which means $\mathbf{v} - \text{proj}_\mathcal{U}(\mathbf{v}) \in \mathcal{U}^\perp$. Moreover, for every $\mathbf{w} \in \mathcal{U}^\perp$ we have

$$
\langle \mathbf{v} - (\mathbf{v} - \text{proj}_\mathcal{U}(\mathbf{v})), \mathbf{w} \rangle = \langle \text{proj}_\mathcal{U}(\mathbf{v}), \mathbf{w} \rangle = 0.
$$

Therefore $\mathbf{v} - \text{proj}_\mathcal{U}(\mathbf{v})$ is the projection of $\mathbf{v}$ on $\mathcal{U}^\perp$. $\square$

Theorem 3.2.29. *For any finite dimensional subspace $\mathcal{U}$ of an inner product space V we have*

$$
V = \mathcal{U} \oplus \mathcal{U}^\perp.
$$

Proof. For every $\mathbf{v} \in V$ we have

$$
\mathbf{v} = \text{proj}_\mathcal{U}(\mathbf{v}) + \text{proj}_{\mathcal{U}^\perp}(\mathbf{v}),
$$

by Theorem 3.2.28. Hence $V = \mathcal{U} + \mathcal{U}^\perp$. If $\mathbf{x} \in \mathcal{U} \cap \mathcal{U}^\perp$, then $\| \mathbf{x} \|^2 = \langle \mathbf{x}, \mathbf{x} \rangle = 0$ and thus $\mathbf{x} = \mathbf{0}$, which means that $V = \mathcal{U} \oplus \mathcal{U}^\perp$. $\square$

In the next theorem we list all basic results on orthogonal projections on finite dimensional subspaces. We assume that the subspace $\mathcal{U}$ is finite dimensional to ensure that the $\text{proj}_\mathcal{U}(\mathbf{v})$ exists for every $\mathbf{v} \in V$. If we replace the assumption that $\mathcal{U}$ is finite dimensional by the assumption that the $\text{proj}_\mathcal{U}(\mathbf{v})$ exists for every $\mathbf{v} \in V$, the theorem remains true.

Theorem 3.2.30. *Let $\mathcal{U}$ be a finite dimensional subspace of an inner product space $\mathcal{V}$. Then*

(a) $\mathbf{p} = \mathrm{proj}_{\mathcal{U}}(\mathbf{v})$ *if and only if* $\langle \mathbf{v} - \mathbf{p}, \mathbf{u} \rangle = 0$ *for every* $\mathbf{u} \in \mathcal{U}$;

(b) $\mathrm{proj}_{\mathcal{U}}(\mathbf{v})$ *is the best approximation to the vector $\mathbf{v}$ by vectors from the subspace $\mathcal{U}$, that is,* $\mathbf{p} = \mathrm{proj}_{\mathcal{U}}(\mathbf{v})$ *if and only if* $\|\mathbf{v} - \mathbf{p}\| < \|\mathbf{v} - \mathbf{u}\|$ *for every* $\mathbf{u} \in \mathcal{U}$ *such that* $\mathbf{u} \neq \mathbf{p}$;

(c) $\mathrm{proj}_{\mathcal{U}} : \mathcal{V} \to \mathcal{V}$ *is a linear transformation;*

(d) $\mathbf{u} = \mathrm{proj}_{\mathcal{U}}(\mathbf{u})$ *for every* $\mathbf{u} \in \mathcal{U}$;

(e) $\mathrm{ran}\,\mathrm{proj}_{\mathcal{U}} = \mathcal{U}$;

(f) $\ker \mathrm{proj}_{\mathcal{U}} = \mathcal{U}^{\perp}$;

(g) $\mathrm{proj}_{\mathcal{U}^{\perp}} = \mathrm{Id} - \mathrm{proj}_{\mathcal{U}}$;

(h) $\mathrm{proj}_{\mathcal{U}}(\mathrm{proj}_{\mathcal{U}}(\mathbf{v})) = \mathrm{proj}_{\mathcal{U}}(\mathbf{v})$ *for every* $\mathbf{v} \in \mathcal{V}$;

(i) $\langle \mathrm{proj}_{\mathcal{U}}(\mathbf{v}), \mathbf{w} \rangle = \langle \mathbf{v}, \mathrm{proj}_{\mathcal{U}}(\mathbf{w}) \rangle$ *for every* $\mathbf{v}, \mathbf{w} \in \mathcal{V}$.

Proof. (a) is the definition of orthogonal projections (Definition 3.2.8);

(b) is the statement in Theorem 3.2.13;

(c) If $\mathbf{v}, \mathbf{w} \in \mathcal{V}$ and $\alpha, \beta \in \mathbb{K}$, then

$$\langle \alpha\mathbf{v} + \beta\mathbf{w} - \alpha\mathrm{proj}_{\mathcal{U}}(\mathbf{v}) + \beta\mathrm{proj}_{\mathcal{U}}(\mathbf{w}), \mathbf{u} \rangle = \alpha\langle \mathbf{v} - \mathrm{proj}_{\mathcal{U}}(\mathbf{v}), \mathbf{u} \rangle + \beta\langle \mathbf{w} - \mathrm{proj}_{\mathcal{U}}(\mathbf{w}), \mathbf{u} \rangle = 0$$

for every $\mathbf{u} \in \mathcal{U}$. Hence

$$\mathrm{proj}_{\mathcal{U}}(\alpha\mathbf{v} + \beta\mathbf{w}) = \alpha\mathrm{proj}_{\mathcal{U}}(\mathbf{v}) + \beta\mathrm{proj}_{\mathcal{U}}(\mathbf{w});$$

(d) For every $\mathbf{u} \in \mathcal{U}$ we have $\langle \mathbf{u} - \mathbf{u}, \mathbf{u} \rangle = 0$, which means that $\mathbf{u} = \mathrm{proj}_{\mathcal{U}}(\mathbf{u})$;

(e) follows from (d) and the definition of the projection;

(f) If $\mathbf{v} \in \ker \mathrm{proj}_{\mathcal{U}}$, then for every $\mathbf{u} \in \mathcal{U}$ we have

$$\langle \mathbf{v}, \mathbf{u} \rangle = \langle \mathbf{v} - \mathrm{proj}_{\mathcal{U}}(\mathbf{v}), \mathbf{u} \rangle = 0,$$

which means that $\mathbf{v} \in \mathcal{U}^{\perp}$. Now, if $\mathbf{v} \in \mathcal{U}^{\perp}$, then $\langle \mathbf{v} - \mathbf{0}, \mathbf{u} \rangle = \langle \mathbf{v}, \mathbf{u} \rangle = 0$ for every $\mathbf{u} \in \mathcal{U}$, which means that $\mathrm{proj}_{\mathcal{U}}(\mathbf{v}) = \mathbf{0}$;

(g) is equivalent to the statement in Theorem 3.2.28;

(h) is a consequence of (d);

(i) For any $\mathbf{v}, \mathbf{w} \in \mathcal{V}$ we have

$$\begin{aligned}
\langle \mathrm{proj}_{\mathcal{U}}(\mathbf{v}), \mathbf{w} \rangle &= \langle \mathrm{proj}_{\mathcal{U}}(\mathbf{v}), \mathbf{w} - \mathrm{proj}_{\mathcal{U}}(\mathbf{w}) + \mathrm{proj}_{\mathcal{U}}(\mathbf{w}) \rangle \\
&= \langle \mathrm{proj}_{\mathcal{U}}(\mathbf{v}), \mathrm{proj}_{\mathcal{U}}(\mathbf{w}) \rangle \\
&= \langle \mathbf{v} - \mathrm{proj}_{\mathcal{U}}(\mathbf{v}) + \mathrm{proj}_{\mathcal{U}}(\mathbf{v}), \mathrm{proj}_{\mathcal{U}}(\mathbf{w}) \rangle \\
&= \langle \mathbf{v}, \mathrm{proj}_{\mathcal{U}}(\mathbf{w}) \rangle.
\end{aligned}$$

$\square$

It turns out that properties (c), (h), and (i) in Theorem 3.2.30 characterize orthogonal projections.

Theorem 3.2.31. *Let V be an inner product space. If $f : V \to V$ is a linear transformation such that*

(a) $f(f(\mathbf{x})) = f(\mathbf{x})$ *for every* $\mathbf{x} \in V$,

(b) $\langle f(\mathbf{x}), \mathbf{y} \rangle = \langle \mathbf{x}, f(\mathbf{y}) \rangle$ *for every* $\mathbf{x}, \mathbf{y} \in V$,

then f is the orthogonal projection on the subspace ran f.

Proof. Assume that $f : V \to V$ is a linear transformation satisfying (a) and (b). If $\mathbf{x}, \mathbf{y} \in V$, then

$$\langle \mathbf{v} - f(\mathbf{x}), f(\mathbf{y}) \rangle = \langle \mathbf{x}, f(\mathbf{y}) \rangle - \langle f(\mathbf{x}), f(\mathbf{y}) \rangle$$
$$= \langle \mathbf{x}, f(\mathbf{y}) \rangle - \langle \mathbf{x}, f(f(\mathbf{y})) \rangle$$
$$= \langle \mathbf{x}, f(\mathbf{y}) \rangle - \langle \mathbf{x}, f(\mathbf{y}) \rangle = 0.$$

This means that

$$f(\mathbf{x}) = \text{proj}_{\text{ran } f}(\mathbf{x})$$

for every $\mathbf{x} \in V$. □

3.2.5 The Gram-Schmidt orthogonalization process and orthonormal bases

In Corollary 3.2.19 we noted that calculating the projection on the subspace $\mathcal{U} = \text{Span}\{\mathbf{u}_1, \ldots, \mathbf{u}_k\}$ is especially simple if $\{\mathbf{u}_1, \ldots, \mathbf{u}_k\}$ is an orthonormal set. We are going to show that every finite dimensional subspace can be spanned by orthonormal vectors. This is accomplished by modifying an arbitrary spanning set by what is called the *Gram-Schmidt process*. We motivate the idea of the Gram-Schmidt process by considering a couple of examples.

Example 3.2.32. Find a vector $\mathbf{v} \in \mathbb{C}^3$ such that

$$\text{Span}\left\{ \begin{bmatrix} 1 \\ 1 \\ 1 \end{bmatrix}, \mathbf{v} \right\} = \text{Span}\left\{ \begin{bmatrix} 1 \\ 1 \\ 1 \end{bmatrix}, \begin{bmatrix} i \\ -1 \\ 1 \end{bmatrix} \right\} \quad \text{and} \quad \left\langle \mathbf{v}, \begin{bmatrix} i \\ -1 \\ 1 \end{bmatrix} \right\rangle = 0.$$

Solution. Let $\mathcal{U} = \text{Span} \left\{ \begin{bmatrix} i \\ -1 \\ 1 \end{bmatrix} \right\}$. By Example 3.2.15, we have

$$\text{proj}_{\mathcal{U}} \left(\begin{bmatrix} 1 \\ 1 \\ 1 \end{bmatrix} \right) = \begin{bmatrix} \frac{1}{3} \\ \frac{1}{3}i \\ -\frac{1}{3}i \end{bmatrix}.$$

Consequently, we can take

$$\mathbf{v} = \begin{bmatrix} 1 \\ 1 \\ 1 \end{bmatrix} - \text{proj}_{\mathcal{U}} \left(\begin{bmatrix} 1 \\ 1 \\ 1 \end{bmatrix} \right) = \begin{bmatrix} \frac{2}{3} \\ 1 - \frac{1}{3}i \\ 1 + \frac{1}{3}i \end{bmatrix}.$$

$\square$

Example 3.2.33. Let $\mathbf{u}_1, \ldots, \mathbf{u}_m$ be nonzero orthogonal vectors in an inner product space $\mathcal{V}$ and let $\mathbf{v}$ be a vector in $\mathcal{V}$ such that $\mathbf{v} \notin \text{Span}\{\mathbf{u}_1, \ldots, \mathbf{u}_m\}$. Find a nonzero vector $\mathbf{u}_{m+1}$ such that

$$\mathbf{u}_{m+1} \in \text{Span}\{\mathbf{u}_1, \ldots, \mathbf{u}_m\}^{\perp}$$

and

$$\text{Span}\{\mathbf{u}_1, \ldots, \mathbf{u}_m, \mathbf{u}_{m+1}\} = \text{Span}\{\mathbf{u}_1, \ldots, \mathbf{u}_m, \mathbf{v}\}.$$

Solution. By Theorem 3.2.18, we have

$$\text{proj}_{\text{Span}\{\mathbf{u}_1, \ldots, \mathbf{u}_m\}}(\mathbf{v}) = \frac{\langle \mathbf{v}, \mathbf{u}_1 \rangle}{\langle \mathbf{u}_1, \mathbf{u}_1 \rangle} \mathbf{u}_1 + \frac{\langle \mathbf{v}, \mathbf{u}_2 \rangle}{\langle \mathbf{u}_2, \mathbf{u}_2 \rangle} \mathbf{u}_2 + \cdots + \frac{\langle \mathbf{v}, \mathbf{u}_m \rangle}{\langle \mathbf{u}_m, \mathbf{u}_m \rangle} \mathbf{u}_m.$$

We take

$$\mathbf{u}_{m+1} = \mathbf{v} - \frac{\langle \mathbf{v}, \mathbf{u}_1 \rangle}{\langle \mathbf{u}_1, \mathbf{u}_1 \rangle} \mathbf{u}_1 - \frac{\langle \mathbf{v}, \mathbf{u}_2 \rangle}{\langle \mathbf{u}_2, \mathbf{u}_2 \rangle} \mathbf{u}_2 - \cdots - \frac{\langle \mathbf{v}, \mathbf{u}_m \rangle}{\langle \mathbf{u}_m, \mathbf{u}_m \rangle} \mathbf{u}_m.$$

Clearly, $\mathbf{u}_{m+1} \neq \mathbf{0}$ and

$$\text{Span}\{\mathbf{u}_1, \ldots, \mathbf{u}_m, \mathbf{v}\} = \text{Span}\{\mathbf{u}_1, \ldots, \mathbf{u}_m, \mathbf{u}_{m+1}\}.$$

Moreover, since

$$\langle \mathbf{u}_{m+1}, \mathbf{u}_j \rangle = \langle \mathbf{v} - \text{proj}_{\text{Span}\{\mathbf{u}_1, \ldots, \mathbf{u}_m\}}(\mathbf{v}), \mathbf{u}_j \rangle = 0$$

for $j = 1, \ldots, m$, we have $\mathbf{u}_{m+1} \in \text{Span}\{\mathbf{u}_1, \ldots, \mathbf{u}_m\}^{\perp}$. $\square$

The method used in the above example leads to the following general result.

Theorem 3.2.34. *For any linearly independent vectors* $\mathbf{u}_1, \ldots, \mathbf{u}_m$ *in an inner product space* $\mathcal{V}$ *there are orthogonal vectors* $\mathbf{v}_1, \ldots, \mathbf{v}_m$ *in* $\mathcal{V}$ *such that*

$$\mathrm{Span}\{\mathbf{u}_1, \ldots, \mathbf{u}_k\} = \mathrm{Span}\{\mathbf{v}_1, \ldots, \mathbf{v}_k\}$$

for every $k \in \{1, \ldots, m\}$.

Proof. Let $\mathcal{U}_k = \mathrm{Span}\{\mathbf{u}_1, \ldots, \mathbf{u}_k\}$ for $k \in \{1, \ldots, m\}$. We define $\mathbf{v}_1 = \mathbf{u}_1$ and then successively

$$\mathbf{v}_k = \mathbf{u}_k - \mathrm{proj}_{\mathcal{U}_{k-1}}(\mathbf{u}_k)$$

for $k \in \{2, \ldots, m\}$. Since $\mathbf{u}_k \notin \mathcal{U}_{k-1}$, we have $\mathbf{v}_k \neq \mathbf{0}$.

If $\mathcal{U}_{k-1} = \mathrm{Span}\{\mathbf{v}_1, \ldots, \mathbf{v}_{k-1}\}$ for some $k \in \{2, \ldots, m\}$, then

$$\mathbf{u}_k = \mathbf{v}_k + \mathrm{proj}_{\mathcal{U}_{k-1}}(\mathbf{u}_k) = \mathbf{v}_k + \mathrm{proj}_{\mathrm{Span}\{\mathbf{v}_1, \ldots, \mathbf{v}_{k-1}\}}(\mathbf{u}_k) \in \mathrm{Span}\{\mathbf{v}_1, \ldots, \mathbf{v}_k\}$$

and consequently

$$\mathcal{U}_k = \mathrm{Span}\{\mathbf{v}_1, \ldots, \mathbf{v}_k\},$$

because $\mathbf{v}_k \in \mathcal{U}_k$ for every $k \in \{1, \ldots, m\}$. This shows by induction that

$$\mathrm{Span}\{\mathbf{u}_1, \ldots, \mathbf{u}_k\} = \mathrm{Span}\{\mathbf{v}_1, \ldots, \mathbf{v}_k\}$$

for every $k \in \{1, \ldots, m\}$.

To finish the proof we note that, by part (a) of Theorem 3.2.30, $\langle \mathbf{v}_k, \mathbf{u} \rangle = 0$ for every $\mathbf{u} \in \mathcal{U}_{k-1}$ and every $k \in \{2, \ldots, m\}$. Hence

$$\langle \mathbf{v}_k, \mathbf{v}_1 \rangle = \cdots = \langle \mathbf{v}_k, \mathbf{v}_{k-1} \rangle = 0,$$

because $\mathbf{v}_1, \ldots, \mathbf{v}_{k-1} \in \mathcal{U}_{k-1}$. $\square$

Note that the above proof describes an effective process of constructing an orthogonal basis of a subspace from an arbitrary basis. This process is called the *Gram-Schmidt orthogonalization process*.

Example 3.2.35. Let $\mathbf{u}_1, \mathbf{u}_2, \mathbf{u}_3, \mathbf{u}_4$ be linearly independent vectors in an inner product space $\mathcal{V}$. Find an orthogonal set $\{\mathbf{v}_1, \mathbf{v}_2, \mathbf{v}_3, \mathbf{v}_4\}$ such that

$$\mathbf{u}_1 = \mathbf{v}_1,$$
$$\mathrm{Span}\{\mathbf{u}_1, \mathbf{u}_2\} = \mathrm{Span}\{\mathbf{v}_1, \mathbf{v}_2\},$$
$$\mathrm{Span}\{\mathbf{u}_1, \mathbf{u}_2, \mathbf{u}_3\} = \mathrm{Span}\{\mathbf{v}_1, \mathbf{v}_2, \mathbf{v}_3\},$$

and

$$\mathrm{Span}\{\mathbf{u}_1, \mathbf{u}_2, \mathbf{u}_3, \mathbf{u}_4\} = \mathrm{Span}\{\mathbf{v}_1, \mathbf{v}_2, \mathbf{v}_3, \mathbf{v}_4\}.$$

Solution. We take

$$\mathbf{v}_1 = \mathbf{u}_1,$$

$$\mathbf{v}_2 = \mathbf{u}_2 - \frac{\langle \mathbf{u}_2, \mathbf{v}_1 \rangle}{\langle \mathbf{v}_1, \mathbf{v}_1 \rangle} \mathbf{v}_1,$$

$$\mathbf{v}_3 = \mathbf{u}_3 - \frac{\langle \mathbf{u}_3, \mathbf{v}_1 \rangle}{\langle \mathbf{v}_1, \mathbf{v}_1 \rangle} \mathbf{v}_1 - \frac{\langle \mathbf{u}_3, \mathbf{v}_2 \rangle}{\langle \mathbf{v}_2, \mathbf{v}_2 \rangle} \mathbf{v}_2,$$

and

$$\mathbf{v}_4 = \mathbf{u}_4 - \frac{\langle \mathbf{u}_4, \mathbf{v}_1 \rangle}{\langle \mathbf{v}_1, \mathbf{v}_1 \rangle} \mathbf{v}_1 - \frac{\langle \mathbf{u}_4, \mathbf{v}_2 \rangle}{\langle \mathbf{v}_2, \mathbf{v}_2 \rangle} \mathbf{v}_2 - \frac{\langle \mathbf{u}_4, \mathbf{v}_3 \rangle}{\langle \mathbf{v}_3, \mathbf{v}_3 \rangle} \mathbf{v}_3.$$

$\square$

In the first section of this chapter we proved that $\|\mathbf{u} + \mathbf{v}\|^2 = \|\mathbf{u}\|^2 + \|\mathbf{v}\|^2$ for any orthogonal vectors $\mathbf{u}$ and $\mathbf{v}$ (the Pythagorean Theorem 3.1.35). This property easily generalizes to any finite set of orthogonal vectors.

Theorem 3.2.36 (The General Pythagorean Theorem). *For any orthogonal vectors $\mathbf{v}_1, \ldots, \mathbf{v}_n$ in an inner product space $\mathcal{V}$ we have*

$$\|\mathbf{v}_1 + \cdots + \mathbf{v}_n\|^2 = \|\mathbf{v}_1\|^2 + \cdots + \|\mathbf{v}_n\|^2.$$

Proof. For any orthogonal vectors $\mathbf{v}_1, \ldots, \mathbf{v}_n$ we have

$$\left\| \sum_{j=1}^{n} \mathbf{v}_j \right\|^2 = \left\langle \sum_{j=1}^{n} \mathbf{v}_j, \sum_{k=1}^{n} \mathbf{v}_k \right\rangle = \sum_{j,k=1}^{n} \langle \mathbf{v}_j, \mathbf{v}_k \rangle = \sum_{j=1}^{n} \langle \mathbf{v}_j, \mathbf{v}_j \rangle = \sum_{j=1}^{n} \|\mathbf{v}_j\|^2.$$

$\square$

If $\mathbf{u}_1, \ldots, \mathbf{u}_m$ are linearly independent vectors, then the vectors $\mathbf{v}_1, \ldots, \mathbf{v}_m$ obtained by the Gram-Schmidt orthogonalization process are also linearly independent and thus they are nonzero vectors. By normalizing vectors $\mathbf{v}_1, \ldots, \mathbf{v}_m$ we obtain an orthonormal set

$$\left\{ \frac{\mathbf{v}_1}{\|\mathbf{v}_1\|}, \ldots, \frac{\mathbf{v}_m}{\|\mathbf{v}_m\|} \right\}.$$

The process of obtaining an orthonormal set from an arbitrary linearly independent set is called the *Gram-Schmidt orthonormalization process.*

Corollary 3.2.37. *For any linearly independent vectors* $\mathbf{u}_1, \ldots, \mathbf{u}_m$ *in an inner product space* V *there are orthonormal vectors* $\mathbf{w}_1, \ldots, \mathbf{w}_m$ *in* V *such that*

$$\mathrm{Span}\{\mathbf{u}_1, \ldots, \mathbf{u}_k\} = \mathrm{Span}\{\mathbf{w}_1, \ldots, \mathbf{w}_k\}$$

for every $k \in \{1, \ldots, m\}$.

Example 3.2.38. We apply the Gram-Schmidt orthonormalization process to the set $\{1, t, t^2\}$ in the vector space of polynomials on the interval $[0, 1]$ with the inner product

$$\langle f, g \rangle = \int_0^1 f(t)\overline{g(t)}dt.$$

First we define $f_0(t) = 1$. Since

$$\|f_0\|^2 = \int_0^1 1 dt = 1,$$

we let $g_0(t) = f_0(t) = 1$.

Next we find f_1:

$$f_1(t) = t - \langle t, 1 \rangle = t - \int_0^1 t dt = t - \frac{1}{2}.$$

Since

$$\|f_1\|^2 = \int_0^1 \left(t - \frac{1}{2}\right)^2 dt = \frac{1}{12},$$

we define

$$g_1(t) = \frac{1}{\|f_1\|} f_1(t) = 2\sqrt{3}\left(t - \frac{1}{2}\right)$$

Now we find f_2:

$$f_2(t) = t^2 - \langle t^2, 1 \rangle - \left\langle t^2, 2\sqrt{3}\left(t - \frac{1}{2}\right)\right\rangle 2\sqrt{3}\left(t - \frac{1}{2}\right)$$

$$= t^2 - \int_0^1 t^2 dt - 12 \int_0^1 t^2 \left(t - \frac{1}{2}\right) dt \left(t - \frac{1}{2}\right)$$

$$= t^2 - \frac{1}{3} - \left(t - \frac{1}{2}\right)$$

$$= t^2 - t + \frac{1}{6}.$$

Since

$$\|f_2\|^2 = \int_0^1 \left(t^2 + t + \frac{1}{6}\right)^2 dt = \frac{1}{180},$$

we define

$$g_2(t) = \frac{1}{\|f_2\|} f_2(t) = 6\sqrt{5}\left(t^2 - t + \frac{1}{6}\right).$$

By applying the Gram-Schmidt orthonormalization process to the set $\{1, t, t^2\}$ we obtain the following orthonormal set

$$\left\{1, 2\sqrt{3}\left(t - \frac{1}{2}\right), 6\sqrt{5}\left(t^2 - t + \frac{1}{6}\right)\right\}.$$

Example 3.2.39. Use the result from Example 3.2.38 to find the best approximation to the function $\cos \pi t$ by quadratic polynomials on the interval $[0, 1]$ with respect to the inner product $\langle f, g \rangle = \int_0^1 f(t)\overline{g(t)}dt$.

Solution. Since

$$\langle \cos \pi t, 1 \rangle = 0, \quad \left\langle \cos \pi t, 2\sqrt{3}\left(t - \frac{1}{2}\right)\right\rangle = -\frac{4\sqrt{3}}{\pi^2},$$

and

$$\left\langle \cos \pi t, 6\sqrt{5}\left(t^2 - t + \frac{1}{6}\right)\right\rangle = 0,$$

we have

$$\mathrm{proj}_{\mathrm{Span}\{1,t,t^2\}}(\cos \pi t) = -\frac{4\sqrt{3}}{\pi^2}2\sqrt{3}\left(t - \frac{1}{2}\right) = -\frac{24}{\pi^2}t + \frac{12}{\pi^2}.$$

$\square$

Example 3.2.40. Consider the vector space $\mathcal{P}_m(\mathbb{R})$ with the inner product

$$\langle f, g \rangle = \int_{-1}^1 f(t)g(t)dt.$$

We apply the Gram-Schmidt orthogonalization process to the polynomials $1, t, \ldots, t^m$ and get polynomials $1, p_1, \ldots, p_m$. Show that

$$p_m \in \mathrm{Span}\{((1 - t^2)^m)^{(m)}\}.$$

Solution. First we find that

$$\int_{-1}^{1} ((1-t^2)^m)^{(m)} t^{m-1} dt = ((1-t^2)^m)^{(m-1)} t^{m-1} \Big|_{-1}^{1}$$

$$- (m-1) \int_{-1}^{1} ((1-t^2)^m)^{(m-1)} t^{m-2} dt$$

$$= -(m-1) \int_{-1}^{1} ((1-t^2)^m)^{(m-1)} t^{m-2} dt.$$

If we continue to integrate by parts, we end up with

$$\int_{-1}^{1} ((1-t^2)^m)^{(m)} t^{m-1} dt = (-1)^{m-1} (m-1)! \int_{-1}^{1} ((1-t^2)^m)' dt = 0.$$

In a similar way we get

$$\int_{-1}^{1} ((1-t^2)^m)^{(m)} t^j dt = 0$$

for every $j \in \{0, \ldots, m-1\}$. Now, since

$$\mathcal{P}_{m-1}(\mathbb{R}) \oplus \mathcal{P}_{m-1}^{\perp}(\mathbb{R}) = \mathcal{P}_m(\mathbb{R}),$$

$\dim \mathcal{P}_m(\mathbb{R}) = m+1$, and $\dim \mathcal{P}_{m-1}(\mathbb{R}) = m$, we have $\dim \mathcal{P}_{m-1}^{\perp}(\mathbb{R}) = 1$. This gives us our result because $((1-t^2)^m)^{(m)} \in \mathcal{P}_{m-1}^{\perp}(\mathbb{R})$. □

From Corollary 3.2.37 it follows that every finite dimensional subspace of an inner product space has an orthonormal spanning set. In other words, every finite dimensional subspace of an inner product space has an orthonormal basis.

Theorem 3.2.41. *Let $\{x_1, \ldots, x_n\}$ be an orthonormal set in an inner product space $\mathcal{V}$. The following conditions are equivalent:*

(a) $\{x_1, \ldots, x_n\}$ *is a basis in $\mathcal{V}$;*

(b) $(\mathrm{Span}\{x_1, \ldots, x_n\})^{\perp} = \{0\}$;

(c) $v = \langle v, x_1 \rangle x_1 + \cdots + \langle v, x_n \rangle x_n$ *for every vector $v \in \mathcal{V}$;*

(d) $\langle v, w \rangle = \langle v, x_1 \rangle \langle x_1, w \rangle + \cdots + \langle v, x_n \rangle \langle x_n w \rangle$ *for every $v, w \in \mathcal{V}$;*

(e) $\|v\|^2 = |\langle v, x_1 \rangle|^2 + \cdots + |\langle v, x_n \rangle|^2$ *for every vector $v \in \mathcal{V}$.*

Proof. Assume $\{x_1, \ldots, x_n\}$ is a basis in $\mathcal{V}$. If $v \in (\mathrm{Span}\{x_1, \ldots, x_n\})^{\perp}$, then

$$v = \alpha_1 x_1 + \cdots + \alpha_n x_n$$

for some $\alpha_1, \ldots, \alpha_n \in \mathbb{K}$ and $\langle \mathbf{v}, \mathbf{x}_j \rangle = 0$ for every $j = 1, \ldots, n$. Since the set $\{\mathbf{x}_1, \ldots, \mathbf{x}_n\}$ is orthonormal, we have

$$0 = \langle \mathbf{v}, \mathbf{x}_j \rangle = \langle \alpha_1 \mathbf{x}_1, \ldots, \alpha_n \mathbf{x}_n, \mathbf{x}_j \rangle$$
$$= \langle \alpha_1 \mathbf{x}_1, \mathbf{x}_j \rangle + \cdots + \langle \alpha_n \mathbf{x}_n, \mathbf{x}_j \rangle = \alpha_j \langle \mathbf{x}_j, \mathbf{x}_j \rangle = \alpha_j,$$

for every $j = 1, \ldots, n$, which means that $\mathbf{v} = \mathbf{0}$. This shows that (a) implies (b).

Now we observe that

$$\langle \mathbf{v} - (\langle \mathbf{v}, \mathbf{x}_1 \rangle \mathbf{x}_1 + \cdots + \langle \mathbf{v}, \mathbf{x}_n \rangle \mathbf{x}_n), \mathbf{x}_j \rangle = 0$$

for every $\mathbf{v} \in V$ and every $j = 1, \ldots, n$, and thus

$$\mathbf{v} - (\langle \mathbf{v}, \mathbf{x}_1 \rangle \mathbf{x}_1 + \cdots + \langle \mathbf{v}, \mathbf{x}_n \rangle \mathbf{x}_n) \in (\text{Span}\{\mathbf{x}_1, \ldots, \mathbf{x}_n\})^{\perp}.$$

Consequently, if $(\text{Span}\{\mathbf{x}_1, \ldots, \mathbf{x}_n\})^{\perp} = \{\mathbf{0}\}$, then

$$\mathbf{v} = \langle \mathbf{v}, \mathbf{x}_1 \rangle \mathbf{x}_1 + \cdots + \langle \mathbf{v}, \mathbf{x}_n \rangle \mathbf{x}_n,$$

for every $\mathbf{v} \in V$. This shows that (b) implies (c).

Since (c) clearly implies (a), the conditions (a), (b), and (c) are equivalent. Now assume that (c) holds and consider arbitrary $\mathbf{v}, \mathbf{w} \in V$. Then

$$\langle \mathbf{v}, \mathbf{w} \rangle = \left\langle \sum_{j=1}^{n} \langle \mathbf{v}, \mathbf{x}_j \rangle \mathbf{x}_j, \sum_{k=1}^{n} \langle \mathbf{w}, \mathbf{x}_k \rangle \mathbf{x}_k \right\rangle$$

$$= \sum_{j,k=1}^{n} \langle \mathbf{v}, \mathbf{x}_j \rangle \overline{\langle \mathbf{w}, \mathbf{x}_k \rangle} \langle \mathbf{x}_j, \mathbf{x}_k \rangle$$

$$= \sum_{j=1}^{n} \langle \mathbf{v}, \mathbf{x}_j \rangle \overline{\langle \mathbf{w}, \mathbf{x}_j \rangle} \langle \mathbf{x}_j, \mathbf{x}_j \rangle$$

$$= \sum_{j=1}^{n} \langle \mathbf{v}, \mathbf{x}_j \rangle \langle \mathbf{x}_j, \mathbf{w} \rangle.$$

Thus (c) implies (d).

To see that (d) implies (e) it suffices to let $\mathbf{w} = \mathbf{v}$ in (d).

To complete the proof we show that (e) implies (b). Indeed, if (e) holds and $\mathbf{v} \in (\text{Span}\{\mathbf{x}_1, \ldots, \mathbf{x}_n\})^{\perp}$, then

$$\|\mathbf{v}\|^2 = |\langle \mathbf{v}, \mathbf{x}_1 \rangle|^2 + \cdots + |\langle \mathbf{v}, \mathbf{x}_n \rangle|^2 = 0,$$

and thus $\mathbf{v} = \mathbf{0}$. $\square$

3.3   The adjoint of a linear transformation

For any $\mathbf{v}_0$ in an inner product space V the function $f(\mathbf{x}) = \langle \mathbf{x}, \mathbf{v}_0 \rangle$ is a linear transformation from V to $\mathbb{K}$, which is an immediate consequence of the definition of the inner product. It turns out that, if V is a finite dimensional inner product space, then every linear transformation from V to $\mathbb{K}$ is of such form.

Theorem 3.3.1 (Representation Theorem). *Let V be a finite dimensional inner product spaces and let $f : V \to \mathbb{K}$ be a linear transformation. Then there exists a unique $\mathbf{v}_f \in V$ such that*

$$f(\mathbf{x}) = \langle \mathbf{x}, \mathbf{v}_f \rangle$$

for every $\mathbf{x} \in V$.

Proof. Let $f : V \to \mathbb{K}$ be a linear transformation. If f is the zero transformation, then clearly $\mathbf{v} = \mathbf{0}$ has the desired property.

Assume $f : V \to \mathbb{K}$ is a nonzero linear transformation. Then $\ker f \neq V$. Let $\mathbf{u}$ be a unit vector in $(\ker f)^{\perp}$. Since $f(\mathbf{u})\mathbf{x} - f(\mathbf{x})\mathbf{u} \in \ker f$ for every $\mathbf{x} \in V$, we have

$$\langle f(\mathbf{u})\mathbf{x} - f(\mathbf{x})\mathbf{u}, \mathbf{u} \rangle = 0,$$

which gives us

$$\langle f(\mathbf{u})\mathbf{x}, \mathbf{u} \rangle = \langle f(\mathbf{x})\mathbf{u}, \mathbf{u} \rangle = f(\mathbf{x})\|\mathbf{u}\|^2 = f(\mathbf{x}).$$

Consequently

$$f(\mathbf{x}) = \langle f(\mathbf{u})\mathbf{x}, \mathbf{u} \rangle = \langle \mathbf{x}, \overline{f(\mathbf{u})}\mathbf{u} \rangle.$$

If we take $\mathbf{v}_f = \overline{f(\mathbf{u})}\mathbf{u}$, then

$$f(\mathbf{x}) = \langle \mathbf{x}, \mathbf{v}_f \rangle$$

for every $\mathbf{x} \in V$.

Now suppose $\mathbf{w}$ is another vector such that $f(\mathbf{x}) = \langle \mathbf{x}, \mathbf{w} \rangle$ for every $\mathbf{x} \in V$. But then

$$\|\mathbf{v}_f - \mathbf{w}\|^2 = \langle \mathbf{v}_f - \mathbf{w}, \mathbf{v}_f - \mathbf{w} \rangle = \langle \mathbf{v}_f - \mathbf{w}, \mathbf{v}_f \rangle - \langle \mathbf{v}_f - \mathbf{w}, \mathbf{w} \rangle$$
$$= f(\mathbf{v}_f - \mathbf{w}) - f(\mathbf{v}_f - \mathbf{w}) = 0$$

and thus $\mathbf{w} = \mathbf{v}_f$. $\qquad\square$

From Theorem 3.3.1 we obtain the following important result.

Theorem 3.3.2. *Let V and W be finite dimensional inner product spaces. For every linear transformation $f : V \to W$ there is a unique linear transformation $g : W \to V$ such that*

$$\langle f(\mathbf{v}), \mathbf{w} \rangle = \langle \mathbf{v}, g(\mathbf{w}) \rangle$$

for every $\mathbf{v} \in V$ and $\mathbf{w} \in W$.

Proof. Let $f : \mathcal{V} \to \mathcal{W}$ be a linear transformation. For every $\mathbf{w} \in \mathcal{W}$ the function $f_{\mathbf{w}} : \mathcal{V} \to \mathbb{K}$ defined by $f_{\mathbf{w}}(\mathbf{v}) = \langle f(\mathbf{v}), \mathbf{w} \rangle$ is linear and thus, by Theorem 3.3.1, there is a unique vector $\mathbf{z_w}$ such that $f_{\mathbf{w}}(\mathbf{v}) = \langle \mathbf{v}, \mathbf{z_w} \rangle$ for every $\mathbf{v} \in \mathcal{V}$. Clearly, the function $f_{\mathbf{w}}$ depends on $\mathbf{w}$ and thus $\mathbf{z_w}$ depends on $\mathbf{w}$. In other words, there is a function $g : \mathcal{W} \to \mathcal{V}$ such that

$$\langle f(\mathbf{v}), \mathbf{w} \rangle = \langle \mathbf{v}, \mathbf{z_w} \rangle = \langle \mathbf{v}, g(\mathbf{w}) \rangle$$

for every $\mathbf{v} \in \mathcal{V}$ and $\mathbf{w} \in \mathcal{W}$. We need to show that g is linear.

If $\mathbf{w}_1, \mathbf{w}_2 \in \mathcal{W}$, then

$$\langle f(\mathbf{v}), \mathbf{w}_1 + \mathbf{w}_2 \rangle = \langle f(\mathbf{v}), \mathbf{w}_1 \rangle + \langle f(\mathbf{v}), \mathbf{w}_2 \rangle = \langle \mathbf{v}, g(\mathbf{w}_1) \rangle + \langle \mathbf{v}, g(\mathbf{w}_2) \rangle$$
$$= \langle \mathbf{v}, g(\mathbf{w}_1) + g(\mathbf{w}_2) \rangle$$

for every $\mathbf{v} \in \mathcal{V}$. By the uniqueness part of Theorem 3.3.1 we have

$$g(\mathbf{w}_1 + \mathbf{w}_2) = g(\mathbf{w}_1) + g(\mathbf{w}_2).$$

Similarly, if $\alpha \in \mathbb{K}$ and $\mathbf{w} \in \mathcal{W}$, then

$$\langle f(\mathbf{v}), \alpha \mathbf{w} \rangle = \overline{\alpha} \langle f(\mathbf{v}), \mathbf{w} \rangle = \overline{\alpha} \langle \mathbf{v}, g(\mathbf{w}) \rangle = \langle \mathbf{v}, \alpha g(\mathbf{w}) \rangle$$

for every $\mathbf{v} \in \mathcal{V}$, which gives us

$$g(\alpha \mathbf{w}) = \alpha g(\mathbf{w}).$$

$\square$

Definition 3.3.3. Let $\mathcal{V}$ and $\mathcal{W}$ be finite dimensional inner product spaces and let $f : \mathcal{V} \to \mathcal{W}$ be a linear transformation. The unique linear transformation $g : \mathcal{W} \to \mathcal{V}$ such that

$$\langle f(\mathbf{v}), \mathbf{w} \rangle = \langle \mathbf{v}, g(\mathbf{w}) \rangle$$

for every $\mathbf{v} \in \mathcal{V}$ and $\mathbf{w} \in \mathcal{W}$ is called the *adjoint* of f and is denoted by f^*.

Note that, if $g = f^*$, then also $f = g^*$. Indeed, if $\langle f(\mathbf{v}), \mathbf{w} \rangle = \langle \mathbf{v}, g(\mathbf{w}) \rangle$ for every $\mathbf{v} \in \mathcal{V}$ and $\mathbf{w} \in \mathcal{W}$, then

$$\langle g(\mathbf{w}), \mathbf{v} \rangle = \overline{\langle \mathbf{v}, g(\mathbf{w}) \rangle} = \overline{\langle f(\mathbf{v}), \mathbf{w} \rangle} = \langle \mathbf{w}, f(\mathbf{v}) \rangle$$

for every $\mathbf{v} \in \mathcal{V}$ and $\mathbf{w} \in \mathcal{W}$.

Theorem 3.3.2 says that for any finite dimensional inner product spaces $\mathcal{V}$ and $\mathcal{W}$, if $f \in \mathcal{L}(\mathcal{V}, \mathcal{W})$, then $f^* \in \mathcal{L}(\mathcal{W}, \mathcal{V})$. We can think of * as an operation from $\mathcal{L}(\mathcal{V}, \mathcal{W})$ to $\mathcal{L}(\mathcal{W}, \mathcal{V})$, that is, $* : \mathcal{L}(\mathcal{V}, \mathcal{W}) \to \mathcal{L}(\mathcal{W}, \mathcal{V})$. The next theorem lists some useful algebraic properties of the operation *.

> **Theorem 3.3.4.** *Let V, W, and X be finite dimensional inner product spaces.*
>
> (a) $(\text{Id})^* = \text{Id}$;
>
> (b) $(gf)^* = f^*g^*$ *for every* $f \in L(V, W)$ *and* $g \in L(W, X)$;
>
> (c) $(f_1 + f_2)^* = f_1^* + f_2^*$ *for every* $f_1, f_2 \in L(V, W)$;
>
> (d) $(\alpha f)^* = \overline{\alpha} f^*$ *for every* $f \in L(V, W)$ *and* $\alpha \in \mathbb{K}$;
>
> (e) $(f^*)^* = f$ *for every* $f \in L(V, W)$.

Proof. (a) For any $\mathbf{v} \in V$ we have

$$\langle \text{Id}(\mathbf{v}), \mathbf{v} \rangle = \langle \mathbf{v}, \mathbf{v} \rangle = \langle \mathbf{v}, \text{Id}(\mathbf{v}) \rangle.$$

(b) For any $\mathbf{v} \in V$, $\mathbf{w} \in W$, and $\mathbf{x} \in X$, we have

$$\langle g(f(\mathbf{v})), \mathbf{x} \rangle = \langle f(\mathbf{v}), g^*(\mathbf{x}) \rangle = \langle \mathbf{v}, f^*(g^*(\mathbf{x})) \rangle.$$

(c) For any $\mathbf{v} \in V$ and $\mathbf{w} \in W$, we have

$$\langle (f_1 + f_2)(\mathbf{v}), \mathbf{w} \rangle = \langle f_1(\mathbf{v}), \mathbf{w} \rangle + \langle f_2(\mathbf{v}), \mathbf{w} \rangle = \langle \mathbf{v}, f_1^*(\mathbf{w}) \rangle + \langle \mathbf{v}, f_2^*(\mathbf{w}) \rangle$$
$$= \langle \mathbf{v}, (f_1^* + f_2^*)(\mathbf{w}) \rangle.$$

(d) For any $\mathbf{v} \in V$, $\mathbf{w} \in W$, and $\alpha \in \mathbb{K}$, we have

$$\langle (\alpha f)(\mathbf{v}), \mathbf{w} \rangle = \langle \alpha f(\mathbf{v}), \mathbf{w} \rangle = \alpha \langle f(\mathbf{v}), \mathbf{w} \rangle = \alpha \langle \mathbf{v}, f^*(\mathbf{w}) \rangle = \langle \mathbf{v}, \overline{\alpha} f^*(\mathbf{w}) \rangle.$$

(e) For any $\mathbf{v} \in V$ and $\mathbf{w} \in W$, we have

$$\langle f(\mathbf{v}), \mathbf{w} \rangle = \langle \mathbf{v}, f^*(\mathbf{w}) \rangle = \langle (f^*)^*(\mathbf{v}), \mathbf{w} \rangle.$$

$\square$

Example 3.3.5. Let $f : \mathbb{C}^3 \to \mathbb{C}^3$ and $g : \mathbb{C}^3 \to \mathbb{C}^3$ be the linear transformations defined by

$$f\left(\begin{bmatrix} x \\ y \\ z \end{bmatrix}\right) = \begin{bmatrix} y \\ z \\ 0 \end{bmatrix} \quad \text{and} \quad g\left(\begin{bmatrix} x \\ y \\ z \end{bmatrix}\right) = \begin{bmatrix} 0 \\ x \\ y \end{bmatrix}.$$

Show that $f^* = g$.

Solution. We have

$$\left\langle f\left(\begin{bmatrix} x_1 \\ y_1 \\ z_1 \end{bmatrix}\right), \begin{bmatrix} x_2 \\ y_2 \\ z_2 \end{bmatrix} \right\rangle = \left\langle \begin{bmatrix} y_1 \\ z_1 \\ 0 \end{bmatrix}, \begin{bmatrix} x_2 \\ y_2 \\ z_2 \end{bmatrix} \right\rangle = y_1 \overline{x_2} + z_1 \overline{y_2}$$

and

$$\left\langle \begin{bmatrix} x_1 \\ y_1 \\ z_1 \end{bmatrix}, g\left(\begin{bmatrix} x_2 \\ y_2 \\ z_2 \end{bmatrix} \right) \right\rangle = \left\langle \begin{bmatrix} x_1 \\ y_1 \\ z_1 \end{bmatrix}, \begin{bmatrix} 0 \\ x_2 \\ y_2 \end{bmatrix} \right\rangle = y_1\overline{x_2} + z_1\overline{y_2}.$$

□

Example 3.3.6. In this example we give an application of the adjoint of a linear operator.

Let V be a finite dimensional inner product space and let $f : V \to V$ be a linear operator. Use the adjoint of f to show that, if $\{a_1, \ldots, a_n\}$ and $\{b_1, \ldots, b_n\}$ are two orthonormal bases in V, then

$$\sum_{j=1}^{n} \|f(a_j)\|^2 = \sum_{j=1}^{n} \|f(b_j)\|^2.$$

Solution. From Theorem 3.2.41 we get

$$\|f(a_j)\|^2 = \sum_{k=1}^{n} |\langle f(a_j), b_k \rangle|^2$$

for every $j = 1, \ldots, n$ and

$$\|f^*(b_k)\|^2 = \sum_{j=1}^{n} |\langle f^*(b_k), a_j \rangle|^2 = \sum_{j=1}^{n} |\langle a_j, f^*(b_k) \rangle|^2$$

for every $k = 1, \ldots, n$. Hence

$$\sum_{j=1}^{n} \|f(a_j)\|^2 = \sum_{j=1}^{n}\sum_{k=1}^{n} |\langle f(a_j), b_k \rangle|^2$$

$$= \sum_{j=1}^{n}\sum_{k=1}^{n} |\langle a_j, f^*(b_k) \rangle|^2 = \sum_{k=1}^{n} \|f^*(b_k)\|^2.$$

In similar way we obtain

$$\sum_{j=1}^{n} \|f(b_j)\|^2 = \sum_{k=1}^{n} \|f^*(b_k)\|^2,$$

which gives us the desired result.   □

Theorem 3.3.7. *Let $\mathcal{V}$ and $\mathcal{W}$ be finite dimensional inner product spaces. A linear transformation $f : \mathcal{V} \to \mathcal{W}$ is invertible if and only if $f^* : \mathcal{W} \to \mathcal{V}$ is invertible and then we have*

$$(f^*)^{-1} = (f^{-1})^*.$$

Proof. From

$$f^{-1}f = \mathrm{Id}_{\mathcal{V}}$$

we get

$$f^*(f^{-1})^* = (f^{-1}f)^* = \mathrm{Id}_{\mathcal{V}}^* = \mathrm{Id}_{\mathcal{V}}$$

and from

$$ff^{-1} = \mathrm{Id}_{\mathcal{W}}$$

we get

$$(f^{-1})^*f^* = (ff^{-1})^* = \mathrm{Id}_{\mathcal{W}}^* = \mathrm{Id}_{\mathcal{W}}.$$

□

In Theorem 2.3.2 we prove that, if $\mathcal{B} = \{\mathbf{v}_1, \ldots, \mathbf{v}_m\}$ and $\mathcal{C} = \{\mathbf{w}_1, \ldots, \mathbf{w}_n\}$ are bases of vector spaces $\mathcal{V}$ and $\mathcal{W}$, respectively, then for every linear transformation $f : \mathcal{V} \to \mathcal{W}$ there is a unique $n \times m$ matrix A such that $f(\mathbf{v}) = A\mathbf{v}$ for all $\mathbf{v} \in \mathcal{V}$. We say that A is the matrix of f relative to the bases $\mathcal{B}$ and $\mathcal{C}$ and write $A = f_{\mathcal{B} \to \mathcal{C}}$.

In the following theorem we use A^* to denote the conjugate transpose of A. If $A = [a_{kj}]$ is an $n \times m$ matrix with complex entries, then the *conjugate transpose* of A is the $m \times n$ matrix defined by $A^* = [\overline{a_{jk}}]$.

Theorem 3.3.8. *Let $\mathcal{V}$ and $\mathcal{W}$ be finite dimensional inner spaces and let $f : \mathcal{V} \to \mathcal{W}$ be a linear transformation. If $\mathcal{B} = \{\mathbf{v}_1, \ldots, \mathbf{v}_m\}$ is an orthonormal basis of $\mathcal{V}$ and $\mathcal{C} = \{\mathbf{w}_1, \ldots, \mathbf{w}_n\}$ an orthonormal basis of $\mathcal{W}$, then*

$$(f^*)_{\mathcal{C} \to \mathcal{B}} = (f_{\mathcal{B} \to \mathcal{C}})^*.$$

Proof. For all $j \in \{1, \ldots, m\}$ and $k \in \{1, \ldots, n\}$ we let $a_{kj} = \langle f(\mathbf{v}_j), \mathbf{w}_k \rangle$. Then

$$f(\mathbf{v}_j) = \langle f(\mathbf{v}_j), \mathbf{w}_1 \rangle \mathbf{w}_1 + \cdots + \langle f(\mathbf{v}_j), \mathbf{w}_n \rangle \mathbf{w}_n = a_{1j}\mathbf{w}_1 + \cdots + a_{nj}\mathbf{w}_n,$$

for every $j \in \{1, \ldots, m\}$. This means that the matrix $A = [a_{kj}]$ is the matrix of f relative to the bases $\mathcal{B}$ and $\mathcal{C}$.

On the other hand, for every $k \in \{1, \ldots, n\}$, we have

$$\begin{aligned} f^*(\mathbf{w}_k) &= \langle f^*(\mathbf{w}_k), \mathbf{v}_1 \rangle \mathbf{v}_1 + \cdots + \langle f^*(\mathbf{w}_k), \mathbf{v}_m \rangle \mathbf{v}_m \\ &= \langle \mathbf{w}_k, f(\mathbf{v}_1) \rangle \mathbf{v}_1 + \cdots + \langle \mathbf{w}_k, f(\mathbf{v}_m) \rangle \mathbf{v}_m \\ &= \overline{\langle f(\mathbf{v}_1), \mathbf{w}_k \rangle} \mathbf{v}_1 + \cdots + \overline{\langle f(\mathbf{v}_m), \mathbf{w}_k \rangle} \mathbf{v}_m \\ &= \overline{a_{k1}} \mathbf{v}_1 + \cdots + \overline{a_{km}} \mathbf{v}_m, \end{aligned}$$

which means that the matrix of f^* relative to the bases $\mathcal{C}$ and $\mathcal{B}$ is the conjugate transpose of the matrix of f relative to the bases $\mathcal{B}$ and $\mathcal{C}$. □

The adjoint of a linear transformation $f \in \mathcal{L}(\mathcal{V}, \mathcal{W})$ is a linear transformation $f^* \in \mathcal{L}(\mathcal{W}, \mathcal{V})$. If $\mathcal{V} = \mathcal{W}$, then $f, f^* \in \mathcal{L}(\mathcal{V}, \mathcal{V}) = \mathcal{L}(\mathcal{V})$ and we can consider properties of the adjoint operation that simply don't make sense when $\mathcal{V} \neq \mathcal{W}$.

Definition 3.3.9. Let $\mathcal{V}$ be a finite dimensional inner product space and let $f : \mathcal{V} \to \mathcal{V}$ be a linear operator.

(a) If $ff^* = f^*f$, then f is called a *normal operator*.

(b) If $f^* = f$, then f is called a *self-adjoint operator*.

Clearly, every self-adjoint operator is normal.

Note that self-adjoint operators can be defined for all inner product spaces (not necessarily finite dimensional): a linear operator $f : \mathcal{V} \to \mathcal{V}$ is self-adjoint if $\langle f(\mathbf{x}), \mathbf{y} \rangle = \langle \mathbf{x}, f(\mathbf{y}) \rangle$ for every $\mathbf{x}, \mathbf{y} \in \mathcal{V}$. More on operators on infinite dimensional inner product spaces can be found in Section 5.

Example 3.3.10. Consider the operator $f \in \mathcal{L}(\mathbb{C}^2, \mathbb{C}^2)$ defined by $f(\mathbf{x}) = A\mathbf{x}$ where

$$A = \begin{bmatrix} i & 1 - i \\ -1 - i & 2i \end{bmatrix}.$$

Show that f is normal but not self-adjoint.

Solution. Since

$$A^* = \begin{bmatrix} -i & -1 + i \\ 1 + i & -2i \end{bmatrix},$$

f is not self-adjoint. On the other hand, since

$$AA^* = A^*A = \begin{bmatrix} 3 & -3 - 3i \\ -3 + 3i & 6 \end{bmatrix},$$

f is a normal operator. □

Theorem 3.3.11. *Let $\mathcal{V}$ be a finite dimensional inner product space. For every linear operator $f : \mathcal{V} \to \mathcal{V}$, the operators ff^*, f^*f and $f + f^*$ are self-adjoint.*

Proof. Since

$$(ff^*)^* = (f^*)^*f^* = ff^*,$$

ff^* is self-adjoint. In the same way we can show that f^*f is self-adjoint.
Since

$$(f + f^*)^* = f^* + (f^*)^* = f^* + f,$$

$f + f^*$ is self-adjoint. □

The composition of two self-adjoint operators need not be self adjoint. The following theorem tells us exactly when it is rue.

Theorem 3.3.12. *Let f and g be self-adjoint operators on a finite dimensional inner product space V. The operator fg is self-adjoint if and only if $fg = gf$.*

Proof. If f and g are self-adjoint, then for every $\mathbf{v}, \mathbf{w} \in V$ we have

$$\langle fg(\mathbf{v}), \mathbf{w} \rangle = \langle g(\mathbf{v}), f(\mathbf{w}) \rangle = \langle \mathbf{v}, gf(\mathbf{w}) \rangle.$$

Consequently, $fg = gf$ if and only if fg is self-adjoint. □

The following useful result is a consequence of the polarization identity.

Theorem 3.3.13. *Let V be a finite dimensional inner product space.*

(a) *If $f : V \to V$ is a self-adjoint operator such that $\langle f(\mathbf{v}), \mathbf{v} \rangle = 0$ for every $\mathbf{v} \in V$, then $f = 0$.*

(b) *If $f_1, f_2 : V \to V$ are self-adjoint operators such that $\langle f_1(\mathbf{v}), \mathbf{v} \rangle = \langle f_2(\mathbf{v}), \mathbf{v} \rangle$ for every $\mathbf{v} \in V$, then $f_1 = f_2$.*

Proof. Let $f : V \to V$ be a self-adjoint operator. First we note that the form $s(\mathbf{v}, \mathbf{w}) = \langle f(\mathbf{v}), \mathbf{w} \rangle$ is sesquilinear. We show that s is hermitian. Indeed,

$$s(\mathbf{v}, \mathbf{w}) = \langle f(\mathbf{v}), \mathbf{w} \rangle = \langle \mathbf{v}, f^*(\mathbf{w}) \rangle = \langle \mathbf{v}, f(\mathbf{w}) \rangle = \overline{\langle f(\mathbf{w}), \mathbf{v} \rangle} = \overline{s(\mathbf{w}, \mathbf{v})}.$$

Now, if $\langle f(\mathbf{v}), \mathbf{v} \rangle = 0$ for every $\mathbf{v} \in V$, then $\langle f(\mathbf{v}), \mathbf{w} \rangle = 0$ for every $\mathbf{v}, \mathbf{w} \in V$, by Theorem 3.1.11 (or Theorem 3.1.6 for a real inner product space). Consequently, $f = 0$, proving part (a).

To prove part (b) we take $f = f_1 - f_2$ and use part (a). □

Example 3.3.14. The above theorem does not hold if we drop the assumption that f is self-adjoint. For example, consider the operator $f \in \mathcal{L}(\mathbb{C}^2, \mathbb{C}^2)$ defined by $f\left(\begin{bmatrix} x \\ y \end{bmatrix}\right) = \begin{bmatrix} -y \\ x \end{bmatrix}$. Then

$$\left\langle f\left(\begin{bmatrix} x \\ y \end{bmatrix}\right), \begin{bmatrix} x \\ y \end{bmatrix}\right\rangle = \left\langle \begin{bmatrix} -y \\ x \end{bmatrix}, \begin{bmatrix} x \\ y \end{bmatrix}\right\rangle = 0$$

for every $\begin{bmatrix} x \\ y \end{bmatrix} \in \mathbb{C}^2$, but $f \neq 0$.

Theorem 3.3.15. *Let $f : \mathcal{V} \to \mathcal{V}$ be a self-adjoint operator on a finite dimensional inner product space $\mathcal{V}$. If $f^k = \mathbf{0}$ for some integer $k \geq 1$, then $f = \mathbf{0}$.*

Proof. If $f^2 = \mathbf{0}$, then for every $\mathbf{v} \in \mathcal{V}$ we have

$$0 = \langle f^2(\mathbf{v}), \mathbf{v} \rangle = \langle f(\mathbf{v}), f(\mathbf{v}) \rangle = \|f(\mathbf{v})\|^2,$$

and thus $f = \mathbf{0}$. If $f^4 = (f^2)^2 = \mathbf{0}$, then $f^2 = \mathbf{0}$ and thus $f = \mathbf{0}$. This way we can show that if $f^{2^n} = \mathbf{0}$ and then $f = \mathbf{0}$. For any other integer $k \geq 1$ we find an integer $n \geq 1$ such that $k \leq 2^n$. Then, if $f^k = \mathbf{0}$, then $f^{2^n} = \mathbf{0}$ and thus $f = \mathbf{0}$. □

The above property may seem obvious, but it's not true for arbitrary linear operators. For example, for the operator $f : \mathbb{C}^2 \to \mathbb{C}^2$ defined as $f(\mathbf{x}) = A\mathbf{x}$, where $A = \begin{bmatrix} 0 & 1 \\ 0 & 0 \end{bmatrix}$, we have $f^2 = \mathbf{0}$, but $f \neq \mathbf{0}$.

We close this section with an important characterization of normal operators.

Theorem 3.3.16. *Let $\mathcal{V}$ be a finite dimensional inner product space. A linear operator $f : \mathcal{V} \to \mathcal{V}$ is normal if and only if $\|f(\mathbf{v})\| = \|f^*(\mathbf{v})\|$ for every $\mathbf{v} \in \mathcal{V}$.*

Proof. Assume f is normal. Then for every $\mathbf{v} \in \mathcal{V}$ we have

$$\|f(\mathbf{v})\|^2 = \langle f(\mathbf{v}), f(\mathbf{v}) \rangle = \langle f^* f(\mathbf{v}), \mathbf{v} \rangle = \langle f f^*(\mathbf{v}), \mathbf{v} \rangle = \langle f^*(\mathbf{v}), f^*(\mathbf{v}) \rangle = \|f^*(\mathbf{v})\|^2.$$

Now assume $\|f(\mathbf{v})\| = \|f^*(\mathbf{v})\|$ for every $\mathbf{v} \in \mathcal{V}$. Since

$$\langle f^* f(\mathbf{v}), \mathbf{v} \rangle = \langle f(\mathbf{v}), f(\mathbf{v}) \rangle = \|f(\mathbf{v})\|^2 = \|f^*(\mathbf{v})\|^2 = \langle f^*(\mathbf{v}), f^*(\mathbf{v}) \rangle = \langle f f^*(\mathbf{v}), \mathbf{v} \rangle,$$

for every $\mathbf{v} \in \mathcal{V}$, we have $f f^* = f^* f$ by Theorems 3.3.11 and 3.3.13. □

3.4 Spectral theorems

Spectral decomposition of matrices is one the most important ideas in matrix linear algebra. Here we generalize this idea to operators on arbitrary finite dimensional inner product spaces.

3.4.1 Spectral theorems for operators on complex inner product spaces

In this section all inner product spaces are assumed to be complex.

Definition 3.4.1. Let V be a complex vector space and let $f : V \to V$ be a linear operator.

(a) $\lambda \in \mathbb{C}$ is called an *eigenvalue* of f if $f(\mathbf{v}) = \lambda \mathbf{v}$ for some nonzero $\mathbf{v} \in V$.

(b) If $\lambda \in \mathbb{C}$ is an eigenvalue of f, then every nonzero vector $\mathbf{v} \in V$ such that $f(\mathbf{v}) = \lambda \mathbf{v}$ is called an *eigenvector* of f corresponding to λ.

The set of all eigenvectors of a linear operator f corresponding to an eigenvalue λ is not a vector subspace of V because the zero vector is not an eigenvector, but if we include the zero vector, then we obtain a subspace that is called the *eigenspace* of f corresponding to λ and is denoted by $\mathcal{E}_\lambda$.

Example 3.4.2. Consider the complex vector space $V = \text{Span}\{e^t, e^{2t}, \ldots, e^{nt}\}$. Show that $1, 2, \ldots, n$ are eigenvalues of the differential operator $\frac{d}{dt}$ on the space V.

Solution. For every $k \in \{1, 2, \ldots, n\}$ we have

$$\frac{d}{dt} e^{kt} = k e^{kt}.$$

This means that k is an eigenvalue of the differential operator $\frac{d}{dt}$ and the function e^{kt} is an eigenvector corresponding to k. □

Example 3.4.3. Consider the real vector space $V = \text{Span}\{1, t, t^2, \ldots, t^n\}$. Show that 0 is the only eigenvalue of the differential operator $\frac{d}{dt}$ on the space V.

Solution. Since $\frac{d}{dt}1 = 0 = 0 \cdot 1$, the number 0 is an eigenvalue of the differential operator $\frac{d}{dt}$ and the constant function 1 is an eigenvector corresponding to 0.

Now suppose there is a $\lambda \neq 0$ that is an eigenvalue of $\frac{d}{dt}$ on $\mathcal{V}$. Then

$$\frac{d}{dt}(z_0 + z_1 t + \cdots + z_k t^k) = \lambda(z_0 + z_1 t + \cdots + z_k t^k)$$

for some $k \leq n$ and some $z_0, z_1, \ldots, z_k \in \mathbb{R}$ such that $z_k \neq 0$. But this means that

$$z_1 + 2z_2 t + \cdots + kz_k t^{k-1} = \lambda(z_0 + z_1 t + \cdots + z_k t^k),$$

which implies $\lambda z_k = 0$, a contradiction. $\qquad\square$

Example 3.4.4. This example uses derivatives of the functions of the form $F : \mathbb{R} \to \mathbb{C}$.

Consider the complex vector space $\mathcal{V} = \mathrm{Span}\{\cos t, \sin t\}$. Show that i and $-i$ are eigenvalues of the differential operator $\frac{d}{dt}$.

Solution. Since

$$\frac{d}{dt}(\cos t + i \sin t) = -\sin t + i \cos t = i(\cos t + i \sin t),$$

i is an eigenvalue of the differential operator $\frac{d}{dt}$ and the function $\cos t + i \sin t$ is an eigenvector corresponding to i. Similarly, since

$$\frac{d}{dt}(\cos t - i \sin t) = -\sin t - i \cos t = -i(\cos t - i \sin t),$$

$-i$ is an eigenvalue of the differential operator $\frac{d}{dt}$ and the function $\cos t - i \sin t$ is an eigenvector corresponding to $-i$. $\qquad\square$

In the next example we are assuming that $f : \mathcal{V} \to \mathcal{V}$ is an operator such that there is an orthonormal basis $\{e_1, \ldots, e_n\}$ of $\mathcal{V}$ consisting of eigenvectors of f. As we will see later in this section, every normal operator has this property.

Example 3.4.5. Let $\mathcal{V}$ be a finite dimensional inner product space and let $f : \mathcal{V} \to \mathcal{V}$ be a linear operator such that there is an orthonormal basis $\{e_1, \ldots, e_n\}$ of $\mathcal{V}$ consisting of eigenvectors of f. Show that for every $\mathbf{v} \in \mathcal{V}$ we have

$$f(\mathbf{v}) = \sum_{j=1}^{n} \lambda_j \langle \mathbf{v}, e_j \rangle e_j,$$

where λ_j is the eigenvalue corresponding to the eigenvector $\mathbf{v}_j$.

Solution. Since $\mathbf{v} = \sum_{j=1}^{n} \langle \mathbf{v}, \mathbf{e}_j \rangle \mathbf{e}_j$ for every $\mathbf{v} \in V$, we have

$$f(\mathbf{v}) = \sum_{j=1}^{n} \langle \mathbf{v}, \mathbf{e}_j \rangle f(\mathbf{e}_j) = \sum_{j=1}^{n} \lambda_j \langle \mathbf{v}, \mathbf{e}_j \rangle \mathbf{e}_j.$$

$\square$

The following theorem gives us two useful descriptions of eigenvalues.

Theorem 3.4.6. *Let V be a finite dimensional complex vector space and let $f : V \to V$ be a linear operator. The following conditions are equivalent:*

(a) $\lambda \in \mathbb{C}$ *is an eigenvalue of f;*

(b) $\ker(f - \lambda \operatorname{Id}) \neq \{\mathbf{0}\}$;

(c) *The operator $f - \lambda \operatorname{Id}$ is not invertible.*

Proof. Equivalence (a) and (b) is an immediate consequence of the definitions. Equivalence (b) and (c) follows from Theorem 2.1.23. $\square$

Definition 3.4.7. *If $f : V \to V$ is a linear operator on a vector space V and $p(z) = a_0 + a_1 z + \cdots + a_m z^m$ is a polynomial, we define*

$$p(f) = a_0 \operatorname{Id} + a_1 f + \cdots + a_m f^m.$$

Since compositions and linear combination of linear operators are linear operators, $p(f)$ is a linear operator. Clearly, if p and q are polynomials, then $(p+q)(f) = p(f) + q(f)$ and $(pq)(f) = p(f)q(f)$.

Theorem 3.4.8. *If V is a nontrivial complex vector space of finite dimension, then every linear operator $f : V \to V$ has an eigenvalue.*

Proof. Let $\dim V = n$. Since $\dim \mathcal{L}(V) = n^2$, the operators $\operatorname{Id}, f, f^2, \ldots, f^{n^2}$ are linearly dependent and thus

$$a_0 \operatorname{Id} + a_1 f + a_2 f^2 + \cdots + a_k f^k = \mathbf{0}$$

for some $k \leq n^2$ and $a_0, a_1, a_2, \ldots, a_k \in \mathbb{C}$ such that $a_k \neq 0$. Now, by the Fundamental Theorem of Algebra, there are complex numbers $z_1, \ldots, z_k$ such that

$$a_0 + a_1 t + \cdots + a_k t^k = a_k (t - z_1) \cdots (t - z_k).$$

Consequently,

$$a_0 \operatorname{Id} + a_1 f + a_2 f^2 + \cdots + a_k f^k = a_k (f - z_1 \operatorname{Id}) \cdots (f - z_k \operatorname{Id}).$$

Since the operator $(f - z_1 \operatorname{Id}) \cdots (f - z_k \operatorname{Id})$ is not invertible, for at least one $j \in \{1, \ldots, k\}$ the operator $f - z_j \operatorname{Id}$ is not invertible, which means that z_j is an eigenvalue of f, as noted in Theorem 3.4.6. □

Theorem 3.4.9. *Let V be a finite dimensional inner product space and let $f : V \to V$ be a normal operator. If λ is an eigenvalue of f, then $\overline{\lambda}$ is an eigenvalue of f^*. Moreover, every eigenvector of f corresponding λ is an eigenvector of f^* corresponding $\overline{\lambda}$.*

Proof. First we note that, if f is a normal operator, then $f - \lambda \operatorname{Id}$ is a normal operator and we have

$$(f - \lambda \operatorname{Id})^* = f^* - \overline{\lambda} \operatorname{Id}.$$

Let $\mathbf{v}$ be an eigenvector of f corresponding to λ. Then, by Theorem 3.3.16, we have

$$0 = \|f(\mathbf{v}) - \lambda \mathbf{v}\| = \|f^*(\mathbf{v}) - \overline{\lambda} \mathbf{v}\|$$

and consequently

$$f^*(\mathbf{v}) = \overline{\lambda} \mathbf{v}.$$

□

Theorem 3.4.10. *Let V be a finite dimensional inner product space inner product space and let $f : V \to V$ be a normal operator. Eigenvectors of f corresponding to different eigenvalues are orthogonal.*

Proof. We need to show that if λ and μ are two distinct eigenvalues of f and $\mathbf{v}$ and $\mathbf{w}$ eigenvectors of f corresponding to λ and μ, respectively, then $\langle \mathbf{v}, \mathbf{w} \rangle = 0$. Indeed, since

$$\lambda \langle \mathbf{v}, \mathbf{w} \rangle = \langle f(\mathbf{v}), \mathbf{w} \rangle = \langle \mathbf{v}, f^*(\mathbf{w}) \rangle = \langle \mathbf{v}, \overline{\mu} \mathbf{w} \rangle = \mu \langle \mathbf{v}, \mathbf{w} \rangle,$$

we have $(\lambda - \mu) \langle \mathbf{v}, \mathbf{w} \rangle = 0$. Consequently, $\langle \mathbf{v}, \mathbf{w} \rangle = 0$, because $\lambda \neq \mu$. □

Note that the property in the above theorem can be expressed as follows: *Eigenspaces of a normal operator corresponding to different eigenvalues are mutually orthogonal subspaces.*

> **Definition 3.4.11.** A subspace $\mathcal{U}$ of a vector space V is called an *invariant space* of a linear operator $f : V \to V$, or simply *f-invariant*, if $f(\mathcal{U}) \subseteq \mathcal{U}$. A subspace $\mathcal{U}$ of an inner product space V is called a *reducing space* of a linear operator $f : V \to V$, if both $\mathcal{U}$ and $\mathcal{U}^\perp$ are f-invariant.

Example 3.4.12. Let $f : \mathbb{R}^3 \to \mathbb{R}^3$ be the linear operator defined by

$$f\left(\begin{bmatrix} x \\ y \\ z \end{bmatrix}\right) = \begin{bmatrix} 2 & 3 & 7 \\ 0 & 8 & 0 \\ 1 & -2 & 3 \end{bmatrix}\begin{bmatrix} x \\ y \\ z \end{bmatrix}.$$

Show that $\mathrm{Span}\left\{\begin{bmatrix} 1 \\ 0 \\ 0 \end{bmatrix}, \begin{bmatrix} 0 \\ 0 \\ 1 \end{bmatrix}\right\}$ is f-invariant.

Proof. We have

$$f\left(\begin{bmatrix} 1 \\ 0 \\ 0 \end{bmatrix}\right) = \begin{bmatrix} 2 \\ 0 \\ 1 \end{bmatrix} \in \mathrm{Span}\left\{\begin{bmatrix} 1 \\ 0 \\ 0 \end{bmatrix}, \begin{bmatrix} 0 \\ 0 \\ 1 \end{bmatrix}\right\}$$

and

$$f\left(\begin{bmatrix} 0 \\ 0 \\ 1 \end{bmatrix}\right) = \begin{bmatrix} 7 \\ 0 \\ 3 \end{bmatrix} \in \mathrm{Span}\left\{\begin{bmatrix} 1 \\ 0 \\ 0 \end{bmatrix}, \begin{bmatrix} 0 \\ 0 \\ 1 \end{bmatrix}\right\}.$$

$\square$

The following two theorems characterize invariant spaces and reducing spaces in terms of projections.

> **Theorem 3.4.13.** *Let V be an inner product space and let $f : V \to V$ be a linear operator. A finite dimensional subspace $\mathcal{U} \subseteq V$ is f-invariant if and only if*
> $$f\,\mathrm{proj}_{\mathcal{U}} = \mathrm{proj}_{\mathcal{U}} f\,\mathrm{proj}_{\mathcal{U}}.$$

Proof. If $\mathcal{U} \subseteq V$ is f-invariant, then $\mathrm{proj}_{\mathcal{U}}(\mathbf{v}) \in \mathcal{U}$ and $f(\mathrm{proj}_{\mathcal{U}}(\mathbf{v})) \in \mathcal{U}$ for every $\mathbf{v} \in \mathcal{V}$. Consequently

$$\text{proj}_{\mathcal{U}}(f(\text{proj}_{\mathcal{U}}(\mathbf{v}))) = f(\text{proj}_{\mathcal{U}}(\mathbf{v})),$$

that is, $f\text{proj}_{\mathcal{U}} = \text{proj}_{\mathcal{U}} f \, \text{proj}_{\mathcal{U}}$.

On the other hand, if $f\text{proj}_{\mathcal{U}} = \text{proj}_{\mathcal{U}} f \, \text{proj}_{\mathcal{U}}$, then

$$f(\mathbf{u}) = f(\text{proj}_{\mathcal{U}}(\mathbf{u})) = \text{proj}_{\mathcal{U}}(f(\text{proj}_{\mathcal{U}}(\mathbf{u}))) \in \mathcal{U}$$

for every $\mathbf{u} \in \mathcal{U}$, which means that $\mathcal{U}$ is f-invariant.   □

Theorem 3.4.14. *Let $\mathcal{V}$ be an inner product space and let $f : \mathcal{V} \to \mathcal{V}$ be a linear operator. A finite dimensional subspace $\mathcal{U} \subseteq \mathcal{V}$ is a reducing subspace of f if and only if*

$$f\text{proj}_{\mathcal{U}} = \text{proj}_{\mathcal{U}} f.$$

Proof. By Theorem 3.4.13, the subspace $\mathcal{U} \subseteq \mathcal{V}$ is a reducing subspace of f if and only if

$$f\text{proj}_{\mathcal{U}} = \text{proj}_{\mathcal{U}} f\text{proj}_{\mathcal{U}} \quad \text{and} \quad f(\text{Id} - \text{proj}_{\mathcal{U}}) = (\text{Id} - \text{proj}_{\mathcal{U}})f(\text{Id} - \text{proj}_{\mathcal{U}}),$$

because $\text{proj}_{\mathcal{U}^\perp} = \text{Id} - \text{proj}_{\mathcal{U}}$. The above is equivalent to

$$f\text{proj}_{\mathcal{U}} = \text{proj}_{\mathcal{U}} f\text{proj}_{\mathcal{U}} \quad \text{and} \quad \text{proj}_{\mathcal{U}} f = \text{proj}_{\mathcal{U}} f\text{proj}_{\mathcal{U}}$$

or simply to

$$f\text{proj}_{\mathcal{U}} = \text{proj}_{\mathcal{U}} f,$$

because the equality $f\text{proj}_{\mathcal{U}} = \text{proj}_{\mathcal{U}} f$ implies

$$\text{proj}_{\mathcal{U}} f\text{proj}_{\mathcal{U}} = \text{proj}_{\mathcal{U}} \text{proj}_{\mathcal{U}} f = \text{proj}_{\mathcal{U}} f.$$

□

Theorem 3.4.15. *Let $\mathcal{V}$ be a finite dimensional inner product space and let $f : \mathcal{V} \to \mathcal{V}$ be a linear operator. If a subspace $\mathcal{U} \subseteq \mathcal{V}$ is an invariant subspace of f, then $\mathcal{U}^\perp$ is an invariant subspace of f^*.*

Proof. Assume $\mathcal{U} \subseteq \mathcal{V}$ is an invariant subspace of f. If $\mathbf{u} \in \mathcal{U}$ and $\mathbf{v} \in \mathcal{U}^\perp$, then

$$\langle f^*(\mathbf{v}), \mathbf{u} \rangle = \langle \mathbf{v}, f(\mathbf{u}) \rangle = 0,$$

because $f(\mathbf{u}) \in \mathcal{U}$. Consequently, $f^*(\mathbf{v}) \in \mathcal{U}^\perp$.   □

Theorem 3.4.16. *Let V be a finite dimensional inner product space and let $f : V \to V$ be a nonzero linear operator. The following conditions are equivalent:*

(a) *f is a normal operator;*

(b) *There are orthonormal vectors $\mathbf{e}_1, \ldots, \mathbf{e}_r \in V$ and nonzero complex numbers $\lambda_1, \ldots, \lambda_r$ such that for every $\mathbf{v} \in V$ we have*

$$f(\mathbf{v}) = \sum_{j=1}^{r} \lambda_j \langle \mathbf{v}, \mathbf{e}_j \rangle \mathbf{e}_j;$$

(c) *There are orthonormal vectors $\mathbf{e}_1, \ldots, \mathbf{e}_r \in V$ and nonzero complex numbers $\lambda_1, \ldots, \lambda_r$ such that*

$$f = \sum_{j=1}^{r} \lambda_j \operatorname{proj}_{\mathbf{e}_j}.$$

Proof. First we note that (b) and (c) are equivalent because $\operatorname{proj}_{\mathbf{e}_j}(\mathbf{v}) = \langle \mathbf{v}, \mathbf{e}_j \rangle \mathbf{e}_j$ for every $\mathbf{v} \in V$.

If $f(\mathbf{v}) = \sum_{j=1}^{r} \lambda_j \langle \mathbf{v}, \mathbf{e}_j \rangle \mathbf{e}_j$ for every $\mathbf{v} \in V$, then $f(\mathbf{e}_j) = \lambda_j \mathbf{e}_j$ for every $j = 1, \ldots, r$ and hence

$$ff^*(\mathbf{v}) = \sum_{j=1}^{r} \lambda_j \langle f^* \mathbf{v}, \mathbf{e}_j \rangle \mathbf{e}_j = \sum_{j=1}^{r} \lambda_j \langle \mathbf{v}, f(\mathbf{e}_j) \rangle \mathbf{e}_j$$

$$= \sum_{j=1}^{r} \lambda_j \langle \mathbf{v}, \lambda_j \mathbf{e}_j \rangle \mathbf{e}_j = \sum_{j=1}^{r} \lambda_j \overline{\lambda_j} \langle \mathbf{v}, \mathbf{e}_j \rangle \mathbf{e}_j = \sum_{j=1}^{r} |\lambda_j|^2 \langle \mathbf{v}, \mathbf{e}_j \rangle \mathbf{e}_j.$$

On the other hand, since

$$\langle f(\mathbf{v}), \mathbf{u} \rangle = \left\langle \sum_{j=1}^{r} \lambda_j \langle \mathbf{v}, \mathbf{e}_j \rangle \mathbf{e}_j, \mathbf{u} \right\rangle = \sum_{j=1}^{r} \lambda_j \langle \mathbf{v}, \mathbf{e}_j \rangle \langle \mathbf{e}_j, \mathbf{u} \rangle$$

$$= \sum_{j=1}^{r} \langle \mathbf{v}, \overline{\lambda_j} \langle \mathbf{u}, \mathbf{e}_j \rangle \mathbf{e}_j \rangle = \left\langle \mathbf{v}, \sum_{j=1}^{r} \overline{\lambda_j} \langle \mathbf{u}, \mathbf{e}_j \rangle \mathbf{e}_j \right\rangle,$$

for every $\mathbf{v}, \mathbf{u} \in V$, we have

$$f^*(\mathbf{v}) = \sum_{j=1}^{r} \overline{\lambda_j} \langle \mathbf{v}, \mathbf{e}_j \rangle \mathbf{e}_j$$

and thus $f^*(\mathbf{e}_j) = \overline{\lambda_j}\mathbf{e}_j$ for every $j = 1,\ldots,r$. Consequently

$$f^*f(\mathbf{v}) = \sum_{j=1}^{r}\overline{\lambda_j}\langle f\mathbf{v},\mathbf{e}_j\rangle\mathbf{e}_j = \sum_{j=1}^{r}\overline{\lambda_j}\langle\mathbf{v},f^*(\mathbf{e}_j)\rangle\mathbf{e}_j$$

$$= \sum_{j=1}^{r}\overline{\lambda_j}\langle\mathbf{v},\overline{\lambda_j}\mathbf{e}_j\rangle\mathbf{e}_j = \sum_{j=1}^{r}\overline{\lambda_j}\lambda_j\langle\mathbf{v},\mathbf{e}_j\rangle\mathbf{e}_j = \sum_{j=1}^{r}|\lambda_j|^2\langle\mathbf{v},\mathbf{e}_j\rangle\mathbf{e}_j.$$

This shows that, if $f(\mathbf{v}) = \sum_{j=1}^{r}\lambda_j\langle\mathbf{v},\mathbf{e}_j\rangle\mathbf{e}_j$ for every $\mathbf{v}\in\mathcal{V}$, then f is a normal operator, that is, (b) implies (a).

To complete the proof we show that (a) implies (c). Assume f is a nonzero normal operator. By Theorem 3.4.8, f has an eigenvalue λ. Let $\mathbf{e}$ be a unit eigenvector of f corresponding to λ. Then, by Theorem 3.4.9, $\overline{\lambda}$ is an eigenvalue of f^* with the same eigenvector $\mathbf{e}$ and thus the subspace $\mathrm{Span}\{\mathbf{e}\}$ is f-invariant and f^*-invariant. Consequently, by Theorem 3.4.15, $\mathrm{Span}\{\mathbf{e}\}^{\perp}$ is f-invariant and f^*-invariant.

Now we argue by induction on the dimension of the range of f.

If $\dim\mathrm{ran}\,f = 1$, then clearly $f = \lambda\mathrm{proj}_{\mathbf{e}}$ and we are done.

Now we assume that $\dim\mathrm{ran}\,f = r$ for some $r > 1$ and that the implication (a) implies (c) is proved for every normal operator g such that $\dim\mathrm{ran}\,g = q < r$.

We denote $\lambda = \lambda_r$ and $\mathbf{e} = \mathbf{e}_r$. Because, as observed above, the subspaces $\mathrm{Span}\{\mathbf{e}_r\}$ and $\mathrm{Span}\{\mathbf{e}_r\}^{\perp}$ are f-invariant and f^*-invariant, the operators f and f^* commute with the projection on $\mathrm{Span}\{\mathbf{e}_r\}$ and we have, by Theorem 3.4.14,

$$f\mathrm{proj}_{\mathbf{e}_r} = \mathrm{proj}_{\mathbf{e}_r}f = \lambda_r\mathrm{proj}_{\mathbf{e}_r} \quad\text{and}\quad f^*\mathrm{proj}_{\mathbf{e}_r} = \mathrm{proj}_{\mathbf{e}_r}f^* = \overline{\lambda_r}\mathrm{proj}_{\mathbf{e}_r}.$$

Consequently,

$$(f-\lambda_r\mathrm{proj}_{\mathbf{e}_r})(f^*-\overline{\lambda_r}\mathrm{proj}_{\mathbf{e}_r}) = ff^*-|\lambda_r|^2\mathrm{proj}_{\mathbf{e}_r} = (f^*-\overline{\lambda_r}\mathrm{proj}_{\mathbf{e}_r})(f-\lambda_r\mathrm{proj}_{\mathbf{e}_r}),$$

which means that $f-\lambda_r\mathrm{proj}_{\mathbf{e}_r}$ is a normal operator. Moreover, since $\mathrm{ran}\,f = \mathrm{Span}\{\mathbf{e}_r\}\oplus f(\mathrm{Span}\{\mathbf{e}_r\}^{\perp})$, we have

$$\dim\mathrm{ran}(f-\lambda_r\mathrm{proj}_{\mathbf{e}_r}) = \dim f(\mathrm{Span}\{\mathbf{e}_r\}^{\perp}) = r-1.$$

By our inductive assumption there are orthonormal vectors $\mathbf{e}_1,\ldots,\mathbf{e}_{r-1}$ and nonzero complex numbers $\lambda_1,\ldots,\lambda_{r-1}$ such that

$$f-\lambda_r\mathrm{proj}_{\mathbf{e}_r} = \sum_{j=1}^{r-1}\lambda_j\mathrm{proj}_{\mathbf{e}_j},$$

which gives us the desired representation $f = \sum_{j=1}^{r}\lambda_j\mathrm{proj}_{\mathbf{e}_j}$. $\square$

Theorem 3.4.17. *Let $\mathcal{V}$ be a finite dimensional inner product space. The operator $f : \mathcal{V}\to\mathcal{V}$ is normal if and only if there is an orthonormal basis of $\mathcal{V}$ consisting of eigenvectors of f.*

Proof. Let $\dim V = n$. The orthonormal vectors $\mathbf{e}_1, \ldots, \mathbf{e}_r$ in Theorem 3.4.16 are eigenvectors of f and we have $\operatorname{ran} f = \operatorname{Span}\{\mathbf{e}_1, \ldots, \mathbf{e}_r\}$. Since $V = \operatorname{ran} f \oplus (\operatorname{ran} f)^\perp$, we have $\dim(\operatorname{ran} f)^\perp = n - r$ and there are orthonormal vectors $\mathbf{e}_{r+1}, \ldots, \mathbf{e}_n$ such that $(\operatorname{ran} f)^\perp = \operatorname{Span}\{\mathbf{e}_{r+1}, \ldots, \mathbf{e}_n\}$.

If $\mathbf{v} \in (\operatorname{ran} f)^\perp$, then $\langle \mathbf{v}, f(f^*(\mathbf{v})) \rangle = 0$. Since f is normal, we have $f(f^*(\mathbf{v})) = f^*(f(\mathbf{v}))$ and thus

$$0 = \langle \mathbf{v}, f(f^*(\mathbf{v})) \rangle = \langle \mathbf{v}, f^*(f(\mathbf{v})) \rangle = \|f(\mathbf{v})\|^2.$$

Hence $f(\mathbf{v}) = \mathbf{0}$ and thus $\mathbf{e}_{r+1}, \ldots, \mathbf{e}_n \in \ker f$, which means that they are eigenvectors corresponding to the eigenvalue 0.

By the Rank-Nullity Theorem we have $\dim \ker f = n - r$ and thus the set $\{\mathbf{e}_{r+1}, \ldots, \mathbf{e}_n\}$ is a basis of $\ker f$. Consequently, $\{\mathbf{e}_1, \ldots, \mathbf{e}_n\}$ is an orthonormal basis of V consisting of eigenvectors of f.

On the other hand, if there is an orthonormal basis $\{\mathbf{e}_1, \ldots, \mathbf{e}_n\}$ of V consisting of eigenvectors of f, then the operator f is normal by Example 3.4.5 and Theorem 3.4.16. $\square$

Example 3.4.18. Let V be a finite dimensional inner product space and let $f : V \to V$ be a normal operator such that for every $\mathbf{v} \in V$ we have

$$f(\mathbf{v}) = \sum_{j=1}^n \lambda_j \langle \mathbf{v}, \mathbf{e}_j \rangle \mathbf{e}_j,$$

where $\{\mathbf{e}_1, \ldots, \mathbf{e}_n\}$ is an orthonormal basis of V consisting of eigenvectors of f and $\lambda_1, \ldots, \lambda_n$ are the corresponding eigenvalues. Show that $p(f)$ is a normal operator and we have

$$p(f)(\mathbf{v}) = \sum_{j=1}^n p(\lambda_j) \langle \mathbf{v}, \mathbf{e}_j \rangle \mathbf{e}_j$$

for any polynomial p.

Solution. Since $\lambda_1, \ldots, \lambda_n$ are eigenvalues of f corresponding to eigenvectors $\mathbf{e}_1, \ldots, \mathbf{e}_n$, we have $f^k(\mathbf{e}_j) = \lambda_j^k \mathbf{e}_j$ for every $j \in \{1, \ldots, n\}$ and every integer $k \geq 1$.

Consequently, $p(f)(\mathbf{e}_j) = p(\lambda_j)\mathbf{e}_j$ for every $j \in \{1, \ldots, n\}$ and consequently the linear operator $p(f)$ is normal according to Theorem 3.4.17. From Example 3.4.5 we get

$$p(f)(\mathbf{v}) = \sum_{j=1}^n p(\lambda_j) \langle \mathbf{v}, \mathbf{e}_j \rangle \mathbf{e}_j.$$

$\square$

A representation of a linear operator $f : \mathcal{V} \to \mathcal{V}$ in the form

$$f = \sum_{j=1}^{r} \lambda_j \mathrm{proj}_{\mathbf{e}_j},$$

as in Theorem 3.4.16, is called a *spectral decomposition* of f.

Example 3.4.19. In Example 3.3.10 we show that the operator $f \in \mathcal{L}(\mathbb{C}^2, \mathbb{C}^2)$ defined by $f(\mathbf{x}) = A\mathbf{x}$ where

$$A = \begin{bmatrix} i & 1-i \\ -1-i & 2i \end{bmatrix}$$

is normal. Find a spectral decomposition of f.

Solution. First we find eigenvalues of f. We need to find values of $\lambda \in \mathbb{C}$ such that the operator $f - \lambda \, \mathrm{Id}$ is not invertible. Since

$$(f - \lambda \, \mathrm{Id})\mathbf{x} = \begin{bmatrix} i - \lambda & 1-i \\ -1-i & 2i - \lambda \end{bmatrix} \mathbf{x}$$

and

$$\det \begin{bmatrix} i - \lambda & 1-i \\ -1-i & 2i - \lambda \end{bmatrix} = (i - \lambda)(2i - \lambda) - (1-i)(-1-i) = \lambda(\lambda - 3i),$$

f has two eigenvalues: 0 and $3i$.

Next we need to find an eigenvector corresponding to $3i$, that is a nonzero vector $\begin{bmatrix} x_1 \\ x_2 \end{bmatrix} \in \mathbb{C}^2$ such that

$$\begin{bmatrix} -2i & 1-i \\ -1-i & -i \end{bmatrix} \begin{bmatrix} x_1 \\ x_2 \end{bmatrix} = \mathbf{0}.$$

Since the vector $\begin{bmatrix} 1 \\ -1+i \end{bmatrix}$ satisfies the above equation, it is an eigenvector corresponding to the eigenvalue $3i$ and

$$f = 3i\,\mathrm{proj}_{\begin{bmatrix} 1 \\ -1+i \end{bmatrix}}$$

is a spectral decomposition of f.

The vector $\begin{bmatrix} 1+i \\ 1 \end{bmatrix}$ is an eigenvector corresponding to 0, but it is not used in the spectral decomposition of f. Note that, as expected, the vectors $\begin{bmatrix} 1 \\ -1+i \end{bmatrix}$ and $\begin{bmatrix} 1+i \\ 1 \end{bmatrix}$ are orthogonal. $\square$

Example 3.4.20. Show that the operator $f \in \mathcal{L}(\mathbb{C}^2, \mathbb{C}^2)$ defined by $f(\mathbf{x}) = A\mathbf{x}$, where

$$ A = \begin{bmatrix} 3+2i & 2-4i \\ 4+2i & 6+i \end{bmatrix}, $$

is normal and find a spectral decomposition of f.

Solution. Since

$$ \det \begin{bmatrix} 3+2i-\lambda & 2-4i \\ 4+2i & 6+i-\lambda \end{bmatrix} = (3+2i-\lambda)(6+i-\lambda) - (4+2i)(2-4i) $$
$$ = 16 + 15i + \lambda^2 - (9+3i)\lambda + 16 - 12i $$
$$ = \lambda^2 - (9+3i)\lambda + 27i $$
$$ = (9-\lambda)(3i-\lambda) $$

the eigenvalues of the operator f are 9 and $3i$.

Next we find that $\begin{bmatrix} 4i-2 \\ -6+2i \end{bmatrix}$ is an eigenvector corresponding to 9 and $\begin{bmatrix} 4i-2 \\ 3-i \end{bmatrix}$ is an eigenvector corresponding to $3i$. The vectors $\mathbf{v}_1 = \begin{bmatrix} 4i-2 \\ -6+2i \end{bmatrix}$ and $\mathbf{v}_2 = \begin{bmatrix} 4i-2 \\ 3-i \end{bmatrix}$ are orthogonal and thus $\{\mathbf{v}_1, \mathbf{v}_2\}$ is an orthogonal basis of $\mathcal{L}(\mathbb{C}^2, \mathbb{C}^2)$ and consequently the operator f is normal. Thus

$$ f = 9\mathrm{proj}_{\mathbf{v}_1} + 3i\mathrm{proj}_{\mathbf{v}_2} $$

is a spectral decomposition of f.

For practical calculations we can use matrices of the projections $\text{proj}_{\mathbf{v}_1}$ and $\text{proj}_{\mathbf{v}_2}$, which gives us

$$f\left(\begin{bmatrix} x_1 \\ x_2 \end{bmatrix}\right) = 9\left(\frac{1}{\sqrt{60}}\begin{bmatrix} 4i-2 \\ -6+2i \end{bmatrix}\right)\left(\frac{1}{\sqrt{60}}\begin{bmatrix} -4i-2 & -6-2i \end{bmatrix}\right)\begin{bmatrix} x_1 \\ x_2 \end{bmatrix}$$

$$+ 3i\left(\frac{1}{\sqrt{30}}\begin{bmatrix} 4i-2 \\ 3-i \end{bmatrix}\right)\left(\frac{1}{\sqrt{30}}\begin{bmatrix} -4i-2 & 3+i \end{bmatrix}\right)\begin{bmatrix} x_1 \\ x_2 \end{bmatrix}$$

$$= 3\begin{bmatrix} 1 & 1-i \\ 1+i & 2 \end{bmatrix}\begin{bmatrix} x_1 \\ x_2 \end{bmatrix} + i\begin{bmatrix} 2 & -1+i \\ -1-i & 1 \end{bmatrix}\begin{bmatrix} x_1 \\ x_2 \end{bmatrix}$$

for every $\begin{bmatrix} x_1 \\ x_2 \end{bmatrix} \in \mathbb{C}^2$.   □

As an immediate consequence of Theorem 3.4.16 we obtain the following result, which is also referred to as the *spectral representation*.

Theorem 3.4.21. *Let* V *be a finite dimensional inner product space and let* $f : V \to V$ *be a normal operator. Then*

$$f = \sum_{j=1}^{q} \lambda_j \text{proj}_{\mathcal{E}_{\lambda_j}},$$

where $\lambda_1, \ldots, \lambda_q$ *are all distinct eigenvalues of* f.

Let V be an inner product space and let $\mathcal{U}_1, \ldots, \mathcal{U}_n$ be subspaces of V such that

$$V = \mathcal{U}_1 \oplus \cdots \oplus \mathcal{U}_r.$$

If the subspaces $\mathcal{U}_1, \ldots, \mathcal{U}_n$ are mutually orthogonal, that is, $\langle \mathbf{x}, \mathbf{y} \rangle = 0$ for any $\mathbf{x} \in \mathcal{U}_j$ and $\mathbf{y} \in \mathcal{U}_k$ with $j \neq k$, then we say that $\mathcal{U}_1 \oplus \cdots \oplus \mathcal{U}_r$ is an *orthogonal decomposition* of the space V.

From Theorem 3.4.21 we obtain the following important result.

Theorem 3.4.22. *Let* $f : V \to V$ *be a normal operator on a finite dimensional inner product space* V. *If* $\lambda_1, \ldots, \lambda_q$ *are all distinct eigenvalues of* f, *then* $\mathcal{E}_{\lambda_1} \oplus \cdots \oplus \mathcal{E}_{\lambda_q}$ *is an orthogonal decomposition of* V.

The decomposition of a normal operator in Theorem 3.4.21 is unique as shown in the next theorem.

Theorem 3.4.23. *Let V be a finite-dimensional inner product space and let $f : V \to V$ be a linear operator. If $U_1 \oplus \cdots \oplus U_r$ is an orthogonal decomposition of V and*

$$f = \sum_{j=1}^{r} \lambda_j \text{proj}_{U_j}$$

for some distinct $\lambda_1, \ldots, \lambda_r \in \mathbb{K}$, then $U_j = \mathcal{E}_{\lambda_j}$ for every $j \in \{1, \ldots, r\}$.

Solution. If $v \in U_j$, then $f(v) = \lambda_j v$ and thus $v \in \mathcal{E}_{\lambda_k}$.

Now, if $v \in \mathcal{E}_{\lambda_k}$, then $f(v) = \lambda_k v = \sum_{j=1}^{r} \lambda_k \text{proj}_{U_j}(v)$ because $\sum_{j=1}^{r} \text{proj}_{U_j} = \text{Id}_V$. Since

$$f(v) = \sum_{j=1}^{r} \lambda_j \text{proj}_{U_j}(v),$$

we have

$$\sum_{j=1}^{r} \lambda_k \text{proj}_{U_j}(v) = \sum_{j=1}^{r} \lambda_j \text{proj}_{U_j}(v)$$

which gives us

$$\sum_{j=1, j \neq k}^{r} |\lambda_k - \lambda_j|^2 \|\text{proj}_{U_j}(v)\|^2 = 0.$$

Consequently, $\text{proj}_{U_j}(v) = 0$ for $j \neq k$. This means that $v = \text{proj}_{U_k}(v)$ and thus $v \in U_k$. $\square$

Now we turn our attention to spectral properties of self-adjoint operators.

Theorem 3.4.24. *Let V be a complex inner product space. All eigenvalues of a self-adjoint operator $f : V \to V$ are real numbers.*

Proof. Let λ be an eigenvalue of a self-adjoint operator $f : V \to V$. If x is an eigenvector of f corresponding to λ, then

$$\lambda \langle x, x \rangle = \langle \lambda x, x \rangle = \langle f(x), x \rangle = \langle x, f(x) \rangle = \langle x, \lambda x \rangle = \overline{\lambda} \langle x, x \rangle.$$

Since $\langle x, x \rangle = \|x\|^2 \neq 0$, we have $\overline{\lambda} = \lambda$. $\square$

The above property characterizes normal operators that are self-adjoint.

Theorem 3.4.25. *Let V be a finite dimensional complex inner product space. A normal operator $f : V \to V$ is self-adjoint if and only if all eigenvalues of f are real.*

Proof. In the proof of Theorem 3.4.16 we have shown that if

$$f(\mathbf{v}) = \sum_{j=1}^{r} \lambda_j \langle \mathbf{v}, \mathbf{e}_j \rangle \mathbf{e}_j$$

then

$$f^*(\mathbf{v}) = \sum_{j=1}^{r} \overline{\lambda_j} \langle \mathbf{v}, \mathbf{e}_j \rangle \mathbf{e}_j$$

Our result is a consequence of these equalities.

□

Example 3.4.26. Find a spectral decomposition of the self-adjoint operator $f \in \mathcal{L}(\mathbb{C}^2, \mathbb{C}^2)$ defined as

$$f\left(\begin{bmatrix} x_1 \\ x_2 \end{bmatrix}\right) = \begin{bmatrix} 33 & 24 - 24i \\ 24 + 24i & 57 \end{bmatrix} \begin{bmatrix} x_1 \\ x_2 \end{bmatrix}.$$

Solution. Since

$$\det \begin{bmatrix} 33 - \lambda & 24 - 24i \\ 24 + 24i & 57 - \lambda \end{bmatrix} = (33 - \lambda)(57 - \lambda) - (24 + 24i)(24 - 24i)$$

$$= (24 + 9 - \lambda)(48 + 9 - \lambda) - 24 \cdot 48$$
$$= (9 - \lambda)(81 - \lambda),$$

the eigenvalues of f are 9 and 81. The vectors $\begin{bmatrix} 1 - i \\ -1 \end{bmatrix}$ and $\begin{bmatrix} 1 - i \\ 2 \end{bmatrix}$ are eigenvectors of f corresponding to the eigenvalues 9 and 81, respectively. Consequently,

$$f = 9\mathrm{proj}_{\begin{bmatrix} 1 - i \\ -1 \end{bmatrix}} + 81\mathrm{proj}_{\begin{bmatrix} 1 - i \\ 2 \end{bmatrix}}$$

is a the spectral decomposition of f.

Since the matrix of the projection on Span $\left\{ \begin{bmatrix} 1 - i \\ -1 \end{bmatrix} \right\}$ is

$$\left(\frac{1}{\sqrt{3}} \begin{bmatrix} 1 - i \\ -1 \end{bmatrix} \right) \left(\frac{1}{\sqrt{3}} \begin{bmatrix} 1 + i & -1 \end{bmatrix} \right) = \frac{1}{3} \begin{bmatrix} 2 & -1 + i \\ -1 - i & 1 \end{bmatrix}$$

and the matrix of the projection on Span $\left\{ \begin{bmatrix} 1 - i \\ 2 \end{bmatrix} \right\}$ is

$$\left(\frac{1}{\sqrt{6}} \begin{bmatrix} 1 - i \\ 2 \end{bmatrix} \right) \left(\frac{1}{\sqrt{6}} \begin{bmatrix} 1 + i & 2 \end{bmatrix} \right) = \frac{1}{3} \begin{bmatrix} 1 & 1 - i \\ 1 + i & 2 \end{bmatrix},$$

we have

$$f\left(\begin{bmatrix} x_1 \\ x_2 \end{bmatrix}\right) = 3 \begin{bmatrix} 2 & -1+i \\ -1-i & 1 \end{bmatrix} \begin{bmatrix} x_1 \\ x_2 \end{bmatrix} + 27 \begin{bmatrix} 1 & 1-i \\ 1+i & 2 \end{bmatrix} \begin{bmatrix} x_1 \\ x_2 \end{bmatrix}$$

for every $\begin{bmatrix} x_1 \\ x_2 \end{bmatrix} \in \mathbb{C}^2$. □

3.4.2   Self-adjoint operators on real inner product spaces

Now we turn our attention to real inner product spaces. Some properties established for complex spaces remain true, but there are some essential differences.

Example 3.4.27. Consider the operator $f \in \mathcal{L}(\mathbb{R}^2, \mathbb{R}^2)$ defined by $f(\mathbf{x}) = A\mathbf{x}$, where

$$A = \begin{bmatrix} 0 & -1 \\ 1 & 0 \end{bmatrix}.$$

Show that f is a normal operator without any eigenvalues.

Solution. First we note that $f^*(\mathbf{x}) = A^T\mathbf{x}$, where $A^T = \begin{bmatrix} 0 & 1 \\ -1 & 0 \end{bmatrix}$. Since

$$\begin{bmatrix} 0 & -1 \\ 1 & 0 \end{bmatrix} \begin{bmatrix} 0 & 1 \\ -1 & 0 \end{bmatrix} = \begin{bmatrix} 0 & 1 \\ -1 & 0 \end{bmatrix} \begin{bmatrix} 0 & -1 \\ 1 & 0 \end{bmatrix} = \begin{bmatrix} 1 & 0 \\ 0 & 1 \end{bmatrix},$$

f is a normal operator. To show that f has no eigenvalues we note that

$$\det \begin{bmatrix} -\lambda & -1 \\ 1 & -\lambda \end{bmatrix} = \lambda^2 + 1$$

and the equation $\lambda^2 + 1 = 0$ has no real solutions. (A complex number cannot be an eigenvalue of an operator on a real vector space.) □

A matrix $A \in \mathcal{M}_{n \times n}(\mathbb{R})$ is called *symmetric* if $A^T = A$. Note that if $A \in \mathcal{M}_{n \times n}(\mathbb{R})$ is symmetric, then the operator $f : \mathbb{R}^n \to \mathbb{R}^n$ defined by $f(\mathbf{x}) = A\mathbf{x}$ is self-adjoint.

Lemma 3.4.28. *If $A \in \mathcal{M}_{n \times n}(\mathbb{R})$ is symmetric, then the operator $f : \mathbb{R}^n \to \mathbb{R}^n$ defined by $f(\mathbf{x}) = A\mathbf{x}$ has an eigenvalue.*

Proof. Let $g : \mathbb{C}^n \to \mathbb{C}^n$ be the linear operator defined by $g(\mathbf{x}) = A\mathbf{x}$. By Theorem 3.4.8, g has an eigenvalue λ. Since A is a symmetric matrix, λ is a real number. Let $\mathbf{z} \in \mathbb{C}^n$ be an eigenvector corresponding to λ. We can write $\mathbf{z} = \mathbf{x} + i\mathbf{y}$ where $\mathbf{x}, \mathbf{y} \in \mathbb{R}^n$. Then

$$A\mathbf{z} = A(\mathbf{x} + i\mathbf{y}) = \lambda(\mathbf{x} + i\mathbf{y}) = \lambda\mathbf{x} + i\lambda\mathbf{y}.$$

Since $A(\mathbf{x} + i\mathbf{y}) = A\mathbf{x} + iA\mathbf{y}$ and λ is a real number, it follows that $A(\mathbf{x}) = \lambda\mathbf{x}$ and $A(\mathbf{y}) = \lambda\mathbf{y}$. Now, because $\mathbf{z} \neq \mathbf{0}$, we have $\mathbf{x} \neq \mathbf{0}$ or $\mathbf{y} \neq \mathbf{0}$. Consequently, λ is an eigenvalue of f.   □

Theorem 3.4.29. *Let $\mathcal{V}$ be a finite dimensional real inner product space. Every self-adjoint operator $f : \mathcal{V} \to \mathcal{V}$ has an eigenvalue.*

Proof. Let $\mathcal{B} = \{\mathbf{v}_1, \ldots, \mathbf{v}_n\}$ be a basis of $\mathcal{V}$ and let A be the $\mathcal{B}$-matrix of f. Since f is self-adjoint, A is a symmetric matrix. Let $\lambda \in \mathbb{R}$ be an eigenvalue of A and let $\mathbf{x} = \begin{bmatrix} x_1 \\ \vdots \\ x_n \end{bmatrix} \in \mathbb{R}^n$ be an eigenvector of A corresponding to the eigenvalue λ. Then

$$A\begin{bmatrix} x_1 \\ \vdots \\ x_n \end{bmatrix} = \lambda \begin{bmatrix} x_1 \\ \vdots \\ x_n \end{bmatrix} = \begin{bmatrix} \lambda x_1 \\ \vdots \\ \lambda x_n \end{bmatrix}$$

and thus

$$f(x_1\mathbf{v}_1 + \cdots + x_n\mathbf{v}_n) = (\lambda x_1)\mathbf{v}_1 + \cdots + (\lambda x_n)\mathbf{v}_n = \lambda(x_1\mathbf{v}_1 + \cdots + x_n\mathbf{v}_n),$$

which means that the real number λ is an eigenvalue of f and $x_1\mathbf{v}_1 + \cdots + x_n\mathbf{v}_n$ is an eigenvector of f corresponding to λ.   □

The following theorem is a version of Theorem 3.4.16 for self-adjoint operators on real inner product spaces. The proof is similar to the proof of Theorem 3.4.16. Note that the above theorem is needed for the result.

Theorem 3.4.30. *Let V be a finite dimensional inner product space over $\mathbb{R}$ and let $f : V \to V$ be a nonzero linear operator. The following conditions are equivalent:*

(a) *f is a self-adjoint operator;*

(b) *There are orthonormal vectors $e_1, \ldots, e_r \in V$ and nonzero real numbers $\lambda_1, \ldots, \lambda_r$ such that for every $\mathbf{v} \in V$ we have*

$$f(\mathbf{v}) = \sum_{j=1}^{r} \lambda_j \langle \mathbf{v}, e_j \rangle e_j;$$

(c) *There are orthonormal vectors $e_1, \ldots, e_r \in V$ and nonzero real numbers $\lambda_1, \ldots, \lambda_r$ such that*

$$f = \sum_{j=1}^{r} \lambda_j \mathrm{proj}_{e_j}.$$

Example 3.4.31. Find all eigenvalues and a spectral decomposition of the operator $f \in \mathcal{L}(\mathbb{R}^3, \mathbb{R}^3)$ defined by $f(\mathbf{x}) = A\mathbf{x}$, where

$$A = \begin{bmatrix} 4 & 1 & 2 \\ 1 & 5 & 1 \\ 2 & 1 & 4 \end{bmatrix}.$$

Solution. Note that A is a symmetric matrix, so f has at least one real eigenvalue.

We have to determine the real numbers λ such that the system

$$\begin{cases} (4 - \lambda)x + y + 2z = 0 \\ x + (5 - \lambda)y + z = 0 \\ 2x + y + (4 - \lambda)z = 0 \end{cases}$$

has nontrivial solutions. By adding the second and third equations to the first one we get

$$\begin{cases} (7 - \lambda)(x + y + z) = 0 \\ x + (5 - \lambda)y + z = 0 \\ 2x + y + (4 - \lambda)z = 0 \end{cases}.$$

If $\lambda = 7$, then the system has nontrivial solutions. If $\lambda \neq 7$, then the system is equivalent to

$$\begin{cases} x + y + z & = 0 \\ x + (5 - \lambda)y + z & = 0 \\ 2x + y + (4 - \lambda)z = 0 \end{cases}$$

or

$$\begin{cases} x + y + z & = 0 \\ (4 - \lambda)y & = 0 \ . \\ 2x + y + (4 - \lambda)z = 0 \end{cases}$$

If $\lambda = 4$, then the system has nontrivial solutions. If $\lambda \neq 4$, then $y = 0$ and the system becomes

$$\begin{cases} x + z & = 0 \\ 2x + (4 - \lambda)z = 0 \end{cases}$$

or

$$\begin{cases} x + z & = 0 \\ (2 - \lambda)z = 0 \end{cases}$$

If $\lambda = 2$, then the system has nontrivial solutions. If $\lambda \neq 2$, then $x = y = z = 0$.

Consequently, the eigenvalues of the matrix $\begin{bmatrix} 4 & 1 & 2 \\ 1 & 5 & 1 \\ 2 & 1 & 4 \end{bmatrix}$ are 7, 4 and 2.

The vectors $\begin{bmatrix} 1 \\ 1 \\ 1 \end{bmatrix}$, $\begin{bmatrix} 1 \\ -2 \\ 1 \end{bmatrix}$, and $\begin{bmatrix} 1 \\ 0 \\ -1 \end{bmatrix}$, are eigenvectors corresponding to the eigenvalues 7, 4 and 2, respectively. If we let

$$\mathbf{v}_1 = \begin{bmatrix} \frac{1}{\sqrt{3}} \\ \frac{1}{\sqrt{3}} \\ \frac{1}{\sqrt{3}} \end{bmatrix}, \quad \mathbf{v}_2 = \begin{bmatrix} \frac{1}{\sqrt{6}} \\ -\frac{2}{\sqrt{6}} \\ \frac{1}{\sqrt{6}} \end{bmatrix}, \quad \mathbf{v}_3 = \begin{bmatrix} \frac{1}{\sqrt{2}} \\ 0 \\ -\frac{1}{\sqrt{2}} \end{bmatrix},$$

then we can write

$$f(\mathbf{x}) = 7\langle \mathbf{x}, \mathbf{v}_1 \rangle \mathbf{v}_1 + 4\langle \mathbf{x}, \mathbf{v}_2 \rangle \mathbf{v}_2 + 2\langle \mathbf{x}, \mathbf{v}_2 \rangle \mathbf{v}_2$$

or

$$f = 7\text{proj}_{\mathbf{v}_1} + 4\text{proj}_{\mathbf{v}_2} + 2\text{proj}_{\mathbf{v}_3}.$$

$\square$

3.4.3   Unitary operators

In this section all inner product spaces are assumed to be complex. We discuss a special type of normal operators, called unitary operators. These operators have interesting geometric and algebraic properties similar to rotations about the origin in $\mathbb{R}^n$.

Definition 3.4.32. Let V be a finite dimensional inner product space. A linear operator $f : V \to V$ is called a *unitary operator* if

$$ff^* = f^*f = \mathrm{Id},$$

where Id is the identity operator on V.

Note that the condition in the above definition implies that f is invertible, ran $f = V$, and $f^{-1} = f^*$.

Theorem 3.4.33. *Let V be a finite dimensional inner product space. If $f : V \to V$ is a unitary operator, then f^* and f^{-1} are unitary.*

Proof. If f is unitary, then

$$f^*(f^*)^* = f^*f = \mathrm{Id} \quad \text{and} \quad (f^*)^*f^* = ff^* = \mathrm{Id},$$

and hence f^* is unitary. Since $f^{-1} = f^*$, f^{-1} is unitary. $\qquad \square$

Example 3.4.34. Consider the vector space $P_n(\mathbb{C})$ with the inner product $\langle f, g \rangle = \int_0^1 f(t)\overline{g(t)}dt$. Show that the linear operator $\Phi : P_n(\mathbb{C}) \to P_n(\mathbb{C})$ defined as $\Phi(f)(t) = f(1-t)$ is a unitary operator.

Solution. Since, for every $f, g \in P_n(\mathbb{C})$, we have

$$\langle \Phi(f), g \rangle = \int_0^1 f(1-t)\overline{g(t)}dt = \int_0^1 f(t)\overline{g(1-t)}dt = \langle f, \Phi(g) \rangle,$$

Φ is self-adjoint. Hence, $\Phi^*\Phi = \Phi\Phi^* = \Phi^2 = \mathrm{Id}$. $\qquad \square$

Definition 3.4.35. Let V be a normed space. A linear operator $f : V \to V$ is called an *isometric operator* or an *isometry* if

$$\|f(\mathbf{x})\| = \|\mathbf{x}\|$$

for every $\mathbf{x} \in V$.

Example 3.4.36. Let z and w be two complex numbers such that $|z| = |w| = 1$. Find a linear isometry $f : \mathbb{C} \to \mathbb{C}$ such that $f(z) = w$.

Solution. The function $f(x) = w\bar{z}x$ has the desired property. Since for every $x \in \mathbb{C}$ we have

$$f(x) = f\left(\frac{x}{z}z\right) = \frac{x}{z}f(z) = \frac{x}{z}w = w\bar{z}x,$$

this solution is unique.

This function f can be interpreted as the rotation of the complex plane about the origin that takes z to w. $\qquad\square$

Example 3.4.37. Let $\mathcal{V}$ be a finite dimensional complex inner product space and let $\{\mathbf{e}_1, \dots, \mathbf{e}_n\}$ be an orthogonal basis of $\mathcal{V}$. Show that if $\lambda_1, \dots, \lambda_n$ are complex numbers such that $|\lambda_1| = \dots = |\lambda_n| = 1$, then the linear operator $f : \mathcal{V} \to \mathcal{V}$ defined by

$$f(\alpha_1\mathbf{e}_1 + \dots + \alpha_n\mathbf{e}_n) = \lambda_1\alpha_1\mathbf{e}_1 + \dots + \lambda_n\alpha_n\mathbf{e}_n$$

is an isometric operator.

Solution. Since the vectors $\mathbf{e}_1, \dots, \mathbf{e}_n$ are orthogonal, by the Pythagorean Theorem, we get

$$
\begin{aligned}
\|f(\alpha_1\mathbf{e}_1 + \dots + \alpha_n\mathbf{e}_n)\|^2 &= \|\lambda_1\alpha_1\mathbf{e}_1 + \dots + \lambda_n\alpha_n\mathbf{e}_n\|^2 \\
&= \|\lambda_1\alpha_1\mathbf{e}_1\|^2 + \dots + \|\lambda_n\alpha_n\mathbf{e}_n\|^2 \\
&= |\lambda_1|^2\|\alpha_1\mathbf{e}_1\|^2 + \dots + |\lambda_n|^2\|\alpha_n\mathbf{e}_n\|^2 \\
&= \|\alpha_1\mathbf{e}_1\|^2 + \dots + \|\alpha_n\mathbf{e}_n\|^2 \\
&= \|\alpha_1\mathbf{e}_1 + \dots + \alpha_n\mathbf{e}_n\|^2.
\end{aligned}
$$

Consequently, $\|f(\mathbf{x})\| = \|\mathbf{x}\|$ for every $\mathbf{x} \in \mathcal{V}$. $\qquad\square$

Theorem 3.4.38. *Let $\mathcal{V}$ be an inner product space. A linear operator $f : \mathcal{V} \to \mathcal{V}$ is isometric if and only if it preserves the inner product, that is,*

$$\langle f(\mathbf{x}), f(\mathbf{y})\rangle = \langle \mathbf{x}, \mathbf{y}\rangle$$

for every $\mathbf{x}, \mathbf{y} \in \mathcal{V}$.

Proof. If $\langle f(\mathbf{x}), f(\mathbf{y})\rangle = \langle \mathbf{x}, \mathbf{y}\rangle$ for every $\mathbf{x}, \mathbf{y} \in \mathcal{V}$, then

$$\|f(\mathbf{x})\|^2 = \langle f(\mathbf{x}), f(\mathbf{x})\rangle = \langle \mathbf{x}, \mathbf{x}\rangle = \|x\|^2$$

for every $\mathbf{x} \in \mathcal{V}$ and thus f is an isometric operator.

Now assume that f is an isometric operator. From the Polarization Identity (Corollary 3.1.34) we get

$$\langle f(\mathbf{x}), f(\mathbf{y}) \rangle$$

$$= \frac{1}{4} \left(\|f(\mathbf{x}) + f(\mathbf{y})\|^2 - \|f(\mathbf{x}) - f(\mathbf{y})\|^2 + i\|f(\mathbf{x}) + if(\mathbf{y})\|^2 - i\|f(\mathbf{x}) - if(\mathbf{y})\|^2 \right)$$

$$= \frac{1}{4} \left(\|f(\mathbf{x} + \mathbf{y})\|^2 - \|f(\mathbf{x} - \mathbf{y})\|^2 + i\|f(\mathbf{x} + i\mathbf{y})\|^2 - i\|f(\mathbf{x} - i\mathbf{y})\|^2 \right)$$

$$= \frac{1}{4} \left(\|\mathbf{x} + \mathbf{y}\|^2 - \|\mathbf{x} - \mathbf{y}\|^2 + i\|\mathbf{x} + i\mathbf{y}\|^2 - i\|\mathbf{x} - i\mathbf{y}\|^2 \right) = \langle \mathbf{x}, \mathbf{y} \rangle$$

for every $\mathbf{x}, \mathbf{y} \in \mathcal{V}$. $\qquad\square$

Theorem 3.4.39. *On a finite dimensional inner product space a linear operator is unitary if and only if it is isometric.*

Proof. Let $\mathcal{V}$ be a finite dimensional inner product space and let $f : \mathcal{V} \to \mathcal{V}$ be a unitary operator. For every $\mathbf{x} \in \mathcal{V}$ we have

$$\|f(\mathbf{x})\|^2 = \langle f(\mathbf{x}), f(\mathbf{x}) \rangle = \langle \mathbf{x}, f^* f(\mathbf{x}) \rangle = \langle \mathbf{x}, \mathbf{x} \rangle = \|\mathbf{x}\|^2,$$

which means that f is an isometric operator.

Now we assume that $\mathcal{V}$ is a finite dimensional inner product space and $f : \mathcal{V} \to \mathcal{V}$ is an isometric operator. Note that the equality $\|f(\mathbf{x})\| = \|\mathbf{x}\|$ implies that f is injective and thus $\dim \ker f = 0$. If $\dim \mathcal{V} = n$, then $\dim \operatorname{ran} f = n$ and hence $\operatorname{ran} f = \mathcal{V}$ and f is invertible. In other words, f is an isomorphism.

By Theorem 3.4.38, for every $\mathbf{x}, \mathbf{y} \in \mathcal{V}$ we have

$$\langle f^* f(\mathbf{x}), \mathbf{y} \rangle = \langle f(\mathbf{x}), f(\mathbf{y}) \rangle = \langle \mathbf{x}, \mathbf{y} \rangle,$$

and hence $f^* f = \operatorname{Id}$. Moreover, since f is an isomorphism, for every $\mathbf{x} \in \mathcal{V}$, there is a $\mathbf{y} \in \mathcal{V}$ such that $\mathbf{x} = f(\mathbf{y})$ and we have

$$ff^*(\mathbf{x}) = ff^* f(\mathbf{y}) = f(\mathbf{y}) = \mathbf{x},$$

because $f^* f = \operatorname{Id}$. Hence $ff^* = \operatorname{Id}$. $\qquad\square$

From Theorems 3.4.38 and 3.4.56 we obtain the following geometric characterization of unitary operators on finite dimensional inner product spaces: unitary operators are linear operators that preserve the norm or the inner product.

Corollary 3.4.40. *Let $\mathcal{V}$ be a finite dimensional inner product space and let $f : \mathcal{V} \to \mathcal{V}$ be a linear operator. The following conditions are equivalent:*

(a) *f is unitary;*

(b) *$\|f(\mathbf{x})\| = \|\mathbf{x}\|$ for every $\mathbf{x} \in \mathcal{V}$;*

(c) *$\langle f(\mathbf{x}), f(\mathbf{y}) \rangle = \langle \mathbf{x}, \mathbf{y} \rangle$ for every $\mathbf{x}, \mathbf{y} \in \mathcal{V}$.*

In the following theorem we characterize unitary operators on $\mathcal{V}$ in terms orthonormal bases in $\mathcal{V}$.

Theorem 3.4.41. *Let $\mathcal{V}$ be a finite dimensional inner product space and let $f : \mathcal{V} \to \mathcal{V}$ be a linear operator. The following conditions are equivalent:*

(a) *f is unitary;*

(b) *f is normal and $|\lambda| = 1$ for every eigenvalue λ of f;*

(c) *There is an orthonormal basis $\{\mathbf{e}_1, \ldots, \mathbf{e}_n\}$ of $\mathcal{V}$ such $\{f(\mathbf{e}_1), \ldots, f(\mathbf{e}_n)\}$ is an orthonormal basis of $\mathcal{V}$;*

(d) *For every orthonormal basis $\{\mathbf{v}_1, \ldots, \mathbf{v}_n\}$ of $\mathcal{V}$, $\{f(\mathbf{v}_1), \ldots, f(\mathbf{v}_n)\}$ is an orthonormal basis of $\mathcal{V}$;*

(e) *$\mathcal{V}$ has an orthonormal basis $\{\mathbf{e}_1, \ldots, \mathbf{e}_n\}$ of eigenvectors of f corresponding to eigenvalues $\lambda_1, \ldots, \lambda_n$ such that $|\lambda_1| = \cdots = |\lambda_n| = 1$;*

(f) *$f(\mathbf{x}) = \sum_{j=1}^{n} \lambda_j \langle \mathbf{x}, \mathbf{e}_j \rangle \mathbf{e}_j$ for every $\mathbf{x} \in \mathcal{V}$, where $\{\mathbf{e}_1, \ldots, \mathbf{e}_n\}$ is an orthonormal basis of $\mathcal{V}$ and $|\lambda_1| = \cdots = |\lambda_n| = 1$.*

Proof. If f is unitary, then f is normal and, by Corollary 3.4.17, there is an orthonormal basis $\{\mathbf{e}_1, \ldots, \mathbf{e}_n\}$ of $\mathcal{V}$ consisting of eigenvectors of f. Let λ_j be the eigenvalue of f corresponding to the eigenvector $\mathbf{e}_j$. Then

$$|\lambda_j| = \|\lambda_j \mathbf{e}_j\| = \|f(\mathbf{e}_j)\| = \|\mathbf{e}_j\| = 1,$$

so (a) implies (b).

Now we assume that f is normal and $|\lambda| = 1$ for every eigenvalue λ of f. By Corollary 3.4.17, there is an orthonormal basis $\{\mathbf{e}_1, \ldots, \mathbf{e}_n\}$ of $\mathcal{V}$ consisting of eigenvectors of f. Let $\lambda_1, \ldots, \lambda_n$ be the corresponding eigenvalues. Since $|\lambda_1| = \cdots = |\lambda_n| = 1$, we have

$$\langle f(\mathbf{e}_j), f(\mathbf{e}_k) \rangle = \langle \lambda_j \mathbf{e}_j, \lambda_k \mathbf{e}_k \rangle = \lambda_j \overline{\lambda_k} \langle \mathbf{e}_j, \mathbf{e}_k \rangle = \delta_{jk}$$

and thus $\{f(\mathbf{e}_1), \dots, f(\mathbf{e}_n)\}$ is an orthonormal basis of $\mathcal{V}$. This shows that (b) implies (c).

Next we assume that there is an orthonormal basis $\{\mathbf{e}_1, \dots, \mathbf{e}_n\}$ of $\mathcal{V}$ such $\{f(\mathbf{e}_1), \dots, f(\mathbf{e}_n)\}$ is an orthonormal basis of $\mathcal{V}$. Let $\{\mathbf{v}_1, \dots, \mathbf{v}_n\}$ be an orthonormal basis of $\mathcal{V}$. Then for all $j \in \{1, \dots, n\}$ we have

$$\mathbf{v}_j = \sum_{m=1}^{n} \alpha_{jm} \mathbf{e}_m,$$

where $\alpha_{jm} = \langle \mathbf{v}_j, \mathbf{e}_m \rangle$. Since

$$
\langle f(\mathbf{v}_j), f(\mathbf{v}_k) \rangle = \left\langle f\left(\sum_{l=1}^{n} \alpha_{jl} \mathbf{e}_l \right), f\left(\sum_{m=1}^{n} \alpha_{km} \mathbf{e}_m \right) \right\rangle
$$

$$
= \sum_{l=1}^{n} \sum_{m=1}^{n} \alpha_{jl} \overline{\alpha_{km}} \langle f(\mathbf{e}_l), f(\mathbf{e}_m) \rangle
$$

$$
= \sum_{l=1}^{n} \sum_{m=1}^{n} \alpha_{jl} \overline{\alpha_{km}} \delta_{lm}
$$

$$
= \sum_{l=1}^{n} \sum_{m=1}^{n} \alpha_{jl} \overline{\alpha_{km}} \langle \mathbf{e}_l, \mathbf{e}_m \rangle
$$

$$
= \left\langle \sum_{l=1}^{n} \alpha_{jl} \mathbf{e}_l, \sum_{m=1}^{n} \alpha_{km} \mathbf{e}_m \right\rangle
$$

$$
= \langle \mathbf{v}_j, \mathbf{v}_k \rangle = \delta_{jk},
$$

$\{f(\mathbf{v}_1), \dots, f(\mathbf{v}_n)\}$ is an orthonormal basis of $\mathcal{V}$, proving that (c) implies (d).

Let $\{\mathbf{v}_1, \dots, \mathbf{v}_n\}$ be an orthonormal basis of $\mathcal{V}$. If $\{f(\mathbf{v}_1), \dots, f(\mathbf{v}_n)\}$ is an orthonormal basis, then

$$
\|f(\mathbf{x})\|^2 = \left\| f\left(\sum_{j=1}^{n} \langle \mathbf{x}, \mathbf{v}_j \rangle \mathbf{v}_j \right) \right\|^2 = \left\| \sum_{j=1}^{n} \langle \mathbf{x}, \mathbf{v}_j \rangle f(\mathbf{v}_j) \right\|^2
$$

$$
= \sum_{j=1}^{n} |\langle \mathbf{x}, \mathbf{v}_j \rangle|^2 \|f(\mathbf{v}_j)\|^2 = \sum_{j=1}^{n} |\langle \mathbf{x}, \mathbf{v}_j \rangle|^2 = \|\mathbf{x}\|^2,
$$

for every $\mathbf{x} \in \mathcal{V}$. Consequently, by Corollary 3.4.40, f is unitary and thus (d) implies (a).

So far we have proved that (a)-(d) are all equivalent. Clearly (b) implies (e) and (e) implies (f).

Finally, if we assume that there is an orthonormal basis $\{\mathbf{e}_1, \dots, \mathbf{e}_n\}$ of $\mathcal{V}$ and $|\lambda_1| = \cdots = |\lambda_n| = 1$ such that

$$
f(\mathbf{x}) = \sum_{j=1}^{n} \lambda_j \langle \mathbf{x}, \mathbf{e}_j \rangle \mathbf{e}_j
$$

for every $\mathbf{x} \in V$, then

$$\|f(\mathbf{x})\|^2 = \left\|\sum_{j=1}^{n} \lambda_j \langle \mathbf{x}, \mathbf{e}_j \rangle \mathbf{e}_j \right\|^2 = \sum_{j=1}^{n} \|\lambda_j \langle \mathbf{x}, \mathbf{e}_j \rangle \mathbf{e}_j \|^2$$

$$= \sum_{j=1}^{n} |\lambda_j|^2 \|\langle \mathbf{x}, \mathbf{e}_j \rangle \mathbf{e}_j\|^2 = \sum_{j=1}^{n} \|\langle \mathbf{x}, \mathbf{e}_j \rangle \mathbf{e}_j\|^2 = \|\mathbf{x}\|^2.$$

By Corollary 3.4.40, f is unitary and thus (f) implies (a), completing the proof of the theorem. □

Example 3.4.42. Let $\{\mathbf{a}, \mathbf{b}, \mathbf{c}\}$ and $\{\mathbf{u}, \mathbf{v}, \mathbf{w}\}$ be two orthogonal bases in an inner product space V. Find an isometry f such that $\mathrm{Span}\{f(\mathbf{a})\} = \mathrm{Span}\{\mathbf{u}\}$, $\mathrm{Span}\{f(\mathbf{b})\}) = \mathrm{Span}\{\mathbf{v}\}$, and $\mathrm{Span}\{f(\mathbf{c})\} = \mathrm{Span}\{\mathbf{w}\}$.

Solution.

$$f(\mathbf{x}) = \left\langle \mathbf{x}, \frac{1}{\|\mathbf{a}\|}\mathbf{a} \right\rangle \frac{1}{\|\mathbf{u}\|}\mathbf{u} + \left\langle \mathbf{x}, \frac{1}{\|\mathbf{b}\|}\mathbf{b} \right\rangle \frac{1}{\|\mathbf{v}\|}\mathbf{v} + \left\langle \mathbf{x}, \frac{1}{\|\mathbf{c}\|}\mathbf{c} \right\rangle \frac{1}{\|\mathbf{w}\|}\mathbf{w}.$$

□

Example 3.4.43. Consider the operator $f \in \mathcal{L}(\mathbb{C}^2, \mathbb{C}^2)$ defined by $f(\mathbf{x}) = A\mathbf{x}$, where

$$A = \frac{1}{5}\begin{bmatrix} 4+i & 2-2i \\ 2-2i & 1+4i \end{bmatrix}.$$

Show that f is unitary and find its spectral decomposition.

Solution. We can verify that $ff^* = f^*f = \mathrm{Id}$ by simple matrix multiplication. Since the roots of the equation

$$(4+i-\lambda)(1+4i-\lambda) - (2-2i)^2 = \lambda^2 - (5+5i)\lambda + 25i = 0$$

are 5 and $5i$, 1 and i are eigenvalues of f and

$$\begin{bmatrix} \frac{2}{\sqrt{5}} \\ \frac{1}{\sqrt{5}} \end{bmatrix} \quad \text{and} \quad \begin{bmatrix} \frac{1}{\sqrt{5}} \\ -\frac{2}{\sqrt{5}} \end{bmatrix}$$

are unit eigenvectors corresponding to 1 and i. Consequently,

$$f\left(\begin{bmatrix} x_1 \\ x_2 \end{bmatrix}\right) = \left\langle \begin{bmatrix} x_1 \\ x_2 \end{bmatrix}, \begin{bmatrix} \frac{2}{\sqrt{5}} \\ \frac{1}{\sqrt{5}} \end{bmatrix} \right\rangle \begin{bmatrix} \frac{2}{\sqrt{5}} \\ \frac{1}{\sqrt{5}} \end{bmatrix} + i \left\langle \begin{bmatrix} x_1 \\ x_2 \end{bmatrix}, \begin{bmatrix} \frac{1}{\sqrt{5}} \\ -\frac{2}{\sqrt{5}} \end{bmatrix} \right\rangle \begin{bmatrix} \frac{1}{\sqrt{5}} \\ -\frac{2}{\sqrt{5}} \end{bmatrix}.$$

□

The operator in the next example is a normal operator that is not unitary.

Example 3.4.44. Consider the operator $f \in \mathcal{L}(\mathbb{C}^2, \mathbb{C}^2)$ defined by $f(\mathbf{x}) = A\mathbf{x}$, where

$$A = \begin{bmatrix} 5i & 3+4i \\ -3+4i & 5i \end{bmatrix}.$$

It is easy to verify that $ff^* = f^*f = \mathbf{0}$, so f is a normal operator, but it is not unitary. Since

$$f\left(\begin{bmatrix} 1-2i \\ -2+i \end{bmatrix}\right) = \begin{bmatrix} 0 \\ 0 \end{bmatrix} \quad \text{and} \quad f\left(\begin{bmatrix} 2+i \\ 1+2i \end{bmatrix}\right) = 10i \begin{bmatrix} 2+i \\ 1+2i \end{bmatrix},$$

f has eigenvalues 0 and $10i$ and $\begin{bmatrix} 1-2i \\ -2+i \end{bmatrix}$ and $\begin{bmatrix} 2+i \\ 1+2i \end{bmatrix}$ are corresponding eigenvectors.

3.4.4 Orthogonal operators on real inner product spaces

In this section we discuss the orthogonal operators on finite dimensional real inner product spaces, that is, operators that preserve the inner product. All inner product spaces considered in this section are assumed to be real.

Definition 3.4.45. Let $\mathcal{V}$ be a real inner product space. A linear operator $f : \mathcal{V} \to \mathcal{V}$ is called an *orthogonal operator* if

$$\langle f(\mathbf{x}), f(\mathbf{y}) \rangle = \langle \mathbf{x}, \mathbf{y} \rangle$$

for every $\mathbf{x}, \mathbf{y} \in \mathcal{V}$.

Theorem 3.4.46. *Let $\mathcal{V}$ be a finite dimensional real inner product space and let $f : \mathcal{V} \to \mathcal{V}$ be a linear operator. The following conditions are equivalent:*

(a) *f is orthogonal;*

(b) *$ff^* = f^*f = \mathrm{Id}$;*

(c) *$\|f(\mathbf{x})\| = \|\mathbf{x}\|$ for every $\mathbf{x} \in \mathcal{V}$.*

Proof. Let $f : V \to V$ be an orthogonal operator. For every $\mathbf{x}, \mathbf{y} \in V$ we have

$$\langle f^* f(\mathbf{x}), \mathbf{y} \rangle = \langle f(\mathbf{x}), f(\mathbf{y}) \rangle = \langle \mathbf{x}, \mathbf{y} \rangle$$

and thus

$$\langle f^* f(\mathbf{x}) - x, \mathbf{y} \rangle = 0.$$

If we take $\mathbf{y} = f^* f(\mathbf{x}) - x$, then we get

$$\|f^* f(\mathbf{x}) - x\|^2 = \langle f^* f(\mathbf{x}) - x, f^* f(\mathbf{x}) - x \rangle = 0.$$

Consequently, $f^* f(\mathbf{x}) - x = \mathbf{0}$, and hence $f^* f = \text{Id}$. Since V is finite dimensional, we also have $f f^* = \text{Id}$, by Theorem 2.2.11. This proves that (a) implies (b).

If $f f^* = f^* f = \text{Id}$, then for every $\mathbf{x} \in V$ we have

$$\|f(\mathbf{x})\|^2 = \langle f(\mathbf{x}), f(\mathbf{x}) \rangle = \langle \mathbf{x}, f^* f(\mathbf{x}) \rangle = \langle \mathbf{x}, \mathbf{x} \rangle = \|\mathbf{x}\|^2,$$

and thus (b) implies (c).

Finally, if $\|f(\mathbf{x})\| = \|\mathbf{x}\|$ for every $\mathbf{x} \in V$, then

$$2 \langle f(\mathbf{x}), f(\mathbf{y}) \rangle = \|f(\mathbf{x} + \mathbf{y})\|^2 - \|f(\mathbf{x})\|^2 - \|f(\mathbf{y})\|^2$$
$$= \|\mathbf{x} + \mathbf{y}\|^2 - \|\mathbf{x}\|^2 - \|\mathbf{y}\|^2 = 2 \langle \mathbf{x}, \mathbf{y} \rangle$$

for every $\mathbf{x}, \mathbf{y} \in V$. This proves that (c) implies (a).   □

While some properties of orthogonal operators on finite dimensional real inner product spaces are the same as properties of unitary operators on finite dimensional complex inner product spaces, there are some essential differences.

Theorem 3.4.47. *Let V be a finite dimensional real inner product space and let $f : V \to V$ be an orthogonal operator. If λ is an eigenvalue of f, then $\lambda = 1$ or $\lambda = -1$.*

Proof. If $\mathbf{v}$ be an eigenvector corresponding to the eigenvalue λ, then $f(\mathbf{v}) = \lambda \mathbf{v}$ and thus

$$|\lambda| \|\mathbf{v}\| = \|\lambda \mathbf{v}\| = \|f(\mathbf{v})\| = \|\mathbf{v}\|.$$

Consequently, $|\lambda| = 1$ because $\mathbf{v} \neq \mathbf{0}$ and, because λ is a real number, $\lambda = 1$ or $\lambda = -1$.   □

Note that the above theorem does not say that every orthogonal operator on a finite dimensional real inner product space has an eigenvalue. Indeed, if $f : \mathbb{R}^2 \to \mathbb{R}^2$ is the operator $f(\mathbf{x}) = A\mathbf{x}$, where

$$A = \begin{bmatrix} 0 & -1 \\ 1 & 0 \end{bmatrix},$$

then

$$\left\langle A\begin{bmatrix} x_1 \\ x_2 \end{bmatrix}, A\begin{bmatrix} y_1 \\ y_2 \end{bmatrix}\right\rangle = \left\langle \begin{bmatrix} -x_2 \\ x_1 \end{bmatrix}, \begin{bmatrix} -y_2 \\ y_1 \end{bmatrix}\right\rangle = x_1 y_1 + x_2 y_2 = \left\langle \begin{bmatrix} x_1 \\ x_2 \end{bmatrix}, \begin{bmatrix} y_1 \\ y_2 \end{bmatrix}\right\rangle,$$

so f is an orthogonal operator. On the other hand, since

$$\det\begin{bmatrix} -\lambda & -1 \\ 1 & -\lambda \end{bmatrix} = \lambda^2 + 1,$$

f has no real eigenvalues.

Lemma 3.4.48. *Let V be a finite dimensional real inner product space and let $f : V \to V$ be an orthogonal operator. If $\mathcal{U}$ is an f-invariant subspace of V, then $\mathcal{U}^\perp$ is also f-invariant.*

Proof. Let $\mathcal{U}$ be an f-invariant subspace of V. Since f is an isomorphism, we have $f(\mathcal{U}) = \mathcal{U}$. Let $\mathbf{v} \in \mathcal{U}^\perp$ and $\mathbf{u} \in \mathcal{U}$. Then there is $\mathbf{w} \in \mathcal{U}$ such that $f(\mathbf{w}) = \mathbf{u}$ and we have

$$\langle f(\mathbf{v}), \mathbf{u}\rangle = \langle f(\mathbf{v}), f(\mathbf{w})\rangle = \langle \mathbf{v}, \mathbf{w}\rangle = 0.$$

Thus $f(\mathbf{v}) \in \mathcal{U}^\perp$. $\qquad\square$

Lemma 3.4.49. *Let V be a finite dimensional real inner product space and let $f : V \to V$ be an orthogonal operator. There is an f-invariant subspace $\mathcal{U} \subseteq V$ such that $\dim \mathcal{U} = 1$ or $\dim \mathcal{U} = 2$.*

Proof. Since the operator $f + f^*$ is self-adjoint, it has an eigenvalue $\lambda \in \mathbb{R}$. Let $\mathbf{v}$ be an eigenvector corresponding to λ. Then

$$(f + f^*)\mathbf{v} = \lambda\mathbf{v}$$

and hence

$$(ff + ff^*)\mathbf{v} = f(\lambda\mathbf{v}).$$

Since $ff^* = \mathrm{Id}$, the above can be written as

$$f^2(\mathbf{v}) + \mathbf{v} = \lambda f(\mathbf{v})$$

or

$$f^2(\mathbf{v}) = \lambda f(\mathbf{v}) - \mathbf{v}.$$

Consequently, $f^2(\mathbf{v}) \in \mathrm{Span}\{\mathbf{v}, f(\mathbf{v})\}$ and thus $\mathcal{U} = \mathrm{Span}\{\mathbf{v}, f(\mathbf{v})\}$ is f-invariant and $\dim \mathcal{U} = 1$ or $\dim \mathcal{U} = 2$. $\qquad\square$

Theorem 3.4.50. *Let V be a finite dimensional real inner product space and let $f : V \to V$ be an orthogonal operator. There are f-invariant subspaces $\mathcal{U}_1, \ldots, \mathcal{U}_n \subseteq V$ such that*

$$V = \mathcal{U}_1 \oplus \cdots \oplus \mathcal{U}_n$$

and $\dim \mathcal{U}_j = 1$ or $\dim \mathcal{U}_j = 2$ for every $j \in \{1, \ldots, n\}$.

Proof. By Lemma 3.4.49 there is an f-invariant subspace $\mathcal{U}_1 \subseteq V$ such that $\dim \mathcal{U}_1 = 1$ or $\dim \mathcal{U}_1 = 2$. We have $V = \mathcal{U}_1 \oplus \mathcal{U}_1^{\perp}$.

Now, by Lemma 3.4.48, $\mathcal{U}_1^{\perp}$ is f-invariant and thus we can define an operator $g : \mathcal{U}_1^{\perp} \to \mathcal{U}_1^{\perp}$ by $g(\mathbf{x}) = f(\mathbf{x})$ for every $\mathbf{x} \in \mathcal{U}_1^{\perp}$. Clearly g is an orthogonal operator. Since $\dim \mathcal{U}_1^{\perp} = \dim V - 1$ or $\dim \mathcal{U}_1^{\perp} = \dim V - 2$, we can apply Lemma 3.4.49 to g and proceed as before. This gives us the desired result by induction. $\qquad\square$

If $f : V \to V$ is an orthogonal operator on a real inner product space V, then by the above theorem, V is a direct sum of f-invariant subspaces $\mathcal{U}_1, \ldots, \mathcal{U}_n$ of dimension 1 or 2. If $\dim \mathcal{U}_j = 1$, then $f(\mathbf{v}) = \mathbf{v}$ or $f(\mathbf{v}) = -\mathbf{v}$ for every $\mathbf{v} \in \mathcal{U}_j$. Now we are going to consider the case when $\dim \mathcal{U}_j = 2$.

Theorem 3.4.51. *Let V be a real inner product space such that $\dim V = 2$ and let $f : V \to V$ be an orthogonal operator. If $\{\mathbf{v}, \mathbf{w}\}$ is an orthonormal basis of V such that*

$$f(\mathbf{v}) = a\mathbf{v} + b\mathbf{w} \quad and \quad f(\mathbf{w}) = c\mathbf{v} + d\mathbf{w},$$

where $a, b, c, d \in \mathbb{R}$, then one of the following conditions holds

(a) *$ad - bc = 1$ and there is a unique number $\theta \in (-\pi, \pi]$ such that*

$$f(\mathbf{v}) = \cos \theta \, \mathbf{v} + \sin \theta \, \mathbf{w} \quad and \quad f(\mathbf{w}) = -\sin \theta \, \mathbf{v} + \cos \theta \, \mathbf{w};$$

(b) *$ad - bc = -1$ and there is an orthonormal basis $\{\mathbf{u}_1, \mathbf{u}_2\}$ of V such that $f(\mathbf{u}_1) = \mathbf{u}_1$ and $f(\mathbf{u}_2) = -\mathbf{u}_2$.*

Note that in the second case $\mathbf{u}_1$ and $\mathbf{u}_2$ are eigenvectors of f corresponding to eigenvalues 1 and -1, respectively.

Proof. From $\|f(\mathbf{v})\|^2 = \|f(\mathbf{w})\|^2 = 1$ and $\langle f(\mathbf{v}), f(\mathbf{w}) \rangle = 0$ we get

$$\begin{cases} a^2 + b^2 = 1 \\ c^2 + d^2 = 1 \\ ac + bd = 0 \end{cases}.$$

If $a \neq 0$, then $c = -\frac{bd}{a} = -\frac{d}{a}b$ and $d = -\frac{d}{a}a$. If we let $t = -\frac{d}{a}$, then we have

$$1 = c^2 + d^2 = t^2(a^2 + b^2) = t^2$$

and thus $t = 1$ or $t = -1$.

If $t = 1$, then $c = -b$, $d = a$, $ad - bc = 1$, and there is a unique $\theta \in (-\pi, \pi]$ such that $a = \cos\theta$ and $b = \sin\theta$.

If $t = -1$, then $c = b$, $d = -a$, and $ad - bc = -1$. Since for any $p, q, x, y \in \mathbb{R}$ we have

$$\langle f(p\mathbf{v} + q\mathbf{w}), x\mathbf{v} + y\mathbf{w} \rangle = apx + bqx + bpy - aqy = \langle p\mathbf{v} + q\mathbf{w}, f(x\mathbf{v} + y\mathbf{w}) \rangle,$$

f is self-adjoint. Because f is orthogonal, if λ is an eigenvalue of f, then $\lambda = 1$ or $\lambda = -1$.

Next we calculate the corresponding eigenvectors.

To solve the equation

$$f(x\mathbf{v} + y\mathbf{w}) - (x\mathbf{v} + y\mathbf{w}) = (x(a-1) + yb)\mathbf{v} + (xb - (a+1)y)\mathbf{w} = 0$$

we need to solve the system

$$\begin{cases} x(a-1) + yb = 0 \\ xb - (a+1)y = 0 \end{cases}.$$

It is easy to verify that $x = -b$ and $y = a - 1$ is a nonzero solution if $b \neq 0$ or $a \neq 1$. The same way we show that the equation

$$f(x\mathbf{v} + y\mathbf{w}) + (x\mathbf{v} + y\mathbf{w}) = (x(a+1) + yb)\mathbf{v} + (xb - (a-1)y)\mathbf{w} = 0$$

has a nonzero solution $x = -b$ and $y = a + 1$ if $b \neq 0$ or $a \neq -1$.

Note that, as expected, the vectors $-b\mathbf{v} + (a-1)\mathbf{w}$ and $-b\mathbf{v} - (a+1)\mathbf{w}$ are orthogonal because $a^2 + b^2 = 1$.

The cases $b = 0$ and $a = 1$ as well as $b = 0$ and $a = -1$ are trivial. Consequently, there is a basis $\{\mathbf{u}_1, \mathbf{u}_2\}$ of V consisting of orthonormal eigenvectors of f such that $f(\mathbf{u}_1) = \mathbf{u}_1$ and $f(\mathbf{u}_2) = -\mathbf{u}_2$.

Finally, if $a = 0$, then $d = 0$ and $b^2 = c^2 = 1$. There are four possibilities:

$$\begin{bmatrix} a & b \\ c & d \end{bmatrix} = \begin{bmatrix} 0 & -1 \\ 1 & 0 \end{bmatrix} = \begin{bmatrix} \cos(\frac{\pi}{2}) & -\sin(\frac{\pi}{2}) \\ \sin(\frac{\pi}{2}) & \cos(\frac{\pi}{2}) \end{bmatrix};$$

$$\begin{bmatrix} a & b \\ c & d \end{bmatrix} = \begin{bmatrix} 0 & 1 \\ -1 & 0 \end{bmatrix} = \begin{bmatrix} \cos(\frac{\pi}{2}) & -\sin(-\frac{\pi}{2}) \\ \sin(-\frac{\pi}{2}) & \cos(\frac{\pi}{2}) \end{bmatrix};$$

$$\begin{bmatrix} a & b \\ c & d \end{bmatrix} = \begin{bmatrix} 0 & 1 \\ 1 & 0 \end{bmatrix};$$

$$\begin{bmatrix} a & b \\ c & d \end{bmatrix} = \begin{bmatrix} 0 & -1 \\ -1 & 0 \end{bmatrix}.$$

In the first two cases $ad - bc = 1$ and in the last two cases $ad - bc = -1$. In all cases the operator f is self-adjoint and has eigenvalues 1 and -1. $\square$

Using Theorem 3.4.51 we can give a more detailed description of the f-invariant subspaces in Theorem 3.4.50.

Theorem 3.4.52. *Let V be a finite dimensional real inner product space and let $f : V \to V$ be an orthogonal operator. There are f-invariant subspaces $\mathcal{U}_1, \ldots, \mathcal{U}_p, \mathcal{W}_1, \ldots, \mathcal{W}_q, \mathcal{X}_1, \ldots, \mathcal{X}_r \subseteq V$, such that*

$$V = \mathcal{U}_1 \oplus \cdots \oplus \mathcal{U}_p \oplus \mathcal{W}_1 \oplus \cdots \oplus \mathcal{W}_q \oplus \mathcal{X}_1 \oplus \cdots \oplus \mathcal{X}_r$$

and

(a) *for every $j \in \{1, \ldots, p\}$, $\dim \mathcal{U}_j = 1$ and there is a nonzero vector $\mathbf{u}_j \in \mathcal{U}_j$ such that $f(\mathbf{u}_j) = \mathbf{u}_j$;*

(b) *for every $k \in \{1, \ldots, q\}$, $\dim \mathcal{W}_k = 1$ and there is a nonzero vector $\mathbf{w}_k \in \mathcal{W}_k$ such that $f(\mathbf{w}_k) = -\mathbf{w}_k$;*

(c) *for every $l \in \{1, \ldots, r\}$, $\dim \mathcal{X}_l = 2$ and there are orthonormal vectors $\mathbf{x}_l, \mathbf{y}_l \in \mathcal{X}_l$ and a unique $\theta_l \in (-\pi, \pi]$ such that*

$$f(\mathbf{x}_l) = \cos \theta_l \, \mathbf{x}_l + \sin \theta_l \, \mathbf{y}_l$$
$$f(\mathbf{y}_l) = -\sin \theta_l \, \mathbf{x}_l + \cos \theta_l \, \mathbf{y}_l.$$

3.4.5 Positive operators

There is some similarity between operators on a complex inner product space and the complex numbers. Self-adjoint operators are like real numbers and unitary operators are like complex numbers of modulus 1. Now we are going to consider operators that behave like nonnegative numbers.

Definition 3.4.53. Let V be an inner product space. A linear operator $f : V \to V$ is called *positive* if

$$\langle f(\mathbf{x}), \mathbf{x} \rangle \geq 0$$

for every $\mathbf{x} \in V$.

If $f : V \to V$ is a positive operator, then $\langle f(\mathbf{x}), \mathbf{x} \rangle$ is a real number for every $\mathbf{x} \in V$ and thus

$$\langle f(\mathbf{x}), \mathbf{x} \rangle = \overline{\langle f(\mathbf{x}), \mathbf{x} \rangle} = \langle \mathbf{x}, f(\mathbf{x}) \rangle,$$

which shows that positive operators are self-adjoint.

Example 3.4.54. Consider the vector space $\mathcal{C}([a, b])$ of complex-valued continuous functions defined on an interval $[a, b]$ with the inner product $\langle f, g \rangle = \int_a^b f(t)\overline{g(t)}\, dt$ and let $\varphi : [a, b] \rightarrow \mathbb{R}$ be a positive continuous function. Show that the operator $\Phi : \mathcal{C}([a, b]) \rightarrow \mathcal{C}([a, b])$ defined by $\Phi(f) = \varphi f$ is a positive operator.

Solution. For all $f \in \mathcal{C}([a, b])$ we have

$$\langle \Phi(f), f \rangle = \int_a^b \varphi(t) f(t)\overline{f(t)}\, dt = \int_a^b \varphi(t)|f(t)|^2\, dt \geq 0.$$

$\square$

Example 3.4.55. Let $\mathcal{V} = \mathbb{C}^\infty$ be the vector space of all infinite sequences of complex numbers with only a finite number of nonzero terms with the inner product defined as

$$\langle (x_1, x_2, \dots), (y_1, y_2, \dots) \rangle = \sum_{j=1}^\infty x_j \overline{y_j}.$$

Consider the linear operator $f : \mathbb{C}^\infty \rightarrow \mathbb{C}^\infty$ defined as

$$f(x_1, x_2, \dots) = (\alpha_1 x_1, \alpha_2 x_2, \dots)$$

where $\alpha_1, \alpha_2, \dots$ are arbitrary complex numbers. Show that f is a positive operator if and only if all α_j's are positive real numbers.

Solution. If all α_j's are positive real numbers, then for every $(x_1, x_2, \dots) \in \mathbb{C}^\infty$ we have

$$\langle f(x_1, x_2, \dots), (x_1, x_2, \dots) \rangle = \langle (\alpha_1 x_1, \alpha_2 x_2, \dots), (x_1, x_2, \dots) \rangle$$

$$= \sum_{j=1}^\infty \alpha_j x_j \overline{x_j} = \sum_{j=1}^\infty \alpha_j |x_j|^2 \geq 0,$$

so f is a positive operator.

Now assume that $\langle f(x_1, x_2, \dots), (x_1, x_2, \dots) \rangle \geq 0$ for every $(x_1, x_2, \dots) \in \mathbb{C}^\infty$.

If, for every integer $j \geq 1$, we denote by $\mathbf{e}_j \in \mathbb{C}^\infty$ the sequence $(x_1, x_2, \dots)$ such that $x_j = 1$ and $x_k = 0$ for $k \neq j$, then we have

$$0 \leq \langle f(\mathbf{e}_j), \mathbf{e}_j \rangle = \alpha_j.$$

$\square$

The operations of adjoint of an operator and conjugate of a complex number have similar algebraic properties. For example, for any complex number z we have $z\bar{z} \geq 0$. In the next theorem we formulate a similar property for linear operators.

Theorem 3.4.56. *Let V be a finite dimensional inner product space and let $f : V \to V$ be a linear operator. The operators ff^* and f^*f are positive.*

Proof. For every $\mathbf{x} \in V$ we have

$$\langle ff^*(\mathbf{x}), \mathbf{x} \rangle = \langle f^*(\mathbf{x}), f^*(\mathbf{x}) \rangle = \|f^*(\mathbf{x})\|^2 \geq 0$$

and

$$\langle f^*f(\mathbf{x}), \mathbf{x} \rangle = \langle f(\mathbf{x}), f(\mathbf{x}) \rangle = \|f(\mathbf{x})\|^2 \geq 0.$$

$\square$

The following theorem is similar to Theorems 3.4.16 and 3.4.30. These three theorems characterize normal, self-adjoint, and positive operators in terms of their eigenvalues.

Theorem 3.4.57. *Let V be a finite dimensional inner product space and let $f : V \to V$ be a nonzero linear operator. The following conditions are equivalent*

(a) *f is a positive operator;*

(b) *There are orthonormal vectors $\mathbf{e}_1, \ldots, \mathbf{e}_r$ and positive numbers $\lambda_1, \ldots, \lambda_r$ such that for every $\mathbf{v} \in V$ we have*

$$f(\mathbf{v}) = \sum_{j=1}^{r} \lambda_j \langle \mathbf{v}, \mathbf{e}_j \rangle \mathbf{e}_j;$$

(c) *There are orthonormal vectors $\mathbf{e}_1, \ldots, \mathbf{e}_r$ and positive numbers $\lambda_1, \ldots, \lambda_r$ such that*

$$f = \sum_{j=1}^{r} \lambda_j \mathrm{proj}_{\mathbf{e}_j}.$$

Proof. If f is a positive operator, then it is self-adjoint and there are orthonormal vectors $\mathbf{e}_1, \ldots, \mathbf{e}_r$ and nonzero real numbers $\lambda_1, \ldots, \lambda_r$ such that $f(\mathbf{v}) = \sum_{j=1}^{r} \lambda_j \langle \mathbf{v}, \mathbf{e}_j \rangle \mathbf{e}_j$ for every vector $\mathbf{v} \in V$. Since, for every λ_j we have

$$0 \leq \langle f(\mathbf{e}_j), \mathbf{e}_j \rangle = \langle \lambda_j \mathbf{e}_j, \mathbf{e}_j \rangle = \lambda_j \langle \mathbf{e}_j, \mathbf{e}_j \rangle = \lambda_j \|\mathbf{e}_j\|^2 = \lambda_j,$$

$\lambda_1, \ldots, \lambda_r$ are positive numbers. This shows that (a) implies (b).

Conditions (b) and (c) are equivalent, because $\text{proj}_{\mathbf{e}_j}(\mathbf{v}) = \langle \mathbf{v}, \mathbf{e}_j \rangle \mathbf{e}_j$.

Now we assume that for every $\mathbf{v} \in V$ we have $f(\mathbf{v}) = \sum_{j=1}^{r} \lambda_j \langle \mathbf{v}, \mathbf{e}_j \rangle \mathbf{e}_j$, where $\{\mathbf{e}_1, \ldots, \mathbf{e}_r\}$ is an orthonormal set and $\lambda_1, \ldots, \lambda_n$ are positive numbers. Then for every $\mathbf{v} \in V$ we have

$$\langle f(\mathbf{v}), \mathbf{v} \rangle = \left\langle \sum_{j=1}^{r} \lambda_j \langle \mathbf{v}, \mathbf{e}_j \rangle \mathbf{e}_j, \mathbf{v} \right\rangle = \sum_{j=1}^{r} \lambda_j \langle \mathbf{v}, \mathbf{e}_j \rangle \langle \mathbf{e}_j, \mathbf{v} \rangle$$

$$= \sum_{j=1}^{r} \lambda_j \langle \mathbf{v}, \mathbf{e}_j \rangle \overline{\langle \mathbf{v}, \mathbf{e}_j \rangle} = \sum_{j=1}^{r} \lambda_j |\langle \mathbf{v}, \mathbf{e}_j \rangle|^2 \geq 0.$$

This shows that (b) implies (a), which completes the proof. □

Example 3.4.58. Consider the operator $f \in \mathcal{L}(\mathbb{C}^3, \mathbb{C}^3)$ defined as $f(\mathbf{x}) = A\mathbf{x}$, where

$$A = \begin{bmatrix} 14 & 2 & 4 \\ 2 & 17 & -2 \\ 4 & -2 & 14 \end{bmatrix}.$$

Show that f is a positive operator.

Solution. Since the matrix A is symmetric, the operator f is normal and thus $\mathbb{C}^3$ has an orthonormal basis consisting of eigenvectors of f. Hence, to show that f is a positive operator, it suffices to show that all eigenvalues of f are nonnegative numbers. If λ is an eigenvalue of f, then the following system has a nontrivial solution.

$$\begin{cases} (14 - \lambda)x + 2y + 4z = 0 \\ 2x + (17 - \lambda)y - 2z = 0 \\ 4x - 2y + (14 - \lambda)z = 0. \end{cases}$$

If we add the first and the third equations, we get

$$(18 - \lambda)(x + z) = 0.$$

It is easy to see that for $\lambda = 18$ the system has nontrivial solutions. If $\lambda \neq 18$, the system is equivalent to the system

$$\begin{cases} x + z & = 0 \\ 2x + (17 - \lambda)y - 2z = 0 \\ 4x - 2y + (14 - \lambda)z = 0. \end{cases}$$

If we let $z = -x$, then we get

$$\begin{cases} 4x + (17 - \lambda)y = 0 \\ (\lambda - 10)x - 2y = 0. \end{cases}$$

We multiply the first equation by $\frac{10-\lambda}{4}$ and add to the second and get

$$\left(\frac{(10-\lambda)}{4}(17-\lambda) - 2 \right) y = 0$$

or

$$\frac{(\lambda^2 - 27\lambda + 162)}{4} y = 0.$$

The roots of the equation $\lambda^2 - 27\lambda + 162 = 0$ are 9 and 18.

If $\lambda \neq 9$ and $\lambda \neq 18$, then the only solution is $x = y = z = 0$. Consequently, 9 and 18 are the only eigenvalues of f. Since these are positive numbers, f is a positive operator.   □

The square root of a positive operator

Every positive real number has a unique positive square root. A similar property holds for positive operators.

Definition 3.4.59. Let $\mathcal{V}$ be an inner product space and let $f : \mathcal{V} \to \mathcal{V}$ be a linear operator. An operator $g : \mathcal{V} \to \mathcal{V}$ is called a *square root* of f if $g^2 = f$.

Example 3.4.60. Let $\mathcal{V} = \mathbb{C}^\infty$ be the vector space of all infinite sequences of complex numbers with only a finite number of nonzero terms with the inner product defined as $\langle (x_1, x_2, \ldots), (y_1, y_2, \ldots) \rangle = \sum_{j=1}^{\infty} x_j \overline{y_j}$. If $f : \mathbb{C}^\infty \to \mathbb{C}^\infty$ is the operator defined as

$$f(x_1, x_2, \ldots) = (\alpha_1 x_1, \alpha_2 x_2, \ldots),$$

where $\alpha_1, \alpha_2, \ldots$ are positive numbers, then the operator $g : \mathbb{C}^\infty \to \mathbb{C}^\infty$ defined as

$$g(x_1, x_2, \ldots) = (\sqrt{\alpha_1} x_1, \sqrt{\alpha_2} x_2, \ldots),$$

is a square root of f.

Note that every operator of the form

$$h(x_1, x_2, \ldots) = ((-1)^{n_1} \sqrt{\alpha_1} x_1, (-1)^{n_2} \sqrt{\alpha_2} x_2, \ldots),$$

where $n_j \in \{1, 2\}$, is a square root of f, but the only square root of f that is a positive operator is the operator g defined above.

Theorem 3.4.61. *Let V be a finite dimensional inner product space and let $f : V \to V$ be a positive operator. There is a unique positive operator $g : V \to V$ such that $g^2 = f$.*

Proof. We offer two different proofs of existence and two different proofs of uniqueness.

The first proof of existence: Without loss of generality we can assume that f is a nonzero positive operator. By Theorem 3.4.57, for every $\mathbf{x} \in V$ we have

$$f(\mathbf{x}) = \sum_{j=1}^{r} \lambda_j \langle \mathbf{x}, \mathbf{e}_j \rangle \mathbf{e}_j,$$

where $\{\mathbf{e}_1, \ldots, \mathbf{e}_r\}$ is an orthonormal set and $\lambda_1, \ldots, \lambda_r$ are positive numbers. It is easy to verify that for the positive operator

$$g(\mathbf{x}) = \sum_{j=1}^{r} \sqrt{\lambda_j} \langle \mathbf{x}, \mathbf{e}_j \rangle \mathbf{e}_j$$

we have $g^2 = f$.

The second proof of existence: Since f is a positive operator, all eigenvalues of f are nonnegative. Let p be a polynomial such that $p(\lambda) = \sqrt{\lambda}$ for every eigenvalue λ of f. Since, by Example 3.4.18, $(p(f))^2 = f$ and $p(f)$ is a positive operator, we can take $g = p(f)$.

The first proof of uniqueness: Let p be a polynomial such that $p(\lambda) = \sqrt{\lambda}$ for every eigenvalue λ of f and let $g = p(f)$. Now assume h is a positive operator such that $h^2 = f$. If μ is any eigenvalue of h and $\mathbf{v}$ is an eigenvector corresponding to μ, then $\mu \geq 0$ and $f(\mathbf{v}) = \mu^2 \mathbf{v}$, which means that μ^2 is an eigenvalue of f and thus $p(\mu^2) = \mu$. Consequently,

$$g(\mathbf{v}) = p(f)(\mathbf{v}) = p(h^2)(\mathbf{v}) = p(\mu^2)(\mathbf{v}) = \mu\mathbf{v} = h(\mathbf{v}).$$

Since $g(\mathbf{v}) = h(\mathbf{v})$ for every eigenvector of h and there is a basis of V of eigenvectors of h, we can conclude that $g = h$.

The second proof of uniqueness: Let p be a polynomial such that $(p(f))^2 = f$ and let h be a positive operator such that $h^2 = f$. Then

$$hf = hh^2 = h^2h = fh$$

and consequently

$$hg = hp(f) = p(f)h = gh,$$

where $g = p(f)$. Since $h^2 - g^2 = \mathbf{0}$, for every $\mathbf{v} \in V$ we have

$$0 = \langle (h-g)\mathbf{v}, (h^2 - g^2)\mathbf{v} \rangle = \langle (h-g)\mathbf{v}, (h+g)(h-g)\mathbf{v} \rangle$$
$$= \langle (h-g)\mathbf{v}, h(h-g)\mathbf{v} \rangle + \langle (h-g)\mathbf{v}, g(h-g)\mathbf{v} \rangle.$$

This gives us

$$\langle h - g)(\mathbf{v}), h(h - g)(\mathbf{v}) \rangle = 0 \quad \text{and} \quad \langle (h - g)(\mathbf{v}), g(h - g)(\mathbf{v}) \rangle = 0,$$

because both h and g are positive operators, and consequently

$$\langle (h - g)(\mathbf{v}), (h - g)(h - g)(\mathbf{v}) \rangle$$
$$= \langle h - g)(\mathbf{v}), h(h - g)(\mathbf{v}) \rangle - \langle (h - g)(\mathbf{v}), g(h - g)(\mathbf{v}) \rangle = 0.$$

Hence $\langle (h - g)^3(\mathbf{v}), \mathbf{v} \rangle = 0$, which gives $(h - g)^3 = \mathbf{0}$. Since $h - g$ is a self-adjoint operator, we conclude that $h - g = \mathbf{0}$, by Theorem 3.3.15. □

The unique positive square root of a positive operator f will be denoted $\sqrt{f}$.

Example 3.4.62. Consider the operator $f \in \mathcal{L}(\mathbb{C}^2, \mathbb{C}^2)$ defined as $f(\mathbf{x}) = P\mathbf{x}$, where

$$P = \begin{bmatrix} 33 & 24 - 24i \\ 24 + 24i & 57 \end{bmatrix}.$$

Show that f is a positive operator and find its positive square root.

Solution. First we find the spectral decomposition of f:

$$f = 9\text{proj}_{\begin{bmatrix} 1 - i \\ -1 \end{bmatrix}} + 81\text{proj}_{\begin{bmatrix} 1 - i \\ 2 \end{bmatrix}},$$

Thus f is a positive operator and

$$\sqrt{f} = 3\text{proj}_{\begin{bmatrix} 1 - i \\ -1 \end{bmatrix}} + 9\text{proj}_{\begin{bmatrix} 1 - i \\ 2 \end{bmatrix}}.$$

Since

$$3\left(\frac{1}{\sqrt{3}} \begin{bmatrix} 1 - i \\ -1 \end{bmatrix} \right) \left(\frac{1}{\sqrt{3}} [1 + i \ -1] \right) = \begin{bmatrix} 2 & i - 1 \\ -1 - i & 1 \end{bmatrix},$$

$$9\left(\frac{1}{\sqrt{6}} \begin{bmatrix} 1 - i \\ 2 \end{bmatrix} \right) \left(\frac{1}{\sqrt{6}} [1 + i \ 2] \right) = \frac{3}{2} \begin{bmatrix} 2 & 2(1 - i) \\ 2(1 + i) & 4 \end{bmatrix},$$

and

$$\begin{bmatrix} 2 & i - 1 \\ -1 - i & 1 \end{bmatrix} + \frac{3}{2} \begin{bmatrix} 2 & 2(1 - i) \\ 2(1 + i) & 4 \end{bmatrix} = \begin{bmatrix} 5 & 2 - 2i \\ 2 + 2i & 7 \end{bmatrix},$$

we have

$$\sqrt{f}\left(\begin{bmatrix} z_1 \\ z_2 \end{bmatrix} \right) = \begin{bmatrix} 5 & 2 - 2i \\ 2 + 2i & 7 \end{bmatrix} \begin{bmatrix} z_1 \\ z_2 \end{bmatrix}.$$

□

Example 3.4.63. In Example 3.4.58 we show that the operator $f \in \mathcal{L}(\mathbb{C}^3, \mathbb{C}^3)$ defined as $f(\mathbf{x}) = A\mathbf{x}$, where

$$A = \begin{bmatrix} 14 & 2 & 4 \\ 2 & 17 & -2 \\ 4 & -2 & 14 \end{bmatrix},$$

is a positive operator. Find the spectral decomposition of $\sqrt{f}$.

Solution. In Example 3.4.58 we found that $\lambda = 18$ and $\lambda = 9$ are the eigenvalues of f. We need to find an orthogonal basis of $\mathbb{C}^3$ consisting of eigenvectors of f.

For $\lambda = 18$ the system

$$\begin{cases} (14 - \lambda)x + 2y + 4z = 0 \\ 2x + (17 - \lambda)y - 2z = 0 \\ 4x - 2y + (14 - \lambda)z = 0 \end{cases}$$

is equivalent to the equation

$$2x - y - 2z = 0.$$

This means that

$$\mathcal{E}_{18} = \left\{ \begin{bmatrix} x \\ 2x - 2z \\ z \end{bmatrix} : x, z \in \mathbb{C} \right\} = \mathrm{Span} \left\{ \begin{bmatrix} 1 \\ 2 \\ 0 \end{bmatrix}, \begin{bmatrix} 0 \\ -2 \\ 1 \end{bmatrix} \right\}.$$

Note that the vectors $\begin{bmatrix} 1 \\ 2 \\ 0 \end{bmatrix}$ and $\begin{bmatrix} 0 \\ -2 \\ 1 \end{bmatrix}$ are not orthogonal. The projection of

the vector $\begin{bmatrix} 1 \\ 2 \\ 0 \end{bmatrix}$ on $\mathrm{Span} \left\{ \begin{bmatrix} 0 \\ -2 \\ 1 \end{bmatrix} \right\}$ is $-\frac{4}{5} \begin{bmatrix} 0 \\ -2 \\ 1 \end{bmatrix}$ and thus the vector

$$\begin{bmatrix} 1 \\ 2 \\ 0 \end{bmatrix} - \left(-\frac{4}{5} \begin{bmatrix} 0 \\ -2 \\ 1 \end{bmatrix} \right) = \frac{1}{5} \begin{bmatrix} 5 \\ 2 \\ 4 \end{bmatrix}$$

is orthogonal to $\begin{bmatrix} 0 \\ -2 \\ 1 \end{bmatrix}$ and $\left\{ \begin{bmatrix} 5 \\ 2 \\ 4 \end{bmatrix}, \begin{bmatrix} 0 \\ -2 \\ 1 \end{bmatrix} \right\}$ is an orthogonal basis of $\mathcal{E}_{18}$.

For $\lambda = 9$ the system

$$\begin{cases} (14 - \lambda)x + 2y + 4z = 0 \\ 2x + (17 - \lambda)y - 2z = 0 \\ 4x - 2y + (14 - \lambda)z = 0 \end{cases}$$

is equivalent to the system

$$\begin{cases} x + z & = 0 \\ 4x + 8y & = 0 \, , \\ -x - 2y & = 0 \end{cases}$$

which means that

$$\mathcal{E}_9 = \left\{ \begin{bmatrix} -2y \\ y \\ 2y \end{bmatrix} : y \in \mathbb{C} \right\} = \text{Span} \left\{ \begin{bmatrix} -2 \\ 1 \\ 2 \end{bmatrix} \right\}.$$

As expected the vector $\begin{bmatrix} -2 \\ 1 \\ 2 \end{bmatrix}$ is orthogonal to the vectors from $\mathcal{E}_{18}$ and

$$\left\{ \begin{bmatrix} 5 \\ 2 \\ 4 \end{bmatrix}, \begin{bmatrix} 0 \\ -2 \\ 1 \end{bmatrix}, \begin{bmatrix} -2 \\ 1 \\ 2 \end{bmatrix} \right\}$$

is an orthogonal basis of eigenvectors of f. Since

$$f = 18\text{proj}_{\begin{bmatrix} 5 \\ 2 \\ 4 \end{bmatrix}} + 18\text{proj}_{\begin{bmatrix} 0 \\ -2 \\ 1 \end{bmatrix}} + 9\text{proj}_{\begin{bmatrix} -2 \\ 1 \\ 2 \end{bmatrix}}$$

we have

$$\sqrt{f} = 3\sqrt{2}\text{proj}_{\begin{bmatrix} 5 \\ 2 \\ 4 \end{bmatrix}} + 3\sqrt{2}\text{proj}_{\begin{bmatrix} 0 \\ -2 \\ 1 \end{bmatrix}} + 3\text{proj}_{\begin{bmatrix} -2 \\ 1 \\ 2 \end{bmatrix}}.$$

$\square$

3.5   Singular value decomposition

Spectral decomposition is formulated for operators on an inner product space, that is, operators $f : \mathcal{V} \to \mathcal{V}$ where $\mathcal{V}$ is an inner product space. Now we are going to discuss a decomposition similar to spectral decomposition for linear transformations between two different inner product spaces. We begin by presenting an example which will motivate the main result of this section.

Example 3.5.1. Let V and W be finite dimensional inner product spaces and let $f : V \to W$ be a linear transformation. Suppose for some orthonormal vectors $v_1, v_2 \in V$ we have

$$f^*f(x) = 49\langle x, v_1 \rangle v_1 + 25\langle x, v_2 \rangle v_2$$

for every $x \in V$. Let $w_1 = \frac{1}{7}f(v_1)$ and $w_2 = \frac{1}{5}f(v_2)$. Show that $\|w_1\| = \|w_2\| = 1$, $\langle w_1, w_2 \rangle = 0$, and

$$f(x) = 7\langle x, v_1 \rangle w_1 + 5\langle x, v_2 \rangle w_2$$

for every $x \in V$.

Solution. Since

$$\|f(v_1)\|^2 = \langle f^*f(v_1), v_1 \rangle = 49 \quad \text{and} \quad \|f(v_2)\|^2 = \langle f^*f(v_2), v_2 \rangle = 25$$

we have

$$\|w_1\| = \left\| \frac{1}{7}f(v_1) \right\| = 1 \quad \text{and} \quad \|w_2\| = \left\| \frac{1}{5}f(v_2) \right\| = 1$$

and

$$\langle w_1, w_2 \rangle = \frac{1}{35}\langle f(v_1), f(v_2) \rangle$$

$$= \frac{1}{35}\langle f^*f(v_1), v_2 \rangle = \frac{1}{35}\langle 49v_1, v_2 \rangle$$

$$= \frac{49}{35}\langle v_1, v_2 \rangle = 0.$$

Now we extend the set $\{v_1, v_2\}$ to $\{v_1, v_2, \dots, v_n\}$, an orthonormal basis of V. Then, for every $x \in V$, we have

$$x = \langle x, v_1 \rangle v_1 + \langle x, v_2 \rangle v_2 + \cdots + \langle x, v_n \rangle v_n$$

and thus

$$f(x) = \langle x, v_1 \rangle f(v_1) + \langle x, v_2 \rangle f(v_2) + \cdots + \langle x, v_n \rangle f(v_n).$$

Note that $\|f(v_j)\|^2 = \langle f^*f(v_j), v_j \rangle = 0$ for $j > 2$. Hence

$$f(x) = \langle x, v_1 \rangle f(v_1) + \langle x, v_2 \rangle f(v_2) = 7\langle x, v_1 \rangle w_1 + 5\langle x, v_2 \rangle w_2.$$

$\square$

Theorem 3.5.2. *Let V and W be finite dimensional inner product spaces. For every nonzero linear transformation $f : V \to W$ there are positive numbers $\sigma_1, \ldots, \sigma_r$, orthonormal vectors $\mathbf{v}_1, \ldots, \mathbf{v}_r \in V$, and orthonormal vectors $\mathbf{w}_1, \ldots, \mathbf{w}_r \in W$ such that*

$$f(\mathbf{x}) = \sum_{j=1}^{r} \sigma_j \langle \mathbf{x}, \mathbf{v}_j \rangle \mathbf{w}_j$$

for every $\mathbf{x} \in V$.

Proof. The operator $f^* f$ is self-adjoint. Since

$$\langle f^* f(\mathbf{x}), \mathbf{x} \rangle = \langle f(\mathbf{x}), f(\mathbf{x}) \rangle = \| f(\mathbf{x}) \|^2 \geq 0,$$

for every $\mathbf{x} \in V$, $f^* f$ is a nonzero positive operator and thus there are orthonormal vectors $\mathbf{v}_1, \ldots, \mathbf{v}_r$ and positive numbers $\lambda_1, \ldots, \lambda_r$ such that

$$f^* f(\mathbf{x}) = \sum_{j=1}^{r} \lambda_j \langle \mathbf{x}, \mathbf{v}_j \rangle \mathbf{v}_j$$

for every $\mathbf{x} \in V$. Now we extend the set $\{\mathbf{v}_1, \ldots, \mathbf{v}_r\}$ to an orthonormal basis $\{\mathbf{v}_1, \ldots, \mathbf{v}_n\}$ of V. Then, for every $\mathbf{x} \in V$, we have

$$f(\mathbf{x}) = f \left(\sum_{j=1}^{n} \langle \mathbf{x}, \mathbf{v}_j \rangle \mathbf{v}_j \right) = \sum_{j=1}^{n} \langle \mathbf{x}, \mathbf{v}_j \rangle f(\mathbf{v}_j).$$

Since, for every $j = 1, \ldots, r$, we have

$$\| f(\mathbf{v}_j) \|^2 = \langle f^* f(\mathbf{v}_j), \mathbf{v}_j \rangle = \lambda_j \| \mathbf{v}_j \|^2 = \lambda_j,$$

and $\| f(\mathbf{v}_j) \|^2 = \langle f^* f(\mathbf{v}_j), \mathbf{v}_j \rangle = 0$ for $j > r$, we can write

$$f(\mathbf{x}) = \sum_{j=1}^{r} \langle \mathbf{x}, \mathbf{v}_j \rangle f(\mathbf{v}_j) = \sum_{j=1}^{r} \| f(\mathbf{v}_j) \| \langle \mathbf{x}, \mathbf{v}_j \rangle \frac{1}{\| f(\mathbf{v}_j) \|} f(\mathbf{v}_j).$$

If we let $\sigma_j = \| f(\mathbf{v}_j) \| = \sqrt{\lambda_j}$ and $\mathbf{w}_j = \frac{1}{\| f(\mathbf{v}_j) \|} f(\mathbf{v}_j)$, for $j \leq r$, then we obtain the desired decomposition of f:

$$f(\mathbf{x}) = \sum_{j=1}^{r} \sigma_j \langle \mathbf{x}, \mathbf{v}_j \rangle \mathbf{w}_j$$

for every $\mathbf{x} \in V$.

Moreover, for every $j, k \in \{1, \ldots, r\}$ we have

$$\langle \mathbf{w}_j, \mathbf{w}_k \rangle = \left\langle \frac{1}{\|f(\mathbf{v}_j)\|} f(\mathbf{v}_j), \frac{1}{\|f(\mathbf{v}_k)\|} f(\mathbf{v}_k) \right\rangle = \frac{1}{\sigma_j \sigma_k} \langle f(\mathbf{v}_j), f(\mathbf{v}_k) \rangle$$

$$= \frac{1}{\sigma_j \sigma_k} \langle f^* f(\mathbf{v}_j), \mathbf{v}_k \rangle = \frac{1}{\sigma_j \sigma_k} \langle \lambda_j \mathbf{v}_j, \mathbf{v}_k \rangle = \frac{\lambda_j}{\sigma_j \sigma_k} \langle \mathbf{v}_j, \mathbf{v}_k \rangle,$$

so the vectors $\mathbf{w}_1, \ldots, \mathbf{w}_r$ are orthonormal.   □

Corollary 3.5.3. *Let V and W be finite dimensional inner product spaces. For every nonzero linear transformation $f : V \to W$ there are orthonormal bases $\mathbf{v}_1, \ldots, \mathbf{v}_n \in V$ and $\mathbf{w}_1, \ldots, \mathbf{w}_m \in W$ and positive numbers $\sigma_1, \ldots, \sigma_r$, for some $r \leq m$ and $r \leq n$, such that*

$$f(\mathbf{v}_j) = \sigma_j \mathbf{w}_j$$

for $j = 1, \ldots, r$ and $f(\mathbf{v}_j) = \mathbf{0}$ for $j = r+1, \ldots, n$, if $n > r$.

Proof. Let $\mathbf{v}_1, \ldots, \mathbf{v}_n \in V$, $\lambda_1, \ldots, \lambda_r, \sigma_1, \ldots, \sigma_r > 0$, and $\mathbf{w}_1, \ldots, \mathbf{w}_r \in W$ be as defined in the proof of Theorem 3.5.2. It suffices to extend $\{\mathbf{w}_1, \ldots, \mathbf{w}_r\}$ to an orthonormal basis of W.   □

The representation of a linear transformation between inner product spaces given in Theorem 3.5.2 is called the *singular value decomposition* of f. If $f : V \to V$ is a positive operator, then its singular value decomposition is the same as its spectral decomposition. If $f : V \to V$ is a normal operator and

$$f(\mathbf{x}) = \sum_{j=1}^{r} \lambda_j \langle \mathbf{x}, \mathbf{v}_j \rangle \mathbf{v}_j$$

is its spectral decomposition with every $\lambda_j \neq 0$, then

$$f(\mathbf{x}) = \sum_{j=1}^{r} |\lambda_j| \langle \mathbf{x}, \mathbf{v}_j \rangle \frac{\lambda_j}{|\lambda_j|} \mathbf{v}_j$$

is its singular value decomposition. Note that, if λ_j is a nonzero real number, then $\frac{\lambda_j}{|\lambda_j|}$ is 1 or -1.

Example 3.5.4. Let $f : V \to W$, $\mathbf{v}_1, \ldots, \mathbf{v}_r \in V$, $\sigma_1, \ldots, \sigma_r > 0$, and $\mathbf{w}_1, \ldots, \mathbf{w}_r \in W$ be as defined in Theorem 3.5.2. Show that

$$f^*(\mathbf{y}) = \sum_{j=1}^{r} \sigma_j \langle \mathbf{y}, \mathbf{w}_j \rangle \mathbf{v}_j$$

for every $\mathbf{y} \in \mathcal{W}$.

Solution.

$$\langle f(\mathbf{x}), \mathbf{y} \rangle = \left\langle \sum_{j=1}^{r} \sigma_j \langle \mathbf{x}, \mathbf{v}_j \rangle \mathbf{w}_j, \mathbf{y} \right\rangle = \sum_{j=1}^{r} \langle \sigma_j \langle \mathbf{x}, \mathbf{v}_j \rangle \mathbf{w}_j, \mathbf{y} \rangle$$

$$= \sum_{j=1}^{r} \sigma_j \langle \mathbf{x}, \mathbf{v}_j \rangle \langle \mathbf{w}_j, \mathbf{y} \rangle = \sum_{j=1}^{r} \sigma_j \langle \mathbf{x}, \overline{\langle \mathbf{w}_j, \mathbf{y} \rangle} \mathbf{v}_j \rangle$$

$$= \sum_{j=1}^{r} \sigma_j \langle \mathbf{x}, \langle \mathbf{y}, \mathbf{w}_j \rangle \mathbf{v}_j \rangle = \left\langle \mathbf{x}, \sum_{j=1}^{r} \sigma_j \langle \mathbf{y}, \mathbf{w}_j \rangle \mathbf{v}_j \right\rangle = \langle \mathbf{x}, f^*(\mathbf{y}) \rangle.$$

□

The following theorem can be interpreted as a form of uniqueness of the singular value decomposition.

Theorem 3.5.5. *Let $\mathcal{V}$ and $\mathcal{W}$ be finite dimensional inner product spaces and let $f : \mathcal{V} \to \mathcal{W}$ be a nonzero linear transformation. If there are positive numbers $\sigma_1, \dots, \sigma_r$, orthonormal vectors $\mathbf{v}_1, \dots, \mathbf{v}_r \in \mathcal{V}$, and orthonormal vectors $\mathbf{w}_1, \dots, \mathbf{w}_r \in \mathcal{W}$ such that*

$$f(\mathbf{x}) = \sum_{j=1}^{r} \sigma_j \langle \mathbf{x}, \mathbf{v}_j \rangle \mathbf{w}_j$$

for every $\mathbf{x} \in \mathcal{V}$, then

$$f^* f(\mathbf{x}) = \sum_{j=1}^{r} \sigma_j^2 \langle \mathbf{x}, \mathbf{v}_j \rangle \mathbf{v}_j$$

for every $\mathbf{x} \in \mathcal{V}$.

Proof. From the result in Example 3.5.4 we get

$$f^* f(\mathbf{x}) = \sum_{j=1}^{r} \sigma_j \langle f(\mathbf{x}), \mathbf{w}_j \rangle \mathbf{v}_j$$

$$= \sum_{j=1}^{r} \sigma_j \left\langle \sum_{k=1}^{r} \sigma_k \langle \mathbf{x}, \mathbf{v}_k \rangle \mathbf{w}_k, \mathbf{w}_j \right\rangle \mathbf{v}_j$$

$$= \sum_{j=1}^{r} \sigma_j^2 \langle \mathbf{x}, \mathbf{v}_j \rangle \mathbf{v}_j$$

for every $\mathbf{x} \in \mathcal{V}$.   □

Example 3.5.6. Let $f : \mathcal{V} \to \mathcal{W}$, $\mathbf{v}_1, \ldots, \mathbf{v}_r \in \mathcal{V}$, $\sigma_1, \ldots, \sigma_r > 0$, and $\mathbf{w}_1, \ldots, \mathbf{w}_r \in \mathcal{W}$ be as defined in Theorem 3.5.2. Show that the linear transformation $f^+ : \mathcal{W} \to \mathcal{V}$ defined as

$$f^+(\mathbf{y}) = \sum_{j=1}^{r} \frac{1}{\sigma_j} \langle \mathbf{y}, \mathbf{w}_j \rangle \mathbf{v}_j$$

is the unique linear transformation g from $\mathcal{W}$ to $\mathcal{V}$ such that the following two conditions are satisfied:

(a) $gf = \text{proj}_{(\ker f)^\perp}$;

(b) $g(\mathbf{y}) = \mathbf{0}$ for every vector $\mathbf{y} \in (\text{ran } f)^\perp$.

Solution. Since, for every $k \in \{1, \ldots, r\}$,

$$f^+ f(\mathbf{v}_k) = f^+ \left(\sum_{j=1}^{r} \sigma_j \langle \mathbf{v}_k, \mathbf{v}_j \rangle \mathbf{w}_j \right) = f^+(\sigma_k \langle \mathbf{v}_k, \mathbf{v}_k \rangle \mathbf{w}_k) = \sigma_k f^+(\mathbf{w}_k)$$

$$= \sigma_k \left(\sum_{j=1}^{r} \frac{1}{\sigma_j} \langle \mathbf{w}_k, \mathbf{w}_j \rangle \mathbf{v}_j \right) = \sigma_k \left(\frac{1}{\sigma_k} \langle \mathbf{w}_k, \mathbf{w}_k \rangle \mathbf{v}_k \right) = \mathbf{v}_k = \text{proj}_{(\ker f)^\perp}(\mathbf{v}_k),$$

we have $f^+ f = \text{proj}_{(\ker f)^\perp}$, because $(\ker f)^\perp = \text{Span}\{\mathbf{v}_1, \ldots, \mathbf{v}_r\}$.

If $\mathbf{y} \in (\text{ran } f)^\perp$, then $\langle \mathbf{y}, \mathbf{w}_j \rangle = 0$ for every $j \in \{1, \ldots, r\}$ and thus

$$f^+(\mathbf{y}) = \sum_{j=1}^{r} \frac{1}{\sigma_j} \langle \mathbf{y}, \mathbf{w}_j \rangle \mathbf{v}_j = \mathbf{0}.$$

Now assume that a linear transformation $g : \mathcal{W} \to \mathcal{V}$ satisfies (a) and (b). Note that $\text{ran } f = \text{Span}\{\mathbf{w}_1, \ldots, \mathbf{w}_r\}$ and $\mathcal{W} = \text{ran } f \oplus (\text{ran } f)^\perp$. If $\mathbf{y} \in \mathcal{W}$, then $\mathbf{y} = \mathbf{y}_1 + \mathbf{y}_2$, where $\mathbf{y}_1 \in \text{ran } f$ and $\mathbf{y}_2 \in (\text{ran } f)^\perp$. Then $\mathbf{y}_1 = f(\mathbf{x})$ for some $\mathbf{x} \in \mathcal{V}$, and consequently

$$g(\mathbf{y}) = g(\mathbf{y}_1 + \mathbf{y}_2) = g(\mathbf{y}_1) + g(\mathbf{y}_2) = gf(\mathbf{x}) + \mathbf{0}$$

$$= \text{proj}_{(\ker f)^\perp}(\mathbf{x}) = \sum_{j=1}^{r} \langle \mathbf{x}, \mathbf{v}_j \rangle \mathbf{v}_j = \sum_{j=1}^{r} \left\langle \mathbf{x}, \frac{1}{\sigma_j^2} f^* f(\mathbf{v}_j) \right\rangle \mathbf{v}_j$$

$$= \sum_{j=1}^{r} \frac{1}{\sigma_j} \left\langle f(\mathbf{x}), \frac{1}{\sigma_j} f(\mathbf{v}_j) \right\rangle \mathbf{v}_j = \sum_{j=1}^{r} \frac{1}{\sigma_j} \langle \mathbf{y}_1, \mathbf{w}_j \rangle \mathbf{v}_j$$

$$= \sum_{j=1}^{r} \frac{1}{\sigma_j} \langle \mathbf{y}, \mathbf{w}_j \rangle \mathbf{v}_j.$$

Therefore, if $g : \mathcal{W} \to \mathcal{V}$ is a linear transformation that satisfies (a) and (b), then $g(\mathbf{y}) = \sum_{j=1}^{r} \frac{1}{\sigma_j} \langle \mathbf{y}, \mathbf{w}_j \rangle \mathbf{v}_j$. $\qquad\square$

Example 3.5.7. Consider the operator $f \in \mathcal{L}(\mathbb{C}^2, \mathbb{C}^2)$ defined as $f(\mathbf{x}) = A\mathbf{x}$, where

$$A = \begin{bmatrix} 5i & 3 + 4i \\ -3 + 4i & 5i \end{bmatrix}.$$

Find the singular value decomposition of f.

Solution. We have

$$A^* = \begin{bmatrix} -5i & -3 - 4i \\ 3 - 4i & -5i \end{bmatrix}$$

and

$$A^* A = \begin{bmatrix} 50 & 40 - 30i \\ 40 + 30i & 50 \end{bmatrix}.$$

The eigenvalues of the matrix $A^* A$ are the roots of the equation

$$(50 - \lambda)^2 - (40 - 30i)(40 + 30i) = \lambda^2 - 100\lambda = 0,$$

that is, $\lambda = 100$ and $\lambda = 0$, and the orthonormal vectors

$$\frac{1}{5\sqrt{2}} \begin{bmatrix} 4 - 3i \\ 5 \end{bmatrix} \quad \text{and} \quad \frac{1}{5\sqrt{2}} \begin{bmatrix} 4 - 3i \\ -5 \end{bmatrix}$$

are corresponding eigenvectors. Since

$$f\left(\begin{bmatrix} 4 - 3i \\ 5 \end{bmatrix}\right) = \begin{bmatrix} 10(3 + 4i) \\ 50i \end{bmatrix}, \quad f\left(\begin{bmatrix} 4 - 3i \\ -5 \end{bmatrix}\right) = \begin{bmatrix} 0 \\ 0 \end{bmatrix}$$

and

$$\left\| \begin{bmatrix} 10(3 + 4i) \\ 50i \end{bmatrix} \right\| = 50\sqrt{2},$$

the singular value decomposition of f is

$$f\left(\begin{bmatrix} x_1 \\ x_2 \end{bmatrix}\right) = 10 \left\langle \begin{bmatrix} x_1 \\ x_2 \end{bmatrix}, \frac{1}{5\sqrt{2}} \begin{bmatrix} 4 - 3i \\ 5 \end{bmatrix} \right\rangle \left(\frac{1}{50\sqrt{2}} \begin{bmatrix} 10(3 + 4i) \\ 50i \end{bmatrix} \right).$$

For practical calculations we can use a simplified form:

$$f\left(\begin{bmatrix} x_1 \\ x_2 \end{bmatrix}\right) = \frac{1}{5} \left\langle \begin{bmatrix} x_1 \\ x_2 \end{bmatrix}, \begin{bmatrix} 4 - 3i \\ 5 \end{bmatrix} \right\rangle \begin{bmatrix} 3 + 4i \\ 5i \end{bmatrix}.$$

$\qquad\square$

Example 3.5.8. Consider the operator $f \in \mathcal{L}(\mathbb{C}^2, \mathbb{C}^2)$ defined as $f(\mathbf{x}) = A\mathbf{x}$, where

$$A = \begin{bmatrix} 3+2i & 2-4i \\ 4+2i & 6+i \end{bmatrix}.$$

Find the singular value decomposition of f.

Solution. First we calculate

$$A^*A = \begin{bmatrix} 33 & 24-24i \\ 24+24i & 57 \end{bmatrix}.$$

The eigenvalues of the matrix A^*A are 9 and 81 and $\begin{bmatrix} 1-i \\ -1 \end{bmatrix}$ and $\begin{bmatrix} 1-i \\ 2 \end{bmatrix}$ are corresponding eigenvectors. Consequently the singular value decomposition of f is

$$f(\mathbf{x}) = 3\langle \mathbf{x}, \mathbf{v}_1 \rangle \mathbf{w}_1 + 9\langle \mathbf{x}, \mathbf{v}_1 \rangle \mathbf{w}_2,$$

where

$$\mathbf{v}_1 = \frac{1}{\sqrt{3}}\begin{bmatrix} 1+i \\ -1 \end{bmatrix}, \quad \mathbf{w}_1 = \frac{1}{\sqrt{3}}\begin{bmatrix} 1+i \\ -i \end{bmatrix}, \quad \mathbf{v}_2 = \frac{1}{\sqrt{6}}\begin{bmatrix} 1+i \\ 2 \end{bmatrix}, \quad \mathbf{w}_2 = \frac{1}{\sqrt{6}}\begin{bmatrix} 1-i \\ 2 \end{bmatrix}.$$

$\square$

Example 3.5.9. Consider the operator $f \in \mathcal{L}(\mathbb{C}^2, \mathbb{C}^4)$ defined as $f(\mathbf{x}) = A\mathbf{x}$, where

$$A = \begin{bmatrix} 3 & 1 \\ 1 & 2 \\ 1 & -3 \\ -1 & -2 \end{bmatrix}.$$

Find the singular value decomposition of f.

Solution. First we find that

$$\begin{bmatrix} 3 & 1 & 1 & -1 \\ 1 & 2 & -3 & -2 \end{bmatrix}\begin{bmatrix} 3 & 1 \\ 1 & 2 \\ 1 & -3 \\ -1 & -2 \end{bmatrix} = \begin{bmatrix} 12 & 4 \\ 4 & 18 \end{bmatrix}.$$

The eigenvalues of the matrix $\begin{bmatrix} 12 & 4 \\ 4 & 18 \end{bmatrix}$ are 10 and 20, and $\begin{bmatrix} 2 \\ -1 \end{bmatrix}$ and $\begin{bmatrix} 1 \\ 2 \end{bmatrix}$ are corresponding eigenvectors. We have

$$\begin{bmatrix} 3 & 1 \\ 1 & 2 \\ 1 & -3 \\ -1 & -2 \end{bmatrix} \begin{bmatrix} 1 \\ 2 \end{bmatrix} = \begin{bmatrix} 5 \\ 5 \\ -5 \\ -5 \end{bmatrix} \quad \text{and} \quad \begin{bmatrix} 3 & 1 \\ 1 & 2 \\ 1 & -3 \\ -1 & -2 \end{bmatrix} \begin{bmatrix} 2 \\ -1 \end{bmatrix} = \begin{bmatrix} 5 \\ 0 \\ 5 \\ 0 \end{bmatrix}.$$

We note that

$$\left\langle \begin{bmatrix} 5 \\ 5 \\ -5 \\ -5 \end{bmatrix}, \begin{bmatrix} 5 \\ 0 \\ 5 \\ 0 \end{bmatrix} \right\rangle = 0.$$

The singular value decomposition of f is

$$f(\mathbf{x}) = \sqrt{20}\langle \mathbf{x}, \mathbf{v}_1 \rangle \mathbf{w}_1 + \sqrt{10}\langle \mathbf{x}, \mathbf{v}_1 \rangle \mathbf{w}_2,$$

where

$$\mathbf{v}_1 = \begin{bmatrix} \frac{1}{\sqrt{5}} \\ \frac{2}{\sqrt{5}} \end{bmatrix}, \quad \mathbf{w}_1 = \begin{bmatrix} \frac{1}{2} \\ \frac{1}{2} \\ -\frac{1}{2} \\ -\frac{1}{2} \end{bmatrix}, \quad \mathbf{v}_2 = \begin{bmatrix} \frac{2}{\sqrt{5}} \\ -\frac{1}{\sqrt{5}} \end{bmatrix}, \quad \mathbf{w}_2 = \begin{bmatrix} \frac{1}{\sqrt{2}} \\ 0 \\ \frac{1}{\sqrt{2}} \\ 0 \end{bmatrix}.$$

$\square$

Example 3.5.10. Let $\mathcal{P}_2([-1,1])$ and $\mathcal{P}_1([-1,1])$ denote real valued polynomials on the interval $[-1,1]$ of degree at most 2 and 1, respectively, with the inner product defined as

$$\langle p(t), q(t) \rangle = \int_{-1}^{1} p(t)q(t)\, dt.$$

Find the singular value decomposition of the differential operator $D = \dfrac{d}{dt}$.

Solution. First we note that $\left\{ \frac{1}{\sqrt{2}}, \sqrt{\frac{3}{2}}t, \sqrt{\frac{5}{2}}\left(\frac{3}{2}t^2 - \frac{1}{2}\right) \right\}$ is an orthonormal basis in $\mathcal{P}_2([-1,1])$ and $\left\{ \frac{1}{\sqrt{2}}, \sqrt{\frac{3}{2}}t \right\}$ is an orthonormal basis in $\mathcal{P}_1([-1,1])$. The matrix of the differential operator with respect to these bases is

$$\begin{bmatrix} 0 & \sqrt{3} & 0 \\ 0 & 0 & \sqrt{15} \end{bmatrix}$$

and we have

$$\begin{bmatrix} 0 & 0 \\ \sqrt{3} & 0 \\ 0 & \sqrt{15} \end{bmatrix} \begin{bmatrix} 0 & \sqrt{3} & 0 \\ 0 & 0 & \sqrt{15} \end{bmatrix} = \begin{bmatrix} 0 & 0 & 0 \\ 0 & 3 & 0 \\ 0 & 0 & 15 \end{bmatrix},$$

so the nonzero eigenvalues of the operator D^*D are 3 and 15. The polynomials $\sqrt{\frac{3}{2}}t$ and $\sqrt{\frac{5}{2}}\left(\frac{3}{2}t^2 - \frac{1}{2}\right)$ are orthonormal eigenvectors of D^*D corresponding to the eigenvalues 3 and 15. Since

$$\frac{D\left(\sqrt{\frac{3}{2}}t\right)}{\left\|D\left(\sqrt{\frac{3}{2}}t\right)\right\|} = \frac{\sqrt{\frac{3}{2}}}{\sqrt{3}} = \sqrt{\frac{1}{2}}$$

and

$$\frac{D\left(\sqrt{\frac{5}{2}}\left(\frac{3}{2}t^2 - \frac{1}{2}\right)\right)}{\left\|D\left(\sqrt{\frac{5}{2}}\left(\frac{3}{2}t^2 - \frac{1}{2}\right)\right)\right\|} = \frac{\frac{3\sqrt{5}}{2}t}{\sqrt{\frac{15}{2}}} = \sqrt{\frac{3}{2}}\,t,$$

the singular value decomposition of the differential operator $D : \mathcal{P}_2([-1,1]) \to \mathcal{P}_1([-1,1])$ is

$$D(p(t)) = \sqrt{3}\left\langle p(t), \sqrt{\frac{3}{2}}t \right\rangle \sqrt{\frac{1}{2}} + \sqrt{15}\left\langle p(t), \sqrt{\frac{5}{2}}\left(\frac{3}{2}t^2 - \frac{1}{2}\right) \right\rangle \sqrt{\frac{3}{2}}t$$

$$= \frac{3}{2}\int_{-1}^{1} tp(t)\, dt + \frac{15}{4}\int_{-1}^{1}(3t^2 - 1)p(t)\, dt\ t.$$

While this result has limited practical applications, it is interesting that on the space $\mathcal{P}_2([-1,1])$ differentiation can be expressed in terms of definite integrals. □

The polar decomposition

We close this chapter with an introduction of another decomposition of operators on a finite dimensional inner product space, called the *polar decomposition*. To define the polar decomposition of an operator we need the notion of a partial isometry.

Definition 3.5.11. Let V be an inner product space and let U be a subspace of V. A linear transformation $f : V \to V$ is called a *partial isometry* with initial space U if the following two conditions are satisfied:

(a) $\|f(\mathbf{x})\| = \|\mathbf{x}\|$ for every $\mathbf{x} \in U$;

(b) $f(\mathbf{x}) = \mathbf{0}$ for every $\mathbf{x} \in U^\perp$.

Example 3.5.12. Let V be an inner product space and let $g : V \to V$ be the linear transformation defined as

$$g(\mathbf{x}) = \sum_{j=1}^{r} \langle \mathbf{x}, \mathbf{a}_j \rangle \mathbf{b}_j,$$

where $\{\mathbf{a}_1, \ldots, \mathbf{a}_r\}$ and $\{\mathbf{b}_1, \ldots, \mathbf{b}_r\}$ are orthonormal sets in V. Show that g is a partial isometry with initial space $\mathrm{Span}\{\mathbf{a}_1, \ldots, \mathbf{a}_r\}$.

Solution. If $\mathbf{x} \in \mathrm{Span}\{\mathbf{a}_1, \ldots, \mathbf{a}_r\}$, then there are numbers $x_1, \ldots, x_r \in \mathbb{K}$ such that

$$\mathbf{x} = x_1 \mathbf{a}_1 + \cdots + x_r \mathbf{a}_r.$$

Then

$$\|\mathbf{x}\|^2 = \|x_1 \mathbf{a}_1 + \cdots + x_r \mathbf{a}_r\|^2 = |x_1|^2 + \cdots + |x_r|^2$$

and, because $\langle \mathbf{x}, \mathbf{a}_j \rangle = x_j$ for all $j \in \{1, \ldots, r\}$, we also have

$$\|g(\mathbf{x})\|^2 = \|x_1 \mathbf{b}_1 + \cdots + x_r \mathbf{b}_r\|^2 = |x_1|^2 + \cdots + |x_r|^2.$$

It is clear that $g(\mathbf{x}) = \mathbf{0}$ for every $\mathbf{x} \in \mathrm{Span}\{\mathbf{a}_1, \ldots, \mathbf{a}_r\}^\perp$. $\qquad\square$

For any $z \in \mathbb{C}$ we have $z\bar{z} = |z|^2$ and thus $|z| = \sqrt{z\bar{z}}$. We use this analogy to define the operator $|f|$ for an arbitrary linear operator on a finite dimensional inner product space V. If $f : V \to V$ is a linear operator, then f^*f is a positive linear operator on V and thus it has a unique positive square root $\sqrt{f^*f}$. We will use the notation

$$|f| = \sqrt{f^*f}.$$

In other words, for any linear operator $f : V \to V$ on a finite dimensional inner product space V there is a unique positive linear operator $|f| : V \to V$ such that $|f|^2 = f^*f$.

Theorem 3.5.13. *Let V be a finite dimensional inner product space and let $f : V \to V$ be a linear transformation. There is a partial isometry $g : V \to V$ with initial space $\text{ran}\,|f|$ such that*

$$f = g|f|.$$

This representation is unique in the following sense: If $f = hp$ where $p : V \to V$ is a positive operator and h is a partial isometry on V with initial space $\text{ran}\,p$, then $p = |f|$ and $g = h$.

Proof. If $\mathbf{u} = |f|(\mathbf{v})$ for some $\mathbf{v} \in V$, then we would like to define $g(\mathbf{u}) = f(\mathbf{v})$, but this does not define g unless we can show that $|f|(\mathbf{v}_1) = |f|(\mathbf{v}_2)$ implies $f(\mathbf{v}_1) = f(\mathbf{v}_2)$. Since

$$\|f(\mathbf{v})\|^2 = \langle f(\mathbf{v}), f(\mathbf{v}) \rangle = \langle f^* f(\mathbf{v}), \mathbf{v} \rangle$$
$$= \langle |f|^2(\mathbf{v}), \mathbf{v} \rangle = \langle |f|(\mathbf{v}), |f|(\mathbf{v}) \rangle = \||f|(\mathbf{v})\|^2,$$

we have $\|f(\mathbf{v})\| = \||f|(\mathbf{v})\|$ and thus

$$\|f(\mathbf{v}_1) - f(\mathbf{v}_2)\| = \|f(\mathbf{v}_1 - \mathbf{v}_2)\| = \||f|(\mathbf{v}_1 - \mathbf{v}_2)\| = \||f|(\mathbf{v}_1) - |f|(\mathbf{v}_2)\|.$$

Consequently $|f|(\mathbf{v}_1) = |f|(\mathbf{v}_2)$ implies $f(\mathbf{v}_1) = f(\mathbf{v}_2)$ and thus g is well-defined. Moreover, $\|g(|f|(\mathbf{v}))\| = \|f(\mathbf{v})\| = \||f|(\mathbf{v})\|$, so g is an isometry on $\text{ran}\,|f|$. Clearly, $g(\mathbf{x}) = \mathbf{0}$ for every $\mathbf{x} \in (\text{ran}\,|f|)^{\perp}$. Therefore, g is a partial isometry with initial space $\text{ran}\,|f|$.

Now assume that $f = hp$ where $p : V \to V$ is a positive operator and h is a partial isometry $g : V \to V$ with initial space $\text{ran}\,p$. Then for every $\mathbf{v} \in V$ we have

$$\langle f^* f(\mathbf{v}), \mathbf{v} \rangle = \langle f(\mathbf{v}), f(\mathbf{v}) \rangle = \langle h(p(\mathbf{v})), h(p(\mathbf{v})) \rangle = \langle p(\mathbf{v}), p(\mathbf{v}) \rangle = \langle p^2(\mathbf{v}), \mathbf{v} \rangle,$$

which gives us $f^* f = p^2$, because $f^* f$ and p^2 are self adjoint, and thus $p = |f|$. Clearly, $g = h$. $\qquad\square$

The representation of a linear transformation $f : V \to V$ in the form presented in Theorem 3.5.13 is called the *polar decomposition* of f. It is somewhat similar to the polar form of a complex number: $z = |z|(\cos\theta + i\sin\theta)$.

Example 3.5.14. Consider the operator $f \in \mathcal{L}(\mathbb{C}^2, \mathbb{C}^2)$ defined as $f(\mathbf{x}) = A\mathbf{x}$, where

$$A = \begin{bmatrix} 3 + 2i & 2 - 4i \\ 4 + 2i & 6 + i \end{bmatrix}.$$

Determine the polar decomposition of f.

Solution. According to Example 3.4.62 we can write

$$|f| = \sqrt{f^*f} = 3\text{proj}_{\begin{bmatrix} 1-i \\ -1 \end{bmatrix}} + 9\text{proj}_{\begin{bmatrix} 1-i \\ 2 \end{bmatrix}}.$$

We have

$$f\left(\begin{bmatrix} 1-i \\ -1 \end{bmatrix}\right) = \begin{bmatrix} 3+3i \\ -3i \end{bmatrix}, \quad f\left(\begin{bmatrix} 1-i \\ 2 \end{bmatrix}\right) = \begin{bmatrix} 9-9i \\ 18 \end{bmatrix}$$

and

$$|f|\left(\begin{bmatrix} 1-i \\ -1 \end{bmatrix}\right) = \begin{bmatrix} 3-3i \\ -3 \end{bmatrix}, \quad |f|\left(\begin{bmatrix} 1-i \\ 2 \end{bmatrix}\right) = \begin{bmatrix} 9-9i \\ 18 \end{bmatrix}.$$

Now we can define an isometry $g : \mathcal{V} \to \mathcal{V}$ such that

$$g\left(\begin{bmatrix} 3-3i \\ -3 \end{bmatrix}\right) = \begin{bmatrix} 3+3i \\ -3i \end{bmatrix} \quad \text{and} \quad g\left(\begin{bmatrix} 9-9i \\ 18 \end{bmatrix}\right) = \begin{bmatrix} 9-9i \\ 18 \end{bmatrix},$$

that is,

$$g(\mathbf{x}) = \langle \mathbf{x}, \mathbf{v}_1 \rangle \mathbf{w}_1 + \langle \mathbf{x}, \mathbf{v}_2 \rangle \mathbf{w}_2,$$

where $\mathbf{x} \in \mathbb{C}^2$,

$$\mathbf{v}_1 = \frac{1}{\sqrt{3}}\begin{bmatrix} 1-i \\ -1 \end{bmatrix}, \quad \mathbf{w}_1 = \frac{1}{\sqrt{3}}\begin{bmatrix} 1+i \\ -i \end{bmatrix}, \quad \mathbf{v}_2 = \frac{1}{\sqrt{6}}\begin{bmatrix} 1-i \\ 2 \end{bmatrix}, \quad \mathbf{w}_2 = \frac{1}{\sqrt{6}}\begin{bmatrix} 1-i \\ 2 \end{bmatrix}.$$

Then $f = g|f|$ which is the polar decomposition of f. □

Example 3.5.15. Let $\mathcal{V}$ be an inner product space and let $f : \mathcal{V} \to \mathcal{V}$ be a linear transformation such that for every $\mathbf{x} \in \mathcal{V}$ we have

$$f^*f(\mathbf{x}) = \sum_{j=1}^{r} \lambda_j \langle \mathbf{x}, \mathbf{a}_j \rangle \mathbf{a}_j,$$

where $\{\mathbf{a}_1, \ldots, \mathbf{a}_r\}$ is an orthonormal set in $\mathcal{V}$ and $\lambda_1, \ldots, \lambda_r$ are positive numbers. Find the polar decomposition of f.

Solution. The linear operator $|f|$ is defined by

$$|f|(\mathbf{x}) = \sqrt{f^*f}(\mathbf{x}) = \sum_{j=1}^{r} \sqrt{\lambda_j} \langle \mathbf{x}, \mathbf{a}_j \rangle \mathbf{a}_j$$

for every $\mathbf{x} \in \mathcal{V}$. In the proof of Theorem 3.5.13 we show that

$$\|f(\mathbf{a}_j)\| = \||f|(\mathbf{a}_j)\| = \sqrt{\lambda_j} > 0$$

for $j \in \{1, \ldots, r\}$, so we can define the operator $g : \mathcal{V} \rightarrow \mathcal{V}$ by

$$g(\mathbf{y}) = \sum_{j=1}^{r} \langle \mathbf{y}, \mathbf{a}_j \rangle \mathbf{b}_j,$$

where $\mathbf{b}_j = \frac{1}{\|f(\mathbf{a}_j)\|} f(\mathbf{a}_j) = \frac{1}{\sqrt{\lambda_j}} f(\mathbf{a}_j)$. The operator g is a partial isometry and it is easy to verify that $f = g|f|$ which is the polar decomposition of f. $\qquad\square$

3.6 Exercises

3.6.1 Definitions and examples

Exercise 3.1. Let $s : \mathcal{V} \times \mathcal{V} \rightarrow \mathbb{C}$ a positive sesquilinear form. Show that $|s(\mathbf{x}, \mathbf{y})|^2 \leq s(\mathbf{x}, \mathbf{x}) s(\mathbf{y}, \mathbf{y})$ for every $\mathbf{x}, \mathbf{y} \in \mathcal{V}$.

Exercise 3.2. Let $\mathcal{V}$ be an inner product space and let $f : \mathcal{V} \rightarrow \mathcal{V}$ be a positive operator (see Definition 3.4.53). Show that $|\langle f(\mathbf{x}), \mathbf{y} \rangle|^2 \leq \langle f(\mathbf{x}), \mathbf{x} \rangle \langle f(\mathbf{y}), \mathbf{y} \rangle$ for every $\mathbf{x}, \mathbf{y} \in \mathcal{V}$.

Exercise 3.3. Let $\mathcal{V}$ be an inner product space and let $f : \mathcal{V} \rightarrow \mathcal{V}$ be a positive operator (see Definition 3.4.53). Show that $\|f(\mathbf{x})\|^3 \leq \langle f(\mathbf{x}), \mathbf{x} \rangle \|f^2(\mathbf{x})\|$ for every $\mathbf{x} \in \mathcal{V}$.

Exercise 3.4. Let $\mathcal{C}^2([0, 1])$ be the space of functions with continues second derivatives. Show that $\langle f, g \rangle = f(0)\overline{g(0)} + f'(0)\overline{g'(0)} + \int_0^1 f''(t)\overline{g''(t)}dt$ is an inner product in $\mathcal{C}^2([0, 1])$.

Exercise 3.5. Let $\mathbf{v}$ be a vector in an inner product space $\mathcal{V}$. Show that the function $s_{\mathbf{v}} : \mathcal{V} \times \mathcal{V} \rightarrow \mathbb{C}$ defined by $s_{\mathbf{v}}(\mathbf{a}, \mathbf{b}) = \langle \mathbf{v}, \mathbf{b} \rangle \mathbf{a}$ is sesquilinear.

Exercise 3.6. Let $\mathcal{V}$ be a finite dimensional inner product space. Show that for every linear operator $f : \mathcal{V} \rightarrow \mathcal{V}$ there is a nonnegative constant α such that $\|f(\mathbf{v})\| \leq \alpha \|\mathbf{v}\|$ for every $\mathbf{v} \in \mathcal{V}$.

Exercise 3.7. Let $\mathcal{V}$ be an inner product space and let $s : \mathcal{V} \times \mathcal{V} \rightarrow \mathbb{C}$ be a sesquilinear form. Show that, if $s(\mathbf{x}, \mathbf{y}) = s(\mathbf{y}, \mathbf{x})$ for every $\mathbf{x}, \mathbf{y} \in \mathcal{V}$, then $s = \mathbf{0}$.

Exercise 3.8. Let $\mathcal{V}$ and $\mathcal{W}$ be inner product spaces and let $\mathbf{v} \in \mathcal{V}$ and $\mathbf{w} \in \mathcal{W}$. Show that the function $\mathbf{w} \otimes \mathbf{v} : \mathcal{V} \rightarrow \mathcal{W}$ defined by $\mathbf{w} \otimes \mathbf{v}(\mathbf{x}) = \langle \mathbf{x}, \mathbf{v} \rangle \mathbf{w}$ is a linear transformation and we have $f \circ (\mathbf{w} \otimes \mathbf{v}) = f(\mathbf{w}) \otimes \mathbf{v}$ for every linear operator $f : \mathcal{W} \rightarrow \mathcal{W}$.

Exercise 3.9. Let $\mathbf{v} \in \mathbb{C}^n$ and $\mathbf{w} \in \mathbb{C}^m$. Find the matrix of the linear transformation $\mathbf{w} \otimes \mathbf{v}$ as defined in Exercise 3.8.

Exercise 3.10. Let $\mathcal{U}$, $\mathcal{V}$, and $\mathcal{X}$ be inner product spaces and let $\mathbf{u} \in \mathcal{U}$, $\mathbf{v}_1, \mathbf{v}_2 \in \mathcal{V}$, and $\mathbf{w} \in \mathcal{W}$. Show, with the notations from Exercise 3.8, that we have $(\mathbf{w} \otimes \mathbf{v}_2)(\mathbf{v}_1 \otimes \mathbf{u}) = \langle \mathbf{v}_1, \mathbf{v}_2 \rangle (\mathbf{w} \otimes \mathbf{u})$.

Exercise 3.11. Show that $\frac{1}{n}|(a_1 + \cdots + a_n)| \leq \sqrt{\frac{1}{n}(a_1^2 + \cdots + a_n^2)}$ for any $a_1, \ldots, a_n \in \mathbb{R}$.

3.6.2   Orthogonal projections

Exercise 3.12. Let $\{\mathbf{v}_1, \ldots, \mathbf{v}_n\}$ be an orthonormal basis of the inner product space $\mathcal{V}$. Show, with the notation from Exercise 3.8, that $\mathrm{Id}_{\mathcal{V}} = \sum_{j=1}^n \mathbf{v}_j \otimes \mathbf{v}_j$.

Exercise 3.13. Let $\mathcal{V}$ and $\mathcal{W}$ be inner product spaces and let $\{\mathbf{v}_1, \ldots, \mathbf{v}_m\}$ and $\{\mathbf{w}_1, \ldots, \mathbf{w}_n\}$ be orthonormal bases in $\mathcal{V}$ and $\mathcal{W}$, respectively. Show that the set $\{\mathbf{w}_k \otimes \mathbf{v}_j : j \in \{1, \ldots, m\}, k \in \{1, \ldots, n\}\}$ is a basis of $\mathcal{L}(\mathcal{W}, \mathcal{V})$ ($\mathbf{w}_k \otimes \mathbf{v}_j$ is defined in Exercise 3.8).

Exercise 3.14. Determine the projection matrix on $\mathrm{Span}\left\{\begin{bmatrix} i \\ -i \end{bmatrix}\right\}$ in $\mathbb{C}^2$ and use it to determine the projection of the vector $\begin{bmatrix} 1 \\ i \end{bmatrix}$ on $\mathrm{Span}\left\{\begin{bmatrix} i \\ -i \end{bmatrix}\right\}$.

Exercise 3.15. Let $\mathcal{V}$ be an inner product space and let $f : \mathcal{V} \to \mathcal{V}$ be an orthogonal projection. Show that $\|f(\mathbf{v})\| = \|\mathbf{v}\|$ implies $f(\mathbf{v}) = \mathbf{v}$ for every $\mathbf{v} \in \mathcal{V}$.

Exercise 3.16. Let $\mathcal{V}$ be an inner product space and let $f : \mathcal{V} \to \mathcal{V}$ be a linear operator. Show that f is an orthogonal projection if and only if $\ker(\mathrm{Id} - f) = \mathrm{ran}\, f = \ker f^{\perp}$.

Exercise 3.17. Let $\mathcal{U}$ and $\mathcal{V}$ be subspaces of a finite-dimensional inner product space $\mathcal{W}$. Show that $\mathrm{proj}_{\mathcal{U}}\mathrm{proj}_{\mathcal{V}} = \mathbf{0}$ if and only if $\langle \mathbf{u}, \mathbf{v} \rangle = 0$ for every $\mathbf{u} \in \mathcal{U}$ and $\mathbf{v} \in \mathcal{V}$.

Exercise 3.18. Let $\mathcal{V}$ be a finite-dimensional inner product space and let $f : \mathcal{V} \to \mathcal{V}$ be a linear operator. Show, with the notation from Exercise 3.8, that the operator f is a nonzero projection if and only if there is an orthonormal set $\{\mathbf{v}_1, \ldots, \mathbf{v}_k\} \subseteq \mathcal{V}$ such that $f = \sum_{j=1}^k \mathbf{v}_j \otimes \mathbf{v}_j$. Determine $\mathrm{ran}\, f$.

Exercise 3.19. Let $\mathcal{V}$ be a finite-dimensional inner product space and let $f : \mathcal{V} \to \mathcal{V}$ be a linear operator. Show that there is an orthonormal basis $\{\mathbf{v}_1, \ldots, \mathbf{v}_n\}$ of $\mathcal{V}$ and numbers x_{jk} for $1 \leq j \leq k \leq n$ such that

$$f(\mathbf{v}_1) = x_{11}\mathbf{v}_1,$$
$$f(\mathbf{v}_2) = x_{12}\mathbf{v}_1 + x_{22}\mathbf{v}_2,$$
$$f(\mathbf{v}_3) = x_{13}\mathbf{v}_1 + x_{23}\mathbf{v}_2 + x_{33}\mathbf{v}_3,$$
$$\vdots$$
$$f(\mathbf{v}_n) = x_{1n}\mathbf{v}_1 + \cdots + x_{n-1,n}\mathbf{v}_{n-1} + x_{nn}\mathbf{v}_n.$$

Exercise 3.20. Consider the vector space of continuous functions defined on the interval $[-1, 1]$ with the inner product $\langle f, g \rangle = \int_{-1}^1 f(t)\overline{g(t)}\, dt$. Find the best approximation of the function t^3 by functions from $\mathrm{Span}\{1, t\}$.

Exercise 3.21. Consider the vector space of continuous functions defined on the interval $[-\pi, \pi]$ with the inner product $\langle f, g \rangle = \frac{1}{2\pi} \int_{-\pi}^{\pi} f(t)\overline{g(t)}\, dt$. Determine the projection of the function t on $\mathrm{Span}\{e^{nit}\}$ for all integers $n \geq 1$.

Exercise 3.22. Let V be an inner product space and let $p_1, \ldots, p_n$ be orthogonal projections such that $\mathrm{Id} = p_1 + \cdots + p_n$. Show that $p_j p_k = \mathbf{0}$ whenever $j \neq k$.

3.6.3 The adjoint of a linear transformation

Exercise 3.23. Let V be a finite-dimensional inner product space and let $f, g : V \to V$ be orthogonal projections. Show that the following conditions are equivalent.

(a) $\mathrm{ran}\, f \subseteq \mathrm{ran}\, g$

(b) $gf = f$

(c) $fg = f$

Exercise 3.24. Let V be an inner product space and let $f, g : V \to V$ be orthogonal projections. If $\mathrm{ran}\, f \subseteq \mathrm{ran}\, g$, show that $g - f$ is an orthogonal projection and $\mathrm{ran}(g - f) = (\mathrm{ran}\, f)^{\perp} \cap \mathrm{ran}\, g$.

Exercise 3.25. Let V be an inner product space and let $f : V \to V$ be a linear operator. If f is invertible and self-adjoint, show that f^{-1} is self-adjoint.

Exercise 3.26. Let V be a finite-dimensional inner product space and let $f : V \to V$ be a self-adjoint operator. Show that $\|(\lambda - f)\mathbf{x}\| \geq |\mathrm{Im}\,\lambda| \|\mathbf{x}\|$, where $\lambda \in \mathbb{K}$ and $\mathbf{x} \in V$.

Exercise 3.27. Let V and W be finite-dimensional inner product spaces and let $f : V \to W$ be a linear transformation. If f is injective, show that $f^* f$ is an isomorphism.

Exercise 3.28. Let V and W be inner product spaces and let $f : V \to W$ be a linear transformation. If f is surjective, show that $f f^*$ is an isomorphism.

Exercise 3.29. Let V and W be inner product spaces and let $\mathbf{v} \in V$ and $\mathbf{w} \in W$. Show, with the notation from Exercise 3.8, that we have $(\mathbf{w} \otimes \mathbf{v})^* = \mathbf{v} \otimes \mathbf{w}$.

Exercise 3.30. Let V and W be inner product spaces and let $\mathbf{v} \in V$ and $\mathbf{w} \in W$. If $f : W \to V$ is a linear transformation, show that $(\mathbf{w} \otimes \mathbf{v}) \circ f = \mathbf{w} \otimes f^*(\mathbf{v})$, where $\otimes$ is defined as in Exercise 3.8.

Exercise 3.31. Let V be a finite-dimensional inner product space and let $f : V \to V$ be a linear operator. Show that $f = \mathbf{0}$ if and only if $f^* f = \mathbf{0}$.

Exercise 3.32. Let V be an inner product space and let $f, g : V \to V$ be linear operators. If f is self-adjoint, show that $g^* f g$ is self-adjoint.

Exercise 3.33. Let V be a finite-dimensional inner product space and let $f : V \to V$ be a linear operator. Show that $\mathrm{Id} + f^* f$ is invertible.

Exercise 3.34. Let V be a finite dimensional inner product spaces and let $f, g : V \to V$ be self-adjoint operators. Show that fg is self-adjoint if and only if $fg = gf$.

Exercise 3.35. Let $\mathcal{V}$ be an inner product space and let $f, g : \mathcal{V} \to \mathcal{V}$ be orthogonal projections. Show that fg is an orthogonal projection if and only if $fg = gf$.

Exercise 3.36. Let $\mathcal{V}$ be an inner product space and let $f, g : \mathcal{V} \to \mathcal{V}$ be orthogonal projections. If fg is an orthogonal projection, show that $\operatorname{ran} fg = \operatorname{ran} f \cap \operatorname{ran} g$.

Exercise 3.37. Let $\mathcal{V}$ be a finite-dimensional inner product space and let $f : \mathcal{V} \to \mathcal{V}$ be a linear operator. If f^*f is an orthogonal projection, show that ff^* is an orthogonal projection.

Exercise 3.38. Let $\mathcal{V}$ be an n-dimensional complex inner product space and let $\{\mathbf{a}_1, \ldots, \mathbf{a}_n\}$ be an orthonormal basis of $\mathcal{V}$. For a linear operator $f : \mathcal{V} \to \mathcal{V}$ we define the *trace* of f, denoted by $\operatorname{tr} f$, as

$$\operatorname{tr} f = \sum_{j=1}^{n} \langle f(\mathbf{a}_j), \mathbf{a}_j \rangle.$$

Show that $\operatorname{tr} f$ does not depend on the choice of the orthonormal basis $\{\mathbf{a}_1, \ldots, \mathbf{a}_n\}$, that is, for any two orthonormal bases $\{\mathbf{a}_1, \ldots, \mathbf{a}_n\}$ and $\{\mathbf{b}_1, \ldots, \mathbf{b}_n\}$ in $\mathcal{V}$ we have $\sum_{j=1}^{n} \langle f(\mathbf{a}_j), \mathbf{a}_j \rangle = \sum_{j=1}^{n} \langle f(\mathbf{b}_j), \mathbf{b}_j \rangle$.

Exercise 3.39. Let $A = (a_{ij})$ be a matrix from $\mathcal{M}_{n \times n}(\mathbb{C})$ and let $f : \mathbb{C}^n \to \mathbb{C}^n$ be the linear operator defined by $f(\mathbf{v}) = A\mathbf{v}$. Show that $\operatorname{tr} f = a_{11} + \cdots + a_{nn}$. (See Exercise 3.38 for the definition of $\operatorname{tr} f$.)

Exercise 3.40. Let $\mathcal{V}$ be an n-dimensional inner product space and let $\mathbf{x}, \mathbf{y} \in \mathcal{V}$. Show that $tr(\mathbf{x} \otimes \mathbf{y}) = \langle \mathbf{x}, \mathbf{y} \rangle$. (See Exercise 3.38 for the definition of $\operatorname{tr} f$ and Exercise 3.8 for the definition of $\mathbf{x} \otimes \mathbf{y}$.)

Exercise 3.41. Let $\mathcal{V}$ be an inner product space and let $f, g : \mathcal{V} \to \mathcal{V}$ be linear operators. Show that $tr(fg) = tr(gf)$. (See Exercise 3.38 for the definition of $\operatorname{tr} f$.)

Exercise 3.42. Let $\mathcal{V}$ be an inner product space. Show that the function $s : \mathcal{L}(\mathcal{V}) \times \mathcal{L}(\mathcal{V}) \to \mathbb{K}$ defined by $s(f, g) = tr(g^*f)$ is an inner product. (See Exercise 3.38 for the definition of $\operatorname{tr} f$.)

Exercise 3.43. Let $\mathcal{V}$ be an inner product space and let $f, g : \mathcal{V} \to \mathcal{V}$ be self-adjoint operators. If $f^2g^2 = \mathbf{0}$, show that $fg = \mathbf{0}$.

3.6.4   Spectral theorems

Exercise 3.44. Let $\mathcal{V}$ be an inner product space and let $f : \mathcal{V} \to \mathcal{V}$ be an invertible linear operator. If λ is an eigenvalue of f, show that $\frac{1}{\lambda}$ is an eigenvalue of f^{-1}.

Exercise 3.45. Let V be a finite-dimensional inner product space and let $f, g : V \to V$ be linear operators. Show that, if $\mathrm{Id} - fg$ is invertible, then $\mathrm{Id} - gf$ is invertible and $(\mathrm{Id} - gf)^{-1} = \mathrm{Id} + g(\mathrm{Id} - fg)^{-1}f$. Then show that, if $\lambda \neq 0$ is an eigenvalue of fg, then λ is an eigenvalue of gf.

Exercise 3.46. Let V be an inner product space and let $f : V \to V$ be a linear operator. Let $g = \frac{1}{2}(f + f^*)$ and $h = \frac{1}{2i}(f - f^*)$. Show that g and h are self-adjoint and $f = g + ih$. Show also that f is normal if and only if $gh = hg$.

Exercise 3.47. Let V be an inner product space and let $f : V \to V$ be a normal operator. If $\{\mathbf{a}_1, \ldots, \mathbf{a}_n\}$ is an orthonormal basis of V and $\lambda_1, \ldots, \lambda_n$ are the eigenvalues of f, show that $\sum_{j=1}^{n} |\lambda_j|^2 = \sum_{j=1}^{n} \|f(\mathbf{a}_j)\|^2$.

Exercise 3.48. Let V be an inner product space and let $f : V \to V$ be a linear operator. Using Exercise 3.19 show that, if the operator f is normal, then there is an orthonormal basis $\{\mathbf{v}_1, \ldots, \mathbf{v}_n\}$ of V consisting of eigenvectors of V.

Exercise 3.49. Let V be a finite-dimensional inner product space and let $f : V \to V$ be a linear operator. Show that if $ff^*f = f^*ff$, then $(ff^* - f^*f)^2 = 0$ and that f is normal.

Exercise 3.50. Let V be a finite-dimensional inner product space and let $f : V \to V$ be a linear operator. Show that f is normal if and only if there is a polynomial p such that $p(f) = f^*$.

Exercise 3.51. Let V be a finite-dimensional inner product space. Show that the function $S : V \times V \to V \times V$ defined by $S(\mathbf{x}, \mathbf{y}) = (-\mathbf{y}, \mathbf{x})$ is a unitary operator and that $S^2 = \mathrm{Id}_{V \times V}$.

Exercise 3.52. Let V be a finite-dimensional inner product space and let $f : V \to V$ be a linear operator. If

$$\Gamma f = \{(\mathbf{x}, f(\mathbf{x})) \in V \times V | \mathbf{x} \in V\} \quad \text{and} \quad \Gamma f^* = \{(\mathbf{x}, f^*(\mathbf{x})) \in V \times V | \mathbf{x} \in V\},$$

show that $(\Gamma(f))^{\perp} = S(\Gamma(f^*))$, where S is defined in Exercise 3.51.

Exercise 3.53. Let V be an n-dimensional inner product space and let $f : V \to V$ be a linear operator. Show that f normal if and only if there is a unitary operator g such that $fg = f^*$.

Exercise 3.54. Let V be a finite-dimensional inner product space and let $f : V \to V$ be a linear operator. If f is positive, show that g^*fg is positive.

Exercise 3.55. Let V be an n-dimensional inner product space and let $f : V \to V$ be a self-adjoint operator. Show that there are positive operators g and h such that $f = g - h$.

Exercise 3.56. Consider the operator $f \in \mathcal{L}(\mathbb{C}^2, \mathbb{C}^2)$ defined as $f(\mathbf{x}) = A\mathbf{x}$, where $A = \begin{bmatrix} 13 & 5i \\ -5i & 13 \end{bmatrix}$. Show that f is positive and determine its spectral decomposition.

Exercise 3.57. Let V and W be inner product spaces and let $f : V \to W$ be a linear operator. Show that f is an isometry if and only if there are orthonormal bases $\{\mathbf{v}_1, \ldots, \mathbf{v}_n\}$ and $\{\mathbf{w}_1, \ldots, \mathbf{w}_n\}$ in V such that $f = \sum_{j=1}^n \mathbf{w}_j \otimes \mathbf{v}_j$, where $\otimes$ is defined as in Exercise 3.8.

3.6.5   Singular value decomposition

Exercise 3.58. Let V be a finite-dimensional inner product space and let $f : V \to V$ be a linear operator. Show that, if f^*f is the projection on a subspace $\mathcal{U}$, then f is a partial isometry with initial space $\mathcal{U}$.

Exercise 3.59. Consider the operator $f \in \mathcal{L}(\mathbb{C}^2, \mathbb{C}^2)$ defined as $f(\mathbf{x}) = A\mathbf{x}$, where $A = \begin{bmatrix} 1 & i \\ i & 1 \end{bmatrix}$. Determine the singular value decomposition of f.

Exercise 3.60. Let V and W be finite dimensional inner product spaces. Show that for every nonzero linear transformation $f : V \to W$ there are positive numbers $\sigma_1, \ldots, \sigma_r$, orthonormal vectors $\mathbf{v}_1, \ldots, \mathbf{v}_r \in V$, and orthonormal vectors $\mathbf{w}_1, \ldots, \mathbf{w}_r \in W$ such that $f = \sum_{j=1}^r \sigma_j \mathbf{w}_j \otimes \mathbf{v}_j$, where $\otimes$ is defined as in Exercise 3.8.

Exercise 3.61. Let V and W be finite-dimensional inner product spaces and let $f : V \to W$ be a linear operator. If $f = \sum_{j=1}^r \alpha_j \mathbf{w}_j \otimes \mathbf{v}_j$, where $\{\mathbf{v}_1, \ldots, \mathbf{v}_r\}$ are orthonormal vectors in V, $\{\mathbf{w}_1, \ldots, \mathbf{w}_r\}$ are orthonormal vectors in W, and $\alpha_1, \ldots, \alpha_r$ are positive numbers such that $\alpha_1 \geq \cdots \geq \alpha_r$ (and $\otimes$ is defined as in Exercise 3.8), show that $f^*f = \sum_{j=1}^r |\alpha_j|^2 \mathbf{v}_j \otimes \mathbf{v}_j$ and consequently $\mathbf{v}_1, \ldots, \mathbf{v}_r$ are eigenvectors of f^*f with corresponding eigenvalues $|\alpha_1|^2, \ldots, |\alpha_r|^2$.

Exercise 3.62. Let V and W be finite dimensional inner product spaces and let $f : V \to W$ be a linear transformation. Let $f = \sum_{j=1}^r \sigma_j \mathbf{w}_j \otimes \mathbf{v}_j$ be the singular value decomposition of f from Exercise 3.60, where $\{\mathbf{v}_1, \ldots, \mathbf{v}_r\}$ are orthonormal vectors in V, $\{\mathbf{w}_1, \ldots, \mathbf{w}_r\}$ are orthonormal vectors in W, and $\sigma_1, \ldots, \sigma_r$ are positive numbers such that $\sigma_1 \geq \cdots \geq \sigma_r$. Show that $\operatorname{ran} f = \operatorname{Span}\{\mathbf{w}_1, \ldots, \mathbf{w}_r\}$.

Exercise 3.63. Let V and W be finite dimensional inner product spaces and let $f : V \to W$ be a linear transformation. Let $f = \sum_{j=1}^r \sigma_j \mathbf{w}_j \otimes \mathbf{v}_j$ be the singular value decomposition of f from Exercise 3.60, where $\{\mathbf{v}_1, \ldots, \mathbf{v}_r\}$ are orthonormal vectors in V, $\{\mathbf{w}_1, \ldots, \mathbf{w}_r\}$ are orthonormal vectors in W, and $\sigma_1, \ldots, \sigma_r$ are positive numbers such that $\sigma_1 \geq \cdots \geq \sigma_r$. Let $\{\mathbf{v}_1, \ldots, \mathbf{v}_r, \ldots, \mathbf{v}_n\}$ be an orthonormal basis of V. Show that $\ker f = \operatorname{Span}\{\mathbf{v}_{r+1}, \ldots, \mathbf{v}_n\}$.

Exercise 3.64. Let V and W be finite-dimensional inner product spaces and let $f : V \to W$ be a linear transformation. Let $f = \sum_{j=1}^r \sigma_j \mathbf{w}_j \otimes \mathbf{v}_j$ be the singular value decomposition of f from Exercise 3.60, where $\{\mathbf{v}_1, \ldots, \mathbf{v}_r\}$ are orthonormal vectors in V, $\{\mathbf{w}_1, \ldots, \mathbf{w}_r\}$ are orthonormal vectors in W, and $\sigma_1, \ldots, \sigma_r$ are positive numbers such that $\sigma_1 \geq \cdots \geq \sigma_r$. Show that $\operatorname{ran} f^* = \operatorname{Span}\{\mathbf{v}_1, \ldots, \mathbf{v}_r\}$.

Exercise 3.65. Let V and W be finite-dimensional inner product spaces and let $f : V \to W$ be a linear transformation. Let $f = \sum_{j=1}^{r} \sigma_j \mathbf{w}_j \otimes \mathbf{v}_j$ be the singular value decomposition of f from Exercise 3.60, where $\{\mathbf{v}_1, \ldots, \mathbf{v}_r\}$ are orthonormal vectors in V, $\{\mathbf{w}_1, \ldots, \mathbf{w}_r\}$ are orthonormal vectors in W, and $\sigma_1, \ldots, \sigma_r$ are positive numbers such that $\sigma_1 \geq \cdots \geq \sigma_r$. Let $\{\mathbf{w}_1, \ldots, \mathbf{w}_r, \ldots, \mathbf{w}_m\}$ be an orthonormal basis of W. Show that $\ker f^* = \mathrm{Span}\{\mathbf{w}_{r+1}, \ldots, \mathbf{w}_m\}$.

Exercise 3.66. Let V and W be finite-dimensional inner product spaces and let $f : V \to W$ be a linear transformation. Let $f = \sum_{j=1}^{r} \sigma_j \mathbf{w}_j \otimes \mathbf{v}_j$ be the singular value decomposition of f from Exercise 3.60, where $\{\mathbf{v}_1, \ldots, \mathbf{v}_r\}$ are orthonormal vectors in V, $\{\mathbf{w}_1, \ldots, \mathbf{w}_r\}$ are orthonormal vectors in W, and $\sigma_1, \ldots, \sigma_r$ are positive numbers such that $\sigma_1 \geq \cdots \geq \sigma_r$. Show that, if a linear transformation $f^+ : W \to V$ is such that $f^+ f = \mathrm{proj}_{\mathrm{ran}\, f^*}$ and $f^+ = \mathbf{0}$ on $(\mathrm{ran}\, f^*)^{\perp}$, then $f^+ = \sum_{j=1}^{r} \frac{1}{\sigma_j} \mathbf{v}_j \otimes \mathbf{w}_j$.

Exercise 3.67. Let V and W be finite-dimensional inner product spaces and let $f : V \to W$ be a linear transformation. Let $f = \sum_{j=1}^{r} \sigma_j \mathbf{w}_j \otimes \mathbf{v}_j$ be the singular value decomposition of f, from Exercise 3.60, where $\{\mathbf{v}_1, \ldots, \mathbf{v}_r\}$ are orthonormal vectors in V, $\{\mathbf{w}_1, \ldots, \mathbf{w}_r\}$ are orthonormal vectors in W, and $\sigma_1, \ldots, \sigma_r$ are positive numbers such that $\sigma_1 \geq \cdots \geq \sigma_r$. Let $f^+ = \sum_{j=1}^{r} \frac{1}{\sigma_j} \mathbf{v}_j \otimes \mathbf{w}_j$ as in exercise 3.66. Show that $f f^+$ is the projection on $\mathrm{ran}\, f$ and $\mathrm{Id}_W - f f^+$ is the projection on $\ker f^*$.

Exercise 3.68. Let V and W be finite-dimensional inner product spaces and let $f : V \to W$ be a linear transformation. Let $f = \sum_{j=1}^{r} \sigma_j \mathbf{w}_j \otimes \mathbf{v}_j$ be the singular value decomposition of f from Exercise 3.60, where $\{\mathbf{v}_1, \ldots, \mathbf{v}_r\}$ are orthonormal vectors in V, $\{\mathbf{w}_1, \ldots, \mathbf{w}_r\}$ are orthonormal vectors in W, and $\sigma_1, \ldots, \sigma_r$ are positive numbers such that $\sigma_1 \geq \cdots \geq \sigma_r$. If $f^+ = \sum_{j=1}^{r} \frac{1}{\sigma_j} \mathbf{v}_j \otimes \mathbf{w}_j$ (as in Exercise 3.66), show that $f^+ f f^+ = f^+$.

Exercise 3.69. Let V and W be finite-dimensional inner product spaces and let $f : V \to W$ be a linear transformation. Let $f = \sum_{j=1}^{r} \sigma_j \mathbf{w}_j \otimes \mathbf{v}_j$ be the singular value decomposition of f from Exercise 3.60, where $\{\mathbf{v}_1, \ldots, \mathbf{v}_r\}$ are orthonormal vectors in V, $\{\mathbf{w}_1, \ldots, \mathbf{w}_r\}$ are orthonormal vectors in W, and $\sigma_1, \ldots, \sigma_r$ are positive numbers such that $\sigma_1 \geq \cdots \geq \sigma_r$. If f is injective, show that $f^* f$ is invertible and $(f^* f)^{-1} f^* = f^+$, where f^+ is defined in Exercise 3.66.

Exercise 3.70. Let V and W be finite-dimensional inner product spaces and let $f : V \to W$ be a linear transformation. Show that $(f^*)^+ = (f^+)^*$, where f^+ is defined in Exercise 3.66.

Exercise 3.71. Let V be a finite-dimensional inner product space and let $f : V \to V$ be a linear operator. Obtain, using Exercise 3.12 and without using the proof of Theorem 3.5.2, the following form of singular value decomposition:

If $\dim V = n$, then there are orthonormal bases $\{\mathbf{v}_1, \ldots, \mathbf{v}_n\}$ and $\{\mathbf{u}_1, \ldots, \mathbf{u}_n\}$ of V and nonnegative numbers $\sigma_1, \ldots, \sigma_n$, such that $f = \sum_{j=1}^{n} \sigma_j \mathbf{u}_j \otimes \mathbf{v}_j$.

Exercise 3.72. Let V be a finite-dimensional inner product space and let $f : V \to V$ be a linear operator. Using Exercise 3.71 show that there is an isometry g such that $f = g|f|$. Note that this is a form of the polar decomposition of f.

Exercise 3.73. Let V be a finite dimensional inner product space and let $f : V \to V$ be a linear operator. Using Exercise 3.71 show that there is an isometry $g : V \to V$ such that $f = \sqrt{ff^*}g$.

Exercise 3.74. Let V and W be finite dimensional inner product spaces and let $f : V \to W$ be a linear transformation. Let $f = \sum_{j=1}^{r} \sigma_j \mathbf{w}_j \otimes \mathbf{v}_j$ be the singular value decomposition of f from Exercise 3.60, where $\{\mathbf{v}_1, \dots, \mathbf{v}_r\}$ are orthonormal vectors in V, $\{\mathbf{w}_1, \dots, \mathbf{w}_r\}$ are orthonormal vectors in W, and $\sigma_1, \dots, \sigma_r$ are positive numbers such that $\sigma_1 \geq \cdots \geq \sigma_r$. Let $\{\mathbf{v}_1, \dots, \mathbf{v}_r, \dots, \mathbf{v}_n\}$ be an orthonormal basis of V. If $f^+ = \sum_{j=1}^{r} \frac{1}{\sigma_j} \mathbf{v}_j \otimes \mathbf{w}_j$, where $\otimes$ is defined as in Exercise 3.8, show that every least square solution $\mathbf{x}$ of the equation $f(\mathbf{x}) = \mathbf{b}$ is of the form $\mathbf{x} = f^+(\mathbf{b}) + x_{r+1}\mathbf{v}_{r+1} + \cdots + x_n\mathbf{v}_n$ where $x_{r+1}, \dots, x_n \in \mathbb{K}$ are arbitrary. Moreover, there is a unique least square solution of minimal length, which is $\mathbf{x} = f^+(\mathbf{b})$.

Chapter 4

Reduction of Endomorphisms

Introduction

The main topic of this chapter is the following question: Given an endomorphism f on a finite-dimensional vector space $\mathcal{V}$ can we find a base of $\mathcal{V}$ such that the matrix of f is simple and easy to work with, that is, diagonal or block-diagonal. This will help us better understand the structure of endomorphisms and provide important tools for applications of linear algebra, for example, to solve differential equations.

At the beginning of the chapter we discuss alternating multilinear forms and determinants of endomorphisms which will give us a practical way of determining the diagonal and block-diagonal matrices of an endomorphism.

Our presentation of determinants is self-contained, that is, it does not use results on determinants from elementary courses.

In the context of this chapter it is customary to use the name *endomorphisms* instead of *linear operators*.

4.1 Eigenvalues and diagonalization

4.1.1 Multilinear alternating forms and determinants

At the beginning of Chapter 3 we introduce bilinear forms. They are defined as functions $f : \mathcal{V} \times \mathcal{V} \to \mathbb{K}$ that are linear in each variable, that is,

the function $f_{\mathbf{x}} : \mathcal{V} \to \mathbb{K}$ defined as $f_{\mathbf{x}}(\mathbf{y}) = f(\mathbf{x}, \mathbf{y})$ is linear for every $\mathbf{x} \in \mathcal{V}$ and

the function $f_{\mathbf{y}} : \mathcal{V} \to \mathbb{K}$ defined as $f_{\mathbf{y}}(\mathbf{x}) = f(\mathbf{x}, \mathbf{y})$ is linear for every $\mathbf{y} \in \mathcal{V}$.

$$\overbrace{}^{n \text{ times}}$$

A similar definition can be given for any function from $\mathcal{V}^n = V \times \cdots \times V$ to $\mathbb{K}$.

Definition 4.1.1. Let $\mathcal{V}$ be a vector space. A function $\mathbf{E} : \mathcal{V}^n \to \mathbb{K}$ is called an *n-linear form* or a *multilinear form* if for every $j \in \{1, \ldots, n\}$ and every $\mathbf{x}_1, \ldots, \mathbf{x}_{j-1}, \mathbf{x}_{j+1}, \ldots, \mathbf{x}_n \in \mathcal{V}$ the function

$$f_{\mathbf{x}_1, \ldots, \mathbf{x}_{j-1}, \mathbf{x}_{j+1}, \ldots, \mathbf{x}_n} : \mathcal{V} \to \mathbb{K}$$

defined as

$$f_{\mathbf{x}_1, \ldots, \mathbf{x}_{j-1}, \mathbf{x}_{j+1}, \ldots, \mathbf{x}_n}(\mathbf{x}) = \mathbf{E}(\mathbf{x}_1, \ldots, \mathbf{x}_{j-1}, \mathbf{x}, \mathbf{x}_{j+1}, \ldots, \mathbf{x}_n)$$

is a linear form.

Example 4.1.2. The function $\mathbf{E} : \mathbb{K}^n \to \mathbb{K}$ defined as

$$\mathbf{E}(x_1, \ldots, x_n) = c x_1 \ldots x_n$$

is an n-linear form for any $c \in \mathbb{K}$.

This example can be generalized in the following way. Let $\mathcal{V}$ be a vector space and let $f_j : \mathcal{V} \to \mathbb{K}$ be a linear function for $j \in \{1, \ldots, n\}$. Then the function $\mathbf{E} : \mathcal{V}^n \to \mathbb{K}$ defined as

$$\mathbf{E}(\mathbf{x}_1, \ldots, \mathbf{x}_n) = f_1(\mathbf{x}_1) \ldots f_n(\mathbf{x}_n)$$

is an n-linear form.

Definition 4.1.3. Let $\mathcal{V}$ be a vector space. An n-linear form $\mathbf{E} : \mathcal{V}^n \to \mathbb{K}$ is called an *alternating n-linear form* (or *alternating multilinear form*) if

$$\mathbf{E}(\mathbf{x}_1, \ldots, \mathbf{x}_n) = 0$$

whenever $\mathbf{x}_j = \mathbf{x}_k$ for some $j \neq k$.

The following property of alternating multilinear forms is often used in calculations. It is equivalent to the condition in the definition of alternating multilinear forms.

> **Theorem 4.1.4.** *Let* $\mathcal{V}$ *be a vector space and let* $\mathbf{E} : \mathcal{V}^n \to \mathbb{K}$ *be an alternating multilinear form. If* $1 \le j < k \le n$, *then*
>
> $$\mathbf{E}(\mathbf{x}_1, \ldots, \mathbf{x}_{j-1}, \mathbf{x}_k, \mathbf{x}_{j+1}, \ldots, \mathbf{x}_{k-1}, \mathbf{x}_j, \mathbf{x}_{k+1} \ldots \mathbf{x}_n) =$$
>
> $$-\mathbf{E}(\mathbf{x}_1, \ldots, \mathbf{x}_{j-1}, \mathbf{x}_j, \mathbf{x}_{j+1}, \ldots, \mathbf{x}_{k-1}, \mathbf{x}_k, \mathbf{x}_{k+1} \ldots \mathbf{x}_n)$$
>
> *for all* $\mathbf{x}_1, \ldots, \mathbf{x}_n \in \mathcal{V}$.

Proof. Since

$$
\begin{aligned}
0 &= \mathbf{E}(\mathbf{x}_1, \ldots, \mathbf{x}_{j-1}, \mathbf{x}_j + \mathbf{x}_k, \mathbf{x}_{j+1}, \ldots, \mathbf{x}_{k-1}, \mathbf{x}_j + \mathbf{x}_k, \mathbf{x}_{k+1}, \ldots, \mathbf{x}_n) \\
&= \mathbf{E}(\mathbf{x}_1, \ldots, \mathbf{x}_{j-1}, \mathbf{x}_j, \mathbf{x}_{j+1}, \ldots, \mathbf{x}_{k-1}, \mathbf{x}_j, \mathbf{x}_{k+1}, \ldots, \mathbf{x}_n) \\
&\quad + \mathbf{E}(\mathbf{x}_1, \ldots, \mathbf{x}_{j-1}, \mathbf{x}_j, \mathbf{x}_{j+1}, \ldots, \mathbf{x}_{k-1}, \mathbf{x}_k, \mathbf{x}_{k+1}, \ldots, \mathbf{x}_n) \\
&\quad + \mathbf{E}(\mathbf{x}_1, \ldots, \mathbf{x}_{j-1}, \mathbf{x}_k, \mathbf{x}_{j+1}, \ldots, \mathbf{x}_{k-1}, \mathbf{x}_j, \mathbf{x}_{k+1}, \ldots, \mathbf{x}_n) \\
&\quad + \mathbf{E}(\mathbf{x}_1, \ldots, \mathbf{x}_{j-1}, \mathbf{x}_k, \mathbf{x}_{j+1}, \ldots, \mathbf{x}_{k-1}, \mathbf{x}_k, \mathbf{x}_{k+1} \ldots \mathbf{x}_n) \\
&= \mathbf{E}(\mathbf{x}_1, \ldots, \mathbf{x}_{j-1}, \mathbf{x}_j, \mathbf{x}_{j+1}, \ldots, \mathbf{x}_{k-1}, \mathbf{x}_k, \mathbf{x}_{k+1}, \ldots, \mathbf{x}_n) \\
&\quad + \mathbf{E}(\mathbf{x}_1, \ldots, \mathbf{x}_{j-1}, \mathbf{x}_k, \mathbf{x}_{j+1}, \ldots, \mathbf{x}_{k-1}, \mathbf{x}_j, \mathbf{x}_{k+1}, \ldots, \mathbf{x}_n),
\end{aligned}
$$

we have

$$
\begin{aligned}
\mathbf{E}(\mathbf{x}_1, \ldots, \mathbf{x}_{j-1}, & \mathbf{x}_k, \mathbf{x}_{j+1}, \ldots, \mathbf{x}_{k-1}, \mathbf{x}_j, \mathbf{x}_{k+1} \ldots \mathbf{x}_n) \\
&= -\mathbf{E}(\mathbf{x}_1, \ldots, \mathbf{x}_{j-1}, \mathbf{x}_j, \mathbf{x}_{j+1}, \ldots, \mathbf{x}_{k-1}, \mathbf{x}_k, \mathbf{x}_{k+1} \ldots \mathbf{x}_n).
\end{aligned}
$$

$\square$

The following three examples indicate that there is a connection between alternating forms and determinants as defined in elementary courses. The full scope of that connection will become clear later in this chapter. In these examples the determinant of a matrix $\begin{bmatrix} \alpha & \beta \\ \gamma & \delta \end{bmatrix} \in \mathcal{M}_{2,2}(\mathbb{K})$ is defined as usual by

$$\det \begin{bmatrix} \alpha & \beta \\ \gamma & \delta \end{bmatrix} = \alpha\delta - \beta\gamma.$$

Example 4.1.5. Let $\mathcal{V}$ be a vector space and let $E : \mathcal{V} \times \mathcal{V} \to \mathbb{K}$ be an alternating bilinear form. Show that for every $\mathbf{v}_1, \mathbf{v}_2 \in \mathcal{V}$ and $\alpha, \beta, \gamma, \delta \in \mathbb{K}$ we have

$$\mathbf{E}(\alpha\mathbf{v}_1 + \beta\mathbf{v}_2, \gamma\mathbf{v}_1 + \delta\mathbf{v}_2) = \det \begin{bmatrix} \alpha & \beta \\ \gamma & \delta \end{bmatrix} \mathbf{E}(\mathbf{v}_1, \mathbf{v}_2). \tag{4.1}$$

Proof. For every $\mathbf{v}_1, \mathbf{v}_2 \in \mathcal{V}$ and $\alpha, \beta, \gamma, \delta \in \mathbb{K}$ we have

$$\mathbf{E}(\alpha\mathbf{v}_1 + \beta\mathbf{v}_2, \gamma\mathbf{v}_1 + \delta\mathbf{v}_2) = \mathbf{E}(\alpha\mathbf{v}_1, \gamma\mathbf{v}_1) + \mathbf{E}(\alpha\mathbf{v}_1, \delta\mathbf{v}_2)$$
$$+ \mathbf{E}(\beta\mathbf{v}_2, \gamma\mathbf{v}_1) + \mathbf{E}(\beta\mathbf{v}_2, \delta\mathbf{v}_2)$$
$$= \alpha\gamma\mathbf{E}(\mathbf{v}_1, \mathbf{v}_1) + \alpha\delta\mathbf{E}(\mathbf{v}_1, \mathbf{v}_2)$$
$$+ \beta\gamma\mathbf{E}(\mathbf{v}_2, \mathbf{v}_1) + \beta\delta\mathbf{E}(\mathbf{v}_2, \mathbf{v}_2)$$
$$= \alpha\delta\mathbf{E}(\mathbf{v}_1, \mathbf{v}_2) + \beta\gamma\mathbf{E}(\mathbf{v}_2, \mathbf{v}_1)$$
$$= \alpha\delta\mathbf{E}(\mathbf{v}_1, \mathbf{v}_2) - \beta\gamma\mathbf{E}(\mathbf{v}_1, \mathbf{v}_2)$$
$$= (\alpha\delta - \beta\gamma)\mathbf{E}(\mathbf{v}_1, \mathbf{v}_2)$$
$$= \det \begin{bmatrix} \alpha & \beta \\ \gamma & \delta \end{bmatrix} \mathbf{E}(\mathbf{v}_1, \mathbf{v}_2).$$

Note that, if $\mathbf{E}$ satisfies (4.1) for every $\mathbf{v}_1, \mathbf{v}_2 \in \mathcal{V}$ and $\alpha, \beta, \gamma, \delta \in \mathbb{K}$, then $\mathbf{E}$ is alternating. $\qquad\square$

Example 4.1.6. Let $\mathcal{V}$ be a vector space and let $E : \mathcal{V} \times \mathcal{V} \to \mathbb{K}$ be an alternating bilinear form. Show that for every $\mathbf{x}, \mathbf{y}, \mathbf{z} \in \mathcal{V}$ and $a_{11}, a_{21}, a_{31}, a_{12}, a_{22}, a_{32} \in \mathbb{K}$ we have

$$\mathbf{E}(a_{11}\mathbf{x} + a_{21}\mathbf{y} + a_{31}\mathbf{z}, a_{12}\mathbf{x} + a_{22}\mathbf{y} + a_{32}\mathbf{z})$$
$$= \det \begin{bmatrix} a_{11} & a_{12} \\ a_{21} & a_{22} \end{bmatrix} \mathbf{E}(\mathbf{x}, \mathbf{y}) + \det \begin{bmatrix} a_{11} & a_{12} \\ a_{31} & a_{32} \end{bmatrix} \mathbf{E}(\mathbf{x}, \mathbf{z})$$
$$+ \det \begin{bmatrix} a_{21} & a_{22} \\ a_{31} & a_{32} \end{bmatrix} \mathbf{E}(\mathbf{y}, \mathbf{z}).$$

Solution.

$$\mathbf{E}(a_{11}\mathbf{x} + a_{21}\mathbf{y} + a_{31}\mathbf{z}, a_{12}\mathbf{x} + a_{22}\mathbf{y} + a_{32}\mathbf{z})$$
$$= \mathbf{E}(a_{11}\mathbf{x}, a_{22}\mathbf{y}) + \mathbf{E}(a_{21}\mathbf{y}, a_{12}\mathbf{x}) + \mathbf{E}(a_{11}\mathbf{x}, a_{32}\mathbf{z})$$
$$+ \mathbf{E}(a_{31}\mathbf{z}, a_{12}\mathbf{x}) + \mathbf{E}(a_{21}\mathbf{y}, a_{32}\mathbf{z}) + \mathbf{E}(a_{31}\mathbf{z}, a_{22}\mathbf{y})$$
$$= \mathbf{E}(a_{11}\mathbf{x} + a_{21}\mathbf{y}, a_{12}\mathbf{x} + a_{22}\mathbf{y}) + \mathbf{E}(a_{11}\mathbf{x} + a_{31}\mathbf{z}, a_{12}\mathbf{x} + a_{32}\mathbf{z})$$
$$+ \mathbf{E}(a_{21}\mathbf{y} + a_{31}\mathbf{z}, a_{22}\mathbf{y} + a_{32}\mathbf{z})$$
$$= \det \begin{bmatrix} a_{11} & a_{12} \\ a_{21} & a_{22} \end{bmatrix} \mathbf{E}(\mathbf{x}, \mathbf{y}) + \det \begin{bmatrix} a_{11} & a_{12} \\ a_{31} & a_{32} \end{bmatrix} \mathbf{E}(\mathbf{x}, \mathbf{z})$$
$$+ \det \begin{bmatrix} a_{21} & a_{22} \\ a_{31} & a_{32} \end{bmatrix} \mathbf{E}(\mathbf{y}, \mathbf{z})$$

$\qquad\square$

Example 4.1.7. Let $\mathbf{x}, \mathbf{y}, \mathbf{z}$ be vectors in a vector space $\mathcal{V}$ and let $\mathbf{E} : \mathcal{V} \times \mathcal{V} \times \mathcal{V} \to \mathbb{K}$ be an alternating multilinear form. Show that

$$\mathbf{E}(a_{11}\mathbf{x} + a_{21}\mathbf{y} + a_{31}\mathbf{z}, a_{12}\mathbf{x} + a_{22}\mathbf{y} + a_{32}\mathbf{z}, a_{13}\mathbf{x} + a_{23}\mathbf{y} + a_{33}\mathbf{z})$$

$$= \left(a_{11} \det \begin{bmatrix} a_{22} & a_{23} \\ a_{32} & a_{33} \end{bmatrix} - a_{21} \det \begin{bmatrix} a_{12} & a_{13} \\ a_{32} & a_{33} \end{bmatrix} + a_{31} \det \begin{bmatrix} a_{12} & a_{13} \\ a_{22} & a_{23} \end{bmatrix} \right) \mathbf{E}(\mathbf{x}, \mathbf{y}, \mathbf{z})$$

$$= - \left(a_{12} \det \begin{bmatrix} a_{21} & a_{23} \\ a_{31} & a_{33} \end{bmatrix} - a_{22} \det \begin{bmatrix} a_{11} & a_{13} \\ a_{31} & a_{33} \end{bmatrix} + a_{32} \det \begin{bmatrix} a_{11} & a_{13} \\ a_{21} & a_{23} \end{bmatrix} \right) \mathbf{E}(\mathbf{x}, \mathbf{y}, \mathbf{z})$$

$$= \left(a_{13} \det \begin{bmatrix} a_{21} & a_{22} \\ a_{31} & a_{32} \end{bmatrix} - a_{23} \det \begin{bmatrix} a_{11} & a_{12} \\ a_{31} & a_{32} \end{bmatrix} + a_{33} \det \begin{bmatrix} a_{11} & a_{12} \\ a_{21} & a_{22} \end{bmatrix} \right) \mathbf{E}(\mathbf{x}, \mathbf{y}, \mathbf{z})$$

for every $a_{11}, a_{21}, a_{31}, a_{12}, a_{22}, a_{32}, a_{13}, a_{23}, a_{33} \in \mathbb{K}$.

Solution. We prove the second equality. The other equalities can be proven in the same way.

We apply the result from Example 4.1.6 to the function $\mathbf{G} : \mathcal{V} \times \mathcal{V} \to \mathbb{K}$ defined by

$$\mathbf{G}(\mathbf{s}, \mathbf{t}) = \mathbf{E}(\mathbf{s}, a_{12}\mathbf{x} + a_{22}\mathbf{y} + a_{32}\mathbf{z}, \mathbf{t}),$$

where $\mathbf{x}, \mathbf{y}, \mathbf{z} \in \mathcal{V}$ are arbitrary but fixed, and obtain

$$\mathbf{E}(a_{11}\mathbf{x} + a_{21}\mathbf{y} + a_{31}\mathbf{z}, a_{12}\mathbf{x} + a_{22}\mathbf{y} + a_{32}\mathbf{z}, a_{13}\mathbf{x} + a_{23}\mathbf{y} + a_{33}\mathbf{z})$$

$$= \det \begin{bmatrix} a_{11} & a_{13} \\ a_{21} & a_{23} \end{bmatrix} \mathbf{E}(\mathbf{x}, a_{12}\mathbf{x} + a_{22}\mathbf{y} + a_{32}\mathbf{z}, \mathbf{y})$$

$$+ \det \begin{bmatrix} a_{11} & a_{13} \\ a_{31} & a_{33} \end{bmatrix} \mathbf{E}(\mathbf{x}, a_{12}\mathbf{x} + a_{22}\mathbf{y} + a_{32}\mathbf{z}, \mathbf{z})$$

$$+ \det \begin{bmatrix} a_{21} & a_{23} \\ a_{31} & a_{33} \end{bmatrix} \mathbf{E}(\mathbf{y}, a_{12}\mathbf{x} + a_{22}\mathbf{y} + a_{32}\mathbf{z}, \mathbf{z})$$

$$= \det \begin{bmatrix} a_{11} & a_{13} \\ a_{21} & a_{23} \end{bmatrix} \mathbf{E}(\mathbf{x}, a_{32}\mathbf{z}, \mathbf{y}) + \det \begin{bmatrix} a_{11} & a_{13} \\ a_{31} & a_{33} \end{bmatrix} \mathbf{E}(\mathbf{x}, a_{22}\mathbf{y}, \mathbf{z})$$

$$+ \det \begin{bmatrix} a_{21} & a_{23} \\ a_{31} & a_{33} \end{bmatrix} \mathbf{E}(\mathbf{y}, a_{12}\mathbf{x}, \mathbf{z})$$

$$= a_{32} \det \begin{bmatrix} a_{11} & a_{13} \\ a_{21} & a_{23} \end{bmatrix} \mathbf{E}(\mathbf{x}, \mathbf{z}, \mathbf{y}) + a_{22} \det \begin{bmatrix} a_{11} & a_{13} \\ a_{31} & a_{33} \end{bmatrix} \mathbf{E}(\mathbf{x}, \mathbf{y}, \mathbf{z})$$

$$+ a_{12} \det \begin{bmatrix} a_{21} & a_{23} \\ a_{31} & a_{33} \end{bmatrix} \mathbf{E}(\mathbf{y}, \mathbf{x}, \mathbf{z})$$

$$= -a_{32} \det \begin{bmatrix} a_{11} & a_{13} \\ a_{21} & a_{23} \end{bmatrix} \mathbf{E}(\mathbf{x}, \mathbf{y}, \mathbf{z}) + a_{22} \det \begin{bmatrix} a_{11} & a_{13} \\ a_{31} & a_{33} \end{bmatrix} \mathbf{E}(\mathbf{x}, \mathbf{y}, \mathbf{z})$$

$$- a_{12} \det \begin{bmatrix} a_{21} & a_{23} \\ a_{31} & a_{33} \end{bmatrix} \mathbf{E}(\mathbf{x}, \mathbf{y}, \mathbf{z})$$

$$= - \left(a_{32} \det \begin{bmatrix} a_{11} & a_{13} \\ a_{21} & a_{23} \end{bmatrix} - a_{22} \det \begin{bmatrix} a_{11} & a_{13} \\ a_{31} & a_{33} \end{bmatrix} + a_{12} \det \begin{bmatrix} a_{21} & a_{23} \\ a_{31} & a_{33} \end{bmatrix} \right) \mathbf{E}(\mathbf{x}, \mathbf{y}, \mathbf{z}).$$

□

Theorem 4.1.8. *Let V be a vector space and let $\mathbf{E} : V^n \to \mathbb{K}$ be an alternating multilinear form. If $\mathbf{x}_1, \ldots, \mathbf{x}_n \in V$ are linearly dependent, then $\mathbf{E}(\mathbf{x}_1, \ldots, \mathbf{x}_n) = 0$.*

Proof. Without loss of generality, we can assume that $\mathbf{x}_1 = \sum\limits_{j=2}^{n} a_j \mathbf{x}_j$. Then

$$\mathbf{E}(\mathbf{x}_1, \ldots, \mathbf{x}_n) = \mathbf{E}\left(\sum_{j=2}^{n} a_j \mathbf{x}_j, \mathbf{x}_2, \ldots, \mathbf{x}_n\right) = \sum_{j=2}^{n} a_j \mathbf{E}(\mathbf{x}_j, \mathbf{x}_2 \ldots, \mathbf{x}_n) = 0,$$

because $\mathbf{E}$ is alternating. $\qquad\qquad\qquad\qquad\qquad\qquad\qquad\qquad\qquad\quad\square$

Theorem 4.1.9. *Let V be a vector space and let $\mathbf{E} : V^n \to \mathbb{K}$ be a multilinear form. For any permutation $\sigma \in \mathfrak{S}_n$ the function $\mathbf{G} : V^n \to \mathbb{K}$ defined by*

$$\mathbf{G}(\mathbf{x}_1, \ldots, \mathbf{x}_n) = \sum_{\sigma \in \mathfrak{S}_n} \epsilon(\sigma) \mathbf{E}(\mathbf{x}_{\sigma(1)}, \ldots, \mathbf{x}_{\sigma(n)})$$

is an alternating multilinear form.

Proof. It is easy to see that $\mathbf{G}$ is a multilinear form.

Now suppose that $\mathbf{x}_j = \mathbf{x}_k$ for some distinct $j, k \in \{1, \ldots, n\}$. Let $\tau = \sigma_{jk} \in \mathfrak{S}_n$, that is the transposition such that $\tau(j) = k$, $\tau(k) = j$, and $\tau(l) = l$ for any $l \in \{1, \ldots, n\}$ different from j and k. First we note that if σ is an even permutation then $\tau\sigma$ is an odd permutation and the function $s : \mathcal{E}_n \to \mathcal{O}_n$ defined by $s(\sigma) = \tau\sigma$ is a bijection.

We have

$$\mathbf{G}(\mathbf{x}_1, \ldots, \mathbf{x}_n) = \sum_{\sigma \in \mathfrak{S}_n} \epsilon(\sigma) \mathbf{E}(\mathbf{x}_{\sigma(1)}, \ldots, \mathbf{x}_{\sigma(n)})$$

$$= \sum_{\sigma \in \mathcal{E}_n} \epsilon(\sigma) \mathbf{E}(\mathbf{x}_{\sigma(1)}, \ldots, \mathbf{x}_{\sigma(n)}) + \sum_{\sigma \in \mathcal{O}_n} \epsilon(\sigma) \mathbf{E}(\mathbf{x}_{\sigma(1)}, \ldots, \mathbf{x}_{\sigma(n)})$$

$$= \sum_{\sigma \in \mathcal{E}_n} \epsilon(\sigma) \mathbf{E}(\mathbf{x}_{\sigma(1)}, \ldots, \mathbf{x}_{\sigma(n)}) + \sum_{\sigma \in \mathcal{E}_n} \epsilon(\tau\sigma) \mathbf{E}(\mathbf{x}_{\tau\sigma(1)}, \ldots, \mathbf{x}_{\tau\sigma(n)})$$

$$= \sum_{\sigma \in \mathcal{E}_n} \epsilon(\sigma) \mathbf{E}(\mathbf{x}_{\sigma(1)}, \ldots, \mathbf{x}_{\sigma(n)}) - \sum_{\sigma \in \mathcal{E}_n} \epsilon(\sigma) \mathbf{E}(\mathbf{x}_{\tau\sigma(1)}, \ldots, \mathbf{x}_{\tau\sigma(n)}).$$

Now we consider three cases:

Case 1: If $\sigma(l) \neq j$ and $\sigma(l) \neq k$, then $\tau\sigma(l) = \sigma(l)$.

Case 2: If $\sigma(l) = j$, then $\tau\sigma(l) = \tau(j) = k$ and $\mathbf{x}_j = \mathbf{x}_{\sigma(l)} = \mathbf{x}_k = \mathbf{x}_{\tau\sigma(l)}$.

Case 3: If $\sigma(l) = k$, then $\tau\sigma(l) = \tau(k) = j$ and $\mathbf{x}_k = \mathbf{x}_{\sigma(l)} = \mathbf{x}_j = \mathbf{x}_{\tau\sigma(l)}$.

Consequently

$$\mathbf{G}(\mathbf{x}_1,\ldots,\mathbf{x}_n) = \sum_{\sigma\in\mathcal{E}_n} \epsilon(\sigma)\mathbf{E}(\mathbf{x}_{\sigma(1)},\ldots,\mathbf{x}_{\sigma(n)}) - \sum_{\sigma\in\mathcal{E}_n} \epsilon(\sigma)\mathbf{E}(\mathbf{x}_{\tau\sigma(1)},\ldots,\mathbf{x}_{\tau\sigma(n)}) = 0$$

and thus $\mathbf{G}$ is an alternating multilinear form. □

Theorem 4.1.10. *Let $\mathcal{V}$ be vector space and let $\mathbf{E} : \mathcal{V}^n \to \mathbb{K}$ be an alternating n-linear form. Then*

$$\mathbf{E}(\mathbf{x}_{\sigma(1)},\ldots,\mathbf{x}_{\sigma(n)}) = \epsilon(\sigma)\mathbf{E}(\mathbf{x}_1,\ldots,\mathbf{x}_n)$$

for any $\sigma \in \mathfrak{S}_n$ and $\mathbf{x}_1,\ldots,\mathbf{x}_n \in \mathcal{V}$.

Proof. Since, by Theorem 4.1.4, we have

$$\mathbf{E}(\mathbf{x}_{\tau(1)},\ldots,\mathbf{x}_{\tau(n)}) = -\mathbf{E}(\mathbf{x}_1,\ldots,\mathbf{x}_n)$$

for every transposition τ, the result follows from Theorem 5.1 in Appendix A. □

Theorem 4.1.11. *Let $\mathcal{V}$ be vector space and let $\mathbf{E} : \mathcal{V}^n \to \mathbb{K}$ be an alternating n-linear form. If*

$$\mathbf{x}_j = a_{1j}\mathbf{v}_1 + \cdots + a_{nj}\mathbf{v}_n,$$

where $\mathbf{v}_1,\ldots,\mathbf{v}_n,\mathbf{x}_1,\ldots,\mathbf{x}_n \in \mathcal{V}$, $a_{kj} \in \mathbb{K}$, and $j,k \in \{1,\ldots,n\}$, then

$$\mathbf{E}(\mathbf{x}_1,\ldots,\mathbf{x}_n) = \left(\sum_{\sigma\in\mathfrak{S}_n} \epsilon(\sigma)a_{\sigma(1),1}\cdots a_{\sigma(n),n}\right)\mathbf{E}(\mathbf{v}_1,\ldots,\mathbf{v}_n).$$

Proof. Since $\mathbf{E}$ is alternating n-linear, we have

$$\mathbf{E}(\mathbf{x}_1,\ldots,\mathbf{x}_n) = \sum_{\sigma\in\mathfrak{S}_n} a_{\sigma(1),1}\cdots a_{\sigma(n),n}\mathbf{E}(\mathbf{v}_{\sigma(1)},\ldots,\mathbf{v}_{\sigma(n)}).$$

Consequently, by Theorem 4.1.10,

$$\mathbf{E}(\mathbf{x}_1,\ldots,\mathbf{x}_n) = \left(\sum_{\sigma\in\mathfrak{S}_n} \epsilon(\sigma)a_{\sigma(1),1}\cdots a_{\sigma(n),n}\right)\mathbf{E}(\mathbf{v}_1,\ldots,\mathbf{v}_n).$$

□

Theorem 4.1.12. *Let $\mathcal{V}$ be an n-dimensional vector space and let $\mathbf{E} : \mathcal{V}^n \to \mathbb{K}$ be a nonzero alternating n-linear form. Then vectors $\mathbf{v}_1,\ldots,\mathbf{v}_n \in \mathcal{V}$ constitute a basis of $\mathcal{V}$ if and only if $\mathbf{E}(\mathbf{v}_1,\ldots,\mathbf{v}_n) \neq 0$.*

Proof. If $\{\mathbf{v}_1, \ldots, \mathbf{v}_n\}$ is a basis of $\mathcal{V}$ and $\mathbf{E}(\mathbf{v}_1, \ldots, \mathbf{v}_n) = 0$, then $\mathbf{E} = \mathbf{0}$ by Theorem 4.1.11. Consequently, if $\{\mathbf{v}_1, \ldots, \mathbf{v}_n\}$ is a basis and $\mathbf{E} \neq \mathbf{0}$, then $\mathbf{E}(\mathbf{v}_1, \ldots, \mathbf{v}_n) \neq 0$.

If the vectors $\mathbf{v}_1, \ldots, \mathbf{v}_n$ are linearly dependent, then $\mathbf{E}(\mathbf{v}_1, \ldots, \mathbf{v}_n) = 0$, by Theorem 4.1.8. Consequently, if $\mathbf{E}(\mathbf{v}_1, \ldots, \mathbf{v}_n) \neq 0$, then $\mathbf{v}_1, \ldots, \mathbf{v}_n$ must be linearly independent, and thus $\{\mathbf{v}_1, \ldots, \mathbf{v}_n\}$ is a basis of $\mathcal{V}$. $\qquad\square$

Example 4.1.13 (Cramer's rule). Let $\mathcal{V}$ be a vector space and let $\{\mathbf{v}_1, \ldots, \mathbf{v}_n\}$ be a basis of $\mathcal{V}$. If $\mathbf{D} : \mathcal{V}^n \to \mathbb{K}$ is a nonzero alternating multilinear form and

$$\mathbf{a} = x_1 \mathbf{v}_1 + \cdots + x_n \mathbf{v}_n,$$

then

$$x_j = \frac{\mathbf{D}(\mathbf{v}_1, \ldots, \mathbf{v}_{j-1}, \mathbf{a}, \mathbf{v}_{j+1}, \ldots, \mathbf{v}_n)}{\mathbf{D}(\mathbf{v}_1, \ldots, \mathbf{v}_n)}$$

for every $j \in \{1, \ldots, n\}$.

Proof. For every $j \in \{1, \ldots, n\}$ we have

$$
\begin{aligned}
\mathbf{D}&(\mathbf{v}_1, \ldots, \mathbf{v}_{j-1}, \mathbf{a}, \mathbf{v}_{j+1}, \ldots, \mathbf{v}_n) \\
&= \mathbf{D}(\mathbf{v}_1, \ldots, \mathbf{v}_{j-1}, x_1\mathbf{v}_1 + \cdots + x_n\mathbf{v}_n, \mathbf{v}_{j+1}, \ldots, \mathbf{v}_n) \\
&= x_1 \mathbf{D}(\mathbf{v}_1, \ldots, \mathbf{v}_{j-1}, \mathbf{v}_1, \mathbf{v}_{j+1}, \ldots, \mathbf{v}_n) \\
&\quad + \cdots + x_j \mathbf{D}(\mathbf{v}_1, \ldots, \mathbf{v}_{j-1}, \mathbf{v}_j, \mathbf{v}_{j+1}, \ldots, \mathbf{v}_n) \\
&\quad + \cdots + x_n \mathbf{D}(\mathbf{v}_1, \ldots, \mathbf{v}_{j-1}, \mathbf{v}_n, \mathbf{v}_{j+1}, \ldots, \mathbf{v}_n) \\
&= x_j \mathbf{D}(\mathbf{v}_1, \ldots, \mathbf{v}_{j-1}, \mathbf{v}_j, \mathbf{v}_{j+1}, \ldots, \mathbf{v}_n).
\end{aligned}
$$

This gives us the desired equality. $\qquad\square$

Theorem 4.1.14. *Let* $\{\mathbf{v}_1, \ldots, \mathbf{v}_n\}$ *be a basis of a vector space* $\mathcal{V}$. *There is an unique alternating n-linear form*

$$\mathbf{D}_{\mathbf{v}_1, \ldots, \mathbf{v}_n} : \mathcal{V}^n \to \mathbb{K}$$

such that $\mathbf{D}_{\mathbf{v}_1, \ldots, \mathbf{v}_n}(\mathbf{v}_1, \ldots, \mathbf{v}_n) = 1$.

Proof. Let $\{\mathbf{v}_1, \ldots, \mathbf{v}_n\}$ be a basis of $\mathcal{V}$. For $\mathbf{x} = a_1 \mathbf{v}_1 + \cdots + a_n \mathbf{v}_n$ and $j \in \{1, \ldots, n\}$ we define

$$l_{\mathbf{v}_j}(\mathbf{x}) = l_{\mathbf{v}_j}(a_1 \mathbf{v}_1 + \cdots + a_n \mathbf{v}_n) = a_j.$$

Clearly, the function $\mathbf{E}_{\mathbf{v}_1, \ldots, \mathbf{v}_n} : \mathcal{V}^n \to \mathbb{K}$ defined by

$$\mathbf{E}_{\mathbf{v}_1, \ldots, \mathbf{v}_n}(\mathbf{x}_1, \ldots, \mathbf{x}_n) = l_{\mathbf{v}_1}(\mathbf{x}_1) \cdots l_{\mathbf{v}_n}(\mathbf{x}_n)$$

is n-linear. According to Theorem 4.1.9 the function $\mathbf{D}_{\mathbf{v}_1,\ldots,\mathbf{v}_n} : \mathcal{V}^n \to \mathbb{K}$ defined by

$$\mathbf{D}_{\mathbf{v}_1,\ldots,\mathbf{v}_n}(\mathbf{x}_1,\ldots,\mathbf{x}_n) = \sum_{\sigma \in \mathfrak{S}_n} \epsilon(\sigma) l_{\mathbf{v}_1}(\mathbf{x}_{\sigma(1)}) \ldots l_{\mathbf{v}_n}(\mathbf{x}_{\sigma(n)})$$

is an alternating n-linear form such that $\mathbf{D}_{\mathbf{v}_1,\ldots,\mathbf{v}_n}(\mathbf{v}_1,\ldots,\mathbf{v}_n) = 1$. The uniqueness is a consequence of Theorem 4.1.11. $\qquad\square$

Note that the sums

$$\sum_{\sigma \in \mathfrak{S}_n} \epsilon(\sigma) a_{\sigma(1),1} \cdots a_{\sigma(n),n} \quad \text{and} \quad \sum_{\sigma \in \mathfrak{S}_n} \epsilon(\sigma) a_{1,\sigma(1)} \cdots a_{n,\sigma(n)}$$

are equal. Indeed, if $a_{jk} \in \mathbb{K}$ for $j, k \in \{1,\ldots,n\}$ and $\sigma, \tau \in \mathfrak{S}_n$, then

$$a_{\sigma(1),1} \cdots a_{\sigma(n),n} = a_{\sigma\tau(1),\tau(1)} \cdots a_{\sigma\tau(n),\tau(n)}$$

and consequently

$$a_{\sigma(1),1} \cdots a_{\sigma(n),n} = a_{\sigma\sigma^{-1}(1),\sigma^{-1}(1)} \cdots a_{\sigma\sigma^{-1}(n),\sigma^{-1}(n)} = a_{1,\sigma^{-1}(1)} \cdots a_{n,\sigma^{-1}(n)}.$$

Hence

$$\sum_{\sigma \in \mathfrak{S}_n} \epsilon(\sigma) a_{\sigma(1),1} \cdots a_{\sigma(n),n} = \sum_{\sigma \in \mathfrak{S}_n} \epsilon(\sigma) a_{1,\sigma^{-1}(1)} \cdots a_{n,\sigma^{-1}(n)}$$

$$= \sum_{\sigma \in \mathfrak{S}_n} \epsilon(\sigma^{-1}) a_{1,\sigma^{-1}(1)} \cdots a_{n,\sigma^{-1}(n)}$$

$$= \sum_{\sigma \in \mathfrak{S}_n} \epsilon(\sigma) a_{1,\sigma(1)} \cdots a_{n,\sigma(n)}.$$

Definition 4.1.15. Let A be an $n \times n$ matrix with entries $a_{j,k}$. The number

$$\sum_{\sigma \in \mathfrak{S}_n} \epsilon(\sigma) a_{\sigma(1),1} \cdots a_{\sigma(n),n}$$

is called the *determinant* of A and is denoted by $\det A$.

Note that from the calculations presented before the above definition it follows that $\det A = \det A^T$.

It is easy to verify that our definition of the determinant agrees with the familiar formulas for 2×2 and 3×3 matrices:

$$\det \begin{bmatrix} a_{11} & a_{12} \\ a_{21} & a_{22} \end{bmatrix} = a_{11}a_{22} - a_{12}a_{21}$$

and

$$\det \begin{bmatrix} a_{11} & a_{12} & a_{13} \\ a_{21} & a_{22} & a_{23} \\ a_{31} & a_{32} & a_{33} \end{bmatrix} = a_{11}a_{22}a_{33} + a_{12}a_{23}a_{31} + a_{13}a_{21}a_{32}$$
$$- a_{13}a_{22}a_{31} - a_{11}a_{23}a_{32} - a_{12}a_{21}a_{33}.$$

Theorem 4.1.16. *Let $\mathcal{V}$ be an n-dimensional vector space and let $\mathbf{D}$: $\mathcal{V}^n \to \mathbb{K}$ be a nonzero alternating n-linear form. For every alternating n-linear form $\mathbf{E} : \mathcal{V}^n \to \mathbb{K}$ we have*

$$\mathbf{E} = \alpha\mathbf{D}$$

for some unique $\alpha \in \mathbb{K}$.

Proof. Let $\{\mathbf{v}_1, \ldots, \mathbf{v}_n\}$ be a basis in $\mathcal{V}$. For every $\mathbf{x}_1, \ldots, \mathbf{x}_n \in \mathcal{V}$ we have

$$\mathbf{x}_j = a_{1j}\mathbf{v}_1 + \cdots + a_{nj}\mathbf{v}_n,$$

where $a_{jk} \in \mathbb{K}$ and $j, k \in \{1, \ldots, n\}$. Now, using Theorem 4.1.11, we obtain

$$\mathbf{E}(\mathbf{x}_1, \ldots, \mathbf{x}_n) = \left(\sum_{\sigma \in \mathfrak{S}_n} \epsilon(\sigma)a_{\sigma(1),1} \cdots a_{\sigma(n),n} \right) \mathbf{E}(\mathbf{v}_1, \ldots, \mathbf{v}_n)$$

$$= \left(\sum_{\sigma \in \mathfrak{S}_n} \epsilon(\sigma)a_{\sigma(1),1} \cdots a_{\sigma(n),n} \right) \mathbf{E}(\mathbf{v}_1, \ldots, \mathbf{v}_n) \frac{\mathbf{D}(\mathbf{v}_1, \ldots, \mathbf{v}_n)}{\mathbf{D}(\mathbf{v}_1, \ldots, \mathbf{v}_n)}$$

$$= \frac{\mathbf{E}(\mathbf{v}_1, \ldots, \mathbf{v}_n)}{\mathbf{D}(\mathbf{v}_1, \ldots, \mathbf{v}_n)} \left(\sum_{\sigma \in \mathfrak{S}_n} \epsilon(\sigma)a_{\sigma(1),1} \cdots a_{\sigma(n),n} \right) \mathbf{D}(\mathbf{v}_1, \ldots, \mathbf{v}_n)$$

$$= \frac{\mathbf{E}(\mathbf{v}_1, \ldots, \mathbf{v}_n)}{\mathbf{D}(\mathbf{v}_1, \ldots, \mathbf{v}_n)} \mathbf{D}(\mathbf{x}_1, \ldots, \mathbf{x}_n).$$

This means that $\mathbf{E} = \alpha\mathbf{D}$ where $\alpha = \frac{\mathbf{E}(\mathbf{v}_1, \ldots, \mathbf{v}_n)}{\mathbf{D}(\mathbf{v}_1, \ldots, \mathbf{v}_n)}$.

Note that since $\mathbf{E}(\mathbf{x}_1, \ldots, \mathbf{x}_n) = \alpha\mathbf{D}(\mathbf{x}_1, \ldots, \mathbf{x}_n)$ for every $\mathbf{x}_1, \ldots, \mathbf{x}_n \in \mathcal{V}$, the constant α does not depend on the choice of a basis in $\mathcal{V}$. Now the uniqueness of α is immediate. $\qquad\square$

Theorem 4.1.17. *Let $\mathcal{V}$ be an n-dimensional vector space and let f : $\mathcal{V} \to \mathcal{V}$ be an endomorphism. There is a unique number $\alpha \in \mathbb{K}$ such that for every $\mathbf{v}_1, \ldots, \mathbf{v}_n \in \mathcal{V}$ and every alternating n-linear form $\mathbf{E} : \mathcal{V}^n \to \mathbb{K}$ we have*

$$\mathbf{E}(f(\mathbf{v}_1)), \ldots, f(\mathbf{v}_n)) = \alpha\mathbf{E}(\mathbf{v}_1, \ldots, \mathbf{v}_n).$$

Proof. Let $\mathbf{D} : \mathcal{V}^n \to \mathbb{K}$ be a nonzero alternating n-linear form. The function $\mathbf{F} : \mathcal{V}^n \to \mathbb{K}$ defined by

$$\mathbf{F}(\mathbf{v}_1, \dots, \mathbf{v}_n) = \mathbf{D}(f(\mathbf{v}_1), \dots, f(\mathbf{v}_n))$$

is an alternating n-linear form. Consequently, by Theorem 4.1.16, there is a number $\alpha \in \mathbb{K}$ such that

$$\mathbf{D}(f(\mathbf{v}_1), \dots, f(\mathbf{v}_n)) = \mathbf{F}(\mathbf{v}_1, \dots, \mathbf{v}_n) = \alpha \mathbf{D}(\mathbf{v}_1, \dots, \mathbf{v}_n).$$

Now, applying Theorem 4.1.16 to an alternating n-linear form $\mathbf{E} : \mathcal{V}^n \to \mathbb{K}$ we obtain a number $\beta \in \mathbb{K}$ such that $\mathbf{E} = \beta \mathbf{D}$. Hence

$$\mathbf{E}(f(\mathbf{v}_1), \dots, f(\mathbf{v}_n)) = \beta \mathbf{D}(f(\mathbf{v}_1), \dots, f(\mathbf{v}_n)) = \beta \alpha \mathbf{D}(\mathbf{v}_1, \dots, \mathbf{v}_n) = \alpha \mathbf{E}(\mathbf{v}_1, \dots, \mathbf{v}_n).$$

It is clear that α is unique. □

Definition 4.1.18. Let $\mathcal{V}$ be an n-dimensional vector space and let $f : \mathcal{V} \to \mathcal{V}$ be an endomorphism. The number $\alpha \in \mathbb{K}$ such that for every $\mathbf{v}_1, \dots, \mathbf{v}_n \in \mathcal{V}$ and every alternating n-linear form $\mathbf{E} : \mathcal{V}^n \to \mathbb{K}$ we have $\mathbf{E}(f(\mathbf{v}_1), \dots, f(\mathbf{v}_n)) = \alpha \mathbf{E}(\mathbf{v}_1, \dots, \mathbf{v}_n)$ is called the *determinant* of f and is denoted by $\det f$.

Using the notation from the above definition we can write that, if f is an endomorphism on an n-dimensional vector space $\mathcal{V}$, then

$$\mathbf{E}(f(\mathbf{v}_1), \dots, f(\mathbf{v}_n)) = \det f \; \mathbf{E}(\mathbf{v}_1, \dots, \mathbf{v}_n)$$

for every $\mathbf{v}_1, \dots, \mathbf{v}_n \in \mathcal{V}$ and every alternating n-linear form $\mathbf{E} : \mathcal{V}^n \to \mathbb{K}$.

Example 4.1.19. Let $\mathcal{V}$ be a 2-dimensional vector space and let $f : \mathcal{V} \to \mathcal{V}$ be an endomorphism. If $\{\mathbf{v}_1, \mathbf{v}_2\}$ is a basis of $\mathcal{V}$ such that

$$f(\mathbf{v}_1) = \alpha \mathbf{v}_1 + \beta \mathbf{v}_2 \quad \text{and} \quad f(\mathbf{v}_2) = \gamma \mathbf{v}_1 + \delta \mathbf{v}_2$$

show that $\det f = \alpha\delta - \beta\gamma$.

Proof. Let $\mathbf{E}$ be an arbitrary alternating bilinear form $\mathbf{E} : \mathcal{V} \times \mathcal{V} \to \mathbb{K}$. Then

$$\mathbf{E}(f(\mathbf{v}_1), f(\mathbf{v}_2)) = \mathbf{E}(\alpha \mathbf{v}_1 + \beta \mathbf{v}_2, \gamma \mathbf{v}_1 + \delta \mathbf{v}_2).$$

Now we continue as in Example 4.1.5. □

Theorem 4.1.20. *Let V be an n-dimensional vector space and let f and g be endomorphisms on V. Then*

$$\det(gf) = \det g \det f.$$

Proof. Let $\mathbf{D} : V^n \to \mathbb{K}$ be a nonzero alternating n-linear form and let $\mathbf{x}_1, \ldots, \mathbf{x}_n \in V$. Then

$$\mathbf{D}(gf(\mathbf{x}_1), \ldots, gf(\mathbf{x}_n)) = \det(gf)\mathbf{D}(\mathbf{x}_1, \ldots, \mathbf{x}_n)$$

and

$$\mathbf{D}(gf(\mathbf{x}_1), \ldots, gf(\mathbf{x}_n)) = \det g \, \mathbf{D}(f(\mathbf{x}_1), \ldots, f(\mathbf{x}_n)) = \det g \, \det f \, \mathbf{D}(\mathbf{x}_1, \ldots, \mathbf{x}_n).$$

$\square$

Theorem 4.1.21. *Let V be an n-dimensional vector space and let $f : V \to V$ be an endomorphism. Then f is invertible if and only if $\det f \neq 0$. If f is invertible, then $\det f^{-1} = \dfrac{1}{\det f}$.*

Proof. Let $\mathbf{D} : V^n \to \mathbb{K}$ be a nonzero alternating n-linear form. If f is not invertible, then there is a nonzero vector $\mathbf{x}_1$ such that $f(\mathbf{x}_1) = \mathbf{0}$. If $\{\mathbf{x}_1, \ldots, \mathbf{x}_n\}$ is a basis of V, then we have

$$\mathbf{D}(f(\mathbf{x}_1), \ldots, f(\mathbf{x}_n)) = \det f \, \mathbf{D}(\mathbf{x}_1, \ldots, \mathbf{x}_n) = 0.$$

Since $\{\mathbf{x}_1, \ldots, \mathbf{x}_n\}$ is a basis of V, we have $\mathbf{D}(\mathbf{x}_1, \ldots, \mathbf{x}_n) \neq 0$. Consequently, $\det f = 0$. This shows that, if $\det f \neq 0$, then f is invertible.

Conversely, if f is invertible, then

$$1 = \det(\mathrm{Id}_V) = \det(f^{-1}f) = \det f^{-1} \det f.$$

Thus $\det f \neq 0$ and we have $\det f^{-1} = \dfrac{1}{\det f}$.

$\square$

Corollary 4.1.22. *Let V be a finite dimensional vector space and let f and g be endomorphisms on V. If f is invertible, then*

$$\det(f^{-1}gf) = \det g.$$

Proof. According to the previous two results we have

$$\det(f^{-1}gf) = \det f^{-1} \det g \det f = \frac{1}{\det f} \det g \det f = \det g.$$

$\square$

Lemma 4.1.23. *Let $\mathcal{V}_1$ and $\mathcal{V}_2$ be finite dimensional vector spaces and let $\mathcal{V} = \mathcal{V}_1 \oplus \mathcal{V}_2$. If $f : \mathcal{V}_1 \to \mathcal{V}_1$ is an endomorphism and $g : \mathcal{V} \to \mathcal{V}$ is the endomorphism defined by*

$$g(\mathbf{v}_1 + \mathbf{v}_2) = f(\mathbf{v}_1) + \mathbf{v}_2$$

for all $\mathbf{v}_1 \in \mathcal{V}_1$ and $\mathbf{v}_2 \in \mathcal{V}_2$, then

$$\det g = \det f.$$

Proof. Let $\{\mathbf{x}_1, \ldots, \mathbf{x}_n\}$ be a basis of the vector space $\mathcal{V}$ such that $\{\mathbf{x}_1, \ldots, \mathbf{x}_p\}$ is a basis of the vector space $\mathcal{V}_1$ and $\{\mathbf{x}_{p+1}, \ldots, \mathbf{x}_n\}$ is a basis of the vector space $\mathcal{V}_2$. Let $\mathbf{D} : \mathcal{V}^n \to \mathbb{K}$ be a nonzero alternating n-linear form. Then

$$\mathbf{D}(g(\mathbf{x}_1), \ldots, g(\mathbf{x}_n)) = \det g \; \mathbf{D}(\mathbf{x}_1, \ldots, \mathbf{x}_n).$$

Now, if $\mathbf{D}_1 : \mathcal{V}_1^p \to \mathbb{K}$ is the alternating p-linear form defined by

$$\mathbf{D}_1(\mathbf{v}_1, \ldots, \mathbf{v}_p) = \mathbf{D}(\mathbf{v}_1, \ldots, \mathbf{v}_p, \mathbf{x}_{p+1}, \ldots, \mathbf{x}_n),$$

then

$$\mathbf{D}_1(f(\mathbf{x}_1), \ldots, f(\mathbf{x}_p)) = \det f \; \mathbf{D}_1(\mathbf{x}_1, \ldots, \mathbf{x}_p) = \det f \mathbf{D}(\mathbf{x}_1, \ldots, \mathbf{x}_p, \mathbf{x}_{p+1}, \ldots, \mathbf{x}_n).$$

Hence $\det g = \det f$ because

$$\mathbf{D}(g(\mathbf{x}_1), \ldots, g(\mathbf{x}_n)) = \mathbf{D}(f(\mathbf{x}_1), \ldots, f(\mathbf{x}_p), \mathbf{x}_{p+1}, \ldots, \mathbf{x}_n) = \mathbf{D}_1(f(\mathbf{x}_1), \ldots, f(\mathbf{x}_p)).$$

$\square$

Theorem 4.1.24. *Let $\mathcal{V}_1$ and $\mathcal{V}_2$ be finite dimensional vector spaces and let $\mathcal{V} = \mathcal{V}_1 \oplus \mathcal{V}_2$. If $f_1 : \mathcal{V}_1 \to \mathcal{V}_1$ and $f_2 : \mathcal{V}_2 \to \mathcal{V}_2$ are endomorphism and $f : \mathcal{V} \to \mathcal{V}$ is the endomorphism defined by*

$$f(\mathbf{v}_1 + \mathbf{v}_2) = f_1(\mathbf{v}_1) + f_2(\mathbf{v}_2)$$

where $\mathbf{v}_1 \in \mathcal{V}_1$ and $\mathbf{v}_2 \in \mathcal{V}_2$, then

$$\det f = \det f_1 \det f_2.$$

Proof. Let $g_1 : \mathcal{V} \to \mathcal{V}$ and $g_2 : \mathcal{V} \to \mathcal{V}$ be defined as

$$g_1(\mathbf{v}_1 + \mathbf{v}_2) = f_1(\mathbf{v}_1) + \mathbf{v}_2 \quad \text{and} \quad g_2(\mathbf{v}_1 + \mathbf{v}_2) = \mathbf{v}_1 + f_2(\mathbf{v}_2)$$

for all $\mathbf{v}_1 \in \mathcal{V}_1$ and $\mathbf{v}_2 \in \mathcal{V}_2$. Then $f = g_1 g_2$ and thus $\det f = \det g_1 \det g_2$, by Theorem 4.1.20. Now from Lemma 4.1.23 we have $\det g_1 = \det f_1$ and $\det g_2 = \det f_2$, which gives us $\det f = \det f_1 \det f_2$. $\square$

4.1.2   Diagonalization

In the remainder of this chapter it will be convenient to identify a number $\alpha \in \mathbb{K}$ with the operator $\alpha \operatorname{Id}$. This convention is quite natural since $(\alpha \operatorname{Id})\mathbf{x} = \alpha\mathbf{x}$, so the α on the right hand side can be interpreted as a number or an operator.

Eigenvalues and eigenvectors were introduced in Chapter 3 in the context of operators on inner product spaces, but the definitions do not require the inner product. For convenience we recall the definitions of eigenvalues, eigenvectors and eigenspaces.

Definition 4.1.25. Let $\mathcal{V}$ be a vector space and let $f : \mathcal{V} \to \mathcal{V}$ be an endomorphism. A number $\lambda \in \mathbb{K}$ is called an *eigenvalue* of f if the equation

$$f(\mathbf{x}) = \lambda\mathbf{x}$$

has a nontrivial solution, that is, a solution $\mathbf{x} \neq \mathbf{0}$.

The following theorem is useful when finding eigenvalues of endomorphism.

Theorem 4.1.26. *Let $\mathcal{V}$ be a vector space and let $f : \mathcal{V} \to \mathcal{V}$ be an endomorphism. Then*

$$\lambda \ \text{is an eigenvalue of } f \quad \text{if and only if} \quad \det(f - \lambda) = 0.$$

Proof. The equivalence is a consequence of Theorem 4.1.21. Indeed, the equation $f(\mathbf{x}) = \lambda\mathbf{x}$ has a solution $\mathbf{x} \neq \mathbf{0}$ if and only if the equation $(f - \lambda)(\mathbf{x}) = \mathbf{0}$ has a solution $\mathbf{x} \neq \mathbf{0}$, which means that the linear transformation $f - \lambda$ is not invertible and this is equivalent to $\det(f - \lambda) = 0$, by Theorem 4.1.21. $\square$

Definition 4.1.27. Let $\mathcal{V}$ be a vector space and let $f : \mathcal{V} \to \mathcal{V}$ be an endomorphism. The polynomial

$$c_f(t) = \det(f - t)$$

is called the *characteristic polynomial* of f and the equation

$$\det(f - t) = 0$$

is called the *characteristic equation* of f.

Example 4.1.28. Let $f : \mathbb{R}^3 \to \mathbb{R}^3$ be the endomorphism defined by

$$f\left(\begin{bmatrix} x \\ y \\ z \end{bmatrix}\right) = \begin{bmatrix} 3 & 1 & 2 \\ 1 & 3 & 2 \\ 1 & 1 & 4 \end{bmatrix}\begin{bmatrix} x \\ y \\ z \end{bmatrix}.$$

Calculate c_f.

Solution. Let $\mathbf{D} : \mathbb{R}^3 \times \mathbb{R}^3 \times \mathbb{R}^3 \to \mathbb{R}$ be a nonzero alternating 3-linear form. Then

$$\mathbf{D}\left((f-t)\left(\begin{bmatrix} 1 \\ 0 \\ 0 \end{bmatrix}\right), (f-t)\left(\begin{bmatrix} 0 \\ 1 \\ 0 \end{bmatrix}\right), (f-t)\left(\begin{bmatrix} 0 \\ 0 \\ 1 \end{bmatrix}\right)\right)$$

$$= \mathbf{D}\left(\begin{bmatrix} 3-t \\ 1 \\ 1 \end{bmatrix}, \begin{bmatrix} 1 \\ 3-t \\ 1 \end{bmatrix}, \begin{bmatrix} 2 \\ 2 \\ 4-t \end{bmatrix}\right)$$

$$= \mathbf{D}\left(\begin{bmatrix} 3-t \\ 1 \\ 1 \end{bmatrix} - \begin{bmatrix} 1 \\ 3-t \\ 1 \end{bmatrix}, \begin{bmatrix} 1 \\ 3-t \\ 1 \end{bmatrix}, \begin{bmatrix} 2 \\ 2 \\ 4-t \end{bmatrix}\right)$$

$$= \mathbf{D}\left(\begin{bmatrix} 2-t \\ t-2 \\ 0 \end{bmatrix}, \begin{bmatrix} 1 \\ 3-t \\ 1 \end{bmatrix}, \begin{bmatrix} 2 \\ 2 \\ 4-t \end{bmatrix}\right)$$

$$= (2-t)\mathbf{D}\left(\begin{bmatrix} 1 \\ -1 \\ 0 \end{bmatrix}, \begin{bmatrix} 1 \\ 3-t \\ 1 \end{bmatrix}, \begin{bmatrix} 2 \\ 2 \\ 4-t \end{bmatrix}\right).$$

Now we proceed as in Example 4.1.7 and get

$$\mathbf{D}\left(\begin{bmatrix} 1 \\ -1 \\ 0 \end{bmatrix}, \begin{bmatrix} 1 \\ 3-t \\ 1 \end{bmatrix}, \begin{bmatrix} 2 \\ 2 \\ 4-t \end{bmatrix}\right) = (t^2 - 8t + 12)\mathbf{D}\left(\begin{bmatrix} 1 \\ 0 \\ 0 \end{bmatrix}, \begin{bmatrix} 0 \\ 1 \\ 0 \end{bmatrix}, \begin{bmatrix} 0 \\ 0 \\ 1 \end{bmatrix}\right).$$

Hence
$$c_f(t) = (2-t)(t^2 - 8t + 12) = (2-t)^2(6-t).$$

□

Example 4.1.29. Let $f : \mathbb{C}^2 \to \mathbb{C}^2$ be the endomorphism defined by

$$f\left(\begin{bmatrix} x \\ y \end{bmatrix}\right) = \begin{bmatrix} 1+2i & 2i \\ 1+i & 3i \end{bmatrix}\begin{bmatrix} x \\ y \end{bmatrix}.$$

Calculate c_f and the eigenvalues of f.

Proof. Let $\mathbf{D} : \mathbb{C}^2 \times \mathbb{C}^2 \to \mathbb{C}$ be a nonzero alternating bilinear form. Proceeding as in Example 4.1.5 we get

$$\mathbf{D}\left(\begin{bmatrix} 1 + 2i - t \\ 1 + i \end{bmatrix}, \begin{bmatrix} 2i \\ 3i - t \end{bmatrix}\right) = ((1 + 2i - t)(3i - t) - 2i(1 + i))\mathbf{D}\left(\begin{bmatrix} 1 \\ 0 \end{bmatrix}, \begin{bmatrix} 0 \\ 1 \end{bmatrix}\right)$$

$$= (t^2 - (1 + 5i)t + i(1 + 4i))\mathbf{D}\left(\begin{bmatrix} 1 \\ 0 \end{bmatrix}, \begin{bmatrix} 0 \\ 1 \end{bmatrix}\right).$$

Hence

$$c_f(t) = t^2 - (1 + 5i)t + i(1 + 4i)$$

and the eigenvalues are i and $1 + 4i$. $\square$

Definition 4.1.30. Let $\mathcal{V}$ be a vector space and let λ be an eigenvalue of an endomorphism $f : \mathcal{V} \to \mathcal{V}$. A vector $\mathbf{x} \neq \mathbf{0}$ is called an *eigenvector* of f corresponding to the eigenvalue λ if $f(\mathbf{x}) = \lambda\mathbf{x}$. The set

$$\mathcal{E}_\lambda = \{\mathbf{x} \in \mathcal{V} : f(\mathbf{x}) = \lambda\mathbf{x}\}$$

is called the *eigenspace* of f corresponding to λ.

It is easy to verify that $\mathcal{E}_\lambda$ is a subspace of $\mathcal{V}$. It consists of all eigenvectors of f corresponding to λ and the zero vector. Note that $\mathcal{E}_0 = \ker f$.

Theorem 4.1.31. *Let $\mathcal{V}$ be an n-dimensional vector space and let $f : \mathcal{V} \to \mathcal{V}$ be an endomorphism. If $\{\mathbf{v}_1, \ldots, \mathbf{v}_n\}$ is a basis of $\mathcal{V}$ such that*

$$f(\mathbf{v}_1) = x_{11}\mathbf{v}_1$$

$$f(\mathbf{v}_2) = x_{12}\mathbf{v}_1 + x_{22}\mathbf{v}_2$$

$$\vdots$$

$$f(\mathbf{v}_n) = x_{1n}\mathbf{v}_1 + x_{2n}\mathbf{v}_2 + \cdots + x_{n-1,n}\mathbf{v}_{n-1} + x_{nn}\mathbf{v}_n,$$

where $x_{jk} \in \mathbb{K}$ for all $j, k \in \{1, \ldots, n\}$ such that $j \leq k$, then

$$c_f(t) = (x_{11} - t) \cdots (x_{nn} - t).$$

Proof. For any nonzero alternating n-linear form $\mathbf{D}: \mathcal{V}^n \to \mathbb{K}$ we have

$$
\begin{aligned}
c_f(t)\mathbf{D}(\mathbf{v}_1, \ldots, \mathbf{v}_n) &= \det(f - t)\mathbf{D}(\mathbf{v}_1, \ldots, \mathbf{v}_n) \\
&= \mathbf{D}(f(\mathbf{v}_1) - t\mathbf{v}_1, f(\mathbf{v}_2) - t\mathbf{v}_2, \ldots, f(\mathbf{v}_n) - t\mathbf{v}_n) \\
&= \mathbf{D}(x_{11}\mathbf{v}_1 - t\mathbf{v}_1, x_{12}\mathbf{v}_1 + x_{22}\mathbf{v}_2 - t\mathbf{v}_2, \ldots, \\
&\qquad x_{1n}\mathbf{v}_1 + x_{2n}\mathbf{v}_2 + \cdots + x_{n-1,n}\mathbf{v}_{n-1} + x_{nn}\mathbf{v}_n - t\mathbf{v}_n) \\
&= \mathbf{D}(x_{11}\mathbf{v}_1 - t\mathbf{v}_1, x_{22}\mathbf{v}_2 - t\mathbf{v}_2, \ldots, x_{nn}\mathbf{v}_n - t\mathbf{v}_n) \\
&= (x_{11} - t) \cdots (x_{nn} - t)\mathbf{D}(\mathbf{v}_1, \ldots, \mathbf{v}_n).
\end{aligned}
$$

□

It turns out that the converse of the above result is also true.

Theorem 4.1.32. *Let $\mathcal{V}$ be an n-dimensional vector space and let $f : \mathcal{V} \to \mathcal{V}$ be an endomorphism such that*

$$c_f(t) = (\lambda_1 - t) \cdots (\lambda_n - t)$$

for some $\lambda_1, \ldots, \lambda_n \in \mathbb{K}$. Then there is a basis $\{\mathbf{v}_1, \ldots, \mathbf{v}_n\}$ of $\mathcal{V}$ such that

$$
\begin{aligned}
f(\mathbf{v}_1) &= x_{11}\mathbf{v}_1 \\
f(\mathbf{v}_2) &= x_{12}\mathbf{v}_1 + x_{22}\mathbf{v}_2 \\
&\quad\vdots \\
f(\mathbf{v}_n) &= x_{1n}\mathbf{v}_1 + x_{2n}\mathbf{v}_2 + \cdots + x_{n-1,n}\mathbf{v}_{n-1} + x_{nn}\mathbf{v}_n,
\end{aligned}
$$

where $x_{jk} \in \mathbb{K}$ for all $j, k \in \{1, \ldots, n\}$ such that $j \leq k$, and

$$x_{11} = \lambda_1, x_{22} = \lambda_2, \ldots, x_{nn} = \lambda_n.$$

Proof. We are going to use induction on n. Clearly, the theorem holds when $n = 1$. Now let $n \geq 2$ and assume that the theorem holds for $n - 1$.

If $c_f(t) = (\lambda_1 - t) \cdots (\lambda_n - t)$, then $\lambda_1, \ldots, \lambda_n$ are eigenvalues of f. Let $\mathbf{v}_1$ be an eigenvector of f corresponding to the eigenvalue λ_1, that is, $f(\mathbf{v}_1) = \lambda_1\mathbf{v}_1$ and $\mathbf{v}_1 \neq \mathbf{0}$. We define $\mathcal{V}_1 = \mathrm{Span}\{\mathbf{v}_1\}$. Let $\mathcal{W}$ be a vector subspace of $\mathcal{V}$ such that $\mathcal{V} = \mathcal{V}_1 \oplus \mathcal{W}$ and let p be the projection of $\mathcal{V}$ on $\mathcal{W}$ along $\mathcal{V}_1$. We denote by $g : \mathcal{W} \to \mathcal{W}$ the endomorphism induced by pf on $\mathcal{W}$.

Let $\{\mathbf{w}_2, \ldots, \mathbf{w}_n\}$ be a basis of $\mathcal{W}$ and let $\mathbf{D} : \mathcal{V}^n \to \mathbb{K}$ be a nonzero alternating n-linear form. For some $y_{1k} \in \mathbb{K}$, where $k \in \{2, \ldots, n\}$, we have

$$f(\mathbf{w}_2) = y_{12}\mathbf{v}_1 + g(\mathbf{w}_2)$$

$$\vdots$$

$$f(\mathbf{w}_n) = y_{1n}\mathbf{v}_1 + g(\mathbf{w}_n).$$

If c_f is the characteristic polynomial of f and c_g is the characteristic polynomial of g, then for some x_{1k}, where $k \in \{2, \ldots, n\}$, we have

$$\begin{aligned}
c_f(t)\mathbf{D}(\mathbf{v}_1, \mathbf{w}_2, \ldots, \mathbf{w}_n) &= \mathbf{D}(f(\mathbf{v}_1) - t\mathbf{v}_1, f(\mathbf{w}_2) - t\mathbf{w}_2, \ldots, f(\mathbf{w}_n) - t\mathbf{w}_n) \\
&= (\lambda_1 - t)\mathbf{D}(\mathbf{v}_1, y_{12}\mathbf{v}_1 + g(\mathbf{w}_2) - t\mathbf{w}_2, \ldots, y_{1n}\mathbf{v}_1 \\
&\quad + g(\mathbf{w}_n) - t\mathbf{w}_n) \\
&= (\lambda_1 - t)\mathbf{D}(\mathbf{v}_1, g(\mathbf{w}_2) - t\mathbf{w}_2, \ldots, g(\mathbf{w}_n) - t\mathbf{w}_n) \\
&= (\lambda_1 - t)c_g(t)\mathbf{D}(\mathbf{v}_1, \mathbf{w}_2, \ldots, \mathbf{w}_n).
\end{aligned}$$

Consequently, $c_f(t) = (\lambda_1 - t)c_g(t)$ and thus $c_g = (\lambda_2 - t) \cdots (\lambda_n - t)$. By our inductive assumption the theorem holds for the endomorphism $g : \mathcal{W} \to \mathcal{W}$ and thus there is a basis $\{\mathbf{v}_2, \ldots, \mathbf{v}_n\}$ of $\mathcal{W}$ and $x_{jk} \in \mathbb{K}$ for $j, k \in \{2, \ldots, n\}$, $j \leq k$, such that

$$\begin{aligned}
g(\mathbf{v}_2) &= x_{22}\mathbf{v}_2 \\
g(\mathbf{v}_3) &= x_{23}\mathbf{v}_2 + x_{33}\mathbf{v}_3
\end{aligned}$$

$$\vdots$$

$$g(\mathbf{v}_n) = x_{2n}\mathbf{v}_2 + \cdots + x_{nn}\mathbf{v}_n,$$

0 where $x_{22} = \lambda_2, \ldots, x_{nn} = \lambda_n$. Consequently, there are $x_{12}, \ldots, x_{1n} \in \mathbb{K}$ such that

$$\begin{aligned}
f(\mathbf{v}_1) &= x_{11}\mathbf{v}_1 \\
f(\mathbf{v}_2) &= x_{12}\mathbf{v}_1 + g(\mathbf{v}_2) = x_{12}\mathbf{v}_1 + x_{22}\mathbf{v}_2
\end{aligned}$$

$$\vdots$$

$$f(\mathbf{v}_n) = x_{1n}\mathbf{v}_1 + g(\mathbf{v}_n) = x_{1n}\mathbf{v}_1 + x_{2n}\mathbf{v}_1 + \cdots + x_{n-1,n}\mathbf{v}_{n-1} + x_{nn}\mathbf{v}_n,$$

where $x_{11} = \lambda_1, x_{22} = \lambda_2, \ldots, x_{nn} = \lambda_n$.   □

Definition 4.1.33. Let $\mathcal{V}$ be a vector space and let $f : \mathcal{V} \to \mathcal{V}$ be an endomorphism. A polynomial p is called an *f-annihilator* if $p(f) = \mathbf{0}$.

Note that in Chapter 3 we define the annihilator of a subset in an inner product space. In that context the annihilator is a subspace. The f-annihilator of an endomorphism f is a polynomial.

Every endomorphism f on a finite dimensional vector space has f-annihilators. Indeed, if $\mathcal{V}$ is an n-dimensional vector space, then the dimension of the vector space $\mathcal{L}(\mathcal{V})$ of all endomorphisms $f : \mathcal{V} \to \mathcal{V}$ is n^2. Consequently, if $f \in \mathcal{L}(\mathcal{V})$, then the endomorphisms $\mathrm{Id}, f, f^2, \ldots, f^{n^2}$ are linearly dependent and thus there are numbers $x_0, x_1, \ldots, x_{n^2} \in \mathbb{K}$, not all equal to 0, such that

$$x_0 \, \mathrm{Id} + x_1 f + x_2 f^2 + \cdots + x_{n^2} f^{n^2} = \mathbf{0}.$$

Example 4.1.34. Let V be a vector space and let $f : V \to V$ be an endomorphism. We suppose that $\dim V = 4$ and that $\mathcal{B} = \{\mathbf{v}_1, \mathbf{v}_2, \mathbf{v}_3, \mathbf{v}_4\}$ is a basis of V. Let

$$\begin{bmatrix} \alpha & 1 & 0 & 0 \\ 0 & \alpha & 1 & 0 \\ 0 & 0 & \alpha & 1 \\ 0 & 0 & 0 & \alpha \end{bmatrix}$$

be the $\mathcal{B}$-matrix of f for some $\alpha \in \mathbb{K}$. Show that $(t - \alpha)^4$ is an annihilator of f.

Proof. Since

$$f(\mathbf{v}_1) = \alpha \mathbf{v}_1, \quad f(\mathbf{v}_2) = \mathbf{v}_1 + \alpha \mathbf{v}_2, \quad f(\mathbf{v}_3) = \mathbf{v}_2 + \alpha \mathbf{v}_3, \quad f(\mathbf{v}_4) = \mathbf{v}_3 + \alpha \mathbf{v}_4,$$

which can be written as

$$(f - \alpha)(\mathbf{v}_1) = \mathbf{0}, \quad (f - \alpha)(\mathbf{v}_2) = \mathbf{v}_1, \quad (f - \alpha)(\mathbf{v}_3) = \mathbf{v}_2, \quad (f - \alpha)(\mathbf{v}_4) = \mathbf{v}_3,$$

we successively obtain

$$(f - \alpha)^2(\mathbf{v}_1) = (f - \alpha)^2(\mathbf{v}_2) = \mathbf{0}, \quad (f - \alpha)^2(\mathbf{v}_3) = (f - \alpha)\mathbf{v}_2,$$
$$(f - \alpha)^2(\mathbf{v}_4) = (f - \alpha)(\mathbf{v}_3),$$

$$(f - \alpha)^3(\mathbf{v}_1) = (f - \alpha)^3(\mathbf{v}_2) = (f - \alpha)^3(\mathbf{v}_3) = \mathbf{0}, \quad (f - \alpha)^3(\mathbf{v}_4) = (f - \alpha)^2(\mathbf{v}_3),$$

and finally

$$(f - \alpha)^4(\mathbf{v}_1) = (f - \alpha)^4(\mathbf{v}_2) = (f - \alpha)^4(\mathbf{v}_3) = (f - \alpha)^4(\mathbf{v}_4) = \mathbf{0}.$$

Consequently the polynomial $(t - \alpha)^4$ is an f-annihilator. $\qquad\square$

It is easy to verify that for the endomorphism f from the above example we have $c_f(t) = (t - \alpha)^4$. It is not a coincidence that $(f - \alpha)^4 = \mathbf{0}$. Actually, this is true for many endomorphisms as stated in the next theorem.

If a polynomial p can be written as $p(t) = (\lambda_1 - t) \cdots (\lambda_n - t)$ for some $\lambda_1, \ldots, \lambda_n \in \mathbb{K}$, then we say that p *splits over* $\mathbb{K}$. Note that every polynomial with complex coefficients splits over $\mathbb{C}$, by the Fundamental Theorem of Algebra, but not every polynomial with real coefficients splits over $\mathbb{R}$.

Theorem 4.1.35 (Cayley-Hamilton). *Let V be an n-dimensional vector space. If $f : V \to V$ is an endomorphism such that its characteristic polynomial c_f splits over $\mathbb{K}$, then*

$$c_f(f) = \mathbf{0}.$$

Proof. Let $\{\mathbf{v}_1, \ldots, \mathbf{v}_n\}$ be a basis of $\mathcal{V}$ such that

$$f(\mathbf{v}_1) = x_{11}\mathbf{v}_1$$
$$f(\mathbf{v}_2) = x_{12}\mathbf{v}_1 + x_{22}\mathbf{v}_2$$

$$\vdots$$

$$f(\mathbf{v}_n) = x_{1n}\mathbf{v}_1 + x_{2n}\mathbf{v}_1 + \cdots + x_{n-1,n}\mathbf{v}_{n-1} + x_{nn}\mathbf{v}_n,$$

where $x_{jk} \in \mathbb{K}$ for all $j, k \in \{1, \ldots, n\}$ such that $j \leq k$. Then $c_f(t) = (x_{11} - t) \cdots (x_{nn} - t)$, by Theorem 4.1.31. We need to show that $c_f(f) = (x_{11} - f) \cdots (x_{nn} - f) = \mathbf{0}$. We will show by induction that

$$(x_{11} - f) \cdots (x_{jj} - f)(\mathbf{x}) = \mathbf{0}$$

for every $\mathbf{x} \in \mathrm{Span}\{\mathbf{v}_1, \ldots, \mathbf{v}_j\}$.

Clearly,

$$(x_{11} - f)(\mathbf{x}) = \mathbf{0}$$

for every vector $\mathbf{x} \in \mathrm{Span}\{\mathbf{v}_1\}$. Now suppose that for some $j \in \{1, \ldots, n-1\}$ we have

$$(x_{11} - f) \cdots (x_{jj} - f)(\mathbf{x}) = \mathbf{0}$$

for every $\mathbf{x} \in \mathrm{Span}\{\mathbf{v}_1, \ldots, \mathbf{v}_j\}$. Since

$$(x_{j+1,j+1} - f)(\mathbf{v}_{j+1}) = x_{j+1,j+1}\mathbf{v}_{j+1} - x_{1,j+1}\mathbf{v}_1 - x_{2,j+1}\mathbf{v}_2 - \cdots - x_{j,j+1}\mathbf{v}_j$$
$$- x_{j+1,j+1}\mathbf{v}_{j+1}$$
$$= -x_{1,j+1}\mathbf{v}_1 - x_{2,j+1}\mathbf{v}_2 - \cdots - x_{j,j+1}\mathbf{v}_j,$$

we have

$$(x_{11} - f) \cdots (x_{jj} - f)(x_{j+1,j+1} - f)(\mathbf{v}_{j+1})$$
$$= (x_{11} - f) \cdots (x_{jj} - f)(-x_{1,j+1}\mathbf{v}_1 - x_{2,j+1}\mathbf{v}_2 - \cdots$$
$$- x_{j,j+1}\mathbf{v}_j) = \mathbf{0},$$

by the inductive assumption. Consequently, $c_f(f)(\mathbf{x}) = (x_{11} - f) \cdots (x_{nn} - f)(\mathbf{x}) = \mathbf{0}$ for every $\mathbf{x} \in \mathcal{V}$. $\square$

Example 4.1.36. Verify Cayley-Hamilton theorem for the endomorphism f in Example 4.1.34.

Solution. Let $\mathbf{D} : \mathcal{V}^4 \to \mathbb{K}$ be a nonzero alternating multilinear form. Then

$$\mathbf{D}(f(\mathbf{v}_1) - t\mathbf{v}_1, f(\mathbf{v}_2) - t\mathbf{v}_2, f(\mathbf{v}_3) - t\mathbf{v}_3, f(\mathbf{v}_4) - t\mathbf{v}_4)$$
$$= \mathbf{D}(\alpha\mathbf{v}_1 - t\mathbf{v}_1, \mathbf{v}_1 + \alpha\mathbf{v}_2 - t\mathbf{v}_2, \mathbf{v}_2 + \alpha\mathbf{v}_3 - t\mathbf{v}_3, \mathbf{v}_3 + \alpha\mathbf{v}_4 - t\mathbf{v}_4)$$
$$= \mathbf{D}(\alpha\mathbf{v}_1 - t\mathbf{v}_1, \alpha\mathbf{v}_2 - t\mathbf{v}_2, \alpha\mathbf{v}_3 - t\mathbf{v}_3, \alpha\mathbf{v}_4 - t\mathbf{v}_4)$$
$$= (\alpha - t)(\alpha - t)(\alpha - t)(\alpha - t)\mathbf{D}(\mathbf{v}_1, \mathbf{v}_2, \mathbf{v}_3, \mathbf{v}_4)$$
$$= (\alpha - t)^4 \mathbf{D}(\mathbf{v}_1, \mathbf{v}_2, \mathbf{v}_3, \mathbf{v}_4).$$

This shows that $c_f(t) = (\alpha - t)^4$. Thus $c_f(f) = \mathbf{0}$, by Example 4.1.34. □

Example 4.1.37. Verify Cayley-Hamilton Theorem for the endomorphism $f : \mathbb{R}^2 \to \mathbb{R}^2$ defined by

$$f\left(\begin{bmatrix} x \\ y \end{bmatrix}\right) = \begin{bmatrix} 2 & -1 \\ 1 & 5 \end{bmatrix} \begin{bmatrix} x \\ y \end{bmatrix}.$$

Solution. Let $\mathbf{D} : \mathbb{R}^2 \times \mathbb{R}^2 \to \mathbb{R}$ be a nonzero alternating bilinear form. Proceeding as in Example 4.1.5 we obtain

$$\mathbf{D}\left(\begin{bmatrix} 2-t \\ -1 \end{bmatrix}, \begin{bmatrix} 1 \\ 5-t \end{bmatrix}\right) = (t^2 - 7t + 11)\mathbf{D}\left(\begin{bmatrix} 1 \\ 0 \end{bmatrix}, \begin{bmatrix} 0 \\ 1 \end{bmatrix}\right).$$

Hence

$$c_f(t) = t^2 - 7t + 11$$

and we have

$$\begin{bmatrix} 2 & -1 \\ 1 & 5 \end{bmatrix}^2 - 7\begin{bmatrix} 2 & -1 \\ 1 & 5 \end{bmatrix} + 11\begin{bmatrix} 1 & 0 \\ 0 & 1 \end{bmatrix} = \begin{bmatrix} 3 & -7 \\ 7 & 24 \end{bmatrix} + \begin{bmatrix} -14 & 7 \\ -7 & -35 \end{bmatrix} + \begin{bmatrix} 11 & 0 \\ 0 & 11 \end{bmatrix} = \begin{bmatrix} 0 & 0 \\ 0 & 0 \end{bmatrix}.$$

□

Example 4.1.38. Let $f : \mathcal{V} \to \mathcal{V}$ be an endomorphism. We suppose that $t^3 + t$ is a f-annihilator. Show that, if λ is an eigenvalue of f, then $\lambda \in \{0, -i, i\}$.

Solution. Let $\mathbf{x}$ be an eigenvector corresponding to an eigenvalue λ of f. Then

$$(f^3 + f)(\mathbf{x}) = (\lambda^3 + \lambda)\mathbf{x} = \mathbf{0}.$$

Since $\mathbf{x} \neq \mathbf{0}$, we have $\lambda^3 + \lambda = 0$, which gives us the desired result. □

Theorem 4.1.39. *Let $\mathcal{V}$ be a finite dimensional vector space and let $f : \mathcal{V} \to \mathcal{V}$ be a nonzero endomorphism.*

(a) *There is a unique monic polynomial m_f of smallest positive degree such that $m_f(f) = \mathbf{0}$;*

(b) *m_f divides every f-annihilator;*

(c) *If λ is an eigenvalue of f then $m_f(\lambda) = 0$.*

Proof. Note that c_f is an f-annihilator. Recall that a monic polynomial is a single-variable polynomial in which the leading coefficient is equal to 1.

Let p be a monic polynomial of smallest positive degree such that $p(f) = \mathbf{0}$ and let q be a polynomial of positive degree such that $q(f) = \mathbf{0}$. Then $q = pa + r$, where a and r are polynomials and $r = 0$ or the degree of r is strictly less than the degree of p. Since

$$q(f) = p(f)a(f) + r(f),$$

we have $r(f) = \mathbf{0}$. Now, because p is a polynomial of smallest positive degree such that $p(f) = \mathbf{0}$, we have $r = 0$ and thus p divides q.

If q is another monic polynomial of smallest positive degree such that $q(f) = \mathbf{0}$, then the equality $q = pa$ implies $a = 1$ and consequently $p = q$.

Finally, if $\mathbf{v}$ is an eigenvector corresponding to the eigenvalue λ, then

$$m_f(f)(\mathbf{v}) = m_f(\lambda)\mathbf{v},$$

which gives us $m_f(\lambda) = 0$ because $\mathbf{v} \neq \mathbf{0}$.   □

Definition 4.1.40. Let $\mathcal{V}$ be a finite dimensional vector space and let $f : \mathcal{V} \to \mathcal{V}$ be a nonzero endomorphism. The unique monic polynomial m_f of smallest positive degree such that $m_f(f) = \mathbf{0}$ is called the *minimal polynomial* of f.

Example 4.1.41. Let $\mathcal{V}$ be a vector space such that $\dim \mathcal{V} = 4$ and let $f : \mathcal{V} \to \mathcal{V}$ be an endomorphism. If $\mathcal{B} = \{\mathbf{v}_1, \mathbf{v}_2, \mathbf{v}_3, \mathbf{v}_4\}$ is a basis of $\mathcal{V}$ and

$$\begin{bmatrix} \alpha & 1 & 0 & 0 \\ 0 & \alpha & 1 & 0 \\ 0 & 0 & \alpha & 1 \\ 0 & 0 & 0 & \alpha \end{bmatrix}$$

is the $\mathcal{B}$-matrix of f for some $\alpha \in \mathbb{K}$, find m_f.

Solution. From Example 4.1.34 we know that the polynomial $(t - \alpha)^4$ is an annihilator of f. We note that $(t - \alpha)^4$ is the minimal polynomial m_f because

$$(f - \alpha)^3(\mathbf{v}_4) = (f - \alpha)^2(\mathbf{v}_3) = (f - \alpha)(\mathbf{v}_2) = \mathbf{v}_1 \neq \mathbf{0}.$$

□

Theorem 4.1.42. *Let $\mathcal{V}_1$ and $\mathcal{V}_2$ be finite dimensional vector spaces and let $\mathcal{V} = \mathcal{V}_1 \oplus \mathcal{V}_2$. If $f_1 : \mathcal{V}_1 \to \mathcal{V}_1$ and $f_2 : \mathcal{V}_2 \to \mathcal{V}_2$ are nonzero endomorphisms and $f : \mathcal{V} \to \mathcal{V}$ is the endomorphism defined by*

$$f(\mathbf{v}_1 + \mathbf{v}_2) = f_1(\mathbf{v}_1) + f_2(\mathbf{v}_2)$$

where $\mathbf{v}_1 \in \mathcal{V}_1$ and $\mathbf{v}_2 \in \mathcal{V}_2$, then

$$m_f = \mathbf{LCM}(m_{f_1}, m_{f_2}).$$

Proof. Let $p = \mathbf{LCM}(m_{f_1}, m_{f_2})$ and let $\mathbf{v}_1 \in \mathcal{V}_1$ and $\mathbf{v}_2 \in \mathcal{V}_2$. Then

$$p(f)(\mathbf{v}_1 + \mathbf{v}_2) = p(f)(\mathbf{v}_1) + p(f)(\mathbf{v}_2) = p(f_1)(\mathbf{v}_1) + p(f_2)(\mathbf{v}_2) = \mathbf{0},$$

because both m_{f_1} and m_{f_2} divide p. This implies that m_f divides p.

Now, since $m_f(f)(\mathbf{v}) = m_f(f_1)(\mathbf{v}) = \mathbf{0}$ for every $\mathbf{v} \in \mathcal{V}_1$, m_{f_1} divides m_f. Similarly, m_{f_2} divides m_f. Consequently p divides m_f. $\square$

Theorem 4.1.43. *Let $\mathcal{V}_1$ and $\mathcal{V}_2$ be finite dimensional vector spaces and let $\mathcal{V} = \mathcal{V}_1 \oplus \mathcal{V}_2$. If $f_1 : \mathcal{V}_1 \to \mathcal{V}_1$ and $f_2 : \mathcal{V}_2 \to \mathcal{V}_2$ are endomorphisms and $f : \mathcal{V} \to \mathcal{V}$ is the endomorphism defined by*

$$f(\mathbf{v}_1 + \mathbf{v}_2) = f_1(\mathbf{v}_1) + f_2(\mathbf{v}_2)$$

where $\mathbf{v}_1 \in \mathcal{V}_1$ and $\mathbf{v}_2 \in \mathcal{V}_2$, then

$$c_f = c_{f_1} c_{f_2}.$$

Proof. This result is a consequence of Theorem 4.1.24. $\square$

The following result is of significant importance for the remainder of this chapter.

Theorem 4.1.44. *Let $\mathcal{V}$ be a finite dimensional vector space and let $f : \mathcal{V} \to \mathcal{V}$ be an endomorphism. If $p_1, \ldots, p_k$ are polynomials such that $\mathbf{GCD}(p_j, p_l) = 1$ for every $j, l \in \{1, \ldots, k\}$ such that $j \neq l$ and the product $p_1 \cdots p_k$ is an f-annihilator, then*

$$\mathcal{V} = \ker p_1(f) \oplus \cdots \oplus \ker p_k(f).$$

Proof. Let $q_j = p_1 \cdots p_{j-1} p_{j+1} \cdots p_k$ for $j \in \{1, \ldots, k\}$. Clearly $\mathbf{GCD}(q_1, \ldots, q_k) = 1$ and thus there are polynomials $a_1, \ldots, a_k$ such that

$$a_1 q_1 + \cdots + a_k q_k = 1.$$

Consequently,

$$a_1(f)q_1(f) + \cdots + a_k(f)q_k(f) = \mathrm{Id}$$

and thus

$$a_1(f)q_1(f)(\mathbf{v}) + \cdots + a_k(f)q_k(f)(\mathbf{v}) = \mathbf{v} \qquad (4.2)$$

for every $\mathbf{v} \in V$. Note that for every $j \in \{1, \ldots, n\}$ we have

$$p_j(f)a_j(f)q_j(f)(\mathbf{v}) = a_j(f)p_j(f)q_j(f)(\mathbf{v}) = a_j(f)p_1(f) \cdots p_k(f)(\mathbf{v}) = \mathbf{0},$$

and thus $a_j(f)q_j(f)(\mathbf{v}) \in \ker p_j(f)$. Hence, by (4.2), we have

$$V = \ker p_1(f) + \cdots + \ker p_k(f).$$

We need to show that the sum is direct.

Let $\mathbf{v}_j \in \ker p_j(f)$ for $j \in \{1, \ldots, n\}$. From (4.2) we get

$$a_1(f)q_1(f)(\mathbf{v}_j) + \cdots + a_k(f)q_k(f)(\mathbf{v}_j) = \mathbf{v}_j.$$

Since $a_l(f)q_l(f)(\mathbf{v}_j) = \mathbf{0}$ for $l \in \{1, \ldots, j-1, j+1, \ldots, k\}$, we have

$$a_1(f)q_1(f)(\mathbf{v}_j) + \cdots + a_k(f)q_k(f)(\mathbf{v}_j) = a_j(f)q_j(f)(\mathbf{v}_j)$$

and thus

$$\mathbf{v}_j = a_j(f)q_j(f)(\mathbf{v}_j). \qquad (4.3)$$

Suppose now that the vectors $\mathbf{v}_j \in \ker p_j(f)$, $j \in \{1, \ldots, k\}$, are such that

$$\mathbf{v}_1 + \cdots + \mathbf{v}_k = \mathbf{0}.$$

Then for every $j \in \{1, \ldots, k\}$ we have

$$\mathbf{0} = a_j(f)q_j(f)(\mathbf{0}) = a_j(f)q_j(f)(\mathbf{v}_1 + \cdots + \mathbf{v}_k) = a_j(f)q_j(f)(\mathbf{v}_j)$$

because $a_j(f)q_j(f)(\mathbf{v}_l) = \mathbf{0}$ for $l \in \{1, \ldots, j-1, j+1, \ldots, k\}$. Hence, by (4.3), we have $\mathbf{v}_j = a_j(f)q_j(f)(\mathbf{v}_j) = \mathbf{0}$, which shows that the sum $\ker p_1(f) + \cdots + \ker p_k(f)$ is direct. $\qquad\qquad\square$

Invariance of a subspace with respect to a linear operator was initially introduced in Chapter 2 in exercises and then repeated in Chapter 3. For convenience we recall that a subspace U of a vector space V is called f-*invariant*, where $f : V \to V$ is an endomorphism, if $f(U) \subseteq U$. Invariance of subspaces will play an important role in this chapter.

It is easy to see that, if V is a vector space and $f : V \to V$ is an endomorphism, then $\ker f$ and $\mathrm{ran}\, f$ are f-invariant subspaces. More generally, if p is a polynomial, then $\ker p(f)$ is an f-invariant subspace.

Theorem 4.1.45. *Let V be an n-dimensional vector space and let $f : V \to V$ be an endomorphism. If*

$$c_f(t) = (\lambda_1 - t)^{r_1} \cdots (\lambda_k - t)^{r_k}$$

for some distinct $\lambda_1, \ldots, \lambda_k \in \mathbb{K}$ and some positive integers $r_1, \ldots, r_k$, then

$$V = \ker(f - \lambda_1)^{r_1} \oplus \cdots \oplus \ker(f - \lambda_k)^{r_k}$$

and

$$c_{f_j}(t) = (t - \lambda_j)^{r_j}$$

for every $j \in \{1, \ldots, k\}$ where $f_j : \ker(f - \lambda_j)^{r_j} \to \ker(f - \lambda_j)^{r_j}$ is the endomorphism induced by f.

Proof. From Theorem 4.1.44 applied to the polynomials $(\lambda_1 - t)^{r_1}, \ldots, (\lambda_k - t)^{r_k}$ we get

$$V = \ker(f - \lambda_1)^{r_1} \oplus \cdots \oplus \ker(f - \lambda_k)^{r_k}.$$

Now, for every $j \in \{1, \ldots, k\}$, the polynomial $(\lambda_1 - t)^{r_1}$ is an f_j-annihilator. This implies, by Theorem 4.1.39, that $m_{f_j} = (t - \lambda_1)^{q_j}$ where q_j is an integer such that $1 \leq q_j \leq r_j$ and $c_{f_j}(t) = (\lambda_j - t)^{s_j}$ where s_j is an integer such that $q_j \leq s_j$.

From Theorem 4.1.43 we get

$$c_f = c_{f_1} \ldots c_{f_k},$$

that is,

$$(\lambda_1 - t)^{r_1} \cdots (\lambda_k - t)^{r_k} = (\lambda_1 - t)^{s_1} \cdots (\lambda_k - t)^{s_k}$$

Hence $r_j = s_j$ for every $j \in \{1, \ldots, k\}$, completing the proof. $\square$

Corollary 4.1.46. *Let V be an n-dimensional vector space and let $f : V \to V$ be an endomorphism. If*

$$c_f(t) = (\lambda_1 - t)^{r_1} \cdots (\lambda_k - t)^{r_k}$$

for some distinct $\lambda_1, \ldots, \lambda_k \in \mathbb{K}$ and some positive integers $r_1, \ldots, r_k$, then for every $j \in \{1, \ldots, k\}$ we have

$$\dim \ker(f - \lambda_j)^{r_j} = r_j.$$

Proof. This follows from the fact that

$$\deg c_{f_j}(t) = \deg(t - \lambda_j)^{r_j} = r_j$$

for every $j \in \{1, \ldots, k\}$. $\square$

Definition 4.1.47. An endomorphism $f : V \to V$ is called diagonalizable if V has a basis consisting of eigenvectors of f.

If V is an n-dimensional vector space and an endomorphism $f : V \to V$ is diagonalizable, then that is there is a basis $\mathcal{B} = \{\mathbf{v}_1, \ldots, \mathbf{v}_n\}$ of V and $\lambda_1, \ldots, \lambda_n \in \mathbb{K}$ such that $f(\mathbf{v}_j) = \lambda_j \mathbf{v}_j$ for every $j \in \{1, \ldots, n\}$. In other words, the $\mathcal{B}$-matrix of f is diagonal:

$$\begin{bmatrix} \lambda_1 & 0 & \cdots & 0 \\ 0 & \lambda_2 & \cdots & 0 \\ \vdots & \vdots & \ddots & \vdots \\ 0 & 0 & \cdots & \lambda_n \end{bmatrix}.$$

The following result is a direct consequence of the definitions.

Theorem 4.1.48. *Let V be an n-dimensional vector space. An endomorphism $f : V \to V$ is diagonalizable if and only if it has n linearly independent eigenvectors.*

If λ is an eigenvalue of an endomorphism $f : V \to V$, then the characteristic polynomial of f can be written as

$$\det(f - t) = (\lambda - t)^r q(t),$$

where $q(t)$ is a polynomial such that $q(\lambda) \neq 0$. The number r is called the *algebraic multiplicity* of the eigenvalue λ.

The dimension of the eigenspace of f corresponding to an eigenvalue λ, that is $\dim \ker(f - \lambda)$, is called the *geometric multiplicity* of λ.

The algebraic multiplicity of an endomorphism need not be the same as the geometric multiplicity. Indeed, consider the endomorphism $f : \mathbb{R}^3 \to \mathbb{R}^3$ defined as $f(\mathbf{x}) = A\mathbf{x}$ where

$$A = \begin{bmatrix} 3 & 1 & 1 \\ 0 & 3 & 0 \\ 0 & 0 & 3 \end{bmatrix}.$$

Since $\det(f - t) = (3 - t)^3$, f has only one eigenvalue $\lambda = 3$ whose algebraic multiplicity is 3. On the other hand, since the dimension of the null space of the matrix

$$\begin{bmatrix} 0 & 1 & 1 \\ 0 & 0 & 0 \\ 0 & 0 & 0 \end{bmatrix}$$

is 2, the geometric multiplicity of the eigenvalue 3 is 2.

> **Theorem 4.1.49.** *The geometric multiplicity is less than or equal to the algebraic multiplicity.*

Proof. Let V be an n-dimensional vector space and let $f : V \to V$ be an endomorphism. If $\dim \ker(f - \lambda) = k$, then there is a basis $\{\mathbf{v}_1, \dots, \mathbf{v}_n\}$ of V such that $\mathbf{v}_1, \dots, \mathbf{v}_k$ are eigenvectors of f corresponding to the eigenvalue λ.

For any nonzero n-linear alternating form $\mathbf{D} : V^n \to \mathbb{K}$ we have

$$
\begin{aligned}
c_f&(t)\mathbf{D}(\mathbf{v}_1, \dots, \mathbf{v}_k, \mathbf{v}_{k+1}, \dots, \mathbf{v}_n) \\
&= \mathbf{D}(f(\mathbf{v}_1) - t\mathbf{v}_1, \dots, f(\mathbf{v}_k) - t\mathbf{v}_k, f(\mathbf{v}_{k+1}) - t\mathbf{v}_{k+1}, \dots, f(\mathbf{v}_n) - t\mathbf{v}_n) \\
&= \mathbf{D}(\lambda\mathbf{v}_1 - t\mathbf{v}_1, \dots, \lambda\mathbf{v}_k - t\mathbf{v}_k, f(\mathbf{v}_{k+1}) - t\mathbf{v}_{k+1}, \dots, f(\mathbf{v}_n) - t\mathbf{v}_n) \\
&= (\lambda - t)^k \mathbf{D}(\mathbf{v}_1, \dots, \mathbf{v}_k, f(\mathbf{v}_{k+1}) - t\mathbf{v}_{k+1}, \dots, f(\mathbf{v}_n) - t\mathbf{v}_n).
\end{aligned}
$$

Consequently, $(\lambda - t)^k$ divides c_f because

$$
\mathbf{D}(\mathbf{v}_1, \dots, \mathbf{v}_k, f(\mathbf{v}_{k+1}) - t\mathbf{v}_{k+1}, \dots, f(\mathbf{v}_n) - t\mathbf{v}_n) = q(t)\mathbf{D}(\mathbf{v}_1, \dots, \mathbf{v}_k, \mathbf{v}_{k+1}, \dots, \mathbf{v}_n),
$$

where q is a polynomial. $\qquad\square$

In Theorem 3.4.10 we show that eigenvectors corresponding to different eigenvalues of a normal operator on an inner product space V are orthogonal. If V is an arbitrary vector space, then we cannot talk about orthogonality of eigenvectors, but we still have linear independence of eigenvectors corresponding to different eigenvalues, as the following theorem states.

> **Theorem 4.1.50.** *Let V be a vector space and let $f : V \to V$ be an endomorphism. If $\mathbf{v}_1, \dots, \mathbf{v}_k \in V$ are eigenvectors of f corresponding to distinct eigenvalues $\lambda_1, \dots, \lambda_k$, then the vectors $\mathbf{v}_1, \dots, \mathbf{v}_k$ are linearly independent.*

First proof. If

$$
x_1\mathbf{v}_1 + \dots + x_k\mathbf{v}_k = \mathbf{0}
$$

for some $x_1, \dots, x_k \in \mathbb{K}$, then

$$
(f - \lambda_1) \cdots (f - \lambda_{k-1})(x_1\mathbf{v}_1 + \dots + x_k\mathbf{v}_k) = \mathbf{0}
$$

and thus

$$
x_k(f - \lambda_1) \cdots (f - \lambda_{k-1})(\mathbf{v}_k) = \mathbf{0}.
$$

Since

$$
\begin{aligned}
(f &- \lambda_1) \cdots (f - \lambda_{k-1})(\mathbf{v}_k) \\
&= ((f - \lambda_k) + (\lambda_k - \lambda_1)) \cdots ((f - \lambda_k) + (\lambda_k - \lambda_{k-1})))(\mathbf{v}_k) \\
&= (\lambda_k - \lambda_1) \cdots (\lambda_k - \lambda_{k-1})\mathbf{v}_k,
\end{aligned}
$$

we get

$$x_k(\lambda_k - \lambda_1) \cdots (\lambda_k - \lambda_{k-1})\mathbf{v}_k = \mathbf{0}.$$

Consequently $x_k = 0$, because $(\lambda_k - \lambda_1) \cdots (\lambda_k - \lambda_{k-1}) \neq 0$ and $\mathbf{v}_k \neq \mathbf{0}$. In the same way we can show that $x_1 = \cdots = x_{k-1} = 0$.   □

Second proof. Since the subspace $\mathrm{Span}\{\mathbf{v}_1, \ldots, \mathbf{v}_k\}$ is f-invariant, without loss of generality we can assume that $V = \mathrm{Span}\{\mathbf{v}_1, \ldots, \mathbf{v}_k\}$. Then the eigenvalues are roots of the polynomial c_f which has the degree equal to $\dim V$ and consequently $k \leq \dim V$. But $V = \mathrm{Span}\{\mathbf{v}_1, \ldots, \mathbf{v}_k\}$, so we must have $k \geq \dim V$. Thus $k = \dim V$ and the vectors $\mathbf{v}_1, \ldots, \mathbf{v}_k$ are linearly independent.   □

Theorem 4.1.51. *Let V be an n-dimensional vector space and let $f : V \to V$ be an endomorphism. The following conditions are equivalent:*

(a) *f is diagonalizable;*

(b) *There are distinct $\lambda_1, \ldots, \lambda_k \in \mathbb{K}$ and positive integers $r_1, \ldots, r_k$ such that*

$$c_f(t) = (\lambda_1 - t)^{r_1} \cdots (\lambda_k - t)^{r_k} \quad and \quad \dim \ker(f - \lambda_j) = r_j;$$

(c) *If $\lambda_1, \ldots, \lambda_k$ are all distinct eigenvalues of f, then*

$$\ker(f - \lambda_1) \oplus \cdots \oplus \ker(f - \lambda_k) = V;$$

(d) *If $\lambda_1, \ldots, \lambda_k$ are all distinct eigenvalues of f, then*

$$n = \dim \ker(f - \lambda_1) + \cdots + \dim \ker(f - \lambda_k);$$

(e) *If $\lambda_1, \ldots, \lambda_k$ are all distinct eigenvalues of f, then*

$$m_f(t) = (t - \lambda_1) \cdots (t - \lambda_k).$$

Proof. First we prove equivalence of (a) and (b). Then we show that (c) implies (d), (d) implies (a), (a) implies (e), and (e) implies (c).

If $f : V \to V$ is diagonalizable, then there is a basis $\{\mathbf{v}_1, \ldots, \mathbf{v}_n\}$ of V such that

$$f(\mathbf{v}_j) = \lambda_j \mathbf{v}_j$$

for some $\lambda_1, \ldots, \lambda_n \in \mathbb{K}$ and all $j \in \{1, \ldots, n\}$. If $\mathbf{D} : V^n \to \mathbb{K}$ is a nonzero n-linear alternating form, then

$$\mathbf{D}(f(\mathbf{v}_1) - t\mathbf{v}_1, \ldots, f(\mathbf{v}_n) - t\mathbf{v}_n) = \mathbf{D}((\lambda_1 - t)\mathbf{v}_1, \ldots, (\lambda_n - t)\mathbf{v}_n)$$
$$= (\lambda_1 - t) \cdots (\lambda_n - t)\mathbf{D}(\mathbf{v}_1, \ldots, \mathbf{v}_n)$$

and thus $c_f(t) = (\lambda_1 - t) \cdots (\lambda_n - t)$. Without loss of generality, we can suppose that $\lambda_1, \ldots, \lambda_k$ are distinct numbers such that $\{\lambda_{k+1}, \ldots, \lambda_n\} \subseteq \{\lambda_1, \ldots, \lambda_k\}$. Consequently,

$$c_f(t) = (\lambda_1 - t)^{r_1} \ldots (\lambda_k - t)^{r_k}$$

where $r_1, \ldots, r_k$ are positive integers such that $r_1 + \cdots + r_k = n$.

Because for every $l \in \{1, \ldots, n\}$ there is a $j \in \{1, \ldots, k\}$ such that $\mathbf{v}_l \in \ker(f - \lambda_j)$, we have

$$\ker(f - \lambda_1) + \cdots + \ker(f - \lambda_k) = \mathcal{V}.$$

Now since for every $j \in \{1, \ldots, k\}$ we have $\ker(f - \lambda_1) \subseteq \ker(f - \lambda_j)^{r_j}$ and, by Corollary 4.1.46, $\dim \ker(f - \lambda_j)^{r_j} = r_j$, we conclude that

$$\dim \ker(f - \lambda_j) = r_j$$

for every $j \in \{1, \ldots, k\}$. Thus (a) implies (b).

Now suppose that we can write

$$c_f(t) = (\lambda_1 - t)^{r_1} \ldots (\lambda_k - t)^{r_k}$$

where $\lambda_1, \ldots, \lambda_k \in \mathbb{K}$ are distinct and $r_1, \ldots, r_k$ are positive integers such that $\dim \ker(f - \lambda_j) = r_j$ for every $j \in \{1, \ldots, k\}$. Then, by Corollary 4.1.46, $\ker(f - \lambda_j) = \ker(f - \lambda_j)^{r_j}$ and, by Theorem 4.1.45, the sum

$$\ker(f - \lambda_1) + \cdots + \ker(f - \lambda_k)$$

is direct and we have

$$\ker(f - \lambda_1) \oplus \cdots \oplus \ker(f - \lambda_k) = \mathcal{V}.$$

Consequently, if $\mathcal{B}_j$ is a basis of $\ker(f - \lambda_j)$ for $j \in \{1, \ldots, k\}$, then $\mathcal{B}_1 \cup \cdots \cup \mathcal{B}_k$ is a basis of $\mathcal{V}$ because $r_1 + \cdots + r_k = n$. Since all elements of $\mathcal{B}_1 \cup \cdots \cup \mathcal{B}_k$ are eigenvectors, f is diagonalizable. Thus (b) implies (a).

Clearly (c) implies (d).

Next suppose that

$$\dim \ker(f - \lambda_1) + \cdots + \dim \ker(f - \lambda_k) = n.$$

Since $\ker(f - \lambda_j) \subseteq \ker(f - \lambda_j)^{r_j}$, it follows from Theorem 4.1.45 that the sum

$$\ker(f - \lambda_1) + \cdots + \ker(f - \lambda_k)$$

is direct. Now we construct a basis of eigenvectors of f as in the proof of the first part ((b) implies (a)). This proves that (d) implies (a).

Next assume that the endomorphism f satisfies (a). Since $(t - \lambda_1) \ldots (t - \lambda_k)$ is an annihilator of f and m_f is a monic polynomial which has all distinct eigenvalues as zeros and divides every annihilator of f, we have

$$m_f(t) = (t - \lambda_1) \ldots (t - \lambda_k),$$

so (a) implies (e).

Finally, (e) implies (c), because if $m_f(t) = (t - \lambda_1) \dots (t - \lambda_k)$, then

$$\ker(f - \lambda_1) \oplus \cdots \oplus \ker(f - \lambda_k) = \mathcal{V},$$

by Theorem 4.1.44.   □

Example 4.1.52. Let $f : \mathbb{R}^2 \to \mathbb{R}^2$ be the endomorphism defined by

$$f\left(\begin{bmatrix} x \\ y \end{bmatrix}\right) = \begin{bmatrix} 1 & 4 \\ -1 & 5 \end{bmatrix} \begin{bmatrix} x \\ y \end{bmatrix}.$$

Show that f is not diagonalizable.

Proof. Let $\mathbf{D} : \mathbb{R}^2 \times \mathbb{R}^2 \to \mathbb{R}$ be a nonzero alternating bilinear form. Proceeding as in Example 4.1.5 we obtain

$$\mathbf{D}\left(\begin{bmatrix} 1-t \\ 4 \end{bmatrix}, \begin{bmatrix} -1 \\ 5-t \end{bmatrix}\right) = (((1-t)(5-t) + 4)\mathbf{D}\left(\begin{bmatrix} 1 \\ 0 \end{bmatrix}, \begin{bmatrix} 0 \\ 1 \end{bmatrix}\right) = (t-3)^2 \mathbf{D}\left(\begin{bmatrix} 1 \\ 0 \end{bmatrix}, \begin{bmatrix} 0 \\ 1 \end{bmatrix}\right).$$

Hence $c_f(t) = (t-3)^2$.

Next we determine the eigenvectors corresponding to the eigenvalue 3. The equation

$$\begin{bmatrix} 1 & 4 \\ -1 & 5 \end{bmatrix} \begin{bmatrix} x \\ y \end{bmatrix} = 3 \begin{bmatrix} x \\ y \end{bmatrix}$$

is equivalent to the equation $x = 2y$. Consequently, the eigenspace corresponding to the eigenvalue 3 is

$$\mathcal{E}_3 = \mathrm{Span}\left\{\begin{bmatrix} 2 \\ 1 \end{bmatrix}\right\}.$$

Because $\dim \mathcal{E}_3 < 2$, the endomorphism f is not diagonalizable.   □

Example 4.1.53. Find f^n if $f : \mathbb{R}^2 \to \mathbb{R}^2$ is the endomorphism defined by

$$f\left(\begin{bmatrix} x \\ y \end{bmatrix}\right) = \begin{bmatrix} 4 & 1 \\ 2 & 3 \end{bmatrix} \begin{bmatrix} x \\ y \end{bmatrix}.$$

Proof. The eigenvalues are given by the equation $(4-t)(3-t) - 2 = 0$ and are 2 and 5. It is easy to verify that $\mathcal{E}_2 = \mathrm{Span}\left\{\begin{bmatrix} 1 \\ -2 \end{bmatrix}\right\}$ and $\mathcal{E}_5 = \mathrm{Span}\left\{\begin{bmatrix} 1 \\ 1 \end{bmatrix}\right\}$.

Since

$$\begin{bmatrix} 4 & 1 \\ 2 & 3 \end{bmatrix} \begin{bmatrix} 1 & 1 \\ -2 & 1 \end{bmatrix} = \begin{bmatrix} 1 & 1 \\ -2 & 1 \end{bmatrix} \begin{bmatrix} 2 & 0 \\ 0 & 5 \end{bmatrix},$$

we have

$$\begin{bmatrix} 4 & 1 \\ 2 & 3 \end{bmatrix} = \begin{bmatrix} 1 & 1 \\ -2 & 1 \end{bmatrix} \begin{bmatrix} 2 & 0 \\ 0 & 5 \end{bmatrix} \begin{bmatrix} 1 & 1 \\ -2 & 1 \end{bmatrix}^{-1}$$

and thus

$$\begin{bmatrix} 4 & 1 \\ 2 & 3 \end{bmatrix}^n = \begin{bmatrix} 1 & 1 \\ -2 & 1 \end{bmatrix} \begin{bmatrix} 2^n & 0 \\ 0 & 5^n \end{bmatrix} \begin{bmatrix} 1 & 1 \\ -2 & 1 \end{bmatrix}^{-1} = \frac{1}{3} \begin{bmatrix} 1 & 1 \\ -2 & 1 \end{bmatrix} \begin{bmatrix} 2^n & 0 \\ 0 & 5^n \end{bmatrix} \begin{bmatrix} 1 & -1 \\ 2 & 1 \end{bmatrix}.$$

Consequently

$$f^n\left(\begin{bmatrix} x \\ y \end{bmatrix}\right) = \frac{1}{3} \begin{bmatrix} 2^n + 2 \cdot 5^n & -2^n + 5^n \\ -2^{n+1} + 2 \cdot 5^n & 2^{n+1} + 5^n \end{bmatrix} \begin{bmatrix} x \\ y \end{bmatrix}.$$

□

Example 4.1.54. Show that the endomorphism $f : \mathbb{R}^3 \to \mathbb{R}^3$ defined by

$$f\left(\begin{bmatrix} x \\ y \\ z \end{bmatrix}\right) = \begin{bmatrix} 3 & 1 & 2 \\ 1 & 3 & 2 \\ 1 & 1 & 4 \end{bmatrix} \begin{bmatrix} x \\ y \\ z \end{bmatrix}$$

is diagonalizable.

Solution. According to Example 4.1.28 we have $c_f(t) = (2 - t)^2(6 - t)$. It is enough to show that $\dim \ker(f - 2) = 2$. The equation

$$\begin{bmatrix} 3 & 1 & 2 \\ 1 & 3 & 2 \\ 1 & 1 & 4 \end{bmatrix} \begin{bmatrix} x \\ y \\ z \end{bmatrix} = 2 \begin{bmatrix} x \\ y \\ z \end{bmatrix}$$

is equivalent to the equation $x + y + 2z = 0$. Consequently,

$$\begin{bmatrix} x \\ y \\ z \end{bmatrix} = \begin{bmatrix} -y - 2z \\ y \\ z \end{bmatrix} = y \begin{bmatrix} -1 \\ 1 \\ 0 \end{bmatrix} + z \begin{bmatrix} -2 \\ 0 \\ 1 \end{bmatrix},$$

which means that

$$\mathcal{E}_2 = \mathrm{Span}\left\{\begin{bmatrix} -1 \\ 1 \\ 0 \end{bmatrix}, \begin{bmatrix} -2 \\ 0 \\ 1 \end{bmatrix}\right\}.$$

Since the vectors $\begin{bmatrix} -1 \\ 1 \\ 0 \end{bmatrix}, \begin{bmatrix} -2 \\ 0 \\ 1 \end{bmatrix}$ are linearly independent, $\dim \mathcal{E}_2 = 2$.   □

Example 4.1.55. Let V be an n-dimensional vector space and let $f : V \to V$ be an endomorphism such that $m_f(t) = t(t+1)$ and $\dim \ker f = k$. Determine c_f.

Solution. The endomorphism f is diagonalizable by Theorem 4.1.51. The only eigenvalues are 0 and -1. Let $\{\mathbf{v}_1, \ldots, \mathbf{v}_n\}$ be a basis of eigenvectors such that $\{\mathbf{v}_1, \ldots, \mathbf{v}_k\}$ is a basis of $\ker f = \mathcal{E}_0$ and $\{\mathbf{v}_{k+1}, \ldots, \mathbf{v}_n\}$ is a basis of $\mathcal{E}_{-1}$. Then

$$f(\mathbf{v}_j) = \mathbf{0} \quad \text{if } 1 \le j \le k$$

and

$$f(\mathbf{v}_j) = -\mathbf{v}_j \quad \text{if } k+1 \le j \le n.$$

Consequently

$$c_f(t) = (-1)^n t^k (1+t)^{n-k}.$$

□

4.2   Jordan canonical form

Diagonalizable endomorphisms have many good properties that are useful in applications. However, we often have to deal with endomorphisms that are not diagonalizable. In this section we study the structure of such endomorphisms.

4.2.1   Jordan canonical form when the characteristic polynomial has one root

We begin by considering two examples.

Example 4.2.1. Let $f : \mathbb{R}^2 \to \mathbb{R}^2$ be the endomorphism defined by

$$f\left(\begin{bmatrix} x \\ y \end{bmatrix}\right) = \begin{bmatrix} 1 & 4 \\ -1 & 5 \end{bmatrix} \begin{bmatrix} x \\ y \end{bmatrix}.$$

Show that f is not diagonalizable and find a basis $\mathcal{B}$ of $\mathbb{R}^2$ such that the matrix of f in $\mathcal{B}$ is $\begin{bmatrix} 3 & 1 \\ 0 & 3 \end{bmatrix}$.

Proof. First we find that $c_f(t) = (t-3)^2$ and $\ker(f-3) = \text{Span}\left\{\begin{bmatrix} 2 \\ 1 \end{bmatrix}\right\}$. This shows that f is not diagonalizable because $\dim \ker(f-3) = 1$.

Next we choose a vector that is not in $\ker(f-3)$, for example $\begin{bmatrix} 1 \\ 0 \end{bmatrix}$. The vector $(f - 3) \begin{bmatrix} 1 \\ 0 \end{bmatrix}$ is in $\ker(f - 3)$ because according to Cayley-Hamilton theorem we have

$$(f - 3)\left((f - 3)\begin{bmatrix} 1 \\ 0 \end{bmatrix}\right) = (f - 3)^2 \begin{bmatrix} 1 \\ 0 \end{bmatrix} = \mathbf{0}.$$

Now it is easy to verify that the set

$$\left\{(f - 3)\begin{bmatrix} 1 \\ 0 \end{bmatrix}, \begin{bmatrix} 1 \\ 0 \end{bmatrix}\right\} = \left\{\begin{bmatrix} 1 & 4 \\ -1 & 5 \end{bmatrix}\begin{bmatrix} 1 \\ 0 \end{bmatrix}, \begin{bmatrix} 1 \\ 0 \end{bmatrix}\right\} = \left\{\begin{bmatrix} -2 \\ -1 \end{bmatrix}, \begin{bmatrix} 1 \\ 0 \end{bmatrix}\right\}$$

is a basis with the desired property.

Note that $m_f(t) = (t - 3)^2$. $\qquad\qquad\qquad\qquad\qquad\qquad\qquad\qquad \square$

Example 4.2.2. Let V be a vector space such that $\dim V = 3$. Let $f : V \to V$ be a linear transformation and let $\alpha \in \mathbb{K}$. If $m_f(t) = (t - \alpha)^3$, show that there is a basis $\mathcal{P}$ of V such that the $\mathcal{P}$-matrix of the linear transformation f is

$$\begin{bmatrix} \alpha & 1 & 0 \\ 0 & \alpha & 1 \\ 0 & 0 & \alpha \end{bmatrix}.$$

Solution. Let $\mathbf{v}$ be a vector in $\ker(f - \alpha)^3 = V$ that is not in $\ker(f - \alpha)^2$. Then $(f - \alpha)\mathbf{v} \in \ker(f - \alpha)^2$ and $(f - \alpha)\mathbf{v} \notin \ker(f - \alpha)$. We will show that

$$\mathcal{P} = \{(f - \alpha)^2\mathbf{v}, (f - \alpha)\mathbf{v}, \mathbf{v}\}$$

is a basis with the desired property.

Let $x_1, x_2, x_3 \in \mathbb{K}$ be such that

$$x_1(f - \alpha)^2\mathbf{v} + x_2(f - \alpha)\mathbf{v} + x_3\mathbf{v} = \mathbf{0}. \tag{4.4}$$

By applying $(f - \alpha)^2$ to (4.4) we get

$$x_3(f - \alpha)^2\mathbf{v} = \mathbf{0}$$

and consequently $x_3 = 0$. Next, by applying $f - \alpha$ to the equality

$$x_1(f - \alpha)^2\mathbf{x} + x_2(f - \alpha)\mathbf{v} = \mathbf{0}$$

we get

$$x_2(f - \alpha)^2\mathbf{v} = \mathbf{0}$$

which gives us $x_2 = 0$. Now (4.4) becomes

$$x_1(f - \alpha)^2\mathbf{v} = \mathbf{0}$$

which gives us $x_1 = 0$. This shows that the vectors $(f - \alpha)^2\mathbf{v}$, $(f - \alpha)\mathbf{v}$, and $\mathbf{v}$ are linearly independent and consequently

$$\mathcal{P} = \{(f - \alpha)^2\mathbf{v}, (f - \alpha)\mathbf{v}, \mathbf{v}\}$$

is a basis of $\mathcal{V}$.

Since

$$
\begin{aligned}
f((f - \alpha)^2\mathbf{v}) &= (f - \alpha)(f - \alpha)^2\mathbf{v} + \alpha(f - \alpha)^2\mathbf{v} \\
&= \alpha(f - \alpha)^2\mathbf{v} \quad \text{(because } (f - \alpha)^3(\mathbf{v}) = \mathbf{0}\text{)}, \\
f((f - \alpha)\mathbf{v}) &= (f - \alpha)(f - \alpha)\mathbf{v} + \alpha(f - \alpha)\mathbf{v} = (f - \alpha)^2\mathbf{v} + \alpha(f - \alpha)\mathbf{v}, \\
f(\mathbf{v}) &= (f - \alpha)\mathbf{v} + \alpha\mathbf{v},
\end{aligned}
$$

the $\mathcal{P}$-matrix of f is $\begin{bmatrix} \alpha & 1 & 0 \\ 0 & \alpha & 1 \\ 0 & 0 & \alpha \end{bmatrix}$.

$\square$

The minimal polynomial of an endomorphism $f : \mathcal{V} \to \mathcal{V}$ on a finite dimensional vector space is defined as the unique monic polynomial m_f of smallest positive degree such that $m_f(f) = \mathbf{0}$, that is, $m_f(f)(\mathbf{v}) = \mathbf{0}$ for every $\mathbf{v} \in \mathcal{V}$. Now we consider a similar property relative to a fixed $\mathbf{v} \in \mathcal{V}$.

Example 4.2.3. Let $f : \mathcal{V} \to \mathcal{V}$ be an endomorphism on a vector space $\mathcal{V}$ and let $\mathbf{x} \in \mathcal{V}$. If $(f - \alpha)^3(\mathbf{x}) = \mathbf{0}$, show that the subspace Span $\{\mathbf{x}, f(\mathbf{x}), f^2(\mathbf{x})\}$ is f-invariant.

Proof. The equality

$$(f - \alpha)^3(\mathbf{x}) = (f^3 - 3\alpha f^2 + 3\alpha^2 f - \alpha^3)(\mathbf{x}) = \mathbf{0}$$

gives us

$$f^3(\mathbf{x}) = (3\alpha f^2 - 3\alpha^2 f + \alpha^3)(\mathbf{x})$$

and then, for every integer $n \geq 3$, we get

$$f^n(\mathbf{x}) = (3\alpha f^{n-1} - 3\alpha^2 f^{n-2} + \alpha^3 f^{n-3})(\mathbf{x}).$$

This implies by induction that for every nonnegative integer n we have

$$f^n(\mathbf{x}) \in \text{Span} \{\mathbf{x}, f(\mathbf{x}), f^2(\mathbf{x})\}.$$

Our result is an immediate consequence of this fact.

$\square$

Theorem 4.2.4. *Let V be a finite dimensional vector space, $f : V \to V$ a nonzero endomorphism, and $\mathbf{v}$ a nonzero vector in V. Then*

(a) *There is a unique monic polynomial $m_{f,\mathbf{v}}$ of smallest positive degree such that $m_{f,\mathbf{v}}(f)(\mathbf{v}) = \mathbf{0}$;*

(b) *$m_{f,\mathbf{v}}$ divides every polynomial p such that $p(f)(\mathbf{v}) = \mathbf{0}$;*

(c) *If the degree of $m_{f,\mathbf{v}}$ is k, then the vectors $\mathbf{v}, f(\mathbf{v}), \ldots, f^{k-1}(\mathbf{v})$ are linearly independent and Span $\{\mathbf{v}, f(\mathbf{v}), \ldots, f^{k-1}(\mathbf{v})\}$ is the smallest f-invariant subspace of V containing $\mathbf{v}$.*

Proof. (a) and (b) can be obtained as in the proof of Theorem 4.1.39.

If the degree of $m_{f,\mathbf{v}}$ is k, then the vectors $\mathbf{v}, f(\mathbf{v}), \ldots, f^{k-1}(\mathbf{v})$ are linearly independent because, if $x_0, \ldots, x_{k-1} \in \mathbb{K}$ are not all 0, then the degree of the polynomial $x_0 + x_1 t + \cdots + x_{k-1} t^{k-1}$ is strictly less than the degree of $m_{f,\mathbf{v}}$.

To show that the subspace Span $\{\mathbf{v}, f(\mathbf{v}), \ldots, f^{k-1}(\mathbf{v})\}$ is f-invariant it is enough to show that $f^n(\mathbf{v})$ is in this subspace for every positive integer n. Indeed, there are polynomials q and r such that

$$t^n = m_{f,\mathbf{v}}(t)q(t) + r(t),$$

where $r = \mathbf{0}$ or $r \neq \mathbf{0}$ and the degree of r is strictly less than the degree of $m_{f,\mathbf{v}}$. Thus $f^n(\mathbf{v}) = r(f)(\mathbf{v})$ and the subspace Span $\{\mathbf{v}, f(\mathbf{v}), \ldots, f^{k-1}(\mathbf{v})\}$ is f-invariant.

Clearly, if $\mathcal{U}$ is an f-invariant subspace of V such that $\mathbf{v} \in \mathcal{U}$, then Span $\{\mathbf{v}, f(\mathbf{v}), \ldots, f^{k-1}(\mathbf{v})\} \subseteq \mathcal{U}$. $\qquad \square$

Definition 4.2.5. Let V be a vector space, $f : V \to V$ a nonzero endomorphism and $\mathbf{v}$ a nonzero vector in V such that the degree of $m_{f,\mathbf{v}}$ is k. The f-invariant subspace Span$\{\mathbf{v}, f(\mathbf{v}), \ldots, f^{k-1}(\mathbf{v})\}$ is called the *cyclic subspace* of f associated with $\mathbf{v}$ and is denoted by $V_{f,\mathbf{v}}$.

Theorem 4.2.6. *Let V be a finite dimensional vector space and let $f : V \to V$ be an endomorphism such that $m_f(t) = (t - \alpha)^k$ for some integer $k \geq 1$ and some $\alpha \in \mathbb{K}$. Then there is a nonzero vector $\mathbf{v} \in V$ such that $m_{f,\mathbf{v}} = m_f$.*

Proof. Let $\mathbf{v} \in V$ be such that $(f - \alpha)^k \mathbf{v} = \mathbf{0}$ and $(f - \alpha)^{k-1} \mathbf{v} \neq \mathbf{0}$. Then $m_f(t) = (t - \alpha)^k = m_{f,\mathbf{v}}(t)$. $\qquad \square$

Theorem 4.2.7. *Let V be a finite dimensional vector space and let f : $V \to V$ be an endomorphism. For a nonzero vector $\mathbf{v} \in V$ the following conditions are equivalent.*

(a) $m_{f,\mathbf{v}} = (t-\alpha)^k$ *for some integer $k \geq 1$ and some $\alpha \in \mathbb{K}$;*

(b) *The set $\mathcal{B} = \{(f-\alpha)^{k-1}\mathbf{v}, \ldots, (f-\alpha)\mathbf{v}, \mathbf{v}\}$ is a basis of $V_{f,\mathbf{v}}$;*

(c) *There is a basis $\mathcal{B} = \{\mathbf{v}_1, \ldots, \mathbf{v}_{k-1}, \mathbf{v}\}$ of an f-invariant subspace U of V such that the $\mathcal{B}$-matrix of the endomorphism $g : U \to U$ induced by f is*

$$\begin{bmatrix} \alpha & 1 & 0 & 0 & \ldots & 0 & 0 \\ 0 & \alpha & 1 & 0 & \ldots & 0 & 0 \\ 0 & 0 & \alpha & 1 & \ldots & 0 & 0 \\ \vdots & \vdots & \vdots & \vdots & \ddots & \vdots & \vdots \\ 0 & 0 & 0 & 0 & \ldots & 1 & 0 \\ 0 & 0 & 0 & 0 & \ldots & \alpha & 1 \\ 0 & 0 & 0 & 0 & \ldots & 0 & \alpha \end{bmatrix}.$$

Proof. Suppose that $m_{f,\mathbf{v}} = (t-\alpha)^k$. Because for every $j \in \{0, \ldots, k-1\}$ we have

$$f^j(\mathbf{v}) = (f - \alpha + \alpha)^j(\mathbf{v}) = (f-\alpha)^j(\mathbf{v}) + j\alpha(f-\alpha)^{j-1}(\mathbf{v}) + \cdots + j\alpha^{j-1}(f-\alpha)(\mathbf{v}) + \alpha^j\mathbf{v},$$

we get

$$V_{f,\mathbf{v}} = \text{Span}\{\mathbf{v}, f(\mathbf{v}), \ldots, f^{k-1}(\mathbf{v})\} \subseteq \text{Span}\{\mathbf{v}, (f-\alpha)\mathbf{v}, \ldots, (f-\alpha)^{k-1}\mathbf{v}\}.$$

Hence

$$V_{f,\mathbf{v}} = \text{Span}\{\mathbf{v}, f(\mathbf{v}), \ldots, f^{k-1}(\mathbf{v})\} = \text{Span}\{\mathbf{v}, (f-\alpha)\mathbf{v}, \ldots, (f-\alpha)^{k-1}\mathbf{v}\},$$

because the vectors $\mathbf{v}, f(\mathbf{v}), \ldots, f^{k-1}(\mathbf{v})$ are linearly independent. Consequently

$$\mathcal{B} = \{(f-\alpha)^{k-1}\mathbf{v}, \ldots, (f-\alpha)\mathbf{v}, \mathbf{v}\}$$

is a basis of $V_{f,\mathbf{v}}$. This proves (a) implies (b).

Suppose now that $\mathcal{B} = \{(f-\alpha)^{k-1}\mathbf{v}, \ldots, (f-\alpha)\mathbf{v}, \mathbf{v}\}$ is a basis of $V_{f,\mathbf{v}}$. In order to find the $\mathcal{B}$-matrix of f we note that

$$f((f-\alpha)^{k-1}(\mathbf{v})) = (f-\alpha)^k(\mathbf{v}) + \alpha(f-\alpha)^{k-1}(\mathbf{v}) = \alpha(f-\alpha)^{k-1}(\mathbf{v})$$

and

$$f((f-\alpha)^j(\mathbf{v})) = (f-\alpha)^{j+1}(\mathbf{v})+\alpha(f-\alpha)^j(\mathbf{v}) = 1\cdot(f-\alpha)^{j+1}(\mathbf{v})+\alpha(f-\alpha)^j(\mathbf{v})$$

for every $j \in \{0,\ldots,k-2\}$. To prove that (b) implies (c) we take $\mathcal{U} = \mathcal{V}_{f,\mathbf{v}}$.

Now suppose that there is a basis $\mathcal{B} = \{\mathbf{v}_1,\ldots,\mathbf{v}_{k-1},\mathbf{v}\}$ of an f-invariant subspace $\mathcal{U}$ of $\mathcal{V}$ such that the $\mathcal{B}$-matrix of the endomorphism $g : \mathcal{U} \to \mathcal{U}$ induced by f is

$$\begin{bmatrix} \alpha & 1 & 0 & 0 & \cdots & 0 & 0 \\ 0 & \alpha & 1 & 0 & \cdots & 0 & 0 \\ 0 & 0 & \alpha & 1 & \cdots & 0 & 0 \\ \vdots & \vdots & \vdots & \vdots & \ddots & \vdots & \vdots \\ 0 & 0 & 0 & 0 & \cdots & 1 & 0 \\ 0 & 0 & 0 & 0 & \cdots & \alpha & 1 \\ 0 & 0 & 0 & 0 & \cdots & 0 & \alpha \end{bmatrix}.$$

Then

$$(f-\alpha)\mathbf{v}_1 = \mathbf{0}$$
$$(f-\alpha)\mathbf{v}_2 = \mathbf{v}_1$$
$$\vdots$$
$$(f-\alpha)\mathbf{v}_{k-1} = \mathbf{v}_{k-2}$$
$$(f-\alpha)\mathbf{v} = \mathbf{v}_{k-1}.$$

Consequently

$$(f-\alpha)^{k-1}\mathbf{v} = \mathbf{v}_1 \neq \mathbf{0} \quad \text{and} \quad (f-\alpha)^k\mathbf{v} = \mathbf{0}.$$

Hence $m_{f,\mathbf{v}} = (t-\alpha)^k$, which proves that (c) implies (a). □

Example 4.2.8. Let $f : \mathcal{P}_3(\mathbb{R}) \to \mathcal{P}_3(\mathbb{R})$ be the endomorphism defined by $f(p) = p'$. Determine $m_f, \mathcal{V}_{f,x^2+1}$ and m_{f,x^2+1}.

Proof. Since

$$f(x^2+1) = 2x, \quad f(2x) = 2, \quad \text{and } f(2) = 0,$$

we have $m_{f,x^2+1} = t^3$, $\mathcal{V}_{f,x^2+1} = \mathcal{P}_2(\mathbb{R})$, and $m_f(t) = t^4$. □

Definition 4.2.9. Let $\alpha \in \mathbb{K}$ and let k be a positive integer. The $k \times k$ matrix

$$\begin{bmatrix} \alpha & 1 & 0 & 0 & \dots & 0 & 0 \\ 0 & \alpha & 1 & 0 & \dots & 0 & 0 \\ 0 & 0 & \alpha & 1 & \dots & 0 & 0 \\ \vdots & \vdots & \vdots & \vdots & \ddots & \vdots & \vdots \\ 0 & 0 & 0 & 0 & \dots & 1 & 0 \\ 0 & 0 & 0 & 0 & \dots & \alpha & 1 \\ 0 & 0 & 0 & 0 & \dots & 0 & \alpha \end{bmatrix}$$

is called a *Jordan block* and is denoted by $J_{\alpha,k}$. The 1×1 matrix $[\alpha]$ is also considered a Jordan block.

Example 4.2.10. Let $f : \mathbb{R}^3 \to \mathbb{R}^3$ be the endomorphism defined by

$$f\left(\begin{bmatrix} x \\ y \\ z \end{bmatrix}\right) = \begin{bmatrix} 2 & 5 & 1 \\ 0 & 5 & -1 \\ -1 & 2 & 5 \end{bmatrix} \begin{bmatrix} x \\ y \\ z \end{bmatrix}.$$

Show that $m_f(t) = (t-4)^3$ and find a basis of $\mathbb{R}^3$ such that the matrix of f in that basis is the Jordan block

$$\begin{bmatrix} 4 & 1 & 0 \\ 0 & 4 & 1 \\ 0 & 0 & 4 \end{bmatrix}.$$

Proof. First, proceeding as in Example 4.1.28, we get $c_f(t) = (4-t)^3$. Hence, by Cayley-Hamilton theorem, we have $(f-4)^3 = \mathbf{0}$.

To determine $\ker(f-4)^2$ we solve the equation

$$\left(\begin{bmatrix} 2 & 5 & 1 \\ 0 & 5 & -1 \\ -1 & 2 & 5 \end{bmatrix} - \begin{bmatrix} 4 & 0 & 0 \\ 0 & 4 & 0 \\ 0 & 0 & 4 \end{bmatrix}\right)^2 \begin{bmatrix} x \\ y \\ z \end{bmatrix} = \begin{bmatrix} 0 \\ 0 \\ 0 \end{bmatrix},$$

that is,

$$\begin{bmatrix} 3 & -3 & -6 \\ 1 & -1 & -2 \\ 1 & -1 & -2 \end{bmatrix} \begin{bmatrix} x \\ y \\ z \end{bmatrix} = \begin{bmatrix} 0 \\ 0 \\ 0 \end{bmatrix},$$

which gives us $x - y - 2z = 0$. Consequently

$$\ker(f-4)^2 = \left\{ \begin{bmatrix} x \\ y \\ z \end{bmatrix} \in \mathbb{R}^3, x - y - 2z = 0 \right\}.$$

Because $\ker(f-4)^2 \neq \mathcal{V}$, we have $m_f(t) = (t-4)^3$.

Since $\begin{bmatrix} 0 \\ 1 \\ 0 \end{bmatrix} \notin \ker(f-4)^2$, according to Theorem 4.2.7, the set

$$\left\{ (f-4)^2 \begin{bmatrix} 0 \\ 1 \\ 0 \end{bmatrix}, (f-4) \begin{bmatrix} 0 \\ 1 \\ 0 \end{bmatrix}, \begin{bmatrix} 0 \\ 1 \\ 0 \end{bmatrix} \right\}$$

$$= \left\{ \begin{bmatrix} 3 & -3 & -6 \\ 1 & -1 & -2 \\ 1 & -1 & -2 \end{bmatrix} \begin{bmatrix} 0 \\ 1 \\ 0 \end{bmatrix}, \begin{bmatrix} -2 & 5 & 1 \\ 0 & 1 & -1 \\ -1 & 2 & 1 \end{bmatrix} \begin{bmatrix} 0 \\ 1 \\ 0 \end{bmatrix}, \begin{bmatrix} 0 \\ 1 \\ 0 \end{bmatrix} \right\}$$

$$= \left\{ \begin{bmatrix} -3 \\ -1 \\ -1 \end{bmatrix}, \begin{bmatrix} 5 \\ 1 \\ 2 \end{bmatrix}, \begin{bmatrix} 0 \\ 1 \\ 0 \end{bmatrix} \right\}$$

is a basis satisfying the required condition. Note that $m_{f, \begin{bmatrix} 0 \\ 1 \\ 0 \end{bmatrix}} = (t-4)^3$. $\square$

In all examples discussed so far the matrix of the linear transformation had one Jordan block. Now we consider two examples of transformations with two Jordan blocks.

In the next several examples we added horizontal and vertical lines in the matrix to visualize and separate different Jordan blocks. The lines have no other mathematical meaning.

Example 4.2.11. Let $f : \mathbb{R}^3 \to \mathbb{R}^3$ be the endomorphism defined by

$$f\left(\begin{bmatrix} x \\ y \\ z \end{bmatrix} \right) = \begin{bmatrix} 5 & 2 & 4 \\ 0 & 7 & 0 \\ -1 & 1 & 9 \end{bmatrix} \begin{bmatrix} x \\ y \\ z \end{bmatrix}.$$

Find a basis of $\mathbb{R}^3$ such that the matrix of f in that basis is $\left[\begin{array}{cc|c} 7 & 1 & 0 \\ 0 & 7 & 0 \\ \hline 0 & 0 & 7 \end{array} \right].$

Proof. First we find that $c_f(t) = (7-t)^3$. We determine $\ker(f-7)$ by solving the equation

$$\left(\begin{bmatrix} 5 & 2 & 4 \\ 0 & 7 & 0 \\ -1 & 1 & 9 \end{bmatrix} - \begin{bmatrix} 7 & 0 & 0 \\ 0 & 7 & 0 \\ 0 & 0 & 7 \end{bmatrix}\right) \begin{bmatrix} x \\ y \\ z \end{bmatrix} = \begin{bmatrix} 0 \\ 0 \\ 0 \end{bmatrix},$$

that is,

$$\begin{bmatrix} -2 & 2 & 4 \\ 0 & 0 & 0 \\ -1 & 1 & 2 \end{bmatrix} \begin{bmatrix} x \\ y \\ z \end{bmatrix} = \begin{bmatrix} 0 \\ 0 \\ 0 \end{bmatrix},$$

an endomorphism which gives us $x - y - 2z = 0$. Consequently

$$\ker(f - 7) = \left\{ \begin{bmatrix} x \\ y \\ z \end{bmatrix} \in \mathbb{R}^3, x - y - 2z = 0 \right\}.$$

Since $\dim \ker(f - 7) = 2$, the endomorphism f is not diagonalizable. It is easy to see that $m_f(t) = (t-7)^2$. Because $\begin{bmatrix} 0 \\ 0 \\ 1 \end{bmatrix} \notin \ker(f-7)$ and $\dim \ker(f-7) = 2$, we have

$$\ker(f - 7) \oplus \operatorname{Span}\left\{ \begin{bmatrix} 0 \\ 0 \\ 1 \end{bmatrix} \right\} = \mathbb{R}^3. \tag{4.5}$$

Now, since $\begin{bmatrix} 1 \\ 1 \\ 0 \end{bmatrix} \in \ker(f - 7)$ and the eigenvectors $\begin{bmatrix} 1 \\ 1 \\ 0 \end{bmatrix}$ and $(f - 7) \begin{bmatrix} 0 \\ 0 \\ 1 \end{bmatrix} = \begin{bmatrix} 4 \\ 0 \\ 2 \end{bmatrix}$ are linearly independent, it is easy to verify, using (4.5), that the set

$$\left\{ (f - 7) \begin{bmatrix} 0 \\ 0 \\ 1 \end{bmatrix}, \begin{bmatrix} 0 \\ 0 \\ 1 \end{bmatrix}, \begin{bmatrix} 1 \\ 1 \\ 0 \end{bmatrix} \right\} = \left\{ \begin{bmatrix} 4 \\ 0 \\ 2 \end{bmatrix}, \begin{bmatrix} 0 \\ 0 \\ 1 \end{bmatrix}, \begin{bmatrix} 1 \\ 1 \\ 0 \end{bmatrix} \right\}$$

is a basis satisfying the required condition.

Note that $m_{f, \begin{bmatrix} 0 \\ 0 \\ 1 \end{bmatrix}} = (t - 7)^2 = m_f$. $\square$

Example 4.2.12. Let V be a vector space such that $\dim V = 5$. Let $f : V \to V$ be an endomorphism and let $\alpha \in \mathbb{K}$. We assume that $c_f(t) = (\alpha - t)^5$ and

$m_f(t) = (t - \alpha)^3$. If

$$\dim \ker(f - \alpha) = 2 \quad \text{and} \quad \dim \ker(f - \alpha)^2 = 4,$$

show that there is a basis $\mathcal{P}$ of V such that the $\mathcal{P}$-matrix of f is

$$\begin{bmatrix} \alpha & 1 & 0 & 0 & 0 \\ 0 & \alpha & 1 & 0 & 0 \\ 0 & 0 & \alpha & 0 & 0 \\ 0 & 0 & 0 & \alpha & 1 \\ 0 & 0 & 0 & 0 & \alpha \end{bmatrix}.$$

Solution. Let $\mathbf{u}$ be a vector in $\ker(f-\alpha)^3$ which is not in $\ker(f-\alpha)^2$. Then the vector $(f-\alpha)\mathbf{u}$ is in $\ker(f-\alpha)^2$ and not in $\ker(f-\alpha)$. Because $\dim \ker(f-\alpha)^2 = 4$ and $\dim \ker(f - \alpha) = 2$, we can choose a vector $\mathbf{v} \in \ker(f - \alpha)^2$ such that $\{(f - \alpha)\mathbf{u}, \mathbf{v}\}$ is a basis of a complement to $\ker(f - \alpha)$ in $\ker(f - \alpha)^2$. We will show that

$$\{(f - \alpha)^2\mathbf{u}, (f - \alpha)\mathbf{u}, \mathbf{u}, (f - \alpha)\mathbf{v}, \mathbf{v}\}$$

is a basis that has the desired property.

First we show that the vectors $(f-\alpha)^2\mathbf{u}, (f-\alpha)\mathbf{u}, \mathbf{u}, (f-\alpha)\mathbf{v}, \mathbf{v}$ are linearly independent. If

$$x_1(f - \alpha)^2\mathbf{u} + x_2(f - \alpha)\mathbf{u} + x_3\mathbf{u} + x_4(f - \alpha)\mathbf{v} + x_5\mathbf{v} = \mathbf{0} \qquad (4.6)$$

for some $x_1, x_2, x_3, x_4, x_5 \in \mathbb{K}$, by applying $(f - \alpha)^2$ to the above equation we obtain

$$x_3(f - \alpha)^2\mathbf{u} = \mathbf{0}$$

and consequently $x_3 = 0$. Next, by applying $f - \alpha$ to (4.6) we get

$$x_2(f - \alpha)^2\mathbf{u} + x_5(f - \alpha)\mathbf{v} = (f - \alpha)(x_2(f - \alpha)\mathbf{u} + x_5\mathbf{v}) = \mathbf{0}$$

which yields that $x_2 = x_5 = 0$ because $\{(f-\alpha)\mathbf{u}, \mathbf{v}\}$ is a basis of a complement to $\ker(f - \alpha)$ in $\ker(f - \alpha)^2$. Now (4.6) becomes

$$x_1(f - \alpha)^2\mathbf{u} + x_4(f - \alpha)\mathbf{v} = (f - \alpha)(x_1(f - \alpha)\mathbf{u} + x_4\mathbf{v}) = \mathbf{0}$$

which gives us $x_1 = x_4 = 0$ again because $\{(f - \alpha)\mathbf{u}, \mathbf{v}\}$ is a basis of a complement to $\ker(f - \alpha)$ in $\ker(f - \alpha)^2$.

We have proved that the vectors $(f - \alpha)^2\mathbf{u}, (f - \alpha)\mathbf{u}, \mathbf{u}, (f - \alpha)\mathbf{v}, \mathbf{v}$ are linearly independent and consequently

$$\mathcal{P} = \{(f - \alpha)^2\mathbf{u}, (f - \alpha)\mathbf{u}, \mathbf{u}, (f - \alpha)\mathbf{v}, \mathbf{v}\}$$

is a basis of $\mathcal{V}$. Since

$$f((f-\alpha)^2\mathbf{u}) = (f-\alpha)(f-\alpha)^2\mathbf{u} + \alpha(f-\alpha)^2\mathbf{u} = \alpha(f-\alpha)^2\mathbf{u}$$
$$f((f-\alpha)\mathbf{u}) = (f-\alpha)(f-\alpha)\mathbf{u} + \alpha(f-\alpha)\mathbf{u} = (f-\alpha)^2\mathbf{u} + \alpha(f-\alpha)\mathbf{u}$$
$$f(\mathbf{u}) = (f-\alpha)\mathbf{u} + \alpha\mathbf{u}$$
$$f((f-\alpha)\mathbf{v}) = (f-\alpha)(f-\alpha)\mathbf{v} + \alpha(f-\alpha)\mathbf{v} = \alpha(f-\alpha)\mathbf{v}$$
$$f(\mathbf{v}) = (f-\alpha)\mathbf{v} + \alpha\mathbf{v}$$

the $\mathcal{P}$-matrix of f is

$$\left[\begin{array}{ccc|cc}
\alpha & 1 & 0 & 0 & 0 \\
0 & \alpha & 1 & 0 & 0 \\
0 & 0 & \alpha & 0 & 0 \\
\hline
0 & 0 & 0 & \alpha & 1 \\
0 & 0 & 0 & 0 & \alpha
\end{array}\right].$$

Note that the assumption that $\dim\ker(f-\alpha)^2 = 4$ is unnecessary because, if $\mathbf{u}$ and $\mathbf{v}$ were linearly independent vectors in a complement of the vector subspace $\ker(f-\alpha)^2$, then we could show, as before, that the vectors

$$(f-\alpha)^2\mathbf{u}, (f-\alpha)\mathbf{u}, \mathbf{u}, (f-\alpha)^2\mathbf{v}, (f-\alpha)\mathbf{v}, \mathbf{v}$$

are linearly independent, which is not possible because $\dim\mathcal{V} = 5$. $\qquad\square$

The next result plays a central role this and the following sections. The proof is difficult in comparison with the other proofs presented in this book so far and it can be skipped at the first reading. On the other hand, understanding the proof, which uses many ideas presented in this book, is a good indication that you understand linear algebra at the level expected in a second course.

Lemma 4.2.13 (Fundamental lemma). *Let $\mathcal{V}$ be an n-dimensional vector space and let $f : \mathcal{V} \to \mathcal{V}$ be an endomorphism. Let $\mathbf{v} \in \mathcal{V}$ be such that*

$$m_f(t) = m_{f,\mathbf{v}}(t) = a_0 + a_1 t + \cdots + a_{k-1}t^{k-1} + t^k,$$

for some $k \in \{1, 2, \ldots, n\}$ and $a_0, a_1, \ldots, a_{k-1} \in \mathbb{K}$. Then there is an f-invariant subspace W such that

$$\mathrm{Span}\,\{\mathbf{v}, f(\mathbf{v}), \ldots, f^{k-1}(\mathbf{v})\} \oplus W = \mathcal{V}.$$

Proof. Let $\{\mathbf{v}, f(\mathbf{v}), \ldots, f^{k-1}(\mathbf{v}), \mathbf{v}_{k+1}, \ldots, \mathbf{v}_n\}$ be a basis of $\mathcal{V}$ and let $g : \mathcal{V} \to \mathbb{K}$ be the linear functional such that $g(f^j(\mathbf{v})) = 0$ for $j < k-1$, $g(f^{k-1}(\mathbf{v})) = 1$, and $g(\mathbf{v}_j) = 0$ for $j > k$. We define

$$W = \{\mathbf{w} \in \mathcal{V} : g(f^j(\mathbf{w})) = 0, j \in \{0, 1, 2, \ldots\}\}.$$

Clearly, $\mathcal{W}$ is a subspace of $\mathcal{V}$. Because for every $j \in \{0, 1, 2, \ldots\}$ and every $\mathbf{w} \in \mathcal{W}$ we have $g(f^j(f(\mathbf{w}))) = g(f^{j+1}(\mathbf{w})) = 0$, the subspace $\mathcal{W}$ is f-invariant.

Now we prove that the sum $\mathrm{Span}\{\mathbf{v}, \ldots, f^{k-1}(\mathbf{v})\} + \mathcal{W}$ is direct. Suppose

$$\mathbf{w} = x_1 \mathbf{v} + x_2 f(\mathbf{v}) + \cdots + x_k f^{k-1}(\mathbf{v}) \in \mathcal{W}$$

for some $x_1, \ldots, x_k \in \mathbb{K}$. Then $g(\mathbf{w}) = x_k = 0$ and thus $g(f(\mathbf{w})) = x_{k-1} = 0$, because

$$f(\mathbf{w}) = x_1 f(\mathbf{v}) + \cdots + x_{k-1} f^{k-1}(\mathbf{v}) + x_k f^k(\mathbf{v}) = x_1 f(\mathbf{v}) + \cdots + x_{k-1} f^{k-1}(\mathbf{v}).$$

Continuing this way we show that $x_k = x_{k-1} = x_{k-2} = \cdots = x_1 = 0$, so the sum is direct.

Now, since $m_f(f) = m_{f,\mathbf{v}}(f) = \mathbf{0}$, we have

$$f^k = -a_0 - a_1 f - \cdots - a_{k-1} f^{k-1}$$

and then, for every integer $l \geq 0$,

$$f^{k+l} = -a_0 f^l - a_1 f^{l+1} - \cdots - a_{k-1} f^{l+k-1},$$

which yields

$$\mathrm{Span}\left\{f^j, j \in \{0, 1, 2, \ldots\}\right\} = \mathrm{Span}\left\{f^j, j \in \{0, 1, \ldots, k-1\}\right\}.$$

Consequently

$$\mathcal{W} = \left\{\mathbf{w} \in \mathcal{V} : g(f^j(\mathbf{w})) = 0, j \in \{0, \ldots, k-1\}\right\}$$
$$= \left\{\mathbf{w} \in \mathcal{V} : (gf^j)(\mathbf{w}) = 0, j \in \{0, \ldots, k-1\}\right\}.$$

Next we show that the functionals $g, gf, \ldots, gf^{k-1}$ are linearly independent. Suppose

$$x_1 g + x_2 gf + \cdots + x_k gf^{k-1} = \mathbf{0}$$

for some $x_1, \ldots, x_k \in \mathbb{K}$. Applying this equality successively to $\mathbf{v}, f(\mathbf{v}), f^2(\mathbf{v}), \ldots, f^{k-1}(\mathbf{v})$ we get

$$g(x_1 \mathbf{v} + \cdots + x_k f^{k-1}(\mathbf{v})) = 0,$$
$$g(x_1 f(\mathbf{v}) + \cdots + x_k f^k(\mathbf{v})) = 0,$$

$$\vdots$$

$$g(x_1 f^{k-1}(\mathbf{v}) + \cdots + x_k f^{2k-2}(\mathbf{v})) = 0,$$

which can be written as

$$g(x_1 \mathbf{v} + \cdots + x_k f^{k-1}(\mathbf{v})) = 0,$$
$$g(f(x_1 \mathbf{v} + \cdots + x_k f^{k-1}(\mathbf{v}))) = 0,$$

$$\vdots$$

$$g(f^{k-1}(x_1 \mathbf{v} + \cdots + x_k f^{k-1}(\mathbf{v}))) = 0.$$

This shows that $x_1 \mathbf{v} + \cdots + x_k f^{k-1}(\mathbf{v}) \in \mathcal{W}$ and thus $x_1 = \cdots = x_k = 0$, because the sum $\mathrm{Span}\{\mathbf{v}, \ldots, f^{k-1}(\mathbf{v})\} + \mathcal{W}$ is direct. Consequently the functionals $g, gf, \ldots, gf^{k-1}$ are linearly independent. By Theorem 2.4.14, we get $\dim \mathcal{W} = \dim \mathcal{V} - k$, which completes the proof. $\qquad\square$

Theorem 4.2.14. *Let V be a finite dimensional vector space and let $f : V \to V$ be an endomorphism. If $m_f(t) = (t - \alpha)^n$ for some integer $n \geq 1$ and $\alpha \in \mathbb{K}$, then there are nonzero vectors $\mathbf{v}_1, \ldots, \mathbf{v}_r \in V$ such that*

$$V = V_{f,\mathbf{v}_1} \oplus \cdots \oplus V_{f,\mathbf{v}_r}$$

and

$$m_{f,\mathbf{v}_1} = (t - \alpha)^{k_1}, \ldots, m_{f,\mathbf{v}_r} = (t - \alpha)^{k_r},$$

where $k_1, \ldots, k_r$ are integers such that $1 \leq k_r \leq \cdots \leq k_1 = n$.

Proof. Let $\mathbf{v}_1 \in V$ be such that $(f - \alpha)^{n-1}(\mathbf{v}_1) \neq \mathbf{0}$. Then $m_{f,\mathbf{v}_1} = (t - \alpha)^n$.

If $V = V_{f,\mathbf{v}_1}$ and we take $k_1 = n$, then we are done. If $V \neq V_{f,\mathbf{v}_1}$, then there is an f-invariant subspace W of dimension $\dim V - n$ such that $V = V_{f,\mathbf{v}_1} \oplus W$, by Lemma 4.2.13. Let $g : W \to W$ be the endomorphism induced by f on W. Clearly, $m_f(t) = (t - \alpha)^n$ is a g-annihilator. Consequently, $m_g(t)$ divides $m_f(t) = (t - \alpha)^n$. This means that $m_g(t) = (t - \alpha)^m$ where $m \leq n$. Now, since $\dim W = \dim V - n < \dim V$, we can finish the proof using induction. $\qquad \square$

From the above theorem and Theorem 4.2.7 we obtain the following important result.

Corollary 4.2.15. *Let V be a finite dimensional vector space and let $f : V \to V$ be an endomorphism. If $m_f(t) = (t - \alpha)^n$ for some integer $n \geq 1$ and $\alpha \in \mathbb{K}$, then there are integers $1 \leq k_r \leq \cdots \leq k_1 = n$ and nonzero vectors $\mathbf{v}_1, \ldots, \mathbf{v}_r$ such that*

$$V = \bigoplus_{j=1}^{r} \operatorname{Span} \left\{ (f - \alpha)^{k_j-1}\mathbf{v}_j, (f - \alpha)^{k_j-2}\mathbf{v}_j, \ldots, (f - \alpha)\mathbf{v}_j, \mathbf{v}_j \right\}$$

where for every $j \in \{1, \ldots, r\}$ we have $(f - \alpha)^{k_j}\mathbf{v}_j = \mathbf{0}$ and $(f - \alpha)^{k_j-1}\mathbf{v}_j \neq \mathbf{0}$. Moreover, the set

$$\mathcal{B} = \bigcup_{j=1}^{r} \left\{ (f - \alpha)^{k_j-1}\mathbf{v}_j, (f - \alpha)^{k_j-2}\mathbf{v}_j, \ldots, (f - \alpha)\mathbf{v}_j, \mathbf{v}_j \right\}$$

is a basis of V and the $\mathcal{B}$-matrix of f is

$$\begin{bmatrix} J_{\alpha,k_1} & & & \\ & J_{\alpha,k_2} & & \mathbf{0} \\ & & \ddots & \\ \mathbf{0} & & & J_{\alpha,k_r} \end{bmatrix}.$$

Definition 4.2.16. Let $\alpha \in \mathbb{K}$. We say that a matrix is in α-*Jordan canonical form* if it has the form

$$\begin{bmatrix} J_{\alpha,k_1} & & & \\ & J_{\alpha,k_2} & & \text{\LARGE 0} \\ & & \ddots & \\ & \text{\LARGE 0} & & J_{\alpha,k_r} \end{bmatrix}$$

where $k_1, \ldots, k_r$ are integers such that $k_1 \geq \cdots \geq k_r \geq 1$ and $J_{\alpha,k_1}, J_{\alpha,k_2}, \ldots, J_{\alpha,k_r}$ are Jordan blocks.

If it is clear from the context what α is, then instead of "α-Jordan canonical" form we simply say "Jordan canonical form".

Here is an example of a matrix in α-Jordan canonical form

$$\left[\begin{array}{ccc|ccc|cc|c} \alpha & 1 & 0 & 0 & 0 & 0 & 0 & 0 & 0 \\ 0 & \alpha & 1 & 0 & 0 & 0 & 0 & 0 & 0 \\ 0 & 0 & \alpha & 0 & 0 & 0 & 0 & 0 & 0 \\ \hline 0 & 0 & 0 & \alpha & 1 & 0 & 0 & 0 & 0 \\ 0 & 0 & 0 & 0 & \alpha & 1 & 0 & 0 & 0 \\ 0 & 0 & 0 & 0 & 0 & \alpha & 0 & 0 & 0 \\ \hline 0 & 0 & 0 & 0 & 0 & 0 & \alpha & 1 & 0 \\ 0 & 0 & 0 & 0 & 0 & 0 & 0 & \alpha & 0 \\ \hline 0 & 0 & 0 & 0 & 0 & 0 & 0 & 0 & \alpha \end{array}\right].$$

In this example $k_1 = k_2 = 3, k_3 = 2, k_4 = 1$ and

$$J_{\alpha,k_1} = J_{\alpha,k_2} = \begin{bmatrix} \alpha & 1 & 0 \\ 0 & \alpha & 1 \\ 0 & 0 & \alpha \end{bmatrix}, \quad J_{\alpha,k_3} = \begin{bmatrix} \alpha & 1 \\ 0 & \alpha \end{bmatrix}, \quad J_{\alpha,k_4} = \begin{bmatrix} \alpha \end{bmatrix}.$$

The result in the following theorem is useful when computing Jordan canonical forms.

Theorem 4.2.17. *Let V be a vector space, $f : V \to V$ an endomorphism, and $\alpha \in \mathbb{K}$. If, for some integer $k \geq 2$, $\mathbf{v}_1, \ldots, \mathbf{v}_m \in \ker(f - \alpha)^k$ are linearly independent vectors such that*

$$\ker(f - \alpha)^k = \ker(f - \alpha)^{k-1} \oplus \mathbb{K}\mathbf{v}_1 \oplus \cdots \oplus \mathbb{K}\mathbf{v}_m,$$

then the vectors $(f-\alpha)\mathbf{v}_1, \ldots, (f-\alpha)\mathbf{v}_m$ are linearly independent vectors in $\ker(f - \alpha)^{k-1}$ and the sum

$$\ker(f - \alpha)^{k-2} + \mathbb{K}(f - \alpha)\mathbf{v}_1 + \cdots + \mathbb{K}(f - \alpha)\mathbf{v}_m$$

is direct.

Proof. If $\mathbf{w} \in \ker(f - \alpha)^{k-2}$ and

$$\mathbf{w} + x_1(f - \alpha)\mathbf{v}_1 + \cdots + x_m(f - \alpha)\mathbf{v}_m = \mathbf{0}$$

for some $x_1, \ldots, x_m \in \mathbb{K}$, then

$$
\begin{aligned}
\mathbf{0} &= (f - \alpha)^{k-2}\mathbf{w} + x_1(f - \alpha)^{k-1}\mathbf{v}_1 + \cdots + x_m(f - \alpha)^{k-1}\mathbf{v}_m \\
&= x_1(f - \alpha)^{k-1}\mathbf{v}_1 + \cdots + x_m(f - \alpha)^{k-1}\mathbf{v}_m \\
&= (f - \alpha)^{k-1}(x_1\mathbf{v}_1 + \cdots + x_m\mathbf{v}_m)
\end{aligned}
$$

and thus

$$x_1\mathbf{v}_1 + \cdots + x_m\mathbf{v}_m = \ker(f - \alpha)^{k-1}.$$

Since $\ker(f - \alpha)^k = \ker(f - \alpha)^{k-1} \oplus \mathbb{K}\mathbf{v}_1 \oplus \cdots \oplus \mathbb{K}\mathbf{v}_m$, we must have $x_1\mathbf{v}_1 + \cdots + x_m\mathbf{v}_m = \mathbf{0}$. Thus $x_1 = \cdots = x_m = 0$, because the vectors $\mathbf{v}_1, \ldots, \mathbf{v}_m$ are linearly independent, and consequently $\mathbf{w} = \mathbf{0}$.   □

Example 4.2.18. Let V be a vector space and let $f : V \to V$ be an endomorphism such that $c_f(t) = (\alpha - t)^9$ and $m_f(t) = (t - \alpha)^3$ for some $\alpha \in \mathbb{K}$. If

$$\dim \ker(f - \alpha) = 4 \quad \text{and} \quad \dim \ker(f - \alpha)^2 = 7,$$

show that there is a basis $\mathcal{B}$ of V such that the $\mathcal{B}$-matrix of f is

$$
A = \left[
\begin{array}{ccc|ccc|ccc}
\alpha & 1 & 0 & 0 & 0 & 0 & 0 & 0 & 0 \\
0 & \alpha & 1 & 0 & 0 & 0 & 0 & 0 & 0 \\
0 & 0 & \alpha & 0 & 0 & 0 & 0 & 0 & 0 \\
\hline
0 & 0 & 0 & \alpha & 1 & 0 & 0 & 0 & 0 \\
0 & 0 & 0 & 0 & \alpha & 1 & 0 & 0 & 0 \\
0 & 0 & 0 & 0 & 0 & \alpha & 0 & 0 & 0 \\
\hline
0 & 0 & 0 & 0 & 0 & 0 & \alpha & 1 & 0 \\
0 & 0 & 0 & 0 & 0 & 0 & 0 & \alpha & 0 \\
0 & 0 & 0 & 0 & 0 & 0 & 0 & 0 & \alpha
\end{array}
\right].
$$

Solution. Let $\mathbf{u}$ and $\mathbf{v}$ be linearly independent vectors in $\mathcal{V}$ such that

$$\ker(f - \alpha)^2 \oplus \mathbb{K}\mathbf{u} \oplus \mathbb{K}\mathbf{v} = \ker(f - \alpha)^3 = \mathcal{V}.$$

According to Theorem 4.2.17 the sum

$$\ker(f - \alpha) + \mathbb{K}(f - \alpha)\mathbf{u} + \mathbb{K}(f - \alpha)\mathbf{v}$$

is direct. Note that

$$\ker(f - \alpha) + \mathbb{K}(f - \alpha)\mathbf{u} + \mathbb{K}(f - \alpha)\mathbf{v} \subseteq \ker(f - \alpha)^2.$$

Because $\dim \ker(f - \alpha) = 4$ and $\dim \ker(f - \alpha)^2 = 7$, there is a $\mathbf{p} \in \ker(f - \alpha)^2$ such that

$$\ker(f - \alpha) \oplus \mathbb{K}(f - \alpha)\mathbf{u} \oplus \mathbb{K}(f - \alpha)\mathbf{v} \oplus \mathbb{K}\mathbf{p} = \ker(f - \alpha)^2.$$

The vectors

$$(f - \alpha)^2\mathbf{u}, (f - \alpha)^2\mathbf{v}, (f - \alpha)\mathbf{p}$$

are in $\ker(f - \alpha)$, that is, are eigenvectors of f and they are linearly independent, by Theorem 4.2.17 (with $k = 2$).

Because $\dim \ker(f - \alpha) = 4$ we can find an eigenvector $\mathbf{q}$ such that

$$\mathbb{K}(f - \alpha)^2\mathbf{u} \oplus \mathbb{K}(f - \alpha)^2\mathbf{v} \oplus \mathbb{K}(f - \alpha)\mathbf{p} \oplus \mathbb{K}\mathbf{q} = \ker(f - \alpha).$$

Consequently,

$$\mathcal{V} = \mathbb{K}(f - \alpha)^2\mathbf{u} \oplus \mathbb{K}(f - \alpha)\mathbf{u} \oplus \mathbb{K}\mathbf{u} \oplus \mathbb{K}(f - \alpha)^2\mathbf{v} \oplus \mathbb{K}(f - \alpha)\mathbf{v} \oplus \mathbb{K}\mathbf{v}$$
$$\oplus \mathbb{K}(f - \alpha)\mathbf{p} \oplus \mathbb{K}\mathbf{p} \oplus \mathbb{K}\mathbf{q}$$

and

$$\mathcal{B} = \{(f - \alpha)^2\mathbf{u}, (f - \alpha)\mathbf{u}, \mathbf{u}, (f - \alpha)^2\mathbf{v}, (f - \alpha)\mathbf{v}, \mathbf{v}, (f - \alpha)\mathbf{p}, \mathbf{p}, \mathbf{q}\}$$

is a basis of $\mathcal{V}$. It is easy to verify that A is the $\mathcal{B}$-matrix of f. □

Example 4.2.19. Let $\mathcal{V}$ be a vector space and let $f : \mathcal{V} \to \mathcal{V}$ be an endomorphism. If $m_f(t) = t^3$ and there are positive integers q, r, s such that $\dim \mathcal{V} = 3q + 2r + s$, $\dim \ker f^2 = 2q + 2r + s$, and $\dim \ker f = q + r + s$, show that there are linearly independent vectors $\mathbf{v}_1, \ldots, \mathbf{v}_q, \mathbf{w}_1, \ldots, \mathbf{w}_r, \mathbf{u}_1, \ldots, \mathbf{u}_s$ such that $\mathcal{V}$ is a direct sum of the following $q + r + s$ cyclic subspaces

$$\mathrm{Span}\left\{f^2(\mathbf{v}_1), f(\mathbf{v}_1), \mathbf{v}_1\right\}, \ldots, \mathrm{Span}\left\{f^2(\mathbf{v}_q), f(\mathbf{v}_q), \mathbf{v}_q\right\}$$

$$\mathrm{Span}\left\{f(\mathbf{w}_1), \mathbf{w}_1\right\}, \ldots, \mathrm{Span}\left\{f(\mathbf{w}_r), \mathbf{w}_r\right\}$$

$$\mathrm{Span}\{\mathbf{u}_1\}, \ldots, \mathrm{Span}\{\mathbf{u}_s\}$$

Solution. There are linearly independent vectors $\mathbf{v}_1, \ldots, \mathbf{v}_q$ such that

$$\mathcal{V} = \ker f^3 = \ker f^2 \oplus \mathbb{K}\mathbf{v}_1 \oplus \cdots \oplus \mathbb{K}\mathbf{v}_q.$$

By Theorem 4.2.17, the sum $\ker f + \mathbb{K}f(\mathbf{v}_1) + \cdots + \mathbb{K}f(\mathbf{v}_q)$ is direct and it is a subspace of $\ker f^2$. Let $\mathbf{w}_1, \ldots, \mathbf{w}_r \in \ker f^2$ be linearly independent vectors such that

$$\ker f^2 = \ker f \oplus \mathbb{K}f(\mathbf{v}_1) \oplus \cdots \oplus \mathbb{K}f(\mathbf{v}_q) \oplus \mathbb{K}\mathbf{w}_1 \oplus \cdots \oplus \mathbb{K}\mathbf{w}_r.$$

Again by Theorem 4.2.17, the vectors $f^2(\mathbf{v}_1), \ldots, f^2(\mathbf{v}_q), f(\mathbf{w}_1), \ldots, f(\mathbf{w}_r)$ are linearly independent vectors in $\ker f$. Now, let $\mathbf{u}_1, \ldots, \mathbf{u}_s$ be linearly independent vectors such that

$$\ker f = \mathbb{K}f^2(\mathbf{v}_1) \oplus \cdots \oplus \mathbb{K}f^2(\mathbf{v}_q) \oplus \mathbb{K}f(\mathbf{w}_1) \oplus \cdots \oplus \mathbb{K}f(\mathbf{w}_r) \oplus \mathbb{K}\mathbf{u}_1 \oplus \cdots \oplus \mathbb{K}\mathbf{u}_s.$$

Consequently, the vector space $\mathcal{V}$ is a direct sum of the following $q + r + s$ cyclic subspaces:

$$\mathrm{Span}\left\{f^2(\mathbf{v}_1), f(\mathbf{v}_1), \mathbf{v}_1\right\}, \ldots, \mathrm{Span}\left\{f^2(\mathbf{v}_q), f(\mathbf{v}_q), \mathbf{v}_q\right\}$$

$$\mathrm{Span}\left\{f(\mathbf{w}_1), \mathbf{w}_1\right\}, \ldots, \mathrm{Span}\left\{f(\mathbf{w}_r), \mathbf{w}_r\right\}$$

$$\mathrm{Span}\{\mathbf{u}_1\}, \ldots, \mathrm{Span}\{\mathbf{u}_s\}$$

$\square$

It is worth noting that Theorems 4.2.14 and 4.2.15 can be obtained from Theorem 4.2.17. In the next theorem we give a slightly different formulation of these results and use Theorem 4.2.17 to prove it.

Theorem 4.2.20. *Let $\mathcal{V}$ be a finite dimensional vector space and let $f : \mathcal{V} \to \mathcal{V}$ be an endomorphism. If $m_f(t) = (t - \alpha)^k$ for some $\alpha \in \mathbb{K}$ and integer $k \geq 1$, then $\mathcal{V}$ is a direct sum of subspaces of the form*

$$\mathrm{Span}\left\{(f - \alpha)^{m-1}\mathbf{v}, \ldots, (f - \alpha)\mathbf{v}, \mathbf{v}\right\}$$

for some integer $m \geq 1$ and $\mathbf{v} \in \mathcal{V}$ such that $m_{f,\mathbf{v}} = (t - \alpha)^m$.

Proof. Recall that $m_f(t) = (t - \alpha)^k$ means that $\mathcal{V} = \ker(f - \alpha)^k$ and $\mathcal{V} \neq \ker(f - \alpha)^{k-1}$. Similarly, $m_{f,\mathbf{v}} = (t - \alpha)^m$ means that $(f - \alpha)^m \mathbf{v} = \mathbf{0}$ and $(f - \alpha)^{m-1}\mathbf{v} \neq \mathbf{0}$.

Let

$$\mathcal{B}_k = \{\mathbf{v}_{k,1}, \ldots, \mathbf{v}_{k,m_k}\}$$

be a basis of a complement $\mathcal{C}_k$ of $\ker(f - \alpha)^{k-1}$ in $\ker(f - \alpha)^k = \mathcal{V}$, that is,

$$\mathcal{V} = \ker(f - \alpha)^k = \ker(f - \alpha)^{k-1} \oplus \mathcal{C}_k.$$

By Theorem 4.2.17, there are vectors $\mathbf{v}_{k-1,1}, \ldots, \mathbf{v}_{k-1,m_{k-1}} \in \ker(f-\alpha)^{k-1}$ such that the set $\mathcal{B}_{k-1}$ consisting of vectors

$$(f-\alpha)\mathbf{v}_{k,1}, \ldots, (f-\alpha)\mathbf{v}_{k,m_k},$$

$$\mathbf{v}_{k-1,1}, \ldots, \mathbf{v}_{k-1,m_{k-1}},$$

is a basis of a complement $\mathcal{C}_{k-1}$ of $\ker(f-\alpha)^{k-2}$ in $\ker(f-\alpha)^{k-1}$. Consequently,

$$\mathcal{V} = \ker(f-\alpha)^{k-2} \oplus \mathcal{C}_{k-1} \oplus \mathcal{C}_k.$$

Next, there are vectors $\mathbf{v}_{k-2,1}, \ldots, \mathbf{v}_{k-2,m_{k-2}} \in \ker(f-\alpha)^{k-2}$ such that the set $\mathcal{B}_{k-2}$ consisting of vectors

$$(f-\alpha)^2\mathbf{v}_{k,1}, \ldots, (f-\alpha)^2\mathbf{v}_{k,m_k},$$

$$(f-\alpha)\mathbf{v}_{k-1,1}, \ldots, (f-\alpha)\mathbf{v}_{k-1,m_{k-1}},$$

$$\mathbf{v}_{k-2,1}, \ldots, \mathbf{v}_{k-2,m_{k-2}},$$

is a basis of a complement $\mathcal{C}_{k-2}$ of $\ker(f-\alpha)^{k-3}$ in $\ker(f-\alpha)^{k-2}$, again by Theorem 4.2.17, and we have

$$\mathcal{V} = \ker(f-\alpha)^{k-3} \oplus \mathcal{C}_{k-2} \oplus \mathcal{C}_{k-1} \oplus \mathcal{C}_k.$$

Continuing as above we eventually obtain vectors $\mathbf{v}_{2,1}, \ldots, \mathbf{v}_{2,m_2} \in \ker(f-\alpha)^2$ such that the set $\mathcal{B}_2$ consisting of vectors

$$(f-\alpha)^{k-2}\mathbf{v}_{k,1}, \ldots, (f-\alpha)^{k-2}\mathbf{v}_{k,m_k},$$

$$(f-\alpha)^{k-3}\mathbf{v}_{k-1,1}, \ldots, (f-\alpha)^{k-3}\mathbf{v}_{k-1,m_{k-1}},$$

$$\vdots$$

$$(f-\alpha)\mathbf{v}_{3,1}, \ldots, (f-\alpha)\mathbf{v}_{3,m_3},$$

$$\mathbf{v}_{2,1}, \ldots, \mathbf{v}_{2,m_2},$$

is a basis of a complement $\mathcal{C}_2$ of $\ker(f-\alpha)$ in $\ker(f-\alpha)^2$.

Finally, there are eigenvectors $\mathbf{v}_{1,1}, \ldots, \mathbf{v}_{1,m_1}$ of f such that the set $\mathcal{B}_1$ consisting of vectors

$$(f-\alpha)^{k-1}\mathbf{v}_{k,1}, \ldots, (f-\alpha)^{k-1}\mathbf{v}_{k,m_k},$$

$$\vdots$$

$$(f-\alpha)\mathbf{v}_{2,1}, \ldots, (f-\alpha)\mathbf{v}_{2,m_2},$$

$$\mathbf{v}_{1,1}, \ldots, \mathbf{v}_{1,m_1},$$

is a basis of the eigenspace $\ker(f-\alpha)$.

Now, since

$$V = \mathcal{C}_k \oplus \mathcal{C}_{k-1} \oplus \cdots \oplus \mathcal{C}_2 \oplus \ker(f - \alpha)$$

the set of the vectors

$$\mathcal{B}_k \cup \mathcal{B}_{k-1} \cup \cdots \cup \mathcal{B}_2 \cup \mathcal{B}_1$$

is a basis of V. This basis contains m_k sets of vectors

$$\{(f - \alpha)^{k-1}\mathbf{v}_{k,1}, \quad (f - \alpha)^{k-2}\mathbf{v}_{k,1}, \quad \ldots, \quad (f - \alpha)\mathbf{v}_{k,1}, \quad \mathbf{v}_{k,1}\}$$
$$\vdots \qquad\qquad\qquad \vdots \qquad\qquad\qquad\qquad \vdots \qquad\qquad \vdots$$
$$\{(f - \alpha)^{k-1}\mathbf{v}_{k,m_k}, \quad (f - \alpha)^{k-2}\mathbf{v}_{k,m_k}, \quad \ldots, \quad (f - \alpha)\mathbf{v}_{k,m_k}, \quad \mathbf{v}_{k,m_k}\}$$

each with k elements, which will generate m_k Jordan blocks $J_{\alpha,k}$, m_{k-1} sets of vectors

$$\{(f - \alpha)^{k-2}\mathbf{v}_{k-1,1}, \quad (f - \alpha)^{k-3}\mathbf{v}_{k-1,1}, \quad \ldots, \quad (f - \alpha)\mathbf{v}_{k-1,1}, \quad \mathbf{v}_{k-1,1}\}$$
$$\vdots \qquad\qquad\qquad \vdots \qquad\qquad\qquad\qquad \vdots \qquad\qquad \vdots$$
$$\{(f - \alpha)^{k-2}\mathbf{v}_{k-1,m_{k-1}}, \quad (f - \alpha)^{k-3}\mathbf{v}_{k-1,m_{k-1}}, \quad \ldots, \quad (f - \alpha)\mathbf{v}_{k-1,m_{k-1}}, \quad \mathbf{v}_{k-1,m_{k-1}}\}$$

each with $k - 1$ elements, which will generate m_{k-1} Jordan blocks $J_{\alpha,k-1}$,

$$\vdots$$

j_2 sets of vectors

$$\{(f - \alpha)\mathbf{v}_{2,1}, \quad \mathbf{v}_{2,1}\}$$
$$\vdots \qquad\qquad \vdots$$
$$\{(f - \alpha)\mathbf{v}_{2,m_2}, \quad \mathbf{v}_{2,m_2}\}$$

each with 2 elements, which will generate m_2 Jordan blocks $J_{\alpha,2}$, and m_1 eigenvectors of f

$$\mathbf{v}_{1,1}$$
$$\vdots$$
$$\mathbf{v}_{1,m_1}$$

which will generate m_1 Jordan blocks $J_{\alpha,1}$.   □

Example 4.2.21. Let V be a vector space and let $f : V \to V$ be an endomorphism. If $c_f(t) = (\alpha - t)^5$ and $m_f(t) = (t - \alpha)^2$ for some $\alpha \in \mathbb{K}$, determine all possible Jordan canonical forms associated with such an endomorphism f.

Solution. It is easy to see that these forms are

$$\begin{bmatrix} \alpha & 1 & 0 & 0 & 0 \\ 0 & \alpha & 0 & 0 & 0 \\ 0 & 0 & \alpha & 1 & 0 \\ 0 & 0 & 0 & \alpha & 0 \\ 0 & 0 & 0 & 0 & \alpha \end{bmatrix} \quad \text{and} \quad \begin{bmatrix} \alpha & 1 & 0 & 0 & 0 \\ 0 & \alpha & 0 & 0 & 0 \\ 0 & 0 & \alpha & 0 & 0 \\ 0 & 0 & 0 & \alpha & 0 \\ 0 & 0 & 0 & 0 & \alpha \end{bmatrix}.$$

In the first case we have $\dim \ker(f - \alpha) = 3$ and we have three Jordan blocks and in the second case we have $\dim \ker(f - \alpha) = 4$ and there are four Jordan blocks. $\square$

Example 4.2.22. Let V be a vector space such that $\dim V = 9$ and let $f : V \to V$ be an endomorphism. If

$$\dim \ker(f - \alpha) = 3, \ \dim \ker(f - \alpha)^2 = 5, \ \dim \ker(f - \alpha)^3 = 7, \ \ker(f - \alpha)^4 = 8,$$

and $m_f(t) = (t - \alpha)^5$, determine the Jordan canonical form of f.

Solution. Following the proof of Theorem 4.2.20 or Theorem 4.2.17 it is easy to verify that the Jordan canonical form of f is

$$\begin{bmatrix} \alpha & 1 & 0 & 0 & 0 & 0 & 0 & 0 & 0 \\ 0 & \alpha & 1 & 0 & 0 & 0 & 0 & 0 & 0 \\ 0 & 0 & \alpha & 1 & 0 & 0 & 0 & 0 & 0 \\ 0 & 0 & 0 & \alpha & 1 & 0 & 0 & 0 & 0 \\ 0 & 0 & 0 & 0 & \alpha & 0 & 0 & 0 & 0 \\ 0 & 0 & 0 & 0 & 0 & \alpha & 1 & 0 & 0 \\ 0 & 0 & 0 & 0 & 0 & 0 & \alpha & 1 & 0 \\ 0 & 0 & 0 & 0 & 0 & 0 & 0 & \alpha & 0 \\ 0 & 0 & 0 & 0 & 0 & 0 & 0 & 0 & \alpha \end{bmatrix}.$$

$\square$

Example 4.2.23. We consider the endomorphism $f : \mathbb{R}^5 \to \mathbb{R}^5$ defined by $f(\mathbf{x}) = A\mathbf{x}$ where

$$A = \begin{bmatrix} -7 & 2 & -14 & -24 & -4 \\ -4 & 3 & -7 & -16 & -3 \\ 7 & -2 & 13 & 22 & 4 \\ -3 & 1 & -5 & -9 & -2 \\ 7 & -2 & 12 & 22 & 5 \end{bmatrix}.$$

Knowing that $c_f(t) = (1 - t)^5$, determine m_f and a basis $\mathcal{B}$ of $\mathbb{R}^5$ such that the $\mathcal{B}$-matrix of f has a Jordan canonical form.

Solution. Since

$$
B = A - \begin{bmatrix} 1 & 0 & 0 & 0 & 0 \\ 0 & 1 & 0 & 0 & 0 \\ 0 & 0 & 1 & 0 & 0 \\ 0 & 0 & 0 & 1 & 0 \\ 0 & 0 & 0 & 0 & 1 \end{bmatrix} = \begin{bmatrix} -8 & 2 & -14 & -24 & -4 \\ -4 & 2 & -7 & -16 & -3 \\ 7 & -2 & 12 & 22 & 4 \\ -3 & 1 & -5 & -10 & -2 \\ 7 & -2 & 12 & 22 & 4 \end{bmatrix},
$$

$$
B^2 = \begin{bmatrix} 2 & 0 & 2 & 4 & 2 \\ 2 & 0 & 2 & 4 & 2 \\ -2 & 0 & -2 & -4 & -2 \\ 1 & 0 & 1 & 2 & 1 \\ -2 & 0 & -2 & -4 & -2 \end{bmatrix},
$$

and

$$
B^3 = \begin{bmatrix} 0 & 0 & 0 & 0 & 0 \\ 0 & 0 & 0 & 0 & 0 \\ 0 & 0 & 0 & 0 & 0 \\ 0 & 0 & 0 & 0 & 0 \\ 0 & 0 & 0 & 0 & 0 \end{bmatrix},
$$

we have $m_f(t) = (t - 1)^3$.

The set

$$
\left\{ \begin{bmatrix} -1 \\ 0 \\ 1 \\ 0 \\ 0 \end{bmatrix}, \begin{bmatrix} -2 \\ 0 \\ 0 \\ 1 \\ 0 \end{bmatrix}, \begin{bmatrix} -1 \\ 0 \\ 0 \\ 0 \\ 1 \end{bmatrix}, \begin{bmatrix} 0 \\ 1 \\ 0 \\ 0 \\ 0 \end{bmatrix} \right\}
$$

is a basis of $\ker(f - 1)^2$ and the set

$$
\left\{ \begin{bmatrix} -2 \\ 1 \\ 1 \\ 0 \\ 1 \end{bmatrix}, \begin{bmatrix} -2 \\ 4 \\ 0 \\ 1 \\ 0 \end{bmatrix} \right\}
$$

is a basis of $\ker(f - 1)$.

It is easy to see that $\begin{bmatrix} 1 \\ 0 \\ 0 \\ 0 \\ 0 \end{bmatrix}$ is not in $\ker(f - 1)^2$ and $\begin{bmatrix} -1 \\ 0 \\ 1 \\ 0 \\ 0 \end{bmatrix} \in \ker(f - 1)^2$

is not in

$$
\mathrm{Span}\left\{(f-1)\begin{bmatrix}1\\0\\0\\0\\0\end{bmatrix},\begin{bmatrix}-2\\1\\1\\0\\1\end{bmatrix},\begin{bmatrix}-2\\4\\0\\1\\0\end{bmatrix}\right\}=\mathrm{Span}\left\{\begin{bmatrix}-8\\-4\\7\\-3\\7\end{bmatrix},\begin{bmatrix}-2\\1\\1\\0\\1\end{bmatrix},\begin{bmatrix}-2\\4\\0\\1\\0\end{bmatrix}\right\}.
$$

Consequently the set

$$
\mathcal{B}=\left\{(f-1)^2\begin{bmatrix}1\\0\\0\\0\\0\end{bmatrix},(f-1)\begin{bmatrix}1\\0\\0\\0\\0\end{bmatrix},\begin{bmatrix}1\\0\\0\\0\\0\end{bmatrix},(f-1)\begin{bmatrix}-1\\0\\1\\0\\0\end{bmatrix},\begin{bmatrix}-1\\0\\1\\0\\0\end{bmatrix}\right\}
$$

$$
=\left\{\begin{bmatrix}2\\2\\-2\\1\\-2\end{bmatrix},\begin{bmatrix}-8\\-4\\7\\-3\\7\end{bmatrix},\begin{bmatrix}1\\0\\0\\0\\0\end{bmatrix},\begin{bmatrix}-6\\-3\\5\\-2\\5\end{bmatrix},\begin{bmatrix}-1\\0\\1\\0\\0\end{bmatrix}\right\}
$$

is a basis of $\mathbb{R}^5$ with the desired properties, and the $\mathcal{B}$-matrix of f is

$$
\left[\begin{array}{ccc|cc}
1 & 1 & 0 & 0 & 0\\
0 & 1 & 1 & 0 & 0\\
0 & 0 & 1 & 0 & 0\\
\hline
0 & 0 & 0 & 1 & 1\\
0 & 0 & 0 & 0 & 1
\end{array}\right].
$$

□

Example 4.2.24. Let $f : \mathcal{V} \to \mathcal{V}$ be an endomorphism on a vector space $\mathcal{V}$ and let $\alpha \in \mathbb{K}$. If $\mathcal{B} = \{v_1, v_2, v_3, v_4, v_5, v_6, v_7, v_8, v_9\}$ is a basis of $\mathcal{V}$ such that the $\mathcal{B}$-matrix of f is

$$
\left[\begin{array}{cccc|ccc|cc}
\alpha & 1 & 0 & 0 & 0 & 0 & 0 & 0 & 0\\
0 & \alpha & 1 & 0 & 0 & 0 & 0 & 0 & 0\\
0 & 0 & \alpha & 1 & 0 & 0 & 0 & 0 & 0\\
0 & 0 & 0 & \alpha & 0 & 0 & 0 & 0 & 0\\
\hline
0 & 0 & 0 & 0 & \alpha & 1 & 0 & 0 & 0\\
0 & 0 & 0 & 0 & 0 & \alpha & 1 & 0 & 0\\
0 & 0 & 0 & 0 & 0 & 0 & \alpha & 0 & 0\\
\hline
0 & 0 & 0 & 0 & 0 & 0 & 0 & \alpha & 0\\
0 & 0 & 0 & 0 & 0 & 0 & 0 & 0 & \alpha
\end{array}\right],
$$

find bases of
$$\ker(f - \alpha), \ker(f - \alpha)^2 \quad \text{and} \quad \ker(f - \alpha)^3$$
and the polynomial m_f.

Solution. Since
$$f(\mathbf{v}_1) = \alpha \mathbf{v}_1,$$
$$f(\mathbf{v}_2) = \mathbf{v}_1 + \alpha \mathbf{v}_2,$$
$$f(\mathbf{v}_3) = \mathbf{v}_2 + \alpha \mathbf{v}_3,$$
$$f(\mathbf{v}_4) = \mathbf{v}_3 + \alpha \mathbf{v}_4,$$
$$f(\mathbf{v}_5) = \alpha \mathbf{v}_5,$$
$$f(\mathbf{v}_6) = \mathbf{v}_5 + \alpha \mathbf{v}_6,$$
$$f(\mathbf{v}_7) = \mathbf{v}_6 + \alpha \mathbf{v}_7,$$
$$f(\mathbf{v}_8) = \alpha \mathbf{v}_8,$$
$$f(\mathbf{v}_9) = \alpha \mathbf{v}_9,$$

$\{\mathbf{v}_1, \mathbf{v}_5, \mathbf{v}_8, \mathbf{v}_9\}$ is a basis of $\ker(f - \alpha)$,
$\{\mathbf{v}_1, \mathbf{v}_2, \mathbf{v}_5, \mathbf{v}_6, \mathbf{v}_8, \mathbf{v}_9\}$ is a basis of $\ker(f - \alpha)^2$,
$\{\mathbf{v}_1, \mathbf{v}_2, \mathbf{v}_3, \mathbf{v}_5, \mathbf{v}_6, \mathbf{v}_7, \mathbf{v}_8, \mathbf{v}_9\}$ is a basis of $\ker(f - \alpha)^3$,
and
$$m_f(t) = (t - \alpha)^4. \qquad \Box$$

4.2.2 Uniqueness of the Jordan canonical form when the characteristic polynomial has one root

In this section we present a formula for the number of Jordan blocks in the Jordan canonical form of an endomorphism with the characteristic polynomial that has only one root. This result gives us a uniqueness theorem for the Jordan canonical form for such endomorphisms. First we need some preliminary results.

Lemma 4.2.25. *Let $\mathcal{U}$ be a finite dimensional vector space and let $g : \mathcal{U} \to \mathcal{U}$ be an endomorphism. If the Jordan canonical form of g is a Jordan $k \times k$ block $J_{\alpha,k}$ for some $\alpha \in \mathbb{K}$, then $\dim \operatorname{ran}(g - \alpha)^m = k - m$ for $1 \le m \le k - 1$ and $\dim \operatorname{ran}(g - \alpha)^m = 0$ for $m \ge k$.*

Proof. Let $\mathcal{B} = \{\mathbf{v}_1, \ldots, \mathbf{v}_k\}$ be a basis of $\mathcal{U}$ such that the $\mathcal{B}$-matrix of g is the Jordan $k \times k$ block $J_{\alpha,k}$. Then
$$(g - \alpha)(\mathbf{v}_1) = \mathbf{0}, \ (g - \alpha)(\mathbf{v}_2) = \mathbf{v}_1, \ \ldots, \ (g - \alpha)(\mathbf{v}_k) = \mathbf{v}_{k-1}.$$
This means that $\operatorname{ran}(g - \alpha) = \operatorname{Span}\{\mathbf{v}_1, \ldots, \mathbf{v}_{k-1}\}$. Since
$$(g-\alpha)^2(\mathbf{v}_1) = \mathbf{0}, \ (g-\alpha)^2(\mathbf{v}_2) = \mathbf{0}, \ (g-\alpha)^2(\mathbf{v}_3) = \mathbf{v}_1, \ \ldots, \ (g-\alpha)^2(\mathbf{v}_k) = \mathbf{v}_{k-2},$$

we have $\operatorname{ran}(g - \alpha)^2 = \operatorname{Span}\{\mathbf{v}_1, \ldots, \mathbf{v}_{k-2}\}$.

Continuing the same way we get $\operatorname{ran}(g - \alpha)^m = \operatorname{Span}\{\mathbf{v}_1, \ldots, \mathbf{v}_{k-m}\}$ for $m \leq k-1$ and $\operatorname{ran}(g-\alpha)^m = \mathbf{0}$ for $m \geq k$. This gives us $\dim \operatorname{ran}(g-\alpha)^m = k-m$ for $1 \leq m \leq k - 1$ and $\dim \operatorname{ran}(g - \alpha)^m = 0$ for $m \geq k$. □

Lemma 4.2.26. *Let V be a finite dimensional vector space and let $f : V \to V$ be an endomorphism. If the diagonal Jordan blocks of the Jordan canonical form of f are $J_{\alpha, n_1}, \ldots, J_{\alpha, n_p}$ for some positive integers $n_1, \ldots, n_p$, then*

$$\dim \operatorname{ran}(f - \alpha)^{k-1} = \{\text{number of Jordan blocks } J_{\alpha,k}\} + \sum_{n_q > k} (n_q - k + 1),$$

for every integer $k \geq 1$.

Proof. We have

$$V = \mathcal{U}_1 \oplus \cdots \oplus \mathcal{U}_p,$$

where, for every $q \in \{1, \ldots, p\}$, $\mathcal{U}_q$ is a vector subspace of V with a basis $\mathcal{B}_q$ such that the $\mathcal{B}_q$-matrix of the endomorphism $f_q : \mathcal{U}_q \to \mathcal{U}_q$ induced by f on $\mathcal{U}_q$ is J_{α, n_q}. For any vectors $\mathbf{v} \in V, \mathbf{u}_1 \in \mathcal{U}_1, \ldots, \mathbf{u}_p \in \mathcal{U}_p$ such that $\mathbf{v} = \mathbf{u}_1 + \cdots + \mathbf{u}_p$, we have

$$f^m(\mathbf{v}) = f_1^m(\mathbf{u}_1) + \cdots + f_p^m(\mathbf{u}_p),$$

for every integer $m \geq 1$. By Lemma 4.2.25,

$$\dim \operatorname{ran}(f_q - \alpha)^{k-1} = 0 \text{ if } n_q < k,$$

$$\dim \operatorname{ran}(f_q - \alpha)^{k-1} = 1 \text{ if } n_q = k,$$

$$\dim \operatorname{ran}(f_q - \alpha)^{k-1} = n_q - (k - 1) = n_q - k + 1 \text{ if } n_q > k.$$

The desired result is a consequence of these equalities. □

Lemma 4.2.27. *Let V be a finite dimensional vector space and let $f : V \to V$ be an endomorphism. If the diagonal Jordan blocks of the Jordan canonical form of f are $J_{\alpha, n_1}, \ldots, J_{\alpha, n_p}$, for some positive integers $n_1, \ldots, n_p$, then*

$$\{\text{number of Jordan blocks } J_{\alpha,k}\}$$
$$= \dim \operatorname{ran}(f - \alpha)^{k-1} + \dim \operatorname{ran}(f - \alpha)^{k+1} - 2 \dim \operatorname{ran}(f - \alpha)^k,$$

for every integer $k \geq 1$.

Proof. By Lemma 4.2.26 we have

$$\dim \mathrm{ran}(f - \alpha)^{k-1} = \{\text{number of Jordan blocks } J_{\alpha,k}\} + \sum_{n_q > k} (n_q - k + 1)$$

and by Lemma 4.2.25 we have

$$\dim \mathrm{ran}(f - \alpha)^k = \sum_{n_q \geq k+1} (n_q - k) = \sum_{n_q > k} (n_q - k)$$

and

$$\dim \mathrm{ran}(f - \alpha)^{k+1} = \sum_{n_q \geq k+2} (n_q - k - 1) = \sum_{n_q \geq k+1} (n_q - k - 1) = \sum_{n_q > k} (n_q - k - 1).$$

Consequently

$$\dim \mathrm{ran}(f - \alpha)^{k-1} + \dim \mathrm{ran}(f - \alpha)^{k+1} - 2 \dim \mathrm{ran}(f - \alpha)^k$$

$$= \{\text{number of Jordan blocks } J_{\alpha,k}\} + \sum_{n_q > k} (n_q - k + 1) + \sum_{n_q > k} (n_q - k - 1)$$

$$- 2 \sum_{n_q > k} (n_q - k)$$

$$= \{\text{number of Jordan blocks } J_{\alpha,k}\}.$$

$\square$

Theorem 4.2.28. *Let V be a finite dimensional vector space and let $f : V \to V$ be an endomorphism. If the diagonal Jordan blocks of the Jordan canonical form of f are $J_{\alpha,n_1}, \ldots, J_{\alpha,n_p}$, for some positive integers $n_1, \ldots, n_p$, then*

$$\{\text{number of Jordan blocks } J_{\alpha,k}\}$$
$$= 2 \dim \ker(f - \alpha)^k - \dim \ker(f - \alpha)^{k-1} - \dim \ker(f - \alpha)^{k+1},$$

for every integer $k \geq 1$.

Proof. This equality is an immediate consequence of Lemma 4.2.27 and the Rank-Nullity Theorem 2.1.28. $\square$

Corollary 4.2.29. *Let V be a finite dimensional vector space and let $f : V \to V$ be an endomorphism. If the minimal polynomial of f is $m_f = (t - \alpha)^n$ for some integer $n \geq 1$, then the number of Jordan blocks in the Jordan canonical form of f is $\dim \ker(f - \alpha)$.*

Proof. According to Theorem 4.2.28 the number of $n \times n$ Jordan blocks $J_{\alpha,n}$ is

$$2 \dim \ker(f - \alpha)^n - \dim \ker(f - \alpha)^{n+1} - \dim \ker(f - \alpha)^{n-1}$$
$$= \dim \ker(f - \alpha)^n - \dim \ker(f - \alpha)^{n-1},$$

the number of $m \times m$ Jordan blocks $J_{\alpha,m}$, where $m \in \{2, \ldots, n-1\}$, is

$$2 \dim \ker(f - \alpha)^m - \dim \ker(f - \alpha)^{m+1} - \dim \ker(f - \alpha)^{m-1},$$

and the number of 1×1 Jordan blocks $J_{\alpha,1}$, that is the blocks which are eigen-values, is

$$2 \dim \ker(f - \alpha) - \dim \ker(f - \alpha)^2 - \dim \ker(f - \alpha)^0$$
$$= 2 \dim \ker(f - \alpha) - \dim \ker(f - \alpha)^2 - \ker \mathrm{Id}$$
$$= 2 \dim \ker(f - \alpha) - \dim \ker(f - \alpha)^2.$$

Consequently the number of Jordan blocks is

$$\dim \ker(f - \alpha)^n - \dim \ker(f - \alpha)^{n-1}$$
$$+ 2 \dim \ker(f - \alpha)^{n-1} - \dim \ker(f - \alpha)^n - \dim \ker(f - \alpha)^{n-2}$$
$$+ \cdots + 2 \dim \ker(f - \alpha)^2 - \dim \ker(f - \alpha)^3 - \dim \ker(f - \alpha)$$
$$+ 2 \dim \ker(f - \alpha) - \dim \ker(f - \alpha)^2$$
$$= \dim \ker(f - \alpha).$$

□

4.2.3 Jordan canonical form when the characteristic polynomial has several roots

We begin by considering an example.

Example 4.2.30. Let V be a 4-dimensional vector space and let $f : V \to V$ be an endomorphism. If the characteristic polynomial of f is $(t - \alpha)^2 (t - \beta)^2$ for two distinct numbers α and β and

$$\dim \ker(f - \alpha) = 1, \ \dim \ker(f - \alpha)^2 = 2, \ \dim \ker(f - \beta) = 2,$$

show that there is a basis $\mathcal{B}$ of V such that the $\mathcal{B}$- matrix of f is

$$A = \begin{bmatrix} \alpha & 1 & 0 & 0 \\ 0 & \alpha & 0 & 0 \\ 0 & 0 & \beta & 0 \\ 0 & 0 & 0 & \beta \end{bmatrix}.$$

Solution. Let $\mathbf{u}$ be a vector in $\ker(f - \alpha)^2$ which is not in $\ker(f - \alpha)$ and let $\{\mathbf{v}, \mathbf{w}\}$ be a basis of $\ker(f - \beta)$. Then $(f - \alpha)\mathbf{u}$ is a nonzero vector in $\ker(f - \alpha)$.

We will show that

$$\mathcal{B} = \{(f-\alpha)\mathbf{u}, \mathbf{u}, \mathbf{v}, \mathbf{w}\}$$

is a basis with the desired properties.

Let

$$x_1(f-\alpha)\mathbf{u} + x_2\mathbf{u} + x_3\mathbf{v} + x_4\mathbf{w} = \mathbf{0} \qquad (4.7)$$

for some $x_1, x_2, x_3, x_4 \in \mathbb{K}$. By applying $(f-\alpha)^2$ to (4.7) we get

$$x_1(f-\alpha)^3\mathbf{u} + x_2(f-\alpha)^2(\mathbf{u}) + x_3(f-\alpha)^2(\mathbf{v}) + x_4(f-\alpha)^2(\mathbf{w}) = \mathbf{0}$$

and thus

$$x_3(f-\alpha)^2(\mathbf{v}) + x_4(f-\alpha)^2(\mathbf{w}) = \mathbf{0}.$$

Hence $x_3 = x_4 = 0$ because

$$(f-\alpha)^2 = (f-\beta)^2 + 2(\beta-\alpha)(f-\beta) + (\beta-\alpha)^2.$$

Next, using the fact that $x_3 = x_4 = 0$ and applying $f-\alpha$ to (4.7), we get $x_2 = 0$ and then $x_1 = 0$. This shows that the vectors $(f-\alpha)\mathbf{u}$, $\mathbf{u}$, $\mathbf{v}$, and $\mathbf{w}$ are linearly independent and consequently $\{(f-\alpha)\mathbf{u}, \mathbf{u}, \mathbf{v}, \mathbf{w}\}$ is a basis of $\mathbb{R}^4$.

Since

$$f((f-\alpha)(\mathbf{u})) = (f-\alpha)^2(\mathbf{u}) + \alpha(f-\alpha)(\mathbf{u}) = \alpha(f-\alpha)(\mathbf{u}),$$
$$f(\mathbf{u}) = (f-\alpha)\mathbf{u} + \alpha\mathbf{u},$$
$$f(\mathbf{v}) = \beta\mathbf{v},$$
$$f(\mathbf{w}) = \beta\mathbf{w},$$

the $\mathcal{B}$-matrix of f is A.   □

The next two results follow from Theorems 4.2.14 and 4.1.45.

Theorem 4.2.31. *Let V be an n-dimensional vector space and let $f : V \to V$ be an endomorphism. If*

$$c_f(t) = (\lambda_1 - t)^{r_1} \cdots (\lambda_k - t)^{r_k}$$

for some distinct $\lambda_1, \ldots, \lambda_k \in \mathbb{K}$ and some positive integers $r_1, \ldots, r_k$, then there are vectors $\mathbf{x}_1, \ldots, \mathbf{x}_m \in V$ such that

$$V = V_{f,\mathbf{x}_1} \oplus \cdots \oplus V_{f,\mathbf{x}_m}$$

and for every $j \in \{1, \ldots, m\}$ there is an $l \in \{1, \ldots, k\}$ and an integer $q_l \leq r_l$ such that

$$m_{f,\mathbf{x}_j}(t) = (t - \lambda_l)^{q_l}.$$

Theorem 4.2.32. *Let V be an n-dimensional vector space and let $f :$ $V \to V$ be an endomorphism. If*

$$c_f(t) = (\lambda_1 - t)^{r_1} \cdots (\lambda_k - t)^{r_k}$$

for some distinct $\lambda_1, \ldots, \lambda_k \in \mathbb{K}$ and some positive integers $r_1, \ldots, r_k$, then there is a basis $\mathcal{B}$ of V such that the $\mathcal{B}$-matrix of f is of the form

$$\begin{bmatrix} J_1 & & & \\ & J_2 & & \mathbf{0} \\ & & \ddots & \\ \mathbf{0} & & & J_r \end{bmatrix}$$

and for every $m \in \{1, \ldots, r\}$ there is an integer $q \in \{1, \ldots, k\}$ such that J_m is in α_q-Jordan canonical form.

Definition 4.2.33. We say that a matrix is in *Jordan canonical form* if it has the form

$$\begin{bmatrix} J_1 & & & \\ & J_2 & & \mathbf{0} \\ & & \ddots & \\ \mathbf{0} & & & J_r \end{bmatrix}$$

and for every $k \in \{1, \ldots, r\}$ there is a number $\alpha_k \in \mathbb{K}$ such that J_k is in α_k-Jordan canonical form.

Example 4.2.34. Let $f : V \to V$ be an endomorphism and let $\alpha, \beta, \gamma, \delta \in \mathbb{K}$ be such that

$$c_f(t) = (t - \alpha)^4 (t - \beta)^2 (t - \gamma)^2 (t - \delta)^2$$

and

$$\dim \ker (f - \alpha I)^3 = 4, \quad \dim \ker (f - \alpha I)^2 = 3, \quad \dim \ker (f - \alpha I) = 2,$$

$$\dim \ker (f - \beta I)^2 = 2, \quad \dim \ker (f - \beta I) = 1,$$

$$\dim \ker (f - \gamma I) = 2,$$

$$\dim \ker (f - \delta I)^2 = 2, \quad \dim \ker (f - \delta I) = 1.$$

By Theorem 4.2.31 there is a basis $\mathcal{B}$ of $\mathcal{V}$ such that the $\mathcal{B}$-matrix of f has following Jordan canonical form

$$\begin{bmatrix} \alpha & 1 & 0 & 0 & 0 & 0 & 0 & 0 & 0 & 0 \\ 0 & \alpha & 1 & 0 & 0 & 0 & 0 & 0 & 0 & 0 \\ 0 & 0 & \alpha & 0 & 0 & 0 & 0 & 0 & 0 & 0 \\ 0 & 0 & 0 & \alpha & 0 & 0 & 0 & 0 & 0 & 0 \\ 0 & 0 & 0 & 0 & \beta & 1 & 0 & 0 & 0 & 0 \\ 0 & 0 & 0 & 0 & 0 & \beta & 0 & 0 & 0 & 0 \\ 0 & 0 & 0 & 0 & 0 & 0 & \gamma & 0 & 0 & 0 \\ 0 & 0 & 0 & 0 & 0 & 0 & 0 & \gamma & 0 & 0 \\ 0 & 0 & 0 & 0 & 0 & 0 & 0 & 0 & \delta & 1 \\ 0 & 0 & 0 & 0 & 0 & 0 & 0 & 0 & 0 & \delta \end{bmatrix}.$$

4.3   The rational form

When considering Jordan forms we assumed that the minimal polynomial of the endomorphism splits over $\mathbb{K}$. In this section we consider endomorphisms such that their minimal polynomial does not necessary split and construct bases for such endomorphisms for which the matrix has a simple form. We begin by considering an example.

Example 4.3.1. Let $\mathcal{V}$ be a 4-dimensional vector space over $\mathbb{R}$ and let $f :$ $\mathcal{V} \to \mathcal{V}$ be an endomorphism. If $m_f(t) = (t^2 + 1)^2$, show that there is a basis $\mathcal{B}$ of $\mathcal{V}$ such that the $\mathcal{B}$-matrix of f is

$$\begin{bmatrix} 0 & 0 & 0 & -1 \\ 1 & 0 & 0 & 0 \\ 0 & 1 & 0 & -2 \\ 0 & 0 & 1 & 0 \end{bmatrix}.$$

Solution. Let $\mathbf{v}$ be a vector from $\mathcal{V}$ such $\mathbf{v} \in \ker(f^2 + 1)^2$ and $\mathbf{v} \notin \ker(f^2 + 1)$. Then $m_{f,\mathbf{v}} = m_f$ and it is easy to see that

$$\mathcal{B} = \left\{ \mathbf{v}, f(\mathbf{v}), f^2(\mathbf{v}), f^3(\mathbf{v}) \right\}$$

is a basis of $\mathcal{V}$. Since $f(\mathbf{v}) = f(\mathbf{v})$, $f(f(\mathbf{v})) = f^2(\mathbf{v})$, $f(f^2(\mathbf{v})) = f^3(\mathbf{v})$ and

$$(f^2 + 1)^2(\mathbf{v}) = f^4(\mathbf{v}) + 2f^2(\mathbf{v}) + 1 = \mathbf{0},$$

we have

$$f(f^3(\mathbf{v})) = f^4(\mathbf{v}) = -1 - 2f^2(\mathbf{v}).$$

Consequently, the $\mathcal{B}$-matrix of f is

$$\begin{bmatrix} 0 & 0 & 0 & -1 \\ 1 & 0 & 0 & 0 \\ 0 & 1 & 0 & -2 \\ 0 & 0 & 1 & 0 \end{bmatrix}.$$

Note that

$$m_{f,\mathbf{v}}(t) = 1 + 2t^2 + t^4$$

and that the entries in the last column in the $\mathcal{B}$-matrix of f are obtained from the coefficients of the polynomial $m_{f,\mathbf{v}}(t)$. □

The following result generalizes the observation in the above example.

Theorem 4.3.2. *Let V be a finite dimensional vector space and let $\mathbf{v} \in V$ be a nonzero vector. If $f : V \to V$ is a nonzero endomorphism such that*

$$m_{f,\mathbf{v}}(t) = a_0 + a_1 t + \cdots + a_{k-1} t^{k-1} + t^k,$$

for some integer $k \geq 1$, then $\mathcal{B} = \{\mathbf{v}, f(\mathbf{v}), \ldots, f^{k-1}(\mathbf{v})\}$ is a basis of $V_{f,\mathbf{v}}$ and the $\mathcal{B}$-matrix of the endomorphism $f_{\mathbf{v}} : V_{f,\mathbf{v}} \to V_{f,\mathbf{v}}$ induced by f is

$$A = \begin{bmatrix} 0 & 0 & 0 & 0 & \cdots & 0 & 0 & -a_0 \\ 1 & 0 & 0 & 0 & \cdots & 0 & 0 & -a_1 \\ 0 & 1 & 0 & 0 & \cdots & 0 & 0 & -a_2 \\ \vdots & \vdots & \vdots & \vdots & \ddots & \vdots & \vdots & \vdots \\ 0 & 0 & 0 & 0 & \cdots & 0 & 0 & -a_{k-3} \\ 0 & 0 & 0 & 0 & \cdots & 1 & 0 & -a_{k-2} \\ 0 & 0 & 0 & 0 & \cdots & 0 & 1 & -a_{k-1} \end{bmatrix}.$$

Proof. We already know that $\mathcal{B}$ is a basis of $V_{f,\mathbf{v}}$. To show that A is the $\mathcal{B}$-matrix of the endomorphism $f_{\mathbf{v}}$ it suffices to observe that

$$f(\mathbf{v}) = f(\mathbf{v})$$
$$f(f(\mathbf{v})) = f^2(\mathbf{v})$$

$$\vdots$$

$$f(f^{k-3}(\mathbf{v})) = f^{k-2}(\mathbf{v})$$
$$f(f^{k-2}(\mathbf{v})) = f^{k-1}(\mathbf{v})$$
$$f(f^{k-1}(\mathbf{v})) = f^k(\mathbf{v}) = -a_0\mathbf{v} - a_1 f(\mathbf{v}) - \cdots - a_{k-1} f^{k-1}(\mathbf{v}).$$

□

The next theorem is a form of converse of the above theorem. Note that it implies that every polynomial is a minimal polynomial of some endomorphism on a finite dimensional vector space.

Theorem 4.3.3. *Let V be a finite dimensional vector space and let $\mathcal{B} = \{v_1, \ldots, v_k\}$ be a basis of V. If $f : V \to V$ is an endomorphism such that*

$$\begin{bmatrix} 0 & 0 & 0 & 0 & \ldots & 0 & 0 & -a_0 \\ 1 & 0 & 0 & 0 & \ldots & 0 & 0 & -a_1 \\ 0 & 1 & 0 & 0 & \ldots & 0 & 0 & -a_2 \\ \vdots & \vdots & \vdots & \vdots & \ddots & \vdots & \vdots & \vdots \\ 0 & 0 & 0 & 0 & \ldots & 0 & 0 & -a_{k-3} \\ 0 & 0 & 0 & 0 & \ldots & 1 & 0 & -a_{k-2} \\ 0 & 0 & 0 & 0 & \ldots & 0 & 1 & -a_{k-1} \end{bmatrix}$$

is the $\mathcal{B}$-matrix of f for some $a_0, a_1, \ldots, a_{k-1} \in \mathbb{K}$, then $v_j = f^{j-1}(v_1)$ for $j \in \{2, \ldots, k\}$ and

$$m_{f,v_1}(t) = m_f(t) = a_0 + a_1 t + \cdots + a_{k-1} t^{k-1} + t^k.$$

Proof. Since

$$f(v_1) = v_2,$$
$$f(v_2) = v_3 = f^2(v_1),$$
$$f(v_3) = v_4 = f^3(v_1),$$
$$\vdots$$
$$f(v_{k-1}) = v_k = f^{k-1}(v_1),$$
$$f(v_k) = f(f^{k-1}(v_1)) = f^k(v_1) = -a_0 v_1 - a_1 f(v_1) - \cdots - a_{k-1} f^{k-1}(v_1),$$

we have

$$a_0 v_1 + a_1 f(v_1) + \cdots + a_{k-1} f^{k-1}(v_1) + f^k(v_1) = 0.$$

Linear independence of vectors $v_1, \ldots, v_k$ implies that

$$m_{f,v_1}(t) = a_0 + a_1 t + \cdots + a_{k-1} t^{k-1} + t^k.$$

Moreover,

$$m_{f,v_1}(v_2) = m_{f,v_1}(f(v_1)) = f(m_{f,v_1}(v_1)) = 0$$
$$\vdots$$
$$m_{f,v_1}(v_k) = m_{f,v_1}(f^{k-1}(v_1)) = f^{k-1}(m_{f,v_1}(v_1)) = 0,$$

which shows that m_{f,v_1} is an annihilator of f and $m_f = m_{f,v_1}$. $\qquad\square$

Definition 4.3.4. Let $a_0, \ldots, a_{k-1} \in \mathbb{K}$. The matrix

$$
\begin{bmatrix}
0 & 0 & 0 & 0 & \cdots & 0 & 0 & -a_0 \\
1 & 0 & 0 & 0 & \cdots & 0 & 0 & -a_1 \\
0 & 1 & 0 & 0 & \cdots & 0 & 0 & -a_2 \\
\vdots & \vdots & \vdots & \vdots & \ddots & \vdots & \vdots & \vdots \\
0 & 0 & 0 & 0 & \cdots & 0 & 0 & -a_{k-3} \\
0 & 0 & 0 & 0 & \cdots & 1 & 0 & -a_{k-2} \\
0 & 0 & 0 & 0 & \cdots & 0 & 1 & -a_{k-1}
\end{bmatrix},
$$

is called the *companion matrix* of the polynomial

$$
p(t) = a_0 + a_1 t + \cdots + a_{k-1} t^{k-1} + t^k
$$

and is denoted by c_p.

In the next example we consider a matrix where instead of Jordan blocks we have companion matrices.

Example 4.3.5. Let V be a vector space such that $\dim V = 5$. If $f : V \to V$ is an endomorphism such that $c_f(t) = (t - \alpha)^5$ and $m_f(t) = (\alpha - t)^3$, for some $\alpha \in \mathbb{K}$, and

$$
\dim \ker(f - \alpha) = 2 \quad \text{and} \quad \dim \ker(f - \alpha)^2 = 4,
$$

show that there is a basis $\mathcal{B}$ of V such that the $\mathcal{B}$-matrix of f is

$$
\begin{bmatrix}
0 & 0 & \alpha^3 & 0 & 0 \\
1 & 0 & -3\alpha^2 & 0 & 0 \\
0 & 1 & 3\alpha & 0 & 0 \\
0 & 0 & 0 & 0 & -\alpha^2 \\
0 & 0 & 0 & 1 & 2\alpha
\end{bmatrix}.
$$

Solution. Let $\mathbf{u}$ and $\mathbf{v}$ be the vectors from Example 4.2.12. We will show that

$$
\mathcal{B} = \{\mathbf{u}, f(\mathbf{u}), f^2(\mathbf{u}), \mathbf{v}, f(\mathbf{v})\}
$$

is a basis of V with the desired property.
    First we note that

$$
\text{Span}\{(f - \alpha)^2 \mathbf{u}, (f - \alpha)\mathbf{u}, \mathbf{u}\} = \text{Span}\{\mathbf{u}, f(\mathbf{u}), f^2(\mathbf{u})\}
$$

and

$$
\text{Span}\{(f - \alpha)\mathbf{v}, \mathbf{v}\} = \text{Span}\{\mathbf{v}, f(\mathbf{v})\}.
$$

Now, since $\{(f-\alpha)^2\mathbf{u}, (f-\alpha)\mathbf{u}, \mathbf{u}, (f-\alpha)\mathbf{v}, \mathbf{v}\}$ is a basis, the set $\{\mathbf{u}, f(\mathbf{u}), f^2(\mathbf{u}), \mathbf{v}, f(\mathbf{v})\}$ is also a basis.

From

$$(f-\alpha)^3(\mathbf{u}) = f^3(\mathbf{u}) - 3\alpha f^2(\mathbf{u}) + 3\alpha^2 f(\mathbf{u}) - \alpha^3\mathbf{u} = \mathbf{0}$$

we get

$$f(f^2(\mathbf{u})) = f^3(\mathbf{u}) = \alpha^3\mathbf{u} - 3\alpha^2 f(\mathbf{u}) + 3\alpha f^2(\mathbf{u}).$$

Since we also have

$$f(f(\mathbf{v})) = f^2(\mathbf{v}) = -\alpha^2\mathbf{v} + 2\alpha f(\mathbf{v}),$$

the $\mathcal{B}$-matrix of f is

$$\begin{bmatrix} 0 & 0 & \alpha^3 & 0 & 0 \\ 1 & 0 & -3\alpha^2 & 0 & 0 \\ 0 & 1 & 3\alpha & 0 & 0 \\ 0 & 0 & 0 & 0 & -\alpha^2 \\ 0 & 0 & 0 & 1 & 2\alpha \end{bmatrix}.$$

$\square$

Theorem 4.3.6. *Let $\mathcal{V}$ be a finite dimensional vector space and let $f : \mathcal{V} \to \mathcal{V}$ be a nonzero endomorphism such that $m_f = pq$, where p and q are two monic polynomials such that $\mathbf{GCD}(p, q) = 1$. If g is the endomorphism induced by f on $\ker p(f)$ and h is the endomorphism induced by f on $\ker q(f)$, then $m_g = p$ and $m_h = q$.*

Proof. Let a and b be polynomials such that $ap + bq = 1$. Since $m_g(g) = \mathbf{0}$, we have $m_g(g)b(g)q(g) = \mathbf{0}$ and thus $m_g(f)b(f)q(f)(\mathbf{x}) = \mathbf{0}$ for every $\mathbf{x} \in \ker p(f)$. Moreover, $m_g(f)b(f)q(f)(\mathbf{y}) = \mathbf{0}$ for every $\mathbf{y} \in \ker q(f)$. Consequently, $m_g(f)b(f)q(f) = \mathbf{0}$, because every vector from $\mathcal{V}$ is of the form $\mathbf{x} + \mathbf{y}$ where $\mathbf{x} \in \ker p(f)$ and $\mathbf{y} \in \ker q(f)$, by Theorem 4.1.44. Clearly, $m_f = pq$ divides $m_g bq = m_g(1 - ap)$. Consequently, p divides $m_g(1 - ap)$, which implies that p divides m_g, because $\mathbf{GCD}(p, 1 - ap) = 1$.

For every $\mathbf{x} \in \ker p(f)$ we have $p(g)(\mathbf{x}) = p(f)(\mathbf{x}) = \mathbf{0}$, which implies that m_g divides p. Since the monic polynomials p and m_g divide each other, we have $p = m_g$. In a similar way we can show that $q = m_h$. $\square$

In Theorem 4.2.7 we show that, if $\mathcal{V}$ is a finite dimensional vector space and $f : \mathcal{V} \to \mathcal{V}$ is an endomorphism such that $m_f(t) = (t - \alpha)^k$ for some $\alpha \in \mathbb{K}$, then there is a vector $\mathbf{v} \in \mathcal{V}$ such that $m_f = m_{f,\mathbf{v}}$. Now we can show that the assumption that $m_f(t) = (t - \alpha)^k$ is unnecessary.

Theorem 4.3.7. *Let V be a finite dimensional vector space. For every nonzero endomorphism $f : V \to V$ there is a nonzero vector $\mathbf{v} \in V$ such that*

$$m_f = m_{f,\mathbf{v}}.$$

Proof. Let $m_f = p_1^{\alpha_1} \dots p_k^{\alpha_k}$ where $p_1, \dots, p_k$ are irreducible monic polynomials and $\alpha_1, \dots, \alpha_k$ are positive integers. By Theorem 4.3.6, for every $j \in \{1, \dots, k\}$ the polynomial $p_j^{\alpha_j}$ is the minimal polynomial of the endomorphism f_j induced by f on $\ker p_j^{\alpha_j}(f)$. By Theorem 4.1.44, we have

$$V = \ker p_1^{\alpha_1} \oplus \cdots \oplus \ker p_k^{\alpha_k}. \tag{4.8}$$

For every $j \in \{1, \dots, k\}$ we choose a vector $\mathbf{v}_j \in \ker p_j^{\alpha_j}(f)$ such that $\mathbf{v}_j \notin \ker p_j^{\alpha_j - 1}(f)$. Because the polynomial p_j is irreducible this means that $m_{f,\mathbf{v}_j} = p_j^{\alpha_j}$. Now, since

$$\mathbf{0} = m_{f,\mathbf{v}_1 + \cdots + \mathbf{v}_k}(f)(\mathbf{v}_1 + \cdots + \mathbf{v}_k)$$
$$= m_{f,\mathbf{v}_1 + \cdots + \mathbf{v}_k}(f)(\mathbf{v}_1) + \cdots + m_{f,\mathbf{v}_1 + \cdots + \mathbf{v}_k}(f)(\mathbf{v}_k)$$

and because $m_{f,\mathbf{v}_1 + \cdots + \mathbf{v}_k}(f)(\mathbf{v}_j) \in \ker p_j^{\alpha_j}$, it follows from (4.8) that $m_{f,\mathbf{v}_1 + \cdots + \mathbf{v}_k}(f)(\mathbf{v}_j) = \mathbf{0}$ for every $j \in \{1, \dots, k\}$. Hence, using the equality $m_{f,\mathbf{v}_j} = p_j^{\alpha_j}$, it follows that the polynomial $p_j^{\alpha_j}$ divides the polynomial $m_{f,\mathbf{v}_1 + \cdots + \mathbf{v}_k}(t)$. Consequently

$$m_{f,\mathbf{v}_1 + \cdots + \mathbf{v}_k} = p_1^{\alpha_1} \dots p_k^{\alpha_k} = m_f$$

because, by definition, $m_{f,\mathbf{v}_1 + \cdots + \mathbf{v}_k}$ divides the product $p_1^{\alpha_1} \dots p_k^{\alpha_k}$. $\qquad \square$

Example 4.3.8. Let V be a vector space such that $\dim V = 4$ and let $f : V \to V$ be an endomorphism. We assume that $V = V_1 \oplus V_2$ where V_1 and V_2 are two invariant subspaces of f. Let $f_1 : V_1 \to V_1$ and $f_2 : V_2 \to V_2$ be the induced endomorphisms. If $m_{f_1} = t^2$ and $m_{f_2} = t^2 - t + 1$, show that there is a basis $\mathcal{B}$ of V such that the $\mathcal{B}$-matrix of f is

$$\begin{bmatrix} 0 & 0 & 0 & 0 \\ 1 & 0 & 0 & 0 \\ 0 & 1 & 0 & -1 \\ 0 & 0 & 1 & 1 \end{bmatrix}.$$

Solution. Let $\mathbf{v} \in V$ be such that

$$m_{f,\mathbf{v}} = m_f = m_{f_1} m_{f_2} = t^4 - t^3 + t^2.$$

Then
$$\mathcal{B} = \{\mathbf{v}, f(\mathbf{v}), f^2(\mathbf{v}), f^3(\mathbf{v})\}$$

is a basis of $\mathcal{V}$. Obviously,

$$f(\mathbf{v}) = f(\mathbf{v})$$
$$f(f(\mathbf{v})) = f^2(\mathbf{v})$$
$$f(f^2(\mathbf{v})) = f^3(\mathbf{v}).$$

Moreover, since

$$f^4(\mathbf{v}) - f^3(\mathbf{v}) + f^2(\mathbf{v}) = \mathbf{0},$$

we have

$$f(f^3(\mathbf{v})) = f^4(\mathbf{v}) = f^3(\mathbf{v}) - f^2(\mathbf{v}).$$

Consequently, the $\mathcal{B}$-matrix of f is

$$\begin{bmatrix} 0 & 0 & 0 & 0 \\ 1 & 0 & 0 & 0 \\ 0 & 1 & 0 & -1 \\ 0 & 0 & 1 & 1 \end{bmatrix}.$$

$\square$

Theorem 4.3.9. *Let $\mathcal{V}$ be a finite dimensional vector space and let $f : \mathcal{V} \to \mathcal{V}$ be a nonzero endomorphism. There are $\mathbf{v}_1, \ldots, \mathbf{v}_n \in \mathcal{V}$ such that*

$$\mathcal{V} = \mathcal{V}_{f,\mathbf{v}_1} \oplus \cdots \oplus \mathcal{V}_{f,\mathbf{v}_n}$$

and $m_{f,\mathbf{v}_j}$ is a multiple of $m_{f,\mathbf{v}_{j+1}}$ for every $j \in \{1, \ldots, n-1\}$.

Proof. By Theorem 4.3.7, there is a vector $\mathbf{v}_1 \in \mathcal{V}$ such that $m_f(t) = m_{f,\mathbf{v}_1}(t)$ and, by Lemma 4.2.13, we have

$$\mathcal{V} = \mathcal{V}_{f,\mathbf{v}_1} \oplus \mathcal{V}_2,$$

where $\mathcal{V}_2$ is an f-invariant subspace of $\mathcal{V}$. Now, there is a vector $\mathbf{v}_2 \in \mathcal{V}_2$ such that $m_{f_2}(t) = m_{f,\mathbf{v}_2}(t)$, where f_2 is the endomorphism induced by f on $\mathcal{V}_2$. Clearly, m_{f_2} divides m_f. Applying Lemma 4.2.13 again we get

$$\mathcal{V} = \mathcal{V}_{f,\mathbf{v}_1} \oplus \mathcal{V}_{f,\mathbf{v}_2} \oplus \mathcal{V}_3,$$

where V_3 is an f-invariant subspace of V. Continuing this way we can produce the desired decomposition of V. $\qquad\square$

Example 4.3.10. Let V, f, $\mathbf{u}$, $\mathbf{v}$, and $\mathbf{w}$ be as defined in Example 4.2.30. Show that

$$V = V_{f,\mathbf{u}+\mathbf{v}} \oplus V_{f,\mathbf{w}}.$$

Solution. Since $m_{f,\mathbf{u}}(t) = (t-\alpha)^2$ and $m_{f,\mathbf{v}}(t) = t - \beta$, we have

$$m_{f,\mathbf{u}+\mathbf{v}} = (t-\alpha)^2(t-\beta) = m_f.$$

Because $V_{f,\mathbf{u}+\mathbf{v}} \subseteq V_{f,\mathbf{u}} \oplus V_{f,\mathbf{v}}$ and because $\dim V_{f,\mathbf{u}+\mathbf{v}} = 3$, we have $V_{f,\mathbf{u}+\mathbf{v}} = V_{f,\mathbf{u}} \oplus V_{f,\mathbf{v}}$. This gives us $V = V_{f,\mathbf{u}} \oplus V_{f,\mathbf{v}} \oplus V_{f,\mathbf{w}} = V_{f,\mathbf{u}+\mathbf{v}} \oplus V_{f,\mathbf{w}}$. $\qquad\square$

In Theorem 4.3.9 we show that for any endomorphism f on a finite dimensional space V there are $\mathbf{v}_1, \ldots, \mathbf{v}_n \in V$ such that $V = V_{f,\mathbf{v}_1} \oplus \cdots \oplus V_{f,\mathbf{v}_n}$ and $m_{f,\mathbf{v}_j}$ is a multiple of $m_{f,\mathbf{v}_{j+1}}$ for every $j \in \{1, \ldots, n-1\}$. In Theorem 4.3.12 below we will show that, while the vectors $\mathbf{v}_1, \ldots, \mathbf{v}_n$ are not unique, the integer n and the polynomials $m_{f,\mathbf{v}_1}, \ldots, m_{f,\mathbf{v}_n}$ are unique. In the proof of Theorem 4.3.12 we use the following lemma.

Lemma 4.3.11. *Let V be a finite dimensional vector space and let $f : V \to V$ be a nonzero endomorphism. If, for a vector $\mathbf{v} \in V$, the polynomial p is a divisor of $m_{f,\mathbf{v}}$, then*

$$\dim p(f)(V_{f,\mathbf{v}}) = \dim V_{f,\mathbf{v}} - \deg p.$$

Proof. We can assume that the polynomial p is monic and that $p \neq m_{f,\mathbf{v}}$. Then $p(f)(V_{f,\mathbf{v}}) = V_{f,p(f)\mathbf{v}}$.

Now, if q is a monic polynomial such that $m_{f,\mathbf{v}} = pq$, then clearly $q = m_{f,p(f)\mathbf{v}}$. This gives us

$$\dim p(f)(V_{f,\mathbf{v}}) = \deg q = \deg m_{f,\mathbf{v}} - \deg p = \dim V_{f,\mathbf{v}} - \deg p.$$

$\qquad\square$

Theorem 4.3.12. *Let* V *be a finite dimensional vector space and let* $f : V \to V$ *be a nonzero endomorphism. If*

$$V = V_{f,\mathbf{v}_1} \oplus \cdots \oplus V_{f,\mathbf{v}_n}$$

for some nonzero vectors $\mathbf{v}_1, \ldots, \mathbf{v}_n \in V$ *such that* $m_{f,\mathbf{v}_1} = m_f$ *and* $m_{f,\mathbf{v}_j}$ *is a multiple of* $m_{f,\mathbf{v}_{j+1}}$ *for every* $j \in \{1, \ldots, n-1\}$ *and*

$$V = V_{f,\mathbf{w}_1} \oplus \cdots \oplus V_{f,\mathbf{w}_q}$$

for some nonzero vectors $\mathbf{w}_1, \ldots, \mathbf{w}_q \in V$ *such that* $m_{f,\mathbf{w}_1} = m_f$ *and* $m_{f,\mathbf{w}_j}$ *is a multiple of* $m_{f,\mathbf{w}_{j+1}}$ *for every* $j \in \{1, \ldots, q-1\}$, *then* $q = n$ *and*

$$m_{f,\mathbf{v}_j} = m_{f,\mathbf{w}_j}$$

for every $j \in \{1, \ldots, n\}$.

Proof. Let k be a positive integer such that $k \le n$ and $k \le q$. Suppose that $m_{f,\mathbf{v}_j} = m_{f,\mathbf{w}_j}$ for every $j \in \{1, \ldots, k-1\}$. Since

$$m_{f,\mathbf{v}_k}(f)(V_{f,\mathbf{v}_j}) = \mathbf{0}$$

for every $j \in \{k, \ldots, n\}$, we have

$$m_{f,\mathbf{v}_k}(f)(V) = m_{f,\mathbf{v}_k}(f)(V_{f,\mathbf{v}_1}) \oplus \cdots \oplus m_{f,\mathbf{v}_k}(f)(V_{f,\mathbf{v}_{k-1}}),$$

which give us

$$\dim m_{f,\mathbf{v}_k}(f)(V) = \dim m_{f,\mathbf{v}_k}(f)(V_{f,\mathbf{v}_1}) + \cdots + \dim m_{f,\mathbf{v}_k}(f)(V_{f,\mathbf{v}_{k-1}}).$$

We also have

$$m_{f,\mathbf{v}_k}(f)(V) = m_{f,\mathbf{v}_k}(f)(V_{f,\mathbf{w}_1}) \oplus \cdots \oplus m_{f,\mathbf{v}_k}(f)(V_{f,\mathbf{w}_q})$$

and thus

$$\dim m_{f,\mathbf{v}_k}(f)(V) = \dim m_{f,\mathbf{v}_k}(f)(V_{f,\mathbf{w}_1}) + \cdots + \dim m_{f,\mathbf{v}_k}(f)(V_{f,\mathbf{w}_q}).$$

By the assumptions and Lemma 4.3.11 we have

$$\dim m_{f,\mathbf{v}_k}(f)(V_{f,\mathbf{v}_j}) = \dim m_{f,\mathbf{v}_k}(f)(V_{f,\mathbf{w}_j})$$

for every $j \in \{1, \ldots, k-1\}$, because $m_{f,\mathbf{v}_k}$ is a divisor of all polynomials $m_{f,\mathbf{w}_1} = m_{f,\mathbf{v}_1}, \ldots, m_{f,\mathbf{w}_{k-1}} = m_{f,\mathbf{v}_{k-1}}$. Hence $\dim m_{f,\mathbf{v}_k}(f)(V_{f,\mathbf{w}_k}) = 0$, which implies that

$$m_{f,\mathbf{v}_k}(f)(V_{f,\mathbf{w}_k}) = \mathbf{0}.$$

This shows that $m_{f,\mathbf{w}_k}$ divides $m_{f,\mathbf{v}_k}$.

The same way we can show that $m_{f,\mathbf{v}_k}$ divides $m_{f,\mathbf{w}_k}$. Consequently, $m_{f,\mathbf{v}_k} = m_{f,\mathbf{w}_k}$, because the polynomials $m_{f,\mathbf{v}_k}$ and $m_{f,\mathbf{w}_k}$ are monic. We finish the proof by induction and using the equality

$$\dim V = \dim V_{f,\mathbf{v}_1} + \cdots + \dim V_{f,\mathbf{v}_n} = \dim V_{f,\mathbf{w}_1} + \cdots + \dim V_{f,\mathbf{w}_q}.$$

□

The polynomials $m_{f,\mathbf{v}_1}, \ldots, m_{f,\mathbf{v}_n}$ in Theorem 4.3.12 are called the *invariant factors* of f.

Example 4.3.13. Let V be a finite dimensional vector space and let $f : V \to V$ be an endomorphism such that the invariant factors of f are

$$(t-1)^3(t-2)^3(t-3), \quad (t-1)^2(t-2)^3, \quad \text{and} \quad (t-1)(t-2).$$

Determine the Jordan blocks from the Jordan canonical form.

Solution.
$$J_{1,3}, J_{1,2}, J_{1,1}, J_{2,3}, J_{2,3}, J_{2,1}, J_{3,1}.$$

□

Definition 4.3.14. With the notation from Theorem 4.3.12 the matrix

$$\begin{bmatrix} C_{m_{f,\mathbf{v}_1}} & & & & \\ & C_{m_{f,\mathbf{v}_2}} & & \mathbf{0} & \\ & & \ddots & & \\ & \mathbf{0} & & & C_{m_{f,\mathbf{v}_n}} \end{bmatrix}$$

is called the *rational form* of f.

Example 4.3.15. Let V be a finite dimensional vector space and let $f : V \to V$ be an endomorphism. If the Jordan canonical form of f is

$$\begin{bmatrix} \alpha & 1 & 0 & 0 \\ 0 & \alpha & 0 & 0 \\ 0 & 0 & \beta & 0 \\ 0 & 0 & 0 & \beta \end{bmatrix},$$

determine the invariant factors of f and the rational form of f.

Solution. The invariant factors are

$$(t - \alpha)^2(t - \beta) = t^3 - (2\alpha + \beta)t^2 + (\alpha^2 + 2\alpha\beta)t - \alpha^2\beta \quad \text{and} \quad t - \beta.$$

Consequently, the rational form is

$$\begin{bmatrix} 0 & 0 & \alpha^2\beta & 0 \\ 1 & 0 & -(\alpha^2 + 2\alpha\beta) & 0 \\ 0 & 1 & 2\alpha + \beta & 0 \\ 0 & 0 & 0 & \beta \end{bmatrix}.$$

□

Example 4.3.16. Let V be a finite dimensional vector space and let $f : V \to V$ be an endomorphism. If the Jordan canonical form of f is

$$\begin{bmatrix} \alpha & 1 & 0 & 0 & 0 & 0 & 0 & 0 & 0 & 0 \\ 0 & \alpha & 1 & 0 & 0 & 0 & 0 & 0 & 0 & 0 \\ 0 & 0 & \alpha & 0 & 0 & 0 & 0 & 0 & 0 & 0 \\ 0 & 0 & 0 & \alpha & 0 & 0 & 0 & 0 & 0 & 0 \\ 0 & 0 & 0 & 0 & \beta & 1 & 0 & 0 & 0 & 0 \\ 0 & 0 & 0 & 0 & 0 & \beta & 0 & 0 & 0 & 0 \\ 0 & 0 & 0 & 0 & 0 & 0 & \gamma & 1 & 0 & 0 \\ 0 & 0 & 0 & 0 & 0 & 0 & 0 & \gamma & 0 & 0 \\ 0 & 0 & 0 & 0 & 0 & 0 & 0 & 0 & \gamma & 0 \\ 0 & 0 & 0 & 0 & 0 & 0 & 0 & 0 & 0 & \gamma \end{bmatrix},$$

determine the invariant factors of f.

Solution. The invariant factors are

$$(t - \alpha)^3(t - \beta)^2(t - \gamma)^2, \quad (t - \alpha)(t - \gamma), \quad \text{and} \quad t - \gamma.$$

□

4.4   Exercises

4.4.1   Diagonalization

Exercise 4.1. Let V be a finite dimensional vector space and let $f : V \to V$ be an endomorphism. If $\mathcal{U}$ and $\mathcal{W}$ are f-invariant subspaces and $V = \mathcal{U} \oplus \mathcal{W}$, show that $\det f = \det f_{\mathcal{U}} \det f_{\mathcal{W}}$.

Exercise 4.2. Let V be a finite dimensional vector space and let $f : V \to V$ be an endomorphism. If p, q, and r are polynomials such that $\mathbf{GCD}(p, q) = r$, show that $\ker p(f) \cap \ker q(f) = \ker r(f)$.

Exercise 4.3. Let V be a finite dimensional vector space and let $f : V \to V$ be an endomorphism. If p and q are polynomials such that $\mathbf{GCD}(p,q) = 1$, show that $\ker(pq)(f) = \ker p(f) \oplus \ker q(f)$.

Exercise 4.4. Let V be a finite dimensional vector space and let $f : V \to V$ be an endomorphism. Suppose that $p_1, \ldots p_n$ are polynomials such that $\mathbf{GCD}(p_j, p_k) = 1$, for $j, k \in \{1, \ldots, n\}$ and $j \neq k$. Use Exercise 4.3 and mathematical induction to prove that

$$\ker(p_1 \ldots p_n)(f) = \ker p_1(f) \oplus \cdots \oplus \ker p_n(f).$$

Exercise 4.5. Let V be a finite dimensional vector space and let $f : V \to V$ be an endomorphism. If $\lambda_1, \ldots, \lambda_k$ are distinct numbers, show that

$$\dim \ker(f - \lambda_1) + \cdots + \dim \ker(f - \lambda_k) = \dim(\ker(f - \lambda_1) + \cdots + \ker(f - \lambda_k)).$$

Exercise 4.6. Let V be a finite dimensional vector space and let $f : V \to V$ be an endomorphism. Suppose that $p_1, \ldots, p_n$ are polynomials such that $\mathbf{GCD}(p_j, p_k) = 1$ for $j, k \in \{1, \ldots, n\}$ and $j \neq k$. Show that for every $j \in \{1, \ldots, n\}$ there is a polynomial q_j such that $q_j(f)$ is the projection of $\ker p_j(f)$ on $\ker(p_1 \ldots p_n)(f)$ along

$$\ker p_1(f) \oplus \cdots \oplus \ker p_{j-1}(f) \oplus \ker p_{j+1}(f) \oplus \cdots \oplus \ker p_n(f).$$

Exercise 4.7. Let V be a finite dimensional vector space and let $f : V \to V$ be an endomorphism. If $p(f) = \mathbf{0}$ for some polynomial p, show that every eigenvalue of f is a root of p.

Exercise 4.8. Let V be a finite dimensional vector space and let $f : V \to V$ be an endomorphism such that $p(f) = \mathbf{0}$ where p is a polynomial which can be written as a product of distinct monic polynomials of degree 1. Show that f is diagonalizable.

Exercise 4.9. Let V be a finite dimensional vector space and let $f : V \to V$ be an endomorphism. Show that, if f^2 is diagonalizable and has distinct positive eigenvalues, then f is diagonalizable.

Exercise 4.10. Let V be a finite dimensional vector space and let $f : V \to V$ be an endomorphism. If $c_f(t) = (5 - t)^2(3 - t)^7$, show that f is diagonalizable if and only if $(f - 5)(f - 3) = \mathbf{0}$.

Exercise 4.11. Let $f : \mathbb{C}^3 \to \mathbb{C}^3$ be the endomorphism defined by

$$f\left(\begin{bmatrix} x \\ y \\ z \end{bmatrix}\right) = \begin{bmatrix} 0 & 0 & 1 \\ 1 & 0 & 0 \\ 0 & 1 & 0 \end{bmatrix} \begin{bmatrix} x \\ y \\ z \end{bmatrix}.$$

Determine c_f and m_f and show that f is diagonalizable.

Exercise 4.12. Let $f : \mathbb{C}^2 \to \mathbb{C}^2$ be the endomorphism defined by

$$f\left(\begin{bmatrix} x \\ y \end{bmatrix}\right) = \begin{bmatrix} 7i & -2 \\ -25 & -7i \end{bmatrix} \begin{bmatrix} x \\ y \end{bmatrix}.$$

Show that f is diagonalizable and determine the eigenspaces of f.

Exercise 4.13. Let V be a finite dimensional vector space, $f : V \to V$ an endomorphism, and $g : V \to V$ an isomorphism. Show that $c_f = c_{gfg^{-1}}$.

Exercise 4.14. Let V be a finite dimensional vector space, $f : V \to V$ an endomorphism, and $g : V \to V$ an isomorphism. Show that $m_f = m_{gfg^{-1}}$.

Exercise 4.15. Let $f : \mathbb{R}^3 \to \mathbb{R}^3$ be the endomorphism defined by

$$f\left(\begin{bmatrix} x \\ y \\ z \end{bmatrix}\right) = \begin{bmatrix} -3 & 43 & -17 \\ -4 & 29 & -10 \\ -8 & 60 & -21 \end{bmatrix} \begin{bmatrix} x \\ y \\ z \end{bmatrix}.$$

Find c_f.

Exercise 4.16. Show that the endomorphism $f : \mathbb{R}^3 \to \mathbb{R}^3$ defined by

$$f\left(\begin{bmatrix} x \\ y \\ z \end{bmatrix}\right) = \begin{bmatrix} -3 & 43 & -17 \\ -4 & 29 & -10 \\ -8 & 60 & -21 \end{bmatrix} \begin{bmatrix} x \\ y \\ z \end{bmatrix}$$

is not diagonalizable and find the eigenspaces of this endomorphism.

Exercise 4.17. Let V be a finite dimensional vector space and let $f : V \to V$ be an endomorphism. Show that 0 is a root of the minimal polynomial m_f if and only if f is not invertible.

Exercise 4.18. Let V be a finite dimensional vector space and let $f : V \to V$ be an endomorphism. If U is an f-invariant subspace of V such that $\dim U = 1$, show that every nonzero vector in U is an eigenvector.

Exercise 4.19. Let V be a finite dimensional vector space, $f, g : V \to V$ endomorphisms, and $\alpha \in \mathbb{K}$. If α is an eigenvalue of f and $g^3 - 3g^2 f + 3g f^2 - f^3 = 0$, show that α is an eigenvalue of g.

Exercise 4.20. Let V be a finite dimensional vector space and let $f, g; V \to V$ be endomorphisms such that $fg = gf$. If $m_f = t^j$ and $m_f = t^k$ for some positive integers j and k, show that $m_{fg} = t^r$ and $m_{f+g} = t^s$ for some positive integers r and s.

Exercise 4.21. Let V be a finite dimensional vector space and let $f : V \to V$ be an endomorphism. If the characteristic polynomial of f is $c_f(t) = 5t^3 + 7t^2 + 2t + 8$, show that f is invertible and find f^{-1}.

Exercise 4.22. Let V and W be vector spaces. Let $f : V \to V$ be an endomorphism and let $g : W \to V$ be an invertible linear transformation. Show that $\lambda \in \mathbb{K}$ is an eigenvalue of f if and only if λ is an eigenvalue of $g^{-1} fg$.

Exercise 4.23. Let $\mathcal{V}$ be a finite dimensional vector space and let $f : \mathcal{V} \to \mathcal{V}$ be an endomorphism. If $\mathcal{U}$ is an f-invariant subspace of $\mathcal{V}$, show that $m_{f_{\mathcal{U}}}$ divides m_f.

Exercise 4.24. Let $\mathcal{V}$ be a finite dimensional vector space, $f : \mathcal{V} \to \mathcal{V}$ an endomorphism, and $\mathcal{U}$ an f-invariant subspace of $\mathcal{V}$. Let $\mathbf{v}, \mathbf{w} \in \mathcal{V}$ be such that $\mathbf{v} + \mathbf{w} \in \mathcal{U}$. If $\mathbf{v}$ and $\mathbf{w}$ are eigenvectors of f corresponding to distinct eigenvalues, show that both vectors $\mathbf{v}$ and $\mathbf{w}$ are in $\mathcal{U}$.

Exercise 4.25. Let $\mathcal{V}$ be a finite dimensional complex vector space and let $f : \mathcal{V} \to \mathcal{V}$ be an endomorphism such that $f^4 + \mathrm{Id} = \mathbf{0}$. Show that f is diagonalizable.

Exercise 4.26. Let $\mathcal{V}$ be a finite dimensional vector space, $f : \mathcal{V} \to \mathcal{V}$ an endomorphism, and λ a nonzero number. If f is invertible, show that λ is an eigenvalue of f if and only if $\frac{1}{\lambda}$ is an eigenvalue of f^{-1}.

Exercise 4.27. Let $\mathcal{V}$ be a finite dimensional vector space, $f : \mathcal{V} \to \mathcal{V}$ an endomorphism, and $\mathbf{v} \in \mathcal{V}$ an eigenvector of f corresponding to an eigenvalue λ. We assume that $\alpha, \beta \in \mathbb{K}$ are such that $\gamma \neq \alpha$ and $\gamma \neq \beta$. If $\mathbf{u} \in \ker((f - \alpha)(f - \beta))$ and $\mathbf{u} + \mathbf{v} = \mathbf{0}$, show that $\mathbf{v} = \mathbf{0}$.

Exercise 4.28. Let $\mathcal{V}$ be a finite dimensional vector space and let $f : \mathcal{V} \to \mathcal{V}$ be an endomorphism. Show that 0 is an eigenvalue of f if and only f is not invertible.

Exercise 4.29. Let $\mathcal{V}$ be a finite dimensional vector space and let $f : \mathcal{V} \to \mathcal{V}$ be an endomorphism. If $\mathcal{U}$ is an f-invariant subspace of $\mathcal{V}$ and f is diagonalizable, show that $f_{\mathcal{U}}$ is diagonalizable.

Exercise 4.30. Let $\mathcal{V}$ be a finite dimensional vector space, $f : \mathcal{V} \to \mathcal{V}$ an endomorphism, and $\alpha \in \mathbb{K}$. If $\dim \mathcal{V} = n$ and there is an integer $k \geq 1$ such that $(f - \alpha)^k = \mathbf{0}$, determine the characteristic polynomial of f.

Exercise 4.31. Let $\mathcal{V}$ be a finite dimensional complex vector space and let $f : \mathcal{V} \to \mathcal{V}$ be an endomorphism such that $f^2 - 2f + 5 = \mathbf{0}$. What are all possible polynomials m_f?

Exercise 4.32. Let $\mathcal{V}$ be a finite dimensional complex vector space and let $f : \mathcal{V} \to \mathcal{V}$ be an endomorphism such that $f^3 - id = \mathbf{0}$. Show that f is diagonalizable.

Exercise 4.33. Let $\mathcal{V}$ be a finite dimensional vector space and let $f : \mathcal{V} \to \mathcal{V}$ be an endomorphism. If $\mathcal{U}$ is an f-invariant subspace of $\mathcal{V}$, $\mathbf{v}, \mathbf{w} \in \mathcal{V}$ are such that $\mathbf{v}, \mathbf{v} + \mathbf{w} \in \mathcal{U}$, and $\mathbf{w}$ is an eigenvector of f corresponding to a nonzero eigenvalue $\alpha \in \mathbb{K}$, show that $\mathbf{w} \in \mathcal{U}$.

Exercise 4.34. Let $\mathcal{V}$ be a finite dimensional complex vector space and let $f : \mathcal{V} \to \mathcal{V}$ be an endomorphism. If $c_f = pq$, where p and q are polynomials such that $\mathbf{GCD}(p, q) = 1$, and g and h are the linear transformations induced by f on $\ker p(f)$ and $\ker q(f)$, respectively, show that there are nonzero $\alpha, \beta \in \mathbb{C}$ such that $p = \alpha c_g$ and $q = \beta c_h$.

4.4.2 Jordan canonical form

Exercise 4.35. Let V be a finite dimensional complex vector space and let $f : V \to V$ be an endomorphism. If $\mathcal{U}$ is an f-invariant subspace of V and $\mathbf{v} \in \mathcal{U}$, show that the cyclic subspace of f associated with $\mathbf{v}$ is a subspace of $\mathcal{U}$.

Exercise 4.36. Show that the endomorphism $f : \mathbb{R}^2 \to \mathbb{R}^2$ corresponding to the matrix $\begin{bmatrix} 3 & 4 \\ -1 & 7 \end{bmatrix}$ is not diagonalizable and find the Jordan canonical form of f and a Jordan basis.

Exercise 4.37. Use Exercise 4.36 to solve the system
$$\begin{cases} x' = 3x + 4y \\ y' = -x + 7y \end{cases}.$$

Exercise 4.38. Let $f : \mathbb{R}^3 \to \mathbb{R}^3$ be the endomorphism defined by
$$f\left(\begin{bmatrix} x \\ y \\ z \end{bmatrix}\right) = \begin{bmatrix} 1 & 1 & 0 \\ -5 & 4 & 1 \\ 1 & 2 & 1 \end{bmatrix} \begin{bmatrix} x \\ y \\ z \end{bmatrix}.$$

Determine the Jordan canonical form of f and a Jordan basis.

Exercise 4.39. Find an endomorphism $f : \mathbb{C}^2 \to \mathbb{C}^2$ with the Jordan canonical form $\begin{bmatrix} i & 1 \\ 0 & i \end{bmatrix}$ and a Jordan basis $\left\{ \begin{bmatrix} 2i \\ -3 \end{bmatrix}, \begin{bmatrix} 1 \\ i \end{bmatrix} \right\}$.

Exercise 4.40. Let $f : \mathbb{R}^3 \to \mathbb{R}^3$ be the endomorphism defined by
$$f\left(\begin{bmatrix} x \\ y \\ z \end{bmatrix}\right) = \begin{bmatrix} -3 & 43 & -17 \\ -4 & 29 & -10 \\ -8 & 60 & -21 \end{bmatrix} \begin{bmatrix} x \\ y \\ z \end{bmatrix}.$$

Using Exercises 4.15 and 4.16 find the Jordan canonical form of f and a Jordan basis.

Exercise 4.41. Let V be a finite dimensional vector space and let $f : V \to V$ be an endomorphism. If $m_f(t) = t^n$ for some integer $n \geq 1$, show that $\{0\} \subsetneq \ker f \subsetneq \cdots \subsetneq \ker f^{n-1} \subsetneq \ker f^n = V$.

Exercise 4.42. Let V be a finite dimensional vector space and let $f : V \to V$ be an endomorphism. If $m \geq 2$ and $\mathcal{S}_m$ is a subspace of V such that $\ker f^{m-1} \cap \mathcal{S}_m = \{0\}$, show that $f(\mathcal{S}_m) \cap \ker f^{m-2} = \{0\}$.

Exercise 4.43. Let V be a finite dimensional vector space and let $f : V \to V$ be an endomorphism. If $m \geq 2$ and $\mathcal{S}_m$ is a subspace of V such that $\ker f^{m-1} \cap \mathcal{S}_m = \{0\}$, show that the function $g : \mathcal{S}_m \to f(\mathcal{S}_m)$ defined by $g(\mathbf{v}) = f(\mathbf{v})$ is an isomorphism.

Exercise 4.44. Let V be a finite dimensional vector space and let $f : V \to V$ be an endomorphism such that $m_f(t) = t^n$ for some integer $n \geq 1$. Using Exercise 4.42, show that for every $m \in \{1, \ldots, n\}$ there is a subspace $\mathcal{S}_m$ of V such that $\ker f^m = \ker f^{m-1} \oplus \mathcal{S}_m$ for $m \in \{1, \ldots, n\}$ and $f(\mathcal{S}_m) \subseteq \mathcal{S}_{m-1}$ for $m \in \{2, \ldots, n\}$.

Exercise 4.45. Give an example of a vector space V and an endomorphism $f : V \to V$ such that $c_f = (t - 2)^4 (t - 5)^4$ and $m_f = (t - 2)^2 (t - 5)^3$.

Exercise 4.46. Let V and W be finite dimensional vector spaces, $f : V \to V$ and $g : W \to W$ endomorphisms, $\mathbf{v} \in V$ and $\mathbf{w} \in W$. If $m_{f,\mathbf{v}} = m_{g,\mathbf{w}}$, show that there is an isomorphism $h : V_{f,\mathbf{v}} \to W_{f,\mathbf{w}}$ such that $g(\mathbf{x}) = h(f(h^{-1}(\mathbf{x})))$ for every $\mathbf{x} \in W_{f,\mathbf{w}}$.

Exercise 4.47. Let V and W be finite dimensional vector spaces and let $f : V \to V$ and $g : W \to W$ be endomorphisms. If f and g have the same Jordan canonical form, show that there is an isomorphism $h : V \to W$ such that $g = hfh^{-1}$.

Exercise 4.48. Let V be a finite dimensional vector space and let $f : V \to V$ be an endomorphism. If $c_f(t) = (\alpha - t)^n$ for some $\alpha \in \mathbb{K}$ and integer $n \geq 1$, show that there is an endomorphism $g : V \to V$ such that $c_f(t) = (-t)^n$ and we have $f = g + \alpha \operatorname{Id}$.

Exercise 4.49. Let V be a vector space such that $\dim V = 8$ and let $f : V \to V$ be an endomorphism such that $m_f(t) = (t - \alpha)^5$ for some $\alpha \in \mathbb{K}$. If

$$\dim \ker(f - \alpha)^4 = 7, \dim \ker(f - \alpha)^3 = 6, \dim \ker(f - \alpha)^2 = 4, \dim \ker(f - \alpha) = 2,$$

find the Jordan canonical form of f and a Jordan basis.

Exercise 4.50. Let V be a finite dimensional vector space and let $f : V \to V$ be an endomorphism. We assume that $m_f = t^4$ and that the Jordan canonical form of f has 7 Jordan blocks $J_{0,4}$, 4 Jordan blocks $J_{0,3}$, one block $J_{0,2}$, and one Jordan block $J_{0,1}$. Determine $\dim \ker f^3$, $\dim \ker f^2$, and $\dim \ker f$.

Exercise 4.51. With the notation from Exercise 4.44 show that $\dim \mathcal{S}_n$ is the number of Jordan blocks $J_{0,n}$ of the Jordan canonical form of f and that for every $m \in \{1, \ldots, n - 1\}$ the dimension $\dim \mathcal{T}_m$ equals the number of Jordan blocks $J_{0,m}$ of the Jordan canonical form of f.

Exercise 4.52. Let V be a finite dimensional vector space and let $f : V \to V$ be an endomorphism. If $c_f = (t - \alpha)^4 (t - \beta)^3$ and $m_f = (t - \alpha)^3 (t - \beta)^2$ for some $\alpha, \beta \in \mathbb{K}$, find all possible Jordan canonical forms of f.

Exercise 4.53. Let V be a finite dimensional vector space and let $f, g : V \to V$ be endomorphisms. If g is invertible, show that f and gfg^{-1} have the same Jordan canonical form.

Exercise 4.54. Let V be a vector space such that $\dim V = 4$ and let $f : V \to V$ be an endomorphism such that $m_f(t) = (t - \alpha)^3$ for some $\alpha \in \mathbb{K}$. Determine the Jordan canonical form of f and explain the construction of a Jordan basis.

Exercise 4.55. Let V be a vector space such that $\dim V = 4$ and let $f : V \to V$ be an endomorphism such that $c_f(t) = (\alpha - t)^4$ and $m_f(t) = (t - \alpha)^2$ for some $\alpha \in \mathbb{K}$. If $\dim \ker(f - \alpha) = 2$, determine the Jordan canonical form of f and explain the construction of a Jordan basis.

Exercise 4.56. Let V be a finite dimensional vector space and let $f : V \to V$ be an endomorphism such that the Jordan canonical form of f is

$$
\left[
\begin{array}{cccc|cc|cc}
\alpha & 1 & 0 & 0 & 0 & 0 & 0 & 0 \\
0 & \alpha & 1 & 0 & 0 & 0 & 0 & 0 \\
0 & 0 & \alpha & 1 & 0 & 0 & 0 & 0 \\
0 & 0 & 0 & \alpha & 0 & 0 & 0 & 0 \\
\hline
0 & 0 & 0 & 0 & \alpha & 1 & 0 & 0 \\
0 & 0 & 0 & 0 & 0 & \alpha & 0 & 0 \\
\hline
0 & 0 & 0 & 0 & 0 & 0 & \alpha & 1 \\
0 & 0 & 0 & 0 & 0 & 0 & 0 & \alpha
\end{array}
\right]
$$

for some $\alpha \in \mathbb{K}$. Determine m_f.

Exercise 4.57. Let V be a finite dimensional vector space and let $f : V \to V$ be an endomorphism such that the Jordan canonical form of f is

$$
\left[
\begin{array}{cc|cc|cccc}
\alpha & 1 & 0 & 0 & 0 & 0 & 0 & 0 \\
0 & \alpha & 0 & 0 & 0 & 0 & 0 & 0 \\
\hline
0 & 0 & \alpha & 1 & 0 & 0 & 0 & 0 \\
0 & 0 & 0 & \alpha & 0 & 0 & 0 & 0 \\
\hline
0 & 0 & 0 & 0 & \beta & 1 & 0 & 0 \\
0 & 0 & 0 & 0 & 0 & \beta & 1 & 0 \\
0 & 0 & 0 & 0 & 0 & 0 & \beta & 0 \\
0 & 0 & 0 & 0 & 0 & 0 & 0 & \beta
\end{array}
\right]
$$

for some $\alpha, \beta \in \mathbb{K}$. Determine m_f.

Exercise 4.58. Let V be a vector space such that $\dim V = 7$ and let $f : V \to V$ be an endomorphism such that $m_f = (t - \alpha)^4$ for some $\alpha \in \mathbb{K}$. Determine all possible Jordan canonical forms of f.

Exercise 4.59. Let V be a vector space such that $\dim V = 25$ and let $f : V \to V$ be an endomorphism such that $m_f(t) = (t - \alpha)^3$ for some $\alpha \in \mathbb{K}$. If $\dim \ker(f - \alpha)^2 = 21$ and $\dim \ker(f - \alpha) = 12$, determine the number of Jordan blocks of the form $\begin{bmatrix} \alpha & 1 \\ 0 & \alpha \end{bmatrix}$ in the Jordan canonical form of f.

Exercise 4.60. Let V be a finite dimensional vector space and let $f : V \to V$ be an endomorphism. If $c_f = (t - \alpha)^4 (t - \beta)^4$ and $m_f = (t - \alpha)^2 (t - \beta)^2$ for some $\alpha, \beta \in \mathbb{K}$, determine all possible Jordan canonical forms of f.

Exercise 4.61. Let V be a finite dimensional vector space and let $f : V \to V$ be a diagonalizable endomorphism. Assume $\mathbf{v}_1 \ldots \mathbf{v}_n$ is a basis of V consisting of eigenvectors of f corresponding to eigenvalues $\lambda_1, \ldots, \lambda_n$. If $\lambda_1, \ldots, \lambda_k$ are the nonzero eigenvalues of f, show that $\{v_1, \ldots, v_k\}$ is a basis of ran f and $\{v_{k+1}, \ldots, v_n\}$ is a basis of ker f.

Exercise 4.62. Let V be a finite dimensional vector space and let $f : V \to V$ be an endomorphism. If $\dim \ker(f - a)^9 = 33$ and $\dim \ker(f - a)^7 = 27$ for some $a \in \mathbb{K}$, determine the possible number of 8×8 Jordan blocks in the Jordan canonical form of f.

Exercise 4.63. Let V be a finite dimensional vector space, $f : V \to V$ an endomorphism, and $a \in \mathbb{K}$. Show that it is not possible to have $\dim V = 8$, $m_f(t) = (t - a)^7$ and $\dim \ker(f - a) = 1$.

4.4.3 Rational form

Exercise 4.64. Let V be a finite dimensional vector space, $f : V \to V$ an endomorphism, and $v \in V$. If $m_f(t) = (t^2 - t + 1)^5$ and that $v \notin \ker(f^2 - f + 1)^3$, what can $m_{f,v}$ be?

Exercise 4.65. Let V be a finite dimensional vector space, $v, w \in V$, and $f : V \to V$ an endomorphism. If $\mathbf{GCD}(m_{f,v}, m_{f,w}) = 1$, show that $m_{f,v+w} = m_{f,v} m_{f,w}$.

Exercise 4.66. Let $f : \mathbb{R}^3 \to \mathbb{R}^3$ be the endomorphism defined by

$$f\left(\begin{bmatrix} x \\ y \\ z \end{bmatrix}\right) = \begin{bmatrix} -3 & 43 & -17 \\ -4 & 29 & -10 \\ -8 & 60 & -21 \end{bmatrix} \begin{bmatrix} x \\ y \\ z \end{bmatrix}.$$

Find a vector $v \in \mathbb{R}^3$ such that $m_{f,v} = (t+1)(t-3)$.

Exercise 4.67. Let $f : \mathbb{R}^3 \to \mathbb{R}^3$ be the endomorphism defined by

$$f\left(\begin{bmatrix} x \\ y \\ z \end{bmatrix}\right) = \begin{bmatrix} -3 & 43 & -17 \\ -4 & 29 & -10 \\ -8 & 60 & -21 \end{bmatrix} \begin{bmatrix} x \\ y \\ z \end{bmatrix}.$$

Find a vector $v \in \mathbb{R}^3$ such that the set $\mathcal{B} = \{v, f(v), f^2(v)\}$ is a basis of $\mathbb{R}^3$ and determine the $\mathcal{B}$-matrix of f.

Exercise 4.68. Let V be a finite dimensional vector space and let $f : V \to V$ be an endomorphism. If the invariant factors of f are $(t - 2)^3 (t - 5)^2$ and $(t - 2)(t - 5)^2$, determine the Jordan canonical form of f

Exercise 4.69. Let V be a finite dimensional vector space and let $f : V \to V$ be an endomorphism. If the minimal polynomials of the Jordan blocks of the Jordan canonical form of f are

$$(t - 1)^5, (t - 1)^3, t - 1, (t - 4)^3, (t - 4)^2, (t - 2)^4, (t - 2)^2, (t - 2)^2, (t - 2),$$

determine the invariant factors of f.

Exercise 4.70. Let V be a vector space such that $\dim V = 8$ and let $f : V \to V$ be an endomorphism. If $m_f = (t^2+1)(t^2+t+1)$, determine all possible invariant factors of f.

Exercise 4.71. Let V be a finite-dimensional vector space, $f : V \to V$ an endomorphism, and $\mathcal{B}$ a basis of V. If the $\mathcal{B}$-matrix of f is

$$
\begin{bmatrix}
0 & 0 & -1 & 0 & 0 \\
1 & 0 & -3 & 0 & 0 \\
0 & 1 & -3 & 0 & 0 \\
0 & 0 & 0 & 0 & -1 \\
0 & 0 & 0 & 1 & -2
\end{bmatrix},
$$

show that this matrix is the rational form of f and determine the Jordan canonical form of f.

Exercise 4.72. Find a vector space V and an endomorphism $f : V \to V$ such that $c_f = (t^2 + t + 1)^3(t^2 + 1)$ and $m_f = (t^2 + t + 1)(t^2 + 1)$.

Exercise 4.73. Let V be a finite-dimensional vector space, $f : V \to V$ an endomorphism, and $\mathcal{B}$ a basis of V. If the $\mathcal{B}$-matrix of f is

$$
\begin{bmatrix}
0 & -1 & 0 & 0 & 0 \\
1 & -1 & 0 & 0 & 0 \\
0 & 0 & 0 & -1 & 0 \\
0 & 0 & 1 & -1 & 0 \\
0 & 0 & 0 & 0 & 0
\end{bmatrix},
$$

determine the rational form of f.

Exercise 4.74. Let V be a vector space such that $\dim V = 8$ and let $f : V \to V$ be an endomorphism. If the Jordan canonical form of f is

$$
\begin{bmatrix}
\alpha & 1 & 0 & 0 & 0 \\
0 & \alpha & 1 & 0 & 0 \\
0 & 0 & \alpha & 0 & 0 \\
0 & 0 & 0 & \beta & 1 \\
0 & 0 & 0 & 0 & \beta
\end{bmatrix},
$$

where $\alpha \neq \beta$, determine the rational form of f .

Exercise 4.75. Let V be a finite-dimensional vector space, $f : V \to V$ an endomorphism, and $\mathcal{B}$ a basis of V. If the $\mathcal{B}$-matrix of f is

$$
\begin{bmatrix}
0 & 0 & 0 & 0 & -1 \\
1 & 0 & 0 & 0 & -5 \\
0 & 1 & 0 & 0 & -10 \\
0 & 0 & 1 & 0 & -10 \\
0 & 0 & 0 & 1 & -5
\end{bmatrix},
$$

determine the minimal polynomial of f.

Exercise 4.76. Let V be a finite-dimensional vector space, $f : V \to V$ an endomorphism, and $\mathcal{B}$ a basis of V. If the $\mathcal{B}$-matrix of f is

$$
\begin{bmatrix}
0 & -3 & 0 & 0 \\
1 & 0 & 0 & 0 \\
0 & 0 & 0 & -1 \\
0 & 0 & 1 & 0
\end{bmatrix},
$$

determine the rational form of f.

Exercise 4.77. Determine a vector space V and an endomorphism $f : V \to V$ such that $m_f = 3 - t - 4t^2 + 2t^3 + t^4$.

Chapter 5

Appendices

Appendix A   Permutations

By a *permutation* on $\{1, 2, \ldots, n\}$ we mean a bijection

$$\sigma : \{1, 2, \ldots, n\} \to \{1, 2, \ldots, n\}.$$

The set of all permutations on $\{1, 2, \ldots, n\}$ will be denoted by $\mathfrak{S}_n$. Note that the identity function $id : \{1, 2, \ldots, n\} \to \{1, 2, \ldots, n\}$ is an element of $\mathfrak{S}_n$.

A permutation $\sigma \in \mathfrak{S}_n$ defines a permutation on any set with n objects. For example, if $\sigma \in \mathfrak{S}_n$ and $(\mathbf{x}_1, \ldots, \mathbf{x}_n)$ is an ordered n-tuple of vectors, then $(\mathbf{x}_{\sigma(1)}, \ldots, \mathbf{x}_{\sigma(n)})$ is the corresponding permutation of $(\mathbf{x}_1, \ldots, \mathbf{x}_n)$.

If we write the first n integers ≥ 1 in the order $1, 2, \ldots, n$, then we may think of a permutation $\sigma \in \mathfrak{S}_n$ as being a reordering of these numbers such that we get $\sigma(1), \sigma(2), \ldots, \sigma(n)$. A permutation $\sigma \in \mathfrak{S}_n$ will be called a transposition if there are distinct $j, k \in \{1, 2, \ldots, n\}$ such that

$$\sigma(j) = k, \ \ \sigma(k) = j, \ \ \text{and} \ \ \sigma(m) = m \ \text{for every} \ m \in \{1, 2, \ldots, n\}$$
$$\text{different from } j \text{ and } k.$$

In other words, a transposition switches two numbers and leaves the remaining numbers where they were. We will use the symbol σ_{jk} to denote the above transposition. Note that $\sigma_{jk}^{-1} = \sigma_{jk}$ for every $j, k \in \{1, 2, \ldots, n\}$.

A transposition of the form $\sigma_{j,j+1}$ is called an elementary transposition.

Theorem 5.1. *Every permutation in $\mathfrak{S}_n$ with $n \geq 2$ is an elementary transposition or a composition of elementary transpositions.*

Proof. We will prove this result using induction on n. Since every permutation in $\sigma \in \mathfrak{S}_2$ is an elementary transposition, the statement is true for $n = 2$. Now assume that the statement is true for every $k < n$ for some $n > 2$.

301

Let $\sigma : \{1, 2, \ldots, n\} \to \{1, 2, \ldots, n\}$. be a permutation. If $\sigma(n) = j$ for some $j < n$, then the permutation

$$\tau = \sigma_{n,n-1} \cdots \sigma_{j+1,j}\sigma$$

satisfies $\tau(n) = n$ and we have

$$\sigma = \sigma_{j+1,j}^{-1} \cdots \sigma_{n,n-1}^{-1}\tau = \sigma_{j+1,j} \cdots \sigma_{n,n-1}\tau.$$

This means that, if τ is a product of elementary permutations, then σ is also a product of elementary permutations.

Now, if $\sigma(n) = n$, then the restriction of σ to the set $\{1, 2, \ldots, n-1\}$ is a product of elementary permutations on $\{1, 2, \ldots, n-1\}$, by the inductive assumption. Because every elementary permutation π on $\{1, 2, \ldots, n-1\}$ can be extended to an elementary permutation π' on $\{1, 2, \ldots, n\}$ by taking $\pi'(j) = j$ for $j < n$ and $\pi'(n) = n$, σ is a product of elementary permutations.   □

Definition 5.2. Let $\sigma \in \mathfrak{S}_n$. A pair of integers $(j, k), j < k$ is called an *inversion* of σ if $\sigma(j) > \sigma(k)$. A permutation with an even number of inversions is called an *even permutation* and a permutation with an odd number of inversions is called an *odd permutation*. We define the sign of a permutation σ, denoted by $\epsilon(\sigma)$, to be 1 if σ is even and -1 if σ is odd, that is,

$$\epsilon(\sigma) = \begin{cases} 1 & \text{if } \sigma \text{ is even} \\ -1 & \text{if } \sigma \text{ is odd} \end{cases}$$

Example 5.3. Show that every transposition is an odd permutation.

Proof. Let σ_{jk} be a transposition where $j < k$. We represent this transposition by

$$(1, \ldots, j-1, k, j+1, \ldots, k-1, j, k+1, \ldots, n).$$

The inversions are

$$(k, j+1), \ldots, (k, k-1), (j+1, j), \ldots, (k-1, j), (k, j).$$

The total number of inversions is the odd number $(2(k-j) - 1)$.   □

Theorem 5.4. *The sign of a product of permutations is the product of the signs of these permutations, that is,*

$$\epsilon(\sigma_1 \cdots \sigma_k) = \epsilon(\sigma_1) \cdots \epsilon(\sigma_k),$$

for any $\sigma_1, \ldots, \sigma_k \in \mathfrak{S}_n$.

Proof. Let $\sigma \in \mathfrak{S}_n$ and let $\tau = \sigma \sigma_{j+1,j}$.

First note that, since $(\tau(j), \tau(j+1)) = (\sigma(j+1), \sigma(j))$, the pair $(\tau(j), \tau(j+1))$ is an inversion if and only if $(\sigma(j), \sigma(j+1))$ is not an inversion. Moreover, $(\tau(p), \tau(q)) = (\sigma(p)), \sigma(q))$ if $p \neq j, p \neq j+1, q \neq j$ and $q \neq j+1$. Since

$$\{(\tau(1), \tau(j)), \ldots, (\tau(j-1), \tau(j))\} = \{(\sigma(1), \sigma(j+1)), \ldots, (\sigma(j-1), \sigma(j+1))\},$$

$$\{(\tau(1), \tau(j+1)), \ldots, (\tau(j-1), \tau(j+1))\} = \{(\sigma(1), \sigma(j)), \ldots, (\sigma(j-1), \sigma(j))\},$$

$$\{(\tau(j), \tau(j+2)), \ldots, (\tau(j), \tau(n))\} = \{(\sigma(j+1), \sigma(j+2)), \ldots, (\sigma(j+1), \sigma(n))\},$$

$$\{(\tau(j+1), \tau(j+2)), \ldots, (\tau(j+1), \tau(n))\} = \{(\sigma(j), \sigma(j+2)), \ldots, (\sigma(j), \sigma(n))\},$$

we have

$$\{(\tau(1), \tau(j)), \ldots, (\tau(j-1), \tau(j)), (\tau(1), \tau(j+1)), \ldots, (\tau(j-1), \tau(j+1))\} =$$

$$\{(\sigma(1), \sigma(j)), \ldots, (\sigma(j-1)), \sigma(j)), (\sigma(1), \sigma(j+1)), \ldots, (\sigma(j-1)), \sigma(j+1))\}$$

and

$$\{(\tau(j), \tau(j+2)), \ldots, (\tau(j), \tau(n)), (\tau(j+1), \tau(j+2)), \ldots, (\tau(j+1), \tau(n))\} =$$

$$\{(\sigma(j), \sigma(j+2)), \ldots, (\sigma(j), \sigma(n)), (\sigma(j+1), \sigma(j+2)), \ldots, (\sigma(j+1), \sigma(n))\}.$$

Hence $\epsilon(\tau) = -\epsilon(\sigma)$, because $(\tau(j), \tau(j+1)) = (\sigma(j+1)), \sigma(j))$.

Since every permutation is a product of elementary transpositions, the result follows by induction. $\square$

Corollary 5.5. *For every $\sigma \in \mathfrak{S}_n$ we have*

$$\epsilon(\sigma^{-1}) = \epsilon(\sigma).$$

Proof. Since

$$\epsilon(\sigma^{-1})\epsilon(\sigma) = \epsilon(\sigma^{-1}\sigma) = \epsilon(\mathrm{Id}) = 1,$$

we have

$$\epsilon(\sigma^{-1}) = (\epsilon(\sigma))^{-1} = \epsilon(\sigma).$$

$\square$

Appendix B   Complex numbers

The set of complex numbers, denoted by $\mathbb{C}$, can be identified with the set $\mathbb{R}^2$ where we define addition as in the vector space $\mathbb{R}^2$, that is

$$(a, b) + (c, d) = (a + c, b + c),$$

and the multiplication by

$$(a, b)(c, d) = (ac - bd, ad + bc).$$

If $z = (x, y)$, then by $-z$ we denote the number $(-x, -y)$. Note that $z + (-z) = (0, 0)$.

If $z = (x, y) \neq (0, 0)$, then by z^{-1} or $\frac{1}{z}$ we denote the number $\left(\frac{x}{x^2 + y^2}, -\frac{y}{x^2 + y^2} \right)$. Note that $zz^{-1} = z^{-1}z = (1, 0)$.

Theorem 5.6. *If $z, z_1, z_2, z_3 \in \mathbb{C}$, then*

(a) $(z_1 + z_2) + z_3 = z_1 + (z_2 + z_3)$;

(b) $z_1 + z_2 = z_2 + z_1$;

(c) $(z_1 z_2) z_3 = z_1 (z_2 z_3)$;

(d) $z_1 z_2 = z_2 z_1$;

(e) $z(0, 0) = 0, 0)$;

(f) $z(1, 0) = z$;

(g) $(z_1 + z_2) z_3 = z_1 z_3 + z_2 z_3$.

All these properties are easy to verify. As an example, we verify the propriety (c). If $z_1 = (a, b)$, $z_2 = (c, d)$, and $z_3 = (e, f)$, then

$$(z_1 z_2) z_3 = (ac - bd, ad + bc)(e, f) = ((ac - bd)e - (ad + bc)f, (ac - bd)f + (ad + bc)e)$$

and

$$z_1(z_2 z_3) = (a, b)(ce - df, cf + de) = (a(ce - df) - b(cf + de), a(cf + de) + b(ce - df).$$

Since

$$(ac - bd)e - (ad + bc)f = a(ce - df) - b(cf + de)$$

and

$$(ac - bd)f + (ad + bc)e = a(cf + de) + b(ce - df),$$

we get $(z_1 z_2) z_3 = z_1 (z_2 z_3)$.

The function $\varphi : \mathbb{R} \to \{(x, 0) : x \in \mathbb{R}\}$, defined by $\varphi(x) = (x, 0)$, is bijective and satisfies

(a) $\varphi(x + y) = \varphi(x) + \varphi(x)$;

(b) $\varphi(xy) = \varphi(x)\varphi(x)$.

This observation makes it possible to identify the real number x with the complex number $(x, 0)$. It is easy to verify that

$$(0, 1)(0, 1) = -1.$$

The complex $(0, 1)$ is denoted by i. Consequently we can write

$$i^2 = -1.$$

If $z = (x, y)$ is a complex number, then

$$z = (x, y) = (x, 0) + (y, 0)(0, 1) = x + yi,$$

which is the standard notation for complex numbers. In this notation the product of the complex numbers (a, b) and (c, d) becomes

$$(a, b)(c, d) = (a + bi)(c + di) = (ab - bc) + (ad + bc)i.$$

A complex number z is called purely imaginary if there is a nonzero real number y such that $z = (0, y) = yi$. The complex conjugate of a number $z = x + yi$ is the number $x - yi$ and it is denoted by $\overline{z}$.

Theorem 5.7. *If $z, z_1, z_2 \in \mathbb{C}$, then*

(a) $\overline{\overline{z}} = z$;

(b) $\overline{z_1 + z_2} = \overline{z_1} + \overline{z_2}$;

(c) $\overline{z_1 z_2} = \overline{z_1}\ \overline{z_2}$.

Proof. Clearly we have $\overline{\overline{z}} = z$ and $\overline{z_1 + z_2} = \overline{z_1} + \overline{z_2}$. To prove $\overline{z_1 z_2} = \overline{z_1}\ \overline{z_2}$ we let $z_1 = x_1 + y_1 i$ and $z_2 = x_2 + y_2 i$. Now

$$\overline{z_1 z_2} = \overline{(x_1 + y_1 i)(x_2 + y_2 i)}$$
$$= \overline{x_1 x_2 - y_1 y_2 + (x_1 y_2 + y_1 x_2)i}$$
$$= x_1 x_2 - y_1 y_2 - (x_1 y_2 + y_1 x_2)i$$

and

$$\overline{z_1}\ \overline{z_2} = \overline{x_1 + y_1 i}\ \overline{x_2 + y_2 i}$$
$$= (x_1 - y_1 i)(x_2 - y_2 i)$$
$$= x_1 x_2 - y_1 y_2 - (x_1 y_2 + y_1 x_2)i.$$

$\square$

Note that a complex number z is real if and only if $z = \bar{z}$ and a complex number z is purely imaginary if and only if $z = -\bar{z}$.

The absolute value of the number $z = x + yi$ is the nonnegative number $\sqrt{x^2 + y^2}$ and is denoted by $|z|$.

Theorem 5.8. *If $z, z_1, z_2 \in \mathbb{C}$, then*

(a) $z \neq 0$ if and only if $|z| \neq 0$;

(b) $z\bar{z} = |z|^2$;

(c) $z^{-1} = \frac{\bar{z}}{|z|^2}$;

(d) $|z_1 + z_2| \leq |z_1| + |z_2|$;

(e) $|z_1 z_2| = |z_1|\,|z_2|$.

Proof. Parts (a), (b), and (c) are direct consequences of the definitions.

To prove $|z_1 + z_2| \leq |z_1| + |z_2|$ we let $z_1 = x_1 + y_1 i$ and $z_2 = x_2 + y_2 i$.

Since both numbers $|z_1 + z_2|$ and $|z_1| + |z_2|$ are nonnegative, the inequality $|z_1 + z_2| \leq |z_1| + |z_2|$ is equivalent to the inequality $|z_1 + z_2|^2 \leq (|z_1| + |z_2|)^2$ which is equivalent the inequality

$$x_1 x_2 + y_1 y_2 \leq \sqrt{x_1^2 + y_1^2}\sqrt{x_2^2 + y_2^2}.$$

The above inequality is a direct consequence of the inequality

$$(x_1 x_2 + y_1 y_2)^2 \leq (x_1^2 + y_1^2)(x_2^2 + y_2^2)$$

which is equivalent to the inequality

$$2x_1 x_2 y_1 y_2 \leq x_1^2 y_2^2 + y_1^2 x_2^2$$

which can be written as

$$0 \leq (x_1 y_2 - y_1 x_2)^2.$$

Now to prove (e) we observe that

$$|z_1 z_2|^2 = \overline{z_1 z_2} z_1 z_2 = \overline{z_1}\ \overline{z_2} z_1 z_2 = \overline{z_1} z_1\ \overline{z_2} z_2 = |z_1|^2 |z_2|^2,$$

which gives us $|z_1 z_2| = |z_1|\,|z_2|$. $\square$

Corollary 5.9. *For every complex number $z \neq 0$, there are unique real number $r > 0$ and a complex number v such that $|v| = 1$ and $z = rv$.*

Proof. We can take $r = |z|$ and $v = \frac{z}{|z|}$, because $|\bar{z}| = |z|$.

If $z = sw$ where $s > 0$ and $|w| = 1$, then $|z| = |s||w| = s$, by Theorem 5.8, and thus $w = \frac{z}{|z|}$. $\square$

Appendix C Polynomials

A polynomial is a function $p : \mathbb{K} \to \mathbb{K}$ defined by

$$p(t) = a_n t^n + a_{n-1} t^{n-1} + \cdots + a_1 t + a_0$$

where n is a nonnegative integer and $a_0, \ldots, a_n \in \mathbb{K}$. We denote the set of all polynomials by $\mathcal{P}(\mathbb{K})$.

If $p, q \in \mathcal{P}(\mathbb{K})$ and $\alpha \in \mathbb{K}$, then we define polynomials $p + q$, pq, and αp by

$$(p + q)(t) = p(t) + q(t), \quad (pq)(t) = p(t)q(t), \quad \text{and} \quad (\alpha p)(t) = \alpha p(t).$$

The numbers $a_0, \ldots, a_n$ are uniquely defined by f. To prove this important fact we first prove the following result which is a form of the Euclidean algorithm.

Theorem 5.10. *Let* $p(t) = a_n t^n + \cdots + a_0$ *and* $s(t) = b_m t^m + \cdots + b_0$ *be polynomials such that* $n \geq m > 0$ *and* $b_m \neq 0$. *Then there are polynomials* $q(t) = c_{n-m} t^{n-m} + \cdots + c_0$ *and* $r(t) = d_k t^k + \cdots + d_0$ *such that* $m > k \geq 0$ *and*

$$p = qs + r.$$

Proof. It is easy to verify that

$$p_1(t) = p(t) - \frac{a_n}{b_m} t^{n-m} s(t) = a'_{n-1} t^{n-1} + \cdots + a'_0$$

for some $a'_{n-1}, \ldots, a'_0 \in \mathbb{K}$.

If $n - 1 < m$, then

$$p(t) = \frac{a_n}{b_m} t^{n-m} s(t) + p_1(t)$$

and we can take $q(t) = \frac{a_n}{b_m} t^{n-m}$ and $r = p_1$.

If $n - 1 \geq m$, then we can show by induction that there are polynomials $q_1(t) = c'_{n-m-1} t^{n-m-1} + \cdots + c'_0$ and $r_1(t) = d'_j t^j + \cdots + d'_0$ such that $m > j \geq 0$ and

$$p_1 = q_1 s + r_1.$$

Consequently,

$$p(t) = \frac{a_n}{b_m} t^{n-m} s(t) + q_1(t) s(t) + r_1(t) = \left(\frac{a_n}{b_m} t^{n-m} + q_1(t) \right) s(t) + r_1(t)$$

and we can take $q(t) = t^{n-m} a_n b_m^{-1} + q_1(t)$ and $r = r_1$. $\quad\square$

Definition 5.11. By a *root* of a nonzero polynomial $p \in \mathcal{P}(\mathbb{K})$ we mean any number $\alpha \in \mathbb{K}$ such that $p(\alpha) = 0$.

Theorem 5.12. *If $\alpha \in \mathbb{K}$ is a root of a polynomial $p(t) = a_n t^n + \cdots + a_0$, then there is a polynomial $q(t) = c_{n-1} t^{n-1} + \cdots + c_0$ such that*

$$p(t) = (t - \alpha) q(t).$$

Proof. According to Theorem 5.10 there are a polynomial $q(t) = c_{n-1} t^{n-1} + \cdots + c_0$ and a number $r \in \mathbb{K}$ such that

$$p(t) = (t - \alpha) q(t) + r(t).$$

Since $r(\alpha) = 0$, we get $p(t) = (t - \alpha) q(t)$.   □

Theorem 5.13. *If a polynomial $p(t) = a_n t^n + \cdots + a_0$ has $n + 1$ distinct roots, then $p(t) = 0$ for every $t \in \mathbb{K}$.*

Proof. Let $\alpha_1, \ldots, \alpha_{n+1}$ be distinct roots of the polynomial p. From Theorem 5.12 we obtain, by induction, that there is a $c \in \mathbb{K}$ such that

$$p(t) = c(t - \alpha_1) \cdots (t - \alpha_n).$$

Since

$$p(\alpha_{n+1}) = c(\alpha_{n+1} - \alpha_1) \cdots (\alpha_{n+1} - \alpha_n) = 0,$$

we get $c = 0$ and consequently $p = 0$.   □

Corollary 5.14. *If $p \in \mathcal{P}(\mathbb{K})$ and $p(t) = 0$ for all $t \in \mathbb{K}$, then $p = 0$.*

Now we are ready to prove the important result mentioned at the beginning, that is, the uniqueness of numbers $a_n, \ldots, a_0$ for a polynomial $p(t) = a_n t^n + \cdots + a_0$.

Theorem 5.15. *Let $p(t) = a_n t^n + \cdots + a_0$ and $q(t) = b_m t^m + \cdots + b_0$. If $p(t) = q(t)$ for all $t \in \mathbb{K}$, then $n = m$ and $a_j = b_j$ for every $j \in \{1, \ldots, n\}$.*

Proof. Since $(p - q)(t) = 0$ for all $t \in \mathbb{K}$, the result follows from Corollary 5.14.   □

If $p(t) = a_n t^n + \cdots + a_0$, then the numbers $a_0, \ldots, a_n$ are called the *coefficients* of p. If p is a nonzero polynomial and $a_n \neq 0$, then a_n is called the *leading coefficient* of the polynomial p and n is called the *degree* of p, denoted by $\deg p$. We do not define the degree of the zero polynomial, that is, the polynomial defined by $p(t) = 0$ for every $t \in \mathbb{K}$. Note that the degree of a polynomial is well defined because of the Theorem 5.15.

Let n be a nonnegative integer. The set of all polynomials of the form $a_0 + a_1 t + \cdots + a_n t^n$ is denoted by $\mathcal{P}_n(\mathbb{K})$. Since $\mathcal{P}_0(\mathbb{K})$ can be identified with $\mathbb{K}$, we often do not distinguish between a constant polynomial α and the number α.

The following theorem is an immediate consequence of the definition of the degree of a polynomial.

Theorem 5.16. *If p and q are nonzero polynomials, then*

$$\deg(p+q) \le \max\{\deg p, \deg q\} \quad and \quad \deg(pq) = \deg p + \deg q.$$

Note that $\deg(\alpha p) = \deg p$ for any nonzero polynomial p and any nonzero number α.

Now we can formulate the final form of Theorem 5.10.

Theorem 5.17. *If $p, s \in \mathcal{P}(\mathbb{K})$ and s is a nonzero polynomial, then there are unique $q, r \in \mathcal{P}(\mathbb{K})$ such that $\deg r < \deg g$ or $r = 0$ and we have*

$$p = qs + r.$$

Proof. The existence of $q, r \in \mathcal{P}(\mathbb{K})$ can be obtained by a slight modification of the proof of Theorem 5.10.

Suppose that we have

$$p = qs + r \quad \text{and} \quad p = q_1 s + r_1,$$

for some $q, r, q_1, r_1 \in \mathcal{P}(\mathbb{K})$ such that $\deg r < \deg s$ or $r = 0$ and $\deg r_1 < \deg s$ or $r_1 = 0$. Then

$$0 = (q - q_1)s = r_1 - r.$$

If $q - q_1 \ne 0$, then $(q - q_1)s \ne 0$, because $\deg(q - q_1) + \deg s = \deg(q - q_1)s$. Consequently, $q - q_1 = 0$ and $r = r_1$. $\qquad\qquad\square$

The polynomial q is called the quotient on division of the polynomial p by the polynomial s and the polynomial r is the remainder on this division.

Probably the student is familiar with the long division algorithm. We do not use it in the next example.

Example 5.18. Let $p(t) = 4t^3 + 5t^2 + 15t + 8$ and $s(t) = t^2 + t + 3$. Find the quotient and the remainder on division of p by s.

Proof. In this case, it is easy to see that $q(t) = at + b$ and $r(t) = ct + d$. We have

$$4t^3+5t^2+15t+8 = (t^2+t+3)(at+b)+ct+d = at^3+(a+b)t^2+(3a+b+c)t+3b+d$$

which yields

$$4 = a;$$
$$5 = a + b;$$
$$15 = 3a + b + c;$$
$$8 = 3b + d.$$

This gives $q(t) = 4t + 11$ and $r(t) = 2t + 5$.   □

A polynomial p is called *monic* if its leading coefficient is 1.

If p and q are polynomials such that there is a polynomial s satisfying $p = qs$, then we say that q divides p or that q is a divisor of p.

Theorem 5.19. *Let* $p_1, \ldots, p_n$ *be nonzero polynomials. There is a unique monic polynomial d such that*

$$\{s_1 p_1 + \cdots + s_n p_n : s_1, \ldots, s_n \in \mathcal{P}(\mathbb{K})\} = \{sd : s \in \mathcal{P}(\mathbb{K})\}.$$

The polynomial d divides all polynomials $p_1, \ldots, p_n$. Moreover, if a nonzero polynomial h divides all polynomials $p_1, \ldots, p_n$, then h divides d.

Proof. Let d be a nonzero polynomial of the smallest degree in the set

$$\mathcal{F} = \{s_1 p_1 + \cdots + s_n p_n : s_1, \ldots, s_n \in \mathcal{P}(\mathbb{K})\}.$$

Then d divides f_j for every $j \in \{1, \ldots, n\}$. Indeed, by Theorem 5.17, for every $j \in \{1, \ldots, n\}$ we have $p_j = q_j d + r_j$, for some polynomials q_j and r_j, and thus $r_j \in \mathcal{F}$. If $r_j \neq 0$ for some $j \in \{1, \ldots, n\}$, then we get a contradiction because $\deg r_j < \deg d$ and d is a nonzero polynomial of the smallest degree in $\mathcal{F}$. Consequently we must have $r_j = 0$ for every $j \in \{1, \ldots, n\}$ which means that for every $j \in \{1, \ldots, n\}$ the polynomial d is a divisor of the polynomial p_j.
     Since

$$s_1 p_1 + \cdots + s_n p_n = (s_1 q_1 + \cdots + s_n q_n)d,$$

we have $\mathcal{F} \subseteq \{sd : s \in \mathcal{P}(\mathbb{K})\}$. On the other hand, because $d \in \mathcal{F}$ there are polynomials $h_1, \ldots, h_n$ such that

$$d = h_1 p_1 + \cdots + h_n p_n \tag{5.1}$$

and consequently $\{sd : s \in \mathcal{P}(\mathbb{K})\} \subseteq \mathcal{F}$.

Now, if d_1 and d_2 are monic polynomials such that

$$\{sd_1 : s \in \mathcal{P}(\mathbb{K})\} = \{sd_2 : s \in \mathcal{P}(\mathbb{K})\},$$

then there are polynomials s_1 and s_2 such that $d_1 = s_1 d_2$ and $d_2 = s_2 d_1$. Then $d_1 = s_1 s_2 d_1$ and thus $s_1 s_2 = 1$. Since d_1 and d_2 are monic polynomials, we conclude that $d_1 = d_2$.

Finally, if a nonzero polynomial h divides all polynomials $p_1, \ldots, p_n$, then h divides d as a consequence of (5.1).   □

As a consequence of Theorem 5.19 and its proof we get the following result.

Theorem 5.20. *Let $p_1, \ldots, p_n$ be nonzero polynomials. There is a unique monic polynomial d such that d divides all polynomials $p_1, \ldots, p_n$ and if a nonzero polynomial h divides all polynomials $p_1, \ldots, p_n$, then h divides d.*

Definition 5.21. Let $p_1, \ldots, p_n$ be nonzero polynomials. The unique monic polynomial d from Theorem 5.20 is called the *greatest common divisor* of the polynomials $p_1, \ldots, p_n$ and is denoted by

$$\mathbf{GCD}(p_1, \ldots, p_n).$$

Note that if

$$\mathbf{GCD}(p_1, \ldots, p_n) = 1,$$

then there are $s_1, \ldots, s_n \in \mathcal{P}(\mathbb{K})$ such that

$$s_1 p_1 + \cdots + s_n p_n = 1.$$

This observation is often used in proofs.

Definition 5.22. A nonzero polynomial $p \in \mathcal{P}(\mathbb{K})$ with $\deg p \geq 1$ is called *irreducible* in $\mathcal{P}(\mathbb{K})$ if $p = fg$ and $f, g \in \mathcal{P}(\mathbb{K})$ implies $f \in \mathbb{K}$ or $g \in \mathbb{K}$.

Note that irreducibility of a polynomial may depend on $\mathbb{K}$. For example, the polynomial $t^2 + 1$ is irreducible in $\mathcal{P}(\mathbb{R})$ but it is not irreducible in $\mathcal{P}(\mathbb{C})$. Clearly, if a polynomial in $\mathcal{P}(\mathbb{R})$ is irreducible in $\mathcal{P}(\mathbb{C})$ then it is irreducible in $\mathcal{P}(\mathbb{R})$.

Example 5.23. Show that the polynomial $at + b$ is irreducible for any $a, b \in \mathbb{K}$ with $a \neq 0$.

Solution. If $at + b = f(t)g(t)$, then $\deg f + \deg g = 1$ and thus $\deg f = 0$ or $\deg g = 0$. Consequently, $f \in \mathbb{K}$ or $g \in \mathbb{K}$.   □

Example 5.24. Show that the polynomial of degree at least 2 which has a root in $\mathbb{K}$ is not irreducible.

Solution. This is an immediate consequence of Theorem 5.12.   □

Example 5.25. Let $a, b, c \in \mathbb{R}$ with $a \neq 0$. Show that the polynomial $at^2 + bt + c$ is irreducible over $\mathbb{R}$ if and only if $4ac - b^2 > 0$.

Solution. If $4ac - b^2 > 0$, then

$$at^2 + bt + c = a\left(t^2 + 2\frac{b}{2a}t + \frac{b^2}{4a^2} + \frac{4ac - b^2}{4a^2}\right) = a\left(\left(t + \frac{b}{2a}\right)^2 + \frac{4ac - b^2}{4a^2}\right).$$

Since

$$\left(t + \frac{b}{2a}\right)^2 + \frac{4ac - b^2}{4a^2} > 0,$$

the polynomial $at^2 + bt + c$ has no real root and consequently is irreducible. If $4ac - b^2 \leq 0$, then

$$\left(t + \frac{b}{2a}\right)^2 = \frac{b^2 - 4ac}{4a^2}$$

and the polynomial $at^2 + bt + c$ has the well known roots

$$-\frac{b}{2a} \pm \frac{\sqrt{b^2 - 4ac}}{2a}$$

and thus is not irreducible.   □

Example 5.26. Let $a, b, c, d \in \mathbb{R}$ with $a \neq 0$. If the polynomial $at^3 + bt^2 + ct + d$ has a complex root $x + yi$ such that $y \neq 0$, show that the polynomial $(t - x)^2 + y^2$ divides $at^3 + bt^2 + ct + d$.

Solution. If $at^3 + bt^2 + ct + d$ has a root $x + yi$ such that $y \neq 0$, then

$$a(x - yi)^3 + b(x - yi)^2 + c(x - yi) + d = \overline{a(x + yi)^3 + b(x + yi)^2 + c(x + yi) + d} = 0.$$

Note that $(t - x - yi)(t - x + yi) = (t - x)^2 + y^2$ is an irreducible polynomial in $\mathcal{P}(\mathbb{R})$. Since we can write

$$at^3 + bt^2 + ct + d = ((t - x)^2 + y^2)q + et + f,$$

where e and f are real numbers, $et + f$ as a polynomial from $\mathcal{P}(\mathbb{C})$ has two different roots $x + yi$ and $x - yi$. Consequently $e = f = 0$, by Theorem 5.13. This shows that the polynomial $(t - x)^2 + y^2$ divides $at^3 + bt^2 + ct + d$. $\square$

Theorem 5.27. *If an irreducible polynomial p divides the product of two polynomials fg then p divides f or p divides g.*

Proof. Suppose p does not divide f. If d is a polynomial such that $p = dp_1$ and $f = df_1$ for some $p_1, f_1 \in \mathcal{P}(\mathbb{K})$, then $d \in \mathbb{K}$. Consequently, $\mathbf{GCD}(p, f) = 1$ and there are polynomials q and h such that

$$qp + hf = 1.$$

Hence

$$qpg + hfg = g.$$

Because p divides qpg and hfg, it must divide g. $\square$

Lemma 5.28. *If an irreducible polynomial p divides the product $q_1 \cdots q_m$ of irreducible polynomials $q_1, \ldots, q_m$, then there are $j \in \{1, \ldots, m\}$ and $c \in \mathbb{K}$ such that $q_j = cp$.*

Proof. Using Theorem 5.27 and induction we can show there is a $j \in 1, \ldots, m$ such that p divides q_j. Because the polynomial q_j is an irreducible polynomial, there is a number $c \in \mathbb{K}$ such that $q_j = cp$. $\square$

Theorem 5.29. *If*

$$p_1 \cdots p_n = q_1 \cdots q_m$$

for some irreducible polynomials $p_1, \ldots, p_n, q_1, \ldots, q_m$, then $n = m$ and there are numbers $c_1, \ldots, c_n$ such $q_1 = c_1 p_1, \ldots, q_n = c_n p_n$.

Proof. Without loss of generality, we can assume that $q_1 = c_1 p_1$ where $c_1 \in \mathbb{K}$, by Lemma 5.28. This gives us

$$p_2 \dots p_n = c_1 q_2 \dots q_m.$$

Because the polynomial $c_1 q_2$ is irreducible can we finish the proof by induction.

$\square$

> **Theorem 5.30.** *Every nonzero polynomial can be written as a product of irreducible polynomials.*

Proof. Let p be a nonzero polynomial such that $\deg p > 0$. If p is irreducible we are done. If not, then we can write $p = fg$ where $\deg f > 0$ and $\deg g > 0$. Now we continue in the same way with the polynomials f and g and finish the proof by induction.

$\square$

Because for every irreducible polynomial p there is a unique monic irreducible polynomial q such that $p = cq$, where $c \in \mathbb{K}$, every polynomial p such that $\deg p > 0$ can be uniquely written as

$$p = cq_1^{m_1} \cdots q_k^{m_k} \tag{5.2}$$

where $c \in \mathbb{K}$ and $q_1, \dots, q_n$ are irreducible monic polynomials.

As a consequence of Theorems 5.29 and 5.30 we get the following result.

> **Theorem 5.31.** *Let $p_1, \dots, p_n$ be nonzero polynomials. There is a unique monic polynomial m such that all polynomials $p_1, \dots, p_n$ divide m and if all polynomials $p_1, \dots, p_n$, divide a nonzero polynomial h then m divides h.*

> **Definition 5.32.** *Let $p_1, \dots, p_n$ be nonzero polynomials. The unique monic polynomial m from Theorem 5.31 is called the lowest common multiple of the polynomials $p_1, \dots, p_n$ and is denoted by*
>
> $$\mathbf{LCM}(p_1, \dots, p_n).$$

The following theorem is called the Fundamental Theorem of Algebra. It is usually proven using complex analysis.

> **Theorem 5.33.** *For every $p \in \mathcal{P}(\mathbb{C})$ such that $\deg p > 0$ there is $z \in \mathbb{C}$ such that $p(z) = 0$.*

From the Fundamental Theorem of Algebra and Theorem 5.12 we obtain the following important result.

Theorem 5.34. *If $p \in \mathcal{P}(\mathbb{C})$ has exactly k distinct roots $\alpha_1, \ldots, \alpha_k$,*

$$p(t) = c(t - \alpha_1)^{m_1} \cdots (t - \alpha_k)^{m_k}$$

for some $c \in \mathbb{C}$ and $m_1, \ldots, m_k$ are integers ≥ 1.

Theorem 5.35. *If $p \in \mathcal{P}(\mathbb{R})$ is irreducible and $\deg p \geq 2$, then*

$$p(t) = at^2 + bt + c$$

where $a, b, c \in \mathbb{R}$, $a \neq 0$, and $4ac - b^2 > 0$.

Proof. If p has a real root, then p is not irreducible.

If $\deg p \geq 3$, then p has a complex root of the form $x + yi$ with $y \neq 0$ and we can show, as in Example 5.26, that the polynomial $(t - x)^2 + y^2$ divides p and consequently p is not irreducible.

To complete the proof we note that a polynomial $p(t) = at^2 + bt + c$ with $a, b, c \in \mathbb{R}$, $a \neq 0$, and $4ac - b^2 > 0$, is irreducible. □

Now we can state a version of Theorem 5.34 for $\mathcal{P}(\mathbb{R})$.

Theorem 5.36. *If $p \in \mathcal{P}(\mathbb{R})$, then*

$$p = cq_1^{m_1} \cdots q_k^{m_k},$$

where $c \in \mathbb{R}$ and for every $j \in \{1, \ldots, k\}$ the polynomial q_j is either of the form $t - \alpha$ for some $\alpha \in \mathbb{R}$ or of the form $t^2 + \beta t + \gamma$ for some $\beta, \gamma \in \mathbb{R}$ such that $\beta^2 - 4\gamma < 0$.

Lagrange interpolation theorem

Theorem 5.37. *For any integer $n \geq 1$ and $\alpha_1, \ldots, \alpha_n, \beta_1, \ldots, \beta_n \in \mathbb{K}$, such that $\alpha_1, \ldots, \alpha_n$ are distinct, there is a polynomial $p \in \mathcal{P}(\mathbb{K})$ such that*

$$p(\alpha_1) = \beta_1, \ldots, p(\alpha_n) = \beta_n.$$

Proof. For every $j \in \{1, \ldots, n\}$ the polynomial

$$q_j(t) = \frac{(t - \alpha_1) \cdots (t - \alpha_{j-1})(t - \alpha_{j+1}) \cdots (t - \alpha_n)}{(\alpha_j - \alpha_1) \cdots (\alpha_j - \alpha_{j-1})(\alpha_j - \alpha_{j+1}) \cdots (\alpha_j - \alpha_n)}$$

satisfies $q_j(\alpha_j) = 1$ and $q_j(\alpha_k) = 0$ for all $k \neq j$. Thus we can take

$$p = \beta_1 q_1 + \cdots + \beta_n q_n.$$

$\square$

The formal derivative of a polynomial

Definition 5.38. By the *derivative* of a polynomial $p(t) = a_n t^n + \cdots + a_0$ we mean the polynomial

$$n a_n t^{n-1} + \cdots + 2 a_2 t + a_1.$$

The derivative of a polynomial p is denoted by p', that is, if $p(t) = a_n t^n + \cdots + a_0$ then $p'(t) = n a_n t^{n-1} + \cdots + 2 a_2 t + a_1$.

Theorem 5.39. *For any $p, q \in \mathcal{P}(\mathbb{K})$ and $\alpha \in \mathbb{K}$ we have*

(a) $(\alpha p)' = \alpha p'$;

(b) $(p + q)' = p' + q'$;

(c) $(pq)' = p'q + pq'$.

Proof. Clearly, we have $(\alpha p)' = \alpha p'$ and $(p + q)' = p' + q'$. To show that $(pq)' = p'q + pq'$ we first note that

$$(t^{m+n})' = (m + n)t^{m+n-1} = mt^{m-1}t^n + nt^m t^{n-1} = (t^m)'t^n + t^m(t^n)',$$

for any positive integers m and n. To finish the proof we use the fact that, if $p_1, p_2, q \in \mathcal{P}(\mathbb{K})$ are polynomials such that $(p_1 q)' = p_1'q + p_1 q'$ and $(p_2 q)' = p_2'q + p_2 q'$, then

$$((p_1 + p_2)q)' = (p_1 q)' + (p_2 q)' = p_1'q + p_1 q' + p_2'q + p_2 q'$$
$$= (p_1' + p_2')q + (p_1 + p_2)q' = (p_1 + p_2)'q + (p_1 + p_2)q'.$$

$\square$

Using Theorem 5.39 and mathematical induction we obtain the following useful result.

> **Corollary 5.40.** *For any $p \in \mathcal{P}(\mathbb{K})$ and any integer $n \geq 1$ we have*
>
> $$(p^n)' = np^{n-1}p'.$$

Appendix D   Infinite dimensional inner product spaces

Many results in Chapter 3 are formulated and proved for finite dimensional inner product spaces. While some of these results remain true in infinite dimensional inner product space, other are not or require additional assumptions. Here we briefly address the issues arising in infinite dimensional inner product spaces.

The important Representation Theorem 3.3.1 is no longer true if we remove the assumption that $\mathcal{V}$ is finite dimensional. Indeed, consider the space $\mathcal{V}$ of all continuous functions on the interval $[0,1]$ with the inner product $\langle f, g \rangle = \int_0^1 f(t)\overline{g(t)}\, dt$ and the function $\Phi : \mathcal{V} \to \mathbb{K}$ defined by

$$\Phi(f) = \int_0^{\frac{1}{2}} f(t)dt.$$

The function Φ is clearly a linear transformation, but there is no continuous function g_0 such that

$$\Phi(f) = \langle f, g_0 \rangle = \int_0^1 f(t)\overline{g_0(t)}dt,$$

for every continuous function f, because then we would have to have $g_0(t) = 1$ for all $t \in \left(0, \frac{1}{2}\right)$ and $g_0(t) = 0$ for all $t \in \left(\frac{1}{2}, 1\right)$.

The Representation Theorem guarantees that every linear transformation between finite dimensional spaces has an adjoint. This is no longer true in infinite dimensional inner product spaces. Indeed, consider the space $\mathcal{V} = \mathbb{C}^\infty$ of all infinite sequences of complex numbers with only a finite number of nonzero terms with the inner product defined as

$$\langle (x_1, x_2, \dots), (y_1, y_2, \dots) \rangle = \sum_{j=1}^{\infty} x_j \overline{y_j}.$$

Note that because all but a finite number of x_j's and y_j's are 0, the summation is always finite and thus we don't have to worry about convergence of the series. Now consider the functions $f : \mathbb{C}^\infty \to \mathbb{C}^\infty$ defined as

$$f((x_1, x_2, \dots)) = \left(\sum_{j=1}^{\infty} x_j, \sum_{j=2}^{\infty} x_j, \dots \right).$$

It is clearly a linear transformation from $\mathbb{C}^\infty$ to $\mathbb{C}^\infty$. Now suppose there is a linear transformation $g : \mathbb{C}^\infty \to \mathbb{C}^\infty$ such that $\langle f(\mathbf{x}), \mathbf{y} \rangle = \langle \mathbf{x}, g(\mathbf{y}) \rangle$ for all $\mathbf{x}, \mathbf{y} \in \mathbb{C}^\infty$. Then for every integer $j \geq 1$ we would have

$$\langle \mathbf{e}_j, g(\mathbf{e}_1) \rangle = \langle f(\mathbf{e}_j), \mathbf{e}_1 \rangle = 1,$$

where $\mathbf{e}_j$ is the sequence that has 1 in the j-th place and zeros everywhere else. But this is not possible because this means that $g(\mathbf{e}_1) = (1, 1, \dots) \notin \mathbb{C}^\infty$.

Since there are linear transformations that do not have adjoints, in every theorem that says something about the adjoint of a transformation we assume that the inner product space is finite dimensional spaces. Many of those theorems remain true in all inner product spaces if we simply assume that the transformations have adjoints. For example, if we assume that all transformations in Theorem 3.3.4 have adjoints, then the theorem is true for all inner product spaces.

The definition of self-adjoint operators is formulated for operators on finite dimensional spaces, but it is not necessary. We can say that a linear transformation $f : V \to V$ is self-adjoint if $\langle f(\mathbf{x}), \mathbf{y} \rangle = \langle \mathbf{x}, f(\mathbf{y}) \rangle$ for all $\mathbf{x}, \mathbf{y} \in V$. Note that this definition makes sense in any inner product spaces and many of the properties of self-adjoint operators proved in this chapter remain true in infinite dimensional inner product spaces and quite often the presented proof does not require any changes.

There are theorems that depend in an essential way on the assumption that the space is of finite dimension. For example, in Theorem 3.4.56 we show that on a finite dimensional inner product space a linear operator is unitary if and only if it is isometric. This is not true in general. Indeed, consider the space $V = \mathbb{C}^\infty$ defined above and the linear operator $f : \mathbb{C}^\infty \to \mathbb{C}^\infty$ defined as

$$f(x_1, x_2, \dots) = (0, x_1, x_2, \dots).$$

Note that this operator has an adjoint:

$$f^*(x_1, x_2, \dots) = (x_2, x_3, \dots).$$

Since

$$\|f(x_1, x_2, \dots)\|^2 = |0|^2 + |x_1|^2 + |x_2|^2 + \dots = |x_1|^2 + |x_2|^2 + \dots = \|(x_1, x_2, \dots)\|^2,$$

we have $\|f(\mathbf{x})\| = \|\mathbf{x}\|$ for every $\mathbf{x} \in \mathbb{C}^\infty$ and thus f is an isometric operator. On the other hand, since $\operatorname{ran} f \neq \mathbb{C}^\infty$ and $f f^* \neq \operatorname{Id}$, f is not a unitary operator.

It would be an excellent way to review Chapter 3 by checking for which theorems the assumption of finite dimensionality is essential.

Bibliography

[1] D. Atanasiu and P. Mikusiński, *Linear algebra, Core topics for the first course*, World Scientific, 2020.

[2] S. Axler, *Linear algebra done right*, 3rd edition, Springer, 2015.

[3] S. K. Berberian, *Linear algebra*, Dover Publications, 2014.

[4] R. Godement, *Cours d'algèbre*, 3rd edition, Hermann, 1966.

[5] J. S. Golan, *The linear algebra a beginning graduate student ought to know*, 3rd edition, Springer, 2009.

[6] M. Houimi, *Algèbre linéaire, algèbre bilinéaire*, Ellipses, 2021.

[7] H. J. Kowalsky and G. Michler, *Lineare algebra*, 12th edition, de Gruyter, 2003.

[8] T. W. Körner, *Vectors pure and applied*, Cambridge University Press, 2013.

[9] S. Lang, *Linear algebra*, 3rd edition, Springer, 1987.

[10] R. Mansuy and R. Mneimné, *Algèbre linéaire. Réduction des endomorphismes*, 2nd edition, Vuibert, 2016.

[11] L. Spence, A. Insel, and S. Friedberg, *Linear algebra*, 5th edition, Pearson, 2018.

[12] S. Weintraub, *A guide to advanced linear algebra*, The Mathematical Association of America, 2011.

[13] H. Woerdeman, *Advanced linear algebra*, Chapman and Hall/CRC, 2015.

Index